Logic, Methodology and Philosophy of Science

Proceedings of the 14th International Congress (Nancy)

Logic and Science Facing the New Technologies

www.ingramcontent.com/pod-product-compliance
Lightning Source LLC
Chambersburg PA
CBHW051123230426
43670CB00007B/652

Logic, Methodology and Philosophy of Science
Proceedings of the 14th International Congress (Nancy)

Logic and Science Facing the New Technologies

Edited by
Peter Schroeder-Heister,
Gerhard Heinzmann,
Wilfrid Hodges,
and
Pierre Edouard Bour

© Individual author and College Publications 2014
All rights reserved.

ISBN 978-1-84890-169-8

College Publications
Scientific Director: Dov Gabbay
Managing Director: Jane Spurr

http://www.collegepublications.co.uk

Printed by Lightning Source, Milton Keynes, UK

All rights reserved. No part of this publication may be reproduced, stored in a retrieval system or transmitted in any form, or by any means, electronic, mechanical, photocopying, recording or otherwise without prior permission, in writing, from the publisher.

CONTENTS

PETER SCHROEDER-HEISTER, GERHARD HEINZMANN,
WILFRID HODGES, PIERRE EDOUARD BOUR
Preface 5

WILFRID HODGES
DLMPS — Tarski's Vision and Ours 9

DAG PRAWITZ
Is There a General Notion of Proof? 27

JEREMY GRAY
"The Soul of the Fact"— Poincaré and Proof 45

JEFFREY BUB
Einstein and Bohr Meet Alice and Bob 65

JUSTIN TATCH MOORE
The Utility of the Uncountable 79

PAULO OLIVA
Unifying Functional Interpretations: Past and Future 97

DAG WESTERSTÅHL
Questions about Compositionality 123

KOSTA DOŠEN
General Proof Theory 149

ARNAUD CARAYOL & CHRISTOF LÖDING
Uniformization in Automata Theory 153

YIANNIS N. MOSCHOVAKIS
On the Church-Turing Thesis and Relative Recursion 179

CARLO CELLUCCI
Explanatory and Non-Explanatory Demonstrations 201

PAUL HUMPHREYS
X-Ray Data and Empirical Content 219

WOLFGANG SPOHN
A Priori Principles of Reason   235

HEATHER E. DOUGLAS
Scientific Integrity in a Politicized World   253

HUGH LACEY
On the Co-Unfolding of Scientific Knowledge and Viable Values   269

YEMIMA BEN-MENAHEM
Scepticism and Verificationism   285

JEAN-PIERRE MARQUIS
Mathematical Abstraction, Conceptual Variation and Identity   299

DALE MILLER
Communicating and Trusting Proofs:
The Case for Foundational Proof Certificates   323

MICHEL MORANGE
The Rise of Post-Genomics and Epigenetics:
Continuities and discontinuities in the history of biological thought 343

MIRIAM SOLOMON
Evidence-Based Medicine and Mechanistic Reasoning
in the Case of Cystic Fibrosis   353

ROGER M. COOKE
Risk Management and Model Uncertainty in Climate Change   363

PETER KROES
Knowledge and the Creation of Physical Phenomena and
Technical Artefacts   385

BENEDIKT LÖWE
Mathematics and the New Technologies
Part I: Philosophical relevance of a changing culture of mathematics   399

PETER KOEPKE
Mathematics and the New Technologies
Part II: Computer-Assisted Formal Mathematics and
Mathematical Practice   409

JEAN PAUL VAN BENDEGEM
Mathematics and the New Technologies
Part III: The Cloud and the Web of Proofs   427

IMRE HRONSZKY
Technological Paradigm Conceptions in the 1980s   441

ARMIN GRUNWALD
Visionary Communication on Techno-Sciences and
Emerging Challenges to Societal Debate:
The Case of Synthetic Biology   469

ELENA PARIOTTI
The Convergence of Law and Ethics:
Promises and Shortcomings   489

APPUKUTTAN NAIR DAMODARAN
Grand Narratives, Local Minds and Natural Disasters:
Community Response to Tsunami in India   499

GERHARD BANSE
Engineering Design between Science and the Art   511

Appendix A: Sections, Plenary Lectures and Special Symposia   527

Appendix B: Contributed Papers   531

Appendix C: Affiliated Symposia   549

Appendix D: Committees, Patronages, Supports and Partners   551

Preface

PETER SCHROEDER-HEISTER, GERHARD HEINZMANN,
WILFRID HODGES, PIERRE EDOUARD BOUR

The 14th *International Congress of Logic, Methodology and Philosophy of Science* was held in July, 19th – 26th, 2011 in Nancy, the historic capital of Lorraine and birthplace of Henri Poincaré. We were very honored that the President of the French Republic, Monsieur Nicolas Sarkozy, generously agreed his patronage.

The LMPS congresses represent the current state of the art and offer new perspectives in its fields. There were 900 registered participants from 56 different countries. They filled 115 sessions consisting of 391 individual talks (among them 6 plenary lectures and 49 invited lectures), 22 symposia (among them 4 special invited symposia), and 13 affiliated meetings and associated events such as 6 public talks—in all nearly 600 papers. These figures reflect the fact that LMPS is not only a place for scientific communication at the highest level, but also a forum for individual and collective research projects to reach a wide international audience.

Concerning the program, there were two innovations:

(a) For the first time in the LMPS history, the Nancy congress had a special topic: *Logic and Science Facing the New Technologies*. It illuminated issues of major significance today: their integration in society. These questions were of great importance not only to LMPS participants, but to our professional and sponsoring partners likewise. Correspondingly, a section of the congress was entirely devoted to "Methodological and Philosophical Issues in Technology". With 16 individual lectures (three invited) and two symposia this special topic made a grand entrance.

(b) We put much emphasis on symposia in the 'non-invited' part of the program. In addition to four symposia with invited speakers which we organized ourselves, and 13 affiliated symposia related to various topics of congress, for which their respective organizers were responsible, we issued a call for contributed symposia in addition to the call for contributed papers, giving researchers the chance to apply as a group of up to 6 people for a short symposium on a selected topic. This call resulted in 18 contributed symposia, some of which were of exceptionally high quality.

The papers of this volume are a selection of invited plenary talks and invited talks given in particular sections. Even though not every invited speaker submitted a paper, most of these sections are represented in this volume. The detailed program of the congress is presented in appendix A.

A selection of contributed papers will appear in issue 18-3 (2014) and 19-1 (2015) of *Philosophia Scientiæ*. The titles of all contributed papers and symposia are listed in appendix B.

We are indebted to many persons and institutions for their integrated efforts to realize this meeting. First and foremost we would like to thank the members of our respective committees, the Local Organizing Committee, and the General Program Committee including its Senior Advisors and Advisors. They all have worked very hard during the past four years, setting up an outstanding and attractive program and staging it in a comfortable surrounding that would make the congress a scientifically and socially enjoyable event. It has been a great pleasure to work with our colleagues and staff in these committees.

We also thank the Executive Committee of the DLMPS for its constant support and encouragement. Claude Debru (Académie des Sciences, Paris) helped us, amongst many other things, with his knowledge of French institutions, for which we are very grateful. Special thanks are also due to the University Nancy 2 and its Presidents, François Le Poultier and Martial Delignon, as well as to the Deans of Nancy's Faculty of Law, Olivier Cachard and Éric Germain, who willingly let us occupy their splendid lecture halls and facilities. Without the generous financial support of the University of Lorraine, of local, national and international organizations, this meeting would not have been possible. To all these partners we express our warm gratitude.

Special thanks also go to the reviewers who helped us by friendly reading the contributions. And, last but not least we would like to thank the editor of *College Publications*, Dov M. Gabbay, for overseeing the publication of this volume, Jane Spurr, for managing the publication process, and Sandrine Avril, who worked on the LATEX layout of this volume.

Peter Schroeder-Heister, Gerhard Heinzmann,
Wilfrid Hodges, Pierre Edouard Bour

Peter Schroeder-Heister
Universität Tübingen
Germany
psh@uni-tuebingen.de

Gerhard Heinzmann
Laboratoire d'Histoire des Sciences et de Philosophie – Archives H.-Poincaré –
Université de Lorraine – CNRS (UMR 7117)
France
gerhard.heinzmann@univ-lorraine.fr

Wilfrid Hodges
British Academy
United Kingdom
wilfrid.hodges@btinternet.com

Pierre Edouard Bour
Laboratoire d'Histoire des Sciences et de Philosophie – Archives H.-Poincaré –
Université de Lorraine – CNRS (UMR 7117)
France
pierre-edouard.bour@univ-lorraine.fr

DLMPS — Tarski's Vision and Ours

WILFRID HODGES

This paper is an edited and updated version of my Presidential Address at the Fourteenth International Congress of Logic, Methodology and Philosophy of Science at Nancy in 2011, in which I reviewed the history and present status of DLMPS as an organisation, and looked to its future.

The title is the title I gave for my talk. Naming individuals enriches history, and Tarski is a natural person to name, both because of his very articulate views about the reasons for doing logic, and also because of his broad and lasting personal influence. In Chapter 10 'Logic and Methodology, Center Stage' of their book, Feferman & Feferman (2004) give a very readable account of Tarski's role in the setting up of DLMPS. But there is a danger that by naming Tarski I diminished the contributions of many other people whose interests combined to shape DLMPS; I hope the paper itself will set the balance straight.

For help of various sorts I thank Anne Fagot-Largeault, Efthymios Nicolaidis, Thomas Piecha, Peter Schroeder-Heister, Paul van Ulsen, Henk Visser, Jan Woleński, and the DLMPS Executive Committee of 2008–11. But none of them should be held responsible for views expressed below.

1 What happened fifty years ago

DLMPS, or to give it its full title, the Division of Logic, Methodology and Philosophy of Science, held its first International Congress in 1960 at Stanford University, California. Starting with the Third International Congress at Amsterdam in 1967, these congresses have taken place every four years. So the 2011 Congress is the nearest thing we have to a celebration of the first half-century of DLMPS congresses.

The editors of the Proceedings of the 1960 Stanford Congress (Ernest Nagel, Patrick Suppes and Alfred Tarski) wrote in their preface (Nagel et al. 1962, vi)

> This was the first International Congress for Logic, Methodology and Philosophy of Science since the International Union of the History of Science and the International Union of the Philosophy

of Science established the International Union of the History and Philosophy of Science on June 3, 1955. The congresses of a related character held prior to the formation of IUHPS were mainly devoted to the philosophy of science. The title of the 1960 Congress reflects its broader coverage; it was in fact the first international congress to include a large number of papers on both mathematical logic and the methodology and philosophy of science.

The editors refer to the establishment of IUHPS, the International Union of the History and Philosophy of Science. In fact DLMPS came into existence as one of the two Divisions of IUHPS, creating a splatter of acronyms as in Figure 1 below. Let me run through this Figure.

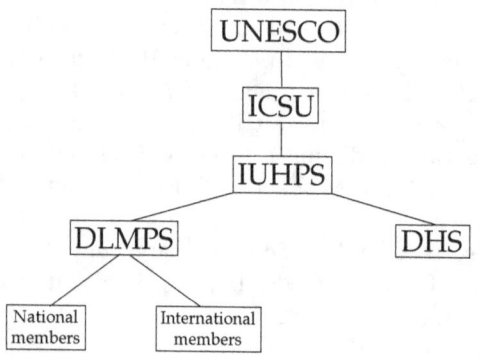

Figure 1. DLMPS in the ICSU family in 1955

1.1 Upwards from ICSU

At the top is UNESCO, the United Nations Educational, Scientific and Cultural Organization, which was born in 1946. During the Second World War there had been discussions between countries on the Allied side with a view to setting up supranational organisations after the war. The creation of the United Nations in 1945 was one result of these discussions. Another was UNESCO, which was attached to the United Nations and thus became funded by and answerable to the national governments ratifying the United Nations Charter. The original plan was for UNESCO to support just Education and Culture; Joseph Needham and Julian Huxley successfully argued that Science should be included too (Greenaway 1996, 71f.).

ICSU, the International Council of Scientific Unions, had existed since 1931 as an international alliance of scientific organisations. It had grown out of collaborations between the scientific academies of some European countries, together with some international scientific projects such as global distance measurements or the establishment of standards. Because of these mixed origins it had two kinds of member: 'national adhering organisations' like the Royal Society, and international scientific unions like the International Union of Pure and Applied Chemistry. The aims of ICSU in 1931 were (in summary):

(1) to coordinate member organisations,

(2) to direct other international scientific activity,

(3) to promote science in countries through their national academies.

At the outset the members of ICSU were forty national members and eight international unions (Greenaway 1996, chap. 3).

In 1946 UNESCO and ICSU formally recognised each other. This meant in practice that UNESCO could call on ICSU for scientific expertise, and ICSU could call on UNESCO for money for the kinds of venture likely to appeal to the United Nations. These arrangements still stand; for example Rio+20, the 2012 United Nations Conference on Sustainable Development held in Rio de Janeiro, had a strong input from UNESCO and ICSU together.

1.2 Downwards from ICSU

The next step down from ICSU in the diagram is IUHPS, the International Union of History and Philosophy of Science. There had been an International Academy of the History of Science as early as 1928. When UNESCO came into being, Needham and others felt that an International Union of the History of Science would be a valuable addition to ICSU. So UNESCO negotiated with the International Academy to convert it into the IUHS, which duly became a member of ICSU in 1947 (Halleux & Severyns 2003).

In 1946, responding to a suggestion of Józef Bocheński who pointed to the recently-formed Association for Symbolic Logic and its associated *Journal of Symbolic Logic*, Ferdinand Gonseth (a Swiss mathematician with interests in philosophy of science and the foundations of mathematics) launched the 'International Society of Logic and Philosophy of Sciences' with an associated journal *Dialectica*. His chief colleagues in this were Paul Bernays, Karl Popper and Karl Dürr. At about the same time, Stanislas Dockx (a Belgian philosopher of science) set up an 'International Academy of Philosophy of Science'. When Gonseth and Dockx became aware that the

International Academy of the History of Science had been converted into a member of ICSU, they decided to pool their efforts so as to create an International Union of the Philosophy of Science (IUPS), which would apply to ICSU for membership. So they called a meeting of interested parties in Brussels in July 1949, where plans were made to set up the IUPS. Besides representatives of UNESCO and ICSU, and Robert Feys representing the Association for Symbolic Logic, the meeting included the logicians Evert Beth and L. E. J. Brouwer together with several leading European philosophers of science. The inaugural meeting of IUPS took place in Paris in October 1949. Sometime between July and September 1949, presumably under pressure from ICSU which wanted to avoid a proliferation of smaller unions, it was agreed that IUHS and IUPS should amalgamate into a single union. In September the executive of IUHS appointed three delegates, and in October IUPS responded with its own three delegates (Gonseth, Dockx and Raymond Bayer), to meet in Paris in 1950 to draw up statutes for a combined IUHPS. In fact it took until 3 June 1955—the date quoted in (1) above—to agree the form of IUHPS, and the new union was admitted to ICSU in August 1955.

The previous paragraph is based on the detailed first-hand account by Dockx (1977). Dockx was writing in honour of Gonseth, and he chose not to mention one embarrassing event. In 1952 there was a coup in IUPS; Gonseth, Dockx and Bayer were all removed from the executive committee, and presumably from the committee to negotiate with IUHS. The new executive consisted of Albert Châtelet, Arend Heyting, Hans Reichenbach, Bocheński and two participants in the July 1949 meeting: Feys and Jean-Louis Destouches. Feys in correspondence gave two reasons for the coup: Gonseth's group wanted to steer UNESCO funds to their own pet projects, and 'they were interested in rather literary forms of "Philosophy of Science"'. Given the commitments made by Gonseth and Dockx in 1949, neither of these two points are likely to have had much direct impact on the negotiations with IUHS. But we know that the Association for Symbolic Logic was unwilling to throw its weight behind the new union until after the coup, so that the coup may have removed a logjam in the negotiations. There was also a perception on the philosophy side that Petre Sergescu, Executive Secretary of IUHS from 1947 till his death in 1954, was against having a combined union. (Van Ulsen (2007), who gives the Feys quotation.)

According to the formula agreed in 1955, IUHS became the Division of History of Science (DHS), IUPS became the Division of Logic, Methodology and Philosophy of Science (DLMPS), and the two divisions together formed the International Union of the History and Philosophy of Science (IUHPS), which became a member of ICSU replacing IUHS.

In Nancy I said that both Divisions seemed to have lost their copies of the IUHPS Statutes by the late 1990s if not earlier—which rather nullified the six years that it had taken to draw up the Statutes in the first place. By 1999 the two Divisions had begun a series of attempts to draw up a Joint Memorandum of Agreement covering various points of collaboration that should really have been covered in the IUHPS Statutes. But in May 2013 Benedikt Loewe discovered a copy of the Statutes, written in French and dated 1962, in an old box containing documents of the German National Committee of the DLMPS. Moves are under way for the two Divisions to agree an updated version of them which will cover the current agreements on collaboration between the Divisions. There is a version online at http://iuhps.net.

Thanks to Benedikt's happy discovery I can replace my previous partial text of the IUHPS Statutes on the aims of the joint Union by a full statement as in the 1962 Statutes:

(1) établir des rapprochements entre les historiens et philosophes des sciences et entre les institutions, sociétés, revues, etc. consacrées à ces disciplines ou à des disciplines connexes;

(2) rassembler les documents utiles au développement de l'Histoire des Sciences et de la Logique, la Méthodologie et la Philosophie des Sciences;

(3) prendre toutes les mesures qu'on croira nécessaires ou utiles pour le développement, la diffusion et l'organisation des études et recherches dans les domaines de l'Histoire des Sciences, de la Philosophie des Sciences et des disciplines connexes;

(4) organiser les Congrès Internationaux d'Histoire des Sciences et les Congrès Internationaux de Philosophie des Sciences, ainsi que des Colloques Internationaux;

(5) contribuer au maintien de l'unité de la science en général et à l'établissement de liens entre les différentes branches du savoir humain;

(6) s'efforcer de favoriser le rapprochement entre historiens, philosophes, savants, soucieux des problèmes de méthode et de fondement que posent leurs disciplines respectives.

This is similar to the aims stated in the DLMPS Statutes on the web at (Status-DLMPS 2011)

We should briefly bring Figure 1 up to date. In 1987 DLMPS changed the name 'National Members' to 'Ordinary Members' because of some political sensitivities. In 1998 ICSU changed its name to 'The International Council for Science', but kept the old acronym. At its General Assembly in Beijing in 2005, DHS added 'and Technology' at the end of its name and became DHST. And finally in 2011 DLMPS joined CIPSH, the Conseil International de la Philosophie et des Sciences Humaines, which in turn is affiliated to UNESCO. In some loose sense CIPSH is to the Humanities as ICSU is to the Sciences. Our sister division DHST had joined CIPSH some years earlier.

2 Pennies from heaven

The institutional structures by themselves don't give many clues about the motivations driving the whole machine. The motivations that chiefly concern us here are money and scientific research. Again it will be helpful to begin the discussion with diagrams (Figures 2 and 3 overleaf). The financial situation today is very different from what it was fifty years ago, so we need diagrams to illustrate both the old situation and the new. These diagrams should be read only as broad indications; one can too easily alter the numbers by adjusting the classifications.

We start with the funds that come to DLMPS from ICSU. UNESCO, which gets its money from countries in the United Nations, makes regular subventions to ICSU. The United States, although it withdrew from funding UNESCO in 1984 and resumed in 2003, continued to make substantial contributions to the ICSU grant fund separately through its National Science Foundation. The US withdrew funding from UNESCO again in 2012, and it remains to be seen how this affects the funding of ICSU (and CIPSH, which is in a similar position to ICSU).

For several decades, ICSU passed on a large part of these subventions as grants to its member unions without close scrutiny. But in 1996 an external assessment (ICSU 2007) recommended that ICSU should be more strategic in its allocations. As a result, since 2002 ICSU has awarded grants by competition and peer review, and only for international multidisciplinary ventures in certain announced priority areas. These changes had a dramatic effect on the funding of Unions, as Figure 2 shows for DLMPS. In fact the only grant from ICSU that came to IUHPS since 2002 and before 2014 was a sum in 2004 to allow DHST to set up databases of bibliographical and archival sources. Figure 3 shows the effect on our outgoings. For a while DLMPS supported only its own meetings and some joint activities with DHST, though since 2012 it has also distributed some small grants to conferences sponsored by members. The money that DLMPS puts into the international congresses held every four years is a small fraction of the cost of these congresses, but it serves to prime the pump. In past decades the sale of Congress Proceedings has brought in some income, but today we no longer expect to make any profit on publications.

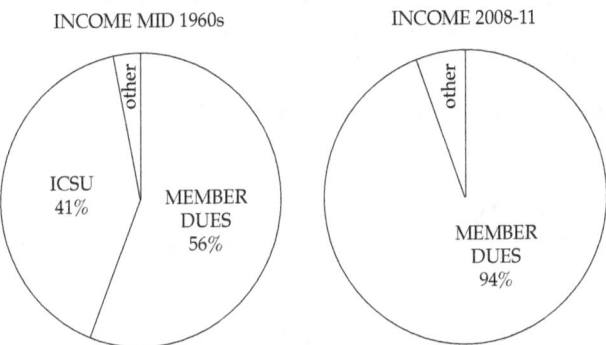

Figure 2. DLMPS income, 1960s and today

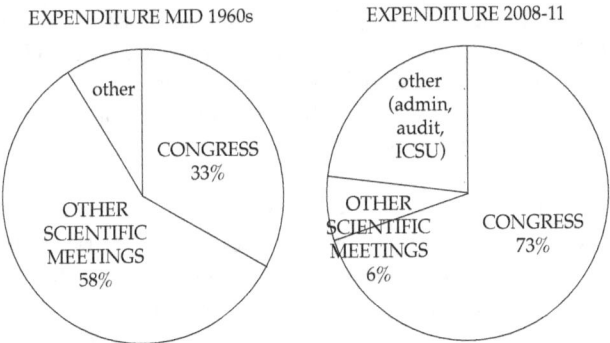

Figure 3. DLMPS expenditure, 1960s and today

Happily the news recently came through, that one of the eight ICSU grants for 2014 was awarded to IUHPS/DLMPS for a project on 'Cultures of Mathematical Research Training'. To quote from the project specification:

> This project aims to mobilize the energies of a currently very active research area (the study of *Practice and Cultures of Mathematics*) to provide the theoretical and empirical resources for designing improvements to the training of the next generations of mathematical researchers and the improvement of research education in developing countries.

The grant application was supported by IMU and its teaching commission ICMI.

As Figure 2 shows, virtually all of our present income comes from our members, both Ordinary and International. Common prudence dictates that we should aim to know what these members reckon they are paying for.

3 Our members and what they pay for

ICSU has no individual members. In its early days it had only two kinds of member: national bodies and international unions. That was partly because ICSU was, so to say, a meta-level association. Its job was to deal with governments or national academies, and to set up and support scientific associations like the international scientific unions. The unions themselves were not meta-level associations in this sense, but they still tended to have structures that copied those of ICSU. The members of a union would be national committees (often administered either by national scientific academies or by national subject societies) and international scientific societies. Our own union IUHPS is a cipher, but its two divisions still both have this style of membership.

The fact that our members represent societies and institutions means that there is a kind of inertia built into our income: institutions that paid this year are likely to carry on paying next year too, because otherwise they would have to make a decision to stop. This could be dangerous for us, because it tends to hide the question whether we are delivering what our members are paying us for. In fact the position is quite complicated and the remarks below are partly guesswork.

3.1 National academies and research councils

About half our members, and two-thirds of our Ordinary Members, are committees of national academies or national research councils. These bodies pass on money from their national governments. Probably most of them reckon that by supporting DLMPS they are supporting science and contributing to the aims of the United Nations as expressed in ICSU. The Canadian National Research Council knows that it is supporting international congresses of DLMPS, and it requests reports from Canadian scientists who attend these congresses; but my impression is that this amount of diligence is very unusual. Some grant-giving bodies ask DLMPS for a copy of our financial report but apparently pay no particular attention to the involvement of logicians or philosophers of science in their countries.

Of course ICSU has its own activities, for example government-level conferences like Rio+20. Let me mention two others that are likely to appeal to national academies. The first is the sharing of expertise between different international scientific bodies. Three recent examples are:

> In 2010 IUHPS was invited to nominate a member for the advisory board of the annual Gruber Cosmology Prize, worth half a million dollars. We nominated a historian of cosmology proposed by a member of DLMPS Council.
>
> In 2011 IUHPS was invited to support the application of the International Council for Industrial and Applied Mathematics to become a Scientific Associate of ICSU. We sent a positive answer, citing the methodological importance of mathematical modelling.
>
> In 2011 ICSU consulted its members for their comments on its draft ICSU Strategic Plan, 2012–2017. Since the Strategic Plan is largely about environmental issues and the integration of science into governmental planning, IUHPS found nothing to say about it. But perhaps DLMPS should have commented on the proposed 'Principle of Universality' for science.

ICSU consultations can be tedious to handle, and often DLMPS is unlikely to have anything to offer. But we could (if membership lists are kept up to date) pass down some consultations to our member societies and national committees. This could help to keep them in touch with the activities of ICSU that they are supporting with their fees.

The second activity of ICSU is its work to protect the free movement of scientists. There is a permanent need for this work, but it was particularly valuable in the days of the Iron Curtain. For example DLMPS consulted ICSU for help in getting visas for East European invitees to the Salzburg Congress in 1983.

Besides these activities, ICSU has committees that rely on the unions for their membership. From 2011 to 2014 Maria Carla Galavotti sat on the ICSU Executive Committee; she was nominated by IUHPS on the proposal of DLMPS. In 2005 Deborah Mayo from DLMPS was one of the authors of the ICSU working group report 'Science and Society: Rights and Responsibilities' (ICSU 2005). In 2008 Susan Lederer became a member of the ICSU Publication Ethics Committee on the proposal of DHST.

3.2 Subject societies

There remain the other half of our members, who are not supported by government-funded bodies. Nearly all of these are supported instead by societies devoted to logic or philosophy of science, or both; some are national and some international. It often seems that random factors have decided whether the societies are primarily devoted to logic or to philosophy of science, and it is possible that we have missed out on support in some countries where the logicians and the philosophers of science were not close to each other. We also have only minimal contact with societies of logicians or philosophers of science in South America. The reasons for this are no doubt partly historical, but we observe that our fellow Division has done much better than us in South America; their next Congress is due to take place in Rio de Janeiro in 2017.

Our supporting societies represent working logicians and philosophers, and they are more likely to support activities that are directly helpful to these working researchers. In the days when ICSU provided grants, these grants often supported smaller meetings and workshops of the kind that researchers relish. Those days are over, and that's a threat to our income. We saw this for example in Britain in the early 1990s, when the government-funded Royal Society and British Academy stopped paying dues for international unions, and the national committees for these unions had to call on scientific societies instead. The British Logic Colloquium at that date was unable to meet its share of the cost, and for a while Britain dropped to a lower category of membership in DLMPS.

The fact that the international scientific unions don't have individual members comes into play here, because it means that there are no DLMPS scientific activities that individual researchers can feel they are involved in. In fact until 2011 DLMPS was an extreme case. There were just two ways in which individuals could be involved with DLMPS. The first was as officers or members of committees, and the second was as participants in congresses or other meetings organised by DLMPS. The officers had a heavy commitment to DLMPS, and the congress organisers an even greater one, but none of the others did. Participants in meetings registered for the meetings and didn't even need to know what DLMPS is. There were the national committees, but in too many cases the committee had lapsed—we found one case where the committee consisted of one person who had died ten years earlier. Sometimes the only task of these committees was to decide who would be delegates at the four-yearly General Assemblies.

Many of the unions have taken steps to involve individuals in actual scientific work. For example the International Union of Radio Science has ten special-subject commissions and a larger number of working groups. The brief of its Commission on Radio Astronomy includes 'observation and interpretation of cosmic radio emissions from the early universe to the present epoch' (URSI 2012). The International Union of Pure and Applied Physics has twenty special-subject Commissions; the Commission on Physics Education goes back to 1960. I think none of these have an open membership, but they do involve quite large numbers of individuals in more than just bureaucracy. Our fellow Division, DHST, has for many years had special-interest commissions; at least some of them have membership open to any interested individuals, and newsletters are circulated to all members. The DHST website lists sixteen commissions, including one on Scientific Instruments and one on Women in Science.

The 2011 General Assembly of DLMPS made a bid to increase the involvement of individual logicians and philosophers of science. It adjusted the Statutes so as to allow commissions in the same style as DHST. It set up four commissions, three of them with open membership. One of those three was the Teaching Commission, which has for many years been a commission of DHST and is now an inter-divisional commission. The other two are new: a Commission on Philosophy of Technology and Engineering Sciences and a Commission on Arabic Logic. Other new commissions are in the pipeline. The aim is for DLMPS to make itself more responsive to the needs of researchers.

4 The name of the Division

When our sister division added 'and Technology' at the end of its name and became DHST, this was a natural step for them to take. The International Committee for the History of Technology had been a Scientific Section of DHS since 1968, and several commissions of DHS already had a strong technology component—for example the Scientific Instruments Commission. So the addition did no more than reflect the facts on the ground.

In June 2008 Claude Debru, on behalf of the French National Committee of History and Philosophy of Science, wrote to DLMPS urging us to go down the same road and add 'technology' to our scope. We put this to the General Assembly in Nancy in 2011, and the result was a pair of resolutions:

> The General Assembly agreed in principle that 'philosophy of science' in the stated scope of the Division should be expanded to 'philosophy of science and technology', and that the Executive Committee

should bring to the 2015 General Assembly proposals for changes in the Statutes and the name of the Division to give effect to this expansion.

The General Assembly asked the Executive Committee to consult with the officers of DHST with a view to changing the name of the Union so as to include technology.

The main reason for proceeding this way was to avoid getting the issue of principle mixed up with debates about the future name of DLMPS. In fact it seemed to many people that just adding T at the end would give a rather monstrous acronym: DLMPST. We tried this acronym on some spell checkers and got back among other things DEMIST, PLUMPEST, ALMOST, DIMMEST and DUMPSITE. Should one or more of the letters be dropped?

4.1 Where did L, M, PS come from?

We know what the organisers of the 1960 Stanford Congress thought these letters stood for (Nagel *et al.* 1962, vi):

> [Stanford] was in fact the first international congress to include a large number of papers on both mathematical logic and the methodology and philosophy of science.

So for the Stanford team, L was for 'mathematical Logic', M was for 'Methodology of science', and PS was for 'Philosophy of Science'.

The name 'Logic, Methodology and Philosophy of Science' could have come from Gonseth back in 1949. Of course if there is evidence against this, then I defer to it; but I know none.

As to mathematical logic: we saw that already in 1947 Gonseth's society was called the International Society of Logic and the Philosophy of Science. Logic was an old interest of Gonseth's. In 1937 he had published a long essay 'Qu'est-ce que la logique?' (1998, 11–94). True, that essay was historical rather than mathematical, and even the chapter on Whitehead and Russell's *Principia Mathematica* hardly contains any formulas. But his essay 'Philosophie Mathématique' (1998, 95–189), published in 1950, is undoubtedly about mathematical logic, including axiomatic set theory and Gödel's incompleteness theorem—even though it does tend to confirm Feys's epithet 'rather literary'. We might add that although some mathematical logicians were certainly repelled by Gonseth's approach to the subject, others found it a stimulus; Gerhard Heinzmann (2001) documents this in the case of Bernays.

As to methodology of science: this phrase goes back to the nineteenth century. In Britain it was popularised by William Hamilton of Edinburgh in his lectures in the 1830s and 1840s (Hamilton 1860, Appendix, 496):

> The Science of Science, or the Methodology of Science—falls into two branches. ... The former—that which treats of those conditions of knowledge which lie in the nature of thought itself—is Logic, properly so called; the latter,—that which treats of those conditions of knowledge which lie in the nature, not of thought itself, but of that which we think about, ... has been called *Heuretic* ... The one owes its systematic development principally to Aristotle, the other to Bacon; [...]

Speaking in Nancy it's appropriate to mention that Henri Poincaré used the phrase in the Introduction to his *Science et Méthode* (1908):

> Je réunis ici diverses études qui se rapportent plus ou moins directement à des questions de méthodologie scientifique.

By the 1940s the notion of scientific methodology was in free circulation among philosophers of science. So it's no surprise that we can document it from Gonseth: *Essai sur la Méthode Axiomatique* (1936); 'une méthodologie dialectique ouverte' (1948); 'La question de la méthode en psychologie' (1949); 'La méthodologie des sciences peut-elle être élevée au rang de discipline scientifique?' (1957); *Essai sur la méthodologie de la recherche* (1964).

In short, the full name 'Logic, methodology and philosophy of science' and the parsing of it in the preface to the 1960 Stanford Proceedings could quite easily have come from Gonseth. This is not to say that they would have meant the same to Gonseth as they did to other members of the Division.

4.2 A name for the future of DLMPS?

The 2011 General Assembly left it to the new Executive to decide on the future name of the Division. It may be superfluous for me to say anything about it here, but I'll make a few remarks anyway.

The two divisions sit together as representing Philosophy of Science and Technology on the one hand and History of Science and Technology on the other. So there is no conceivable case for dropping the PS. The situation is different for both the L and the M, but for different reasons.

In the case of M, there is a case for dropping it straight away. The case is that it no longer represents anything distinctive about DLMPS. In the mid 20th century it was common to distinguish methodology from traditional philosophical areas like epistemology and ontology. By advocating 'methodology of science' one would be supporting philosophy of science but distancing oneself from metaphysics. For example, Herbert Feigl published in 1954 a paper with the title 'Scientific method without metaphysical presuppositions' (1954). His opening words were:

> As the title of this article indicates, I contend that there are no philosophical postulates of science, i.e., that the scientific method can be explicated and justified without metaphysical presuppositions about the order or structure of nature.

On this interpretation the only reason for retaining the M would be to bracket off certain aspects of the philosophy of science that some people don't want to be associated with. That doesn't strike me as an adequate reason.

Feigl's usage of 'method' or 'methodology' was not the only one. Tarski had a distinctive view of the matter. His fullest account of it is in the Introduction to the 1941 English version of his book *Introduction to Logic and to the Methodology of Deductive Sciences* (1941), and it appears unaltered at least up to the 1961 edition, though it has been shortened in the posthumous 1994 edition.

Tarski distinguishes between 'methodology of deductive sciences' and 'methodology of empirical sciences'. Methodology of deductive sciences is what Tarski elsewhere calls metamathematics (for example 1983, 342). It is a part of logic, and a part that Tarski strongly associates himself with. Methodology of empirical sciences 'constitutes an important domain of scientific research', and logic is valuable for it. But: 'logical concepts and methods have not, up to the present, found any specific or fertile applications in this domain' (Tarski 1941, xiii). Tarski comments that this could be a permanent and necessary feature of the subject. He continues:

> It should be added that, in striking opposition to the high development of the empirical sciences themselves, the methodology of these sciences can hardly boast of comparably definite achievements—despite the great efforts that have been made. Even the preliminary task of clarifying the concepts involved in this domain has not yet been carried out in a satisfactory way. Consequently, a course in the methodology of empirical sciences must have a quite different character from one in logic and must be largely confined to evaluations and criticisms of tentative gropings and unsuccessful efforts. (Tarski 1941, xiv)

Tarski doesn't spell out what he regards as the tasks of the methodology of empirical sciences—indeed he suggests that some concepts need to be clarified before we can do that properly. But the comparison with metamathematics sends a strong message. A methodologist of an empirical science should ideally aim to find a suitable formal language in which to carry out the science, with suitable meanings for the primitive terms. Then she should look for suitable axioms. Here part of her task will be to find appropriate criteria for the suitability of the axioms. As Tarski explains in (1944, 366)

> one of the main problems of the methodology of empirical science consists in establishing conditions under which an empirical theory or hypothesis should be regarded as acceptable.

He offers his truth definition as a help here, which suggests that he has in mind a methodologist using a formal metatheory. The oral remarks of Tarski in 1953 reported in (Feferman & Feferman 2004, 250f.) point in the same direction.

Tarski makes a few further remarks about 'the methodology of empirical science' in (1944), but I don't think they help us much here. What is helpful, and perhaps unexpected, is 19 of (Tarski 1944) in which he vigorously dissociates himself from attacks on 'metaphysical elements'.

> When listening to discussions in this subject, sometimes one gets the impression that the term "metaphysical" has lost any objective meaning, and is merely used as a kind of professional philosophical invective. (Tarski 1944, 363)

So he uses a very different language from that of Feigl above.

To my eye, not a single one of the papers on particular empirical sciences in the Proceedings of the 1960 Stanford Proceedings (Nagel *et al.* 1962) is written under the paradigm that Tarski has in mind above. From his remarks in 1941, I doubt that this would have surprised Tarski himself. And given the general usage of the word 'methodology', it seems unlikely that Tarski would have expected many people outside a group of loyal followers to interpret the M in DLMPS in line with his own account of 'the methodology of empirical sciences'. So even a deference to Tarski would hardly give us reason to insist on keeping the M.

By contrast the word 'logic' certainly does mark a major area within the scope of DLMPS. DLMPS Congresses continue to attract top quality speakers in all branches of mathematical logic. Two of the international members of DLMPS are specifically devoted to logic, and several national members have a particular interest in it. Since logic is not a subset of philosophy of science, or indeed of philosophy at all, it follows that as things are at present, there is no question of dropping the L from DLMPS.

But the world moves on. Around 1950 some logicians—Bocheński in particular (Van Ulsen 2007)—wanted an affiliation of 'logic' to ICSU in order to get a wider recognition for modern logic. In this they succeeded magnificently. But logic today gets incomparably more recognition from its role in computer science than it does from the title of DLMPS. Logicians now have so many international outlets that they depend on DLMPS much less than a few decades ago, and this trend will probably continue.

Also in 1955 mathematical logic had stronger links with foundations than it does today. For example mathematical model theory, which was still finding its feet in 1955, is now a branch of mathematics like any other; it has interesting foundations but it is not itself a contribution to foundations. So the links between mathematical logic and philosophy of science grow weaker.

There are already signs that mathematical (as opposed to philosophical) logic may eventually part company from DLMPS. The trend is for fewer papers in mathematical logic to be submitted to DLMPS Congresses. It seems very likely that DLMPS congresses will continue to attract philosophical work that uses mathematical logic, but less of the straight mathematics will find its way there. The General Assembly in Nancy was the first one to which the Association for Symbolic Logic sent no delegates; this was certainly an unintended accident and not a policy decision, but there is a message in the accident.

My own reaction would be to let rivers find their own natural course. The L in DLMPS should be secure for some decades to come.

Bibliography

Dockx, S. (1977). Note historique concernant la fondation de l'Union internationale de Philosophie des Sciences. *Dialectica, 31,* 35–38, doi:10.1111/j.1746-8361.1977.tb01350.x.

Feferman, A. B. & Feferman, S. (2004). *Alfred Tarski: Life and Logic.* Cambridge: Cambridge University Press.

Feigl, H. (1954). Scientific method without metaphysical presuppositions. *Philosophical Studies, 5,* 17–29, doi:10.1007/BF02223254.

Gonseth, F. (1936). *Les Mathématiques et la Réalité: Essai sur la méthode axiomatique.* Paris: Alcan.

Gonseth, F. (1948). Remarque sur l'idée de complémentarité. *Dialectica, 2*(3–4), 413–420, doi:10.1111/j.1746-8361.1948.tb00710.x.

Gonseth, F. (1949). La question de la méthode en psychologie. *Dialectica, 3,* 324–337, doi:10.1111/j.1746-8361.1949.tb00874.x.

Gonseth, F. (1957). La méthodologie des sciences peut-elle être élevée au rang de discipline scientifique? *Dialectica, 11,* 9–20, doi:10.1111/j.1746-8361.1957.tb00348.x.

Gonseth, F. (1964). *Le Problème du temps: Essai sur la méthodologie de la recherche.* Neuchâtel: Griffon.

Gonseth, F. (1998). *Logique et philosophie mathématiques.* Paris: Hermann.

Greenaway, F. (1996). *Science International: A History of the International Council of Scientific Unions.* Cambridge: Cambridge University Press.

Halleux, R. & Severyns, B. (2003). Twenty-five years of international institutions. *Llull: Revista de la Sociedad Española de Historia de las Ciencias*, 26, 315–321, Warning: several statements about 'the Union' in this paper are in fact true only of DHS(T), for example the list of officers and the list of commissions.

Hamilton, W. (1860). *Lectures on Metaphysics and Logic Vol. II. Logic.* Boston: Gould and Lincoln.

Heinzmann, G. (2001). Paul Bernays et la philosophie ouverte. In *Logic and Set Theory in 20th Century Switzerland*, Gasser, J. & Volken, H., eds., PhilSwiss Schriften zur Philosophie I, Bern: PhilSwiss, 19–29.

ICSU (2005). Science and Society: Rights and Responsibilities. A Strategic Review. Tech. rep., ICSU website at www.icsu.org, Paris.

ICSU (2007). Review of the ICSU Grants Programme, 2001–2006. Tech. rep., ICSU website at www.icsu.org, Paris.

Nagel, E., Suppes, P., & Tarski, A. (Eds.) (1962). *Logic, Methodology and Philosophy of Science, Proceedings of the 1960 International Congress.* Stanford: Stanford University Press.

Poincaré, H. (1908). *Science et Méthode.* Paris: Flammarion.

Status-DLMPS (2011). Statutes of the Division of Logic, Methodology and Philosophy of Science. Tech. rep., DLMPS website at www.dlmpst.org/pages/statutes.php.

Tarski, A. (1941). *Introduction to Logic and to the Methodology of the Deductive Sciences.* New York: Oxford University Press.

Tarski, A. (1944). The semantic conception of truth: and the foundations of semantics. *Philosophy and Phenomenological Research*, 4, 341–376.

Tarski, A. (1983). *Logic, Semantics, Metamathematics.* Indianapolis: Hackett.

URSI (2012). Commission J: Radio Astronomy. Tech. rep., International Union of Radio Science website at www.ursi.org.

Van Ulsen, P. (2007). The birth pangs of DLMPS. Tech. rep., DLMPS website at www.dlmpst.org/pages/history.php.

Wilfrid Hodges
Herons Brook, Sticklepath
Okehampton EX20 2PY
England
wilfrid.hodges@btinternet.com

Is There a General Notion of Proof?

DAG PRAWITZ

1 Introduction

The question raised in the title of my talk—Is there a general notion of proof?—may seem a bit strange. Of course, one may say, there is a general concept of deductive proof. We have been familiar with it since the time of ancient Greek. Aristotle had things to say about it, and the general concept has stayed essentially the same since the Greeks, even if today we have more to add.

However, some people may react in a quite opposite direction. With a few exceptions to which I shall soon return, modern logic and philosophy of logic do not deal with a general notion of deductive proof. With the birth of modern logic attention has instead been directed towards formal proofs, and it is clear that one cannot arrive to a general concept of proof in that way. In the main stream of contemporary logic, the interest in even formal proofs has been quite attenuated; often they are seen only as a way of stating facts about the recursive enumerability of the true sentences of a formal language.

Although formal proofs may be seen as representing real proofs or *contentual* proofs, as they are sometimes called, we know from Gödel's incompleteness result that a general concept of proof, if there is such a concept at all, cannot be characterized formally: for any sufficiently rich system of formal proofs that represent contentual proofs, there is an unprovable sentence which is not only true but is intuitively provable. Classically, we have a general concept of computable function, and we discuss general concepts of logical consequence and truth, but there is nothing comparable when we come to the notion of proof. The attitude is often that it makes little sense to speak about proofs in general.

Opinions are in this way divided on the question raised in the title. For my own part, I see the question as a challenge. I share the common opinion that mathematics since the time of the Greeks is distinguished by its deductive character, and I think that to account for this in more detail we should be able to say something about the general concept of deductive proof. But it is a fact that this topic is remarkably neglected.

Proof theory should have something to say about the concept of proof, one could think, but, as Kosta Dozen pointed out in his introduction to the symposium on general proof theory at this congress, its focus has usually been the consistency of mathematics. Nevertheless, when Hilbert coined the term "proof theory", he seemed to have had in mind a much broader field of study. Somewhat bombastically, he depicted the new field by saying:

> we must make the specific mathematical proof itself the object of investigation, just as the astronomer must take his position into account, the physicist must take care of his theory of instruments, and the philosopher criticizes Reason. (Hilbert 1917)

2 The epistemic nature of proofs

If one really wants to study proofs as instruments of the mathematician, one has to take seriously what proofs are instruments for, and obviously the first point of a proof is to acquire knowledge. To make justice to what is essential about proofs, we must therefore acknowledge their epistemic nature.

It is also clear that one cannot get to know something without acting mentally in some way. This means that a proof is first of all an action, the act of proving something.[1] A proof act may be verbally recorded. We then get a representation of the proof act, which may also be called a proof in a transferred sense. It is what we meet in a mathematical paper, and it may serve as an instruction for the reader to carry out the same generic proof act as the author of the paper had carried out.

By a proof in the sense of a proof act, one gets to know that the sentence proved is true. This may seem a banality, but it is far from banal to account for how and why a proof succeeds in giving such knowledge, and this I see as the main conceptual problem about proofs.

To give such an account comes essentially to the same as explaining why a proof is able to justify the assertion that is proved; I take for granted the idea that "to know that p" involves "to be justified in holding p" to be true.

Of course, it is not only in mathematics that assertions are supposed to be justified in some way; a speaker is normally expected to have some ground for what she says. What is particular about mathematics in this respect is a combination of two ideas: assertions are expected to have conclusive grounds, and deductive proofs are supposed to deliver them. The problem is to explain how proofs can have this power.

---

[1] To my knowledge, among contemporary writers, Martin-Löf (1985) and Sundholm (1998) have especially stressed this aspect of proofs.

The term "proof" is commonly used veridically like "know" or "see". Mistakes are of course possible in deductive as well as in empirical matters. But if we have claimed to have a proof of a sentence and the sentence turns out to be false, then we say that we did not really have a proof. Even if there is only a gap in an alleged proof so that it does not provide us with a justification of its last assertion, we say again that we did not have a real proof. In other words, it is a conceptual truth that a proof delivers a justification; it is a part of how we use the notion of proof—for something to be a proof it should give a conclusive ground for the assertion it claims to prove. This common usage does not relieve us from the task of explaining how proofs can have this epistemic power or, in other words, how there can be something that falls under what we call proof.

There are many other aspects of proofs that have intrigued mathematicians and philosophers. For instance, Henri Poincaré stressed features that make one see what he called "the soul of a fact", and blamed logicians for neglecting them in a concern for rigour. Nevertheless he also said:

> In mathematics rigor is not everything, but without it there would be nothing. (Poincaré 1910)[2]

The rigour of proofs can be identified, I think, with they giving justifications or conclusive grounds for their assertions. This aspect of proofs, on which I shall concentrate here, is thus even to Poincaré the most fundamental one, without which "there would be nothing". Being only concerned with conclusive grounds, I shall usually drop the attribute "conclusive" when talking about grounds.

3   Inference and inference assertion

Proof acts are usually compound, that is, they are made up of a number of other acts, inference acts, linked to each other. Problems about proofs may thus be restated as problems about inferences, and again inferences are first of all acts.

The importance of taking inferences acts seriously may be seen as the main lesson of Lewis Carroll's (1895) well-known tale about Achilles and the Tortoise. As we recall, the Tortoise is questioning an inference in Euclid whose premisses he accepts, and Achilles is to compel him to accept the conclusion too. We may slightly modify the tale by assuming that the task of Achilles is instead to show the Tortoise that he is justified in asserting the conclusion since, as he agrees, he is justified in asserting the premisses.

---

[2] Originally stated in an address delivered at the general session of the Fourth International Congress of Mathematics, Rome, April 6-11, 1908.

Achilles strategy is to point out that if the premisses are true then the conclusion must also be true. The Tortoise accepts this implication as an additional justified premiss but asks why he now becomes justified in asserting the conclusion. Achilles repeats his strategy and points out that if the original premisses and the added one are both true, then the conclusion must be true. The Tortoise again accepts this new implication and so it goes on, Achilles getting nowhere by adding more and more premisses. The point is that the Tortoise cannot get justified in making the assertion in question unless he acts by performing an inference. It does not help that he accepts implications.

What is then an inference act? We commonly announce an inference by stating its premisses and conclusion. For instance, we make a construction C and verify that applied to a prime number it yields a greater prime number. From this we infer that there are infinitely many primes. We may announce this inference by stating: "C applied to a prime number yields a greater prime number. Hence, there are infinitely many prime numbers." I shall call such a statement an *inference assertion*. Of course, we may make such an inference without publically announcing it. We then judge that there are infinitely many prime numbers, and take this judgement to be supported by our verification concerning C. When I speak of inference assertions I include silent assertions of that kind.

An inference act thus involves the making of an inference assertion, which is a kind of speech act, but more complex. It consists not only of a number of assertions but also of the claim that one of them, the conclusion, is supported by other ones, the premisses. A lesson from Lewis Carroll is thus that one should not confuse an inference assertion with simply asserting the implication formed by taking the conjunction of the premisses as the antecedent and the conclusion as the consequent; in an inference assertion, the premisses as well as the conclusion are asserted, and in addition it is claimed that the conclusion is supported or inferred from the premisses. The latter is typically indicated by inserting some word like "hence" or "therefore", if the conclusion is stated after the premisses, or by words like "because" or "since", if the conclusion is stated first and the premisses afterwards.

The premisses need not be asserted categorically. They may be assumptions or assertions made under some assumptions. The conclusion may then also be an assertion made under assumptions, or as happens in reductio ad absurdum and implication introduction, the inference discharges assumptions, so that the conclusion is asserted categorically or under fewer assumptions than the premisses. The structure of an inference assertion may thus be more complex than first exemplified. We must even allow that the sentence that is asserted in a premiss or in the conclusion is open, as when we say: "Assume that $\sqrt{2}$ equals a rational number n/m. Then, 2 equals n^2/m^2".

Does an inference act only amount to making an inference assertion? Since one often takes an inference to be individuated by just its premisses and conclusion, one may be inclined to answer yes to this question. But I think that this would be a mistake. It seems fairly obvious that a person who is justified in asserting A does not get justified in asserting B by just making the speech act:

A. Hence B.

But it is less obvious what precisely is contained in an inference above an inference assertion.

The inference figures studied in proof theory, which depict the premisses and conclusion of an inference, represent generic acts of inference assertions. But if it is right that an inference contains something more than an inference assertion, a full representation of an inference cannot consist of just an inference figure but should contain some additional element. I shall return to this question.

4   Legitimate inference

In contrast to the notion of proof, the notion of inference is usually not used veridically.[3] We speak of correct and incorrect, or valid and invalid, inferences. What is it then for an inference to be correct? The obvious, general answer is that it should deliver a justification for its conclusion given that the premisses are justified. I have called such an inference *legitimate* (Prawitz 2011). If we define a proof as a chain of legitimate inferences, it is conceptually guaranteed that a proof delivers a justification for its ultimate assertion provided that its initial premisses are justified.

The crucial issue can now be stated as the question what it is that gives an inference legitimacy. Can we give a criterion for an inference to be legitimate or in some general way characterize the legitimate inferences so as to explain their epistemic power?

It is obvious that logical validity of an inference as it is usually defined in analogy with Bolzano's and Tarski's concept of logical consequence is totally inadequate as a criterion for the legitimacy of an inference. That

---

[3] In the first version of this paper read at the CLMPS in Nancy, I suggested that it would be better to change this terminology and use also the term inference veridically. I argued that it is unnatural to say that one inferred a conclusion from premisses for which one had grounds but yet did not get a ground for the conclusion. I now think that there are other considerations that may make one more doubtful about this proposal. In any case, I am now sticking to the usual terminology in this respect.

the definition pays attention only to the premises and conclusion of the inference is not in itself what makes it inadequate as such a criterion. But, clearly, it is in general insufficient for legitimacy that the inference is *in fact* truth-preserving under all interpretations of the non-logical terms of the sentences involved. For instance, if the premises are the axioms of a theory and the conclusion is a difficult, not yet established theorem, no one would consider the inference to be a legitimate step in a proof, despite its validity, and it would remain so even if the validity becomes known.

It is noteworthy that when one wants to distinguish between bad and good inferences, for instance in elementary textbooks of logic, one has only this very inadequate concept of validity at hand in the main stream of contemporary logic. Aristotle was in fact more advanced. Already at the beginning of *Prior Analytics*, he introduces the notion of *perfect syllogism*, which has the same drift as legitimacy. In his words, as translated by Ross:

> A perfect syllogism is one which needs nothing other than the premises to make the conclusion evident. (Ross 1949)

Aristotle did not try to explain what it is that makes a syllogism perfect, for which we cannot blame him. But it is time now, after more than 2 000 years, to try to make progress in this respect. To get a substantial concept of proof seems to require that we can say something informative about what an inference is and, in particular, what a legitimate inference is.

5 Gentzen's idea about justification of inferences

I turn now to what I take to be the two most important exceptions to the general lack of interest in the topic that I am discussing. One is Gentzen's (1935) ideas about justification of inferences, which I think contain an embryo to a general concept of proof. It has two main ingredients that can be summarized in the form of two principles that are implicit in Gentzen's work. The first says that certain direct or canonical means of proving an assertion are given with the meaning of the asserted sentence, and the second that other indirect or non-canonical means of proving assertions are justified when it is shown that they can be transformed to direct means.

The canonical means consist in inferences that are instances of what Gentzen called introduction rules. Their justification is merely a reflection of they being constitutive of the meaning of the logical constants in question; the introduction rules "represent, as it were, the 'definitions' of the symbols concerned" (Gentzen 1969, 80). An argument or piece of reasoning that ends with an introduction inference is said to be in *canonical form* (Prawitz 1974), and is considered to represent a proof provided that the arguments for the premises do; the inference is in other words taken to be legitimate in virtue of the meaning of the conclusion. The idea that the meaning of a logical constant is given by certain inference rules, in other words, by taking certain forms of inference as legitimate may be seen as a special case of Wittgenstein's more general and vaguer thesis, stated at about the same time, that it is the use of a symbol that determines its meaning. It has been taken up in another way by what has become known as *inferentialism*, saying roughly that all inferences that a language community accepts and that cannot be decomposed into simpler inferences determine the meaning that the words involved have in that language.

In contrast, according to Gentzen it is only the introduction inferences that are meaning constitutive. Other inferences must therefore be justified in a different way. In particular, what Gentzen called an elimination inference for a logical constant, where one of the premisses, the major premiss, has this constant as its outer symbol, has to be justified according to the second principle. To this end there must be a *reduction procedure* such that any argument ending with this inference where the major premiss is inferred by an introduction inference is transformed to another argument for the same conclusion using only the immediate sub-arguments of the given argument.

Such reduction procedures were first explicitly defined to show that a proof in Gentzen's system for natural deduction can be reduced to a certain normal form (Prawitz 1965). But their relevance for our present topic is that they show that an elimination inference uses the major premiss only "in the sense afforded it by the [corresponding] introduction", as Gentzen says—the inference is in other words semantically justified.

Gentzen's ideas were presented with reference to the particular system of natural deduction that he developed but they are not necessarily limited to any particular formal system. To be turned into a general concept of proof they need of course to be further elaborated and generalized. One such attempt has resulted in a notion called *valid argument*.[4] One of its main features is that a non-canonical argument for a closed assertion not depending on assumptions is valid if and only if there are associated reduction procedures that take it to a valid canonical argument for the same assertion. However, as it stands it cannot be said to amount to a general notion of proof. A valid argument does not give us a ground for its conclusion unless we know that the argument is valid, and when this knowledge comes from a proof, it is this proof rather than the argument itself that gives a ground

---

[4]First presented at the 4th CLMPS (Prawitz 1973) and later discussed and modified by among others Dummett (1991) and Schroeder-Heister (2006).

6 The notion of proof within the intuitionistic tradition

The other main exception to the general neglect of the notion of proof that I want to take up is found within the intuitionistic tradition, where notions of proof play an essential semantic role. I am not concerned here with the conflict between classical and intuitionistic logic, and keep it open whether similar ideas about proofs could be developed with respect to the classical meaning of sentences. One may ask if, as far as intuitionism is concerned, there is not already a general concept of proof that satisfies what I am asking for. However, there is not one unambiguous view of proofs within intuitionism, so we should look at some of the different proposals.

When in the beginning of the 1930's Heyting (1930; 1931; 1934) was engaged in the foundation and logic of intuitionism, he explained an intuitionistic proposition as expressing the intention of a construction that satisfies certain conditions and an assertion as affirming the realization of this intention. He furthermore identified a proof of a proposition with the realization of the intention expressed by the proposition.

For instance, the meaning of a proposition $a \to b$ is explained by saying that it expresses "the intention of a construction which from any proof of a leads to a proof of b". To assert $a \to b$ is thus to affirm that this intention has been realized, in other words, that the intended construction has been found, and to prove $a \to b$ is to find such a construction that joined with a proof of a leads to a proof of b.

A proof of a proposition is thus seen as an act, the act of realizing the intention expressed by the proposition. Furthermore, it is an act that justifies the assertion of the proposition, since what is affirmed is just that the intention has been realized or, in other words, that a proof has taken place. Proofs are on this view not only what justify our assertions, but also what one affirms to exist when making an assertion.

As seen, Heyting's view of proofs agrees in two important respects with what I said initially about proofs: proofs are acts that justify assertions. But it breaks with the traditional view of proofs as chains of inferences. In a later survey paper, Heyting (1958) remarked that corresponding to the inference steps of a proof as traditionally viewed there are steps of the mathematical construction that constitutes an intuitionistic proof.

Although Heyting's remarks about proofs are sketchy, his semantic view of propositions and assertions explains to some extent how proofs can have the epistemic power of justifying assertions. One can ask whether having made a construction that in fact satisfies what is required of the construction intended by a proposition means that one also knows that the obtained construction satisfies the required conditions, and if not, whether it is really sufficient to have found the intended construction in order to be justified in making the assertion in question. These are questions that Heyting does not enter into. Nor does he say much about the nature of the constructions and the steps by which they are constructed. One could expect that questions similar to the ones I have raised about inference steps could be asked about construction steps.

There is also a certain ambiguity in Heyting's use of the notion of proof. A proof is not only seen as an act but also as a mathematical construction that could itself be treated mathematically.

Writers on intuitionism after Heyting have often been explaining the meaning of propositions or sentences not in terms of constructions but in terms of what they have called proofs, whose status has sometimes been unclear. When Kreisel at the 1st CLMPS in 1960 (Kreisel 1962) tried to lay a foundation of the logic of intuitionism, he saw proofs as just objects. He then naturally became concerned with the epistemic force of proofs and saw a need to supplement what Heyting had said about proofs of implications and universal quantifications. A proof of an implication $p \to q$ was required to consist of a pair (f, d) where f is a function that transforms any proof of p into a proof of q, and d is a proof of the fact that f has this property; the latter proof was supposed to consist in an application of a decision method to avoid a regress.

Kreisel's work inspired Troelstra (1977) to speak of the Brouwer-Heyting-Kreisel interpretation of intuitionism, abbreviated the BHK-interpretation, where he followed Kreisel in taking proofs of implications and universal quantifications to consist of pairs (f, d), but where d was now said to consist not of a proof but of the insight that f satisfied the required property. Later Troelstra & van Dalen (1987) presented what they also called the BHK-interpretation of intuitionism, but where K now stood for Kolmogorov and where the supplementation of proofs of implications and universal quantifications inspired by Kreisel was dropped.

Let us recall how Troelstra's and van Dalen's well-known interpretation runs. Their aim is to explain the use of logical operations in a constructive context by telling what forms proofs of logically compound statements take in terms of proofs of the constituents, which is done by stating the following clauses:

1. A proof of $A \wedge B$ is given by presenting a proof of A and a proof of B.

2. A proof of $A \vee B$ is given by presenting either a proof of A or a proof of B (plus the stipulation that we want to regard the proof presented as evidence for $A \vee B$).

3. A proof of $A \to B$ is a construction which permits us to transform any proof of A into a proof of B.

4. Absurdity $\bot$ has no proof.

5. A proof of $\forall x A(x)$ is a construction which transforms a proof of $d \in D$ (D the intended range of the variable x) into a proof of $A(d)$.

6. A proof of $\exists x A(x)$ is given by providing $d \in D$ and a proof $A(d)$.

The interpretation is intentionally quite informal, but its account of proofs seems to differ in essential respects from how Heyting saw it and to mix different lines of thought. Clauses 1, 2 and 6 tell how proofs are formed or "given", and there is an indication that a proof is seen as evidence for what it is a proof of. They may be understood as saying that a proof in these cases is formed by applying one of Gentzen's introduction rules. But if so, they characterize only what was called canonical proofs in connection with Gentzen's ideas of proofs and not proofs in general; we must of course acknowledge that there are proofs of conjunctions, disjunctions, and existential quantifications that are not in canonical form.

Clauses 3 and 5 on the other hand identify proofs in these cases with certain mathematical objects, namely the constructions in terms of which Heyting explained respective proposition, and not with the acts of constructing these objects, which is what Heyting took as proofs. One may wonder about the epistemic force of these objects. In what way do they constitute evidence for what they are said to be proofs of?

However, one may instead understand what is said in the BHK-interpretation as differing from Heyting's explanations in merely terminological respects: what Heyting called intended construction is now simply called proof instead, and the point of the clauses 1-5 is to give a recursive account of these intended constructions.

If so, it would be preferable to make explicit how the construction intended by a compound proposition or statement is built up by applying operations to constructions intended by the constituents. For instance, it may be said that a construction of (or intended by) a conjunction $A \wedge B$ is a pair whose elements are constructions of the conjuncts. It is not important that we use exactly the operation of pairing to form a construction of a conjunction. What matters is only that for each of the clauses 1-3 and 5-6 there is a certain operation with the help of which the construction is formed. Since they will have a structural affinity with Gentzen's introduction rules, we may call them *introduction operations* and name them $\wedge I, \vee I$, and so on. Note that there are two disjunction introductions, $\vee I_1$ and $\vee I_2$, and that the two operations $\to I$ and $\forall I$ are variable binding and are therefore written $(\to Ix)$ and $(\forall Ix)$. We then get the following clauses, which should be supplemented with indications of types and with a clause for atomic statements:

1') α is a construction of $A_1 \wedge A_2$ iff $\alpha = \wedge I(\alpha_1, \alpha_2)$ for some constructions α_i of A_i ($i = 1, 2$).

2') α is a construction of $A_1 \vee A_2$ iff $\alpha = \vee I_i(\beta)$ for some construction β of A_i ($i = 1$ or 2).

3') α is a construction of $A \to B$ iff $\alpha = (\to Ix)\Phi(x)$ for some $\Phi(x)$ such that $\Phi(\beta)$ is construction of B if β is a construction of A.

4') α is a construction of $\forall x A(x)$ iff $\alpha = (\forall Ix)\Phi(x)$ for some $\Phi(x)$ such that $\Phi(d)$ is construction of $A(d)$ if $d \in D$ (D the intended range of the variable x).

5') α is a construction of $\exists x A(x)$ iff $\alpha = \exists I(\beta, d)$ for some β and $d \in D$ (D the intended range of the variable x) such that β is a construction of $A(d)$.[5]

There are two ways to read these clauses. One is to see them as defining an abstract notion of construction; they may be denoted by terms that can be introduced later. They should then be supplemented with identity stipulations, saying for instance that $\wedge I(\alpha_1, \alpha_2) = \wedge I(\beta_1, \beta_2)$, if and only if, $\alpha_1 = \beta_1$ and $\alpha_2 = \beta_2$. $\Phi(x)$ stands in this case for a function from constructions or individuals, respectively, to constructions.

Alternatively, the clauses may be seen as steps in building up a language of terms standing for constructions. In this case they are to be understood as describing the canonical forms of such terms, and have to be supplemented by introducing also non-canonical terms. $\Phi(x)$ is in this case an open term in the language, and it is important to note that $\Phi(\beta)$ or $\Phi(d)$ need not be terms in canonical form. The language may be seen as that of an extended lambda calculus. If it is supplemented with operations analogous to Gentzen elimination rules, it will be structurally similar to Gentzen's system of natural deduction; a well-known fact known as the Curry-Howard isomorphism.

The BHK-interpretation of Troelstra and van Dalen is best seen in this way as a recursive characterization of the constructions intended by compound propositions or sentences in predicate logic in the sense of Heyting—

---

[5] Already Kreisel Kreisel (1962) made explicit operations of this kind by which the various constructions are formed. Clauses of more or less the form exhibited here have since then been used by a number of authors, see for instance Prawitz (1970), Howard (1980), and Martin-Löf (1984, 12–13), or (1994).

supplemented with operations for how to form compound constructions from constructions of the constituents—and not as proofs that give evidence.

The conclusion that the proofs referred to in intuitionistic meaning explanations are mathematical objects and not epistemic proofs was drawn some time ago by Per Martin-Löf (1998). In his type-theory there is a clear distinction between on the one hand proofs in the sense of constructions of propositions and on the other hand proofs in the sense of demonstrations of assertions or judgements. Nowadays he usually calls the former *proof-objects* and the latter *demonstrations*. The demonstrations establish that a proof-object is a proof-object of a particular proposition or an object of a particular type. They proceed by inferences in the traditional way, and the question what makes them legitimate remains.

In the works of Gentzen and Heyting there are implicit suggestions based on their semantic ideas for a general concept of proof that could clarify how and why a proof has epistemic force, but as far as I can see there has been no successful attempt to work out such a concept.

7 Grounds

It is the epistemic character of their meaning explanations that makes Gentzen's and Heyting's ideas promising for the project to explain the epistemic force of proofs, in my opinion. That Gentzen's explanations are in epistemic terms is plain since he explains the meaning of sentences in terms of inference rules. It is less clear in what sense the intuitionistic explanations in terms of constructions are epistemic. Martin-Löf (1998) lists the notion of construction among non-epistemic concepts. Sundholm wants to deny that they have any epistemic significance whatever, saying for instance:

> A proof-object is a mathematical object like any other, say a function in a Banach space whence, from an epistemological point of view, it is of no more forcing than such objects. (Sundholm 1998, 194)

I want to argue to the contrary that the proof-object in terms of which the meaning of a sentence A is explained intuitionistically constitutes a ground for asserting A. By saying that something is a ground for an assertion, I mean that it is sufficient to be in possession of it in order to be justified in making the assertion. That the proof-objects are grounds in this sense is something that Martin-Löf seems to agree with, saying "to have the right to make a judgement of the form 'A is true', you must know a proof of A" (Martin-Löf 1998, 112); in this quote "proof" stands for what Martin-Löf now calls proof-object.

The crucial question is what it is to be in possession of a ground or know a proof-object. Note that grounds are objects that it is possible to get in possession of. Sundholm (1994) makes the reasonable suggestion that a proof-object of a proposition A is a truth-maker for A. When one adopts the position of realism, one cannot assume that the entities that are taken as truth-makers of propositions are always possible to get in possession of. But in contrast, as Heyting was keen to emphasize, the intention expressed by a proposition is not to be understood transcendentally "as an imagined state of affair existing independently of us, but as an imagined possible experience" (Heyting 1931, 113). The intended constructions are in other words thought of as something we can experience or get to know, and this is what makes it possible for intuitionistic truth-makers to be at the same time grounds for assertions. But we have to account for how we get in possession of them.

We can know an object only under some description of it, and to get in possession of a construction α we have to form a term that denotes α. It may be in canonical or in non-canonical form. For instance, if one has formed a term $\wedge I(t_1, t_2)$, and knows that t_i denotes a ground for asserting A_i ($i = 1, 2$), I shall say that one is in *direct possession* of a ground for asserting $A_1 \wedge A_2$. Provided that one knows the meaning of $A_1 \wedge A_2$, one then knows that $\wedge I(t_1, t_2)$ denotes a ground for asserting $A_1 \wedge A_2$.

When we have formed a term in non-canonical form that denotes a proof-object, it is more difficult to say what has to be required in order to be in *indirect possession* of a ground for the assertion in question. For instance, we can define two operations $\wedge E_1$ and $\wedge E_2$ by the equations $\wedge E_i(\wedge I(\alpha_1, \alpha_2)) = \alpha_i$ ($i = 1$ or 2) and form a non-canonical term $\wedge E_1(u)$. If one knows that u denotes a proof-object of $A_1 \wedge A_2$ and knows how $\wedge E_1$ is defined, I shall say that one is in indirect possession of a ground for A_1. I think that it is because of the nature of the definition of the operation $\wedge E_1$ that it is right to say so in this case.

The operations $\wedge E_1$ and $\wedge E_2$ are of the same nature as the reduction procedure for conjunction that Gentzen saw as justifying the inference conjunction elimination. It is essential to characterize this common nature of such reduction procedures and certain operations on constructions if one is to argue for the view that one can be in indirect possession of a ground for the assertion of a sentence A by having formed a non-canonical term t denoting a proof-object α of A without having actually proved that t denotes α. But I have to leave this problem here.

8 The concept of inference reconstructed

To arrive at a general concept of proof where proofs are seen as chains of inferences, the main problem as I see it is to say how the performance of an inference can result in a ground for asserting the conclusion. It cannot be enough to prove that the inference is justified in the sense that there exists a ground for the conclusion, because then it is this proof rather than the performance of the inference in itself that gives a ground for the conclusion.

To make an advance concerning this problem, we have to reconsider the concept of inference. A kind of synthesis of ideas from Heyting and Gentzen can help us here. We should now return to the question that I raised in the beginning of the paper what the difference may be between performing an inference and making an inference assertion. What more is involved when we perform an inference? If we accept the view discussed in the previous section that to be justified in asserting a sentence is to be in possession of a construction of the sentence, and can make sufficiently clear what it is to be in possession of a ground, the natural answer is, I want to suggest, that the performance of an inference involves in addition to an inference assertion an operation on the grounds that one considers oneself to have for the premisses. When successful, the operation results in one getting in possession of a ground for the conclusion.

Examples of operations that one can apply are firstly the primitive ones referred to when saying in clauses 1'—6' what counts as constructions of sentences in predicate logic and secondly operations that can be defined on these constructions such as $\wedge E_1$ and $\wedge E_2$. In the first case we can get in direct possession and in the second case in indirect possession of a ground.

For instance, to make an inference by conjunction introduction is to support the assertion of a sentence $A_1 \wedge A_2$ by two premisses asserting A_1 and A_2 and to apply in addition the operation $\wedge I$ to alleged grounds for the premisses. To make an inference by the first form of conjunction elimination is to support the assertion of a sentence A_1 by a premiss asserting $A_1 \wedge A_2$ and to apply in addition the operation $\wedge E_1$ to an alleged ground for the premiss. If these operations are applied to real grounds for the premisses, one thereby gets in possession of a ground for the conclusion.

The proposal is thus that an inference is to be individuated not only by its premisses and conclusion but also by an operation applicable to grounds for the premisses. For instance, modus ponens (implication elimination) is defined not only by saying that it is an inference with premisses of the form A and $A \to B$ and a conclusion B but also by giving an operation $\to E$ applicable to grounds for the premisses and defined by the equation

$$\to E[(\to Ix)\Phi(x), \alpha] = \Phi(\alpha).$$

An inference can now be defined as *deductively valid* when the result of applying its operation to grounds for the premisses is a ground for the conclusion.

When the notions of inference is reconstructed in this way it is a conceptual truth that when one performs a deductively valid inference as now defined one gets in possession of a ground for the conclusion provided that one was in possession of grounds for the premisses. We can then arrive at a general notion of deductive proof by defining it as a chain of deductively valid inferences.

9 Logically valid inferences

The suggested definition of deductively valid inference makes an inference valid in virtue of the meaning of the involved sentences. This is a consequence of the leading idea that the meaning of a sentence is given in terms of what counts as ground for asserting the sentence. For this reason, one could speak of *analytic validity* instead of deductive validity.

Let me end by noting that from this notion of validity one can easily define a notion of logical validity by applying the same general idea that Bolzano and Tarski used when defining the concept of logical consequence. We can simply define an inference as *logically valid* when it remains deductively (or analytically) valid under any (re-)interpretation of the non-logical terms occurring in the sentences involved. In contrast to Bolzano's and Tarski's definition, this definition does not refer to the truth-values of the sentences involved but to the notion deductively valid inference, which is a more basic notion than the notion of logically valid inference.

Bibliography

Carroll, L. (1895). What the Tortoise said to Achilles. *Mind*, *4*(14), 278–280, doi:10.1093/mind/IV.14.278.

Dummett, M. (1991). *The Logical Basis of Metaphysics*. London: Duckworth.

Gentzen, G. (1935). Untersuchungen über das logische Schließen. I. *Mathematische Zeitschrift*, *39*(1), 176–210, 405–431, doi:10.1007/BF01201353.

Gentzen, G. (1969). *The Collected Papers by Gerhard Gentzen*, Szabo, M. E., ed., Amsterdam: North Holland.

Heyting, A. (1930). Sur la logique intuitionniste. *Académie Royale de Belgique, Bulletins de la Classe des Sciences*, *16*, 957–963.

Heyting, A. (1931). Die intuitionistische Grundlegung der Mathematik. *Erkenntnis*, *2*(1), 106–115, doi:10.1007/BF02028143.

Heyting, A. (1934). *Mathematische Grundlagenforschung, Intuitionismus, Beweistheorie*. Berlin: Springer.

Heyting, A. (1958). Intuitionism in mathematics. In *Philosophy in the Mid-Century*, Klibansky, R., ed., Florence: La Nuova Italia, 101–115.

Hilbert, D. (1917). Axiomatisches Denken. *Mathematische Annalen*, 78(1), 405–415, doi:10.1007/BF01457115.

Howard, W. A. (1980). The formulas-as-types notion of construction. In *To H. B. Curry: Essays on Combinatory Logic, Lambda Calculus, and Formalism*, Seldin, J. P. & Hindley, J. R., eds., London: Academic Press, 479–490.

Kreisel, G. (1962). Foundations of intuitionistic logic. In *Logic, Methodology and Philosophy of Science*, Nagel, E., ed., Stanford: Stanford University Press, 198–212.

Martin-Löf, P. (1984). *Intuitionistic Type Theory*. Napoli: Bibliopolis.

Martin-Löf, P. (1985). On the meanings of the logical constants and the justifications of the logical laws. In *Atti degli Incontri di Logica Matematica*, vol. 2, Università di Siena: Scuola di Specializzazione in Logica Matematica, Dipartimento di Matematica, 203–281, republished in *Nordic Journal of Philosophical Logic*, 1,11–60, 1996.

Martin-Löf, P. (1994). Analytic and synthetic judgements in type theory. In *Kant and Contemporary Epistemology, The University of Western Ontario Series in Philosophy of Science*, vol. 54, Parrini, P., ed., Dordrecht: Kluwer Academic Publishers, 87–99, doi:10.1007/978-94-011-0834-8_5.

Martin-Löf, P. (1998). Truth and knowability: on the principles C and K of Michael Dummett. In *Truth in Mathematics*, Dales, H. G. & Oliveri, G., eds., Oxford: Clarendon Press, 105–114.

Poincaré, H. (1910). *The Future of Mathematics*. Washington: U. S. Government.

Prawitz, D. (1965). *Natural Deduction. A Proof-Theoretical Study*. Stockholm: Almqvist & Wiksell, republished by Dover Publications, New York, 2006.

Prawitz, D. (1970). Constructive semantics. In *Proceedings of the 1st Scandinavian Logic Symposium Åbo 1968*, Uppsala: Filosofiska Föreningen och Filosofiska Institutionen vid Uppsala Universitet, 96–114.

Prawitz, D. (1973). Towards a foundation of a general proof theory. In *Logic, Methodology, and Philosophy of Science IV*, Suppes, P., ed., Amsterdam: North-Holland, 225–250.

Prawitz, D. (1974). On the idea of a general proof theory. *Synthese*, 27(1–2), 63–77, doi:10.1007/BF00660889.

Prawitz, D. (2011). Proofs and perfect syllogisms. In *Logic and Knowledge*, Cellucci, C., ed., Newcastle upon Tyne: Cambridge Scholars Publishing, 385–402.

Ross, W. D. (1949). *Aristotle Prior and Posterior Analytics*. Oxford: Oxford University Press.

Schroeder-Heister, P. (2006). Validity concepts in proof-theoretic semantics. *Synthese*, *148*(3), 525–571, doi:10.1007/s11229-004-6296-1.

Sundholm, G. (1994). Existence, proof and truth-making: A perspective on the intuitionistic conception of truth. *Topoi*, *13*(2), 117–126, doi: 10.1007/BF00763510.

Sundholm, G. (1998). Proof as acts and proofs as objects. *Theoria*, *54*, 187–216.

Troelstra, A. S. (1977). Aspects of constructive mathematics. In *Handbook of Mathematical Logic*, Barwise, J., ed., Amsterdam: North-Holland, 973–1052.

Troelstra, A. S. & van Dalen, D. (1987). *Constructivism in Mathematics*. Amsterdam: North-Holland.

Dag Prawitz
Department of Philosophy
Stockholm University
Sweden
dag.prawitz@philosophy.su.se

"The Soul of the Fact"— Poincaré and Proof

JEREMY GRAY

ABSTRACT. Throughout his life Poincaré reflected on how to be a productive mathematician and physicist. Many of his popular essays were influential, and remain interesting today, because they argue for his opinions about the nature of mathematics and science. But his work has acquired a reputation for being impressionistic and lacking in rigour, and while there is some justification for this there is more to be said for the view that Poincaré always sought to advance one's understanding of a problem or a topic. This could be done in various ways, he suggested, chief among them being the identification of "the soul of the fact", the key concept that enabled the best way to organise one's ideas. Poincaré's sense of human understanding was focussed on its capacity to create new knowledge, and can be illuminated from a perspective that places it close to what Wittgenstein later advocated.

1 Poincaré and rigour

For Poincaré, the uninteresting part of proof was rigour, the interesting part was the role a proof plays in understanding a piece of mathematics. As he put it in *L'Avenir*, his address to the International Congress of Mathematicians (ICM) of 1908:

> Rigour is not everything—but without it there is nothing. (Poincaré 1908a, 932)[1]

Nonetheless, he cared about rigour, as his correspondence with Fuchs in 1880 demonstrated, and as does his work on asymptotic series.

The correspondence with Fuchs (see Poincaré 1921b) began on 29 May 1880 as soon as Poincaré had submitted his essay on differential equations in the complex domain for the prize of the Paris Académie des Sciences.

---

[1] A full English translation appears in (Gray 2012b).

This is the competition later won by Georges Halphen that was the occasion for Poincaré to discover the theory of automorphic functions (see Poincaré 1997). But in May Poincaré was still considering the subject from an entirely analytic point of view, and his questions to Fuchs were about the analytic continuation of the quotient of two independent solutions of a linear differential equation. This was standard research material of the day, and one that Fuchs was the acknowledged expert in, but we have the somewhat comic sight of Poincaré explaining the subtleties of analytical continuation to the older man. What he saw, and Fuchs had missed, was an insight into the global nature of the image defined by the quotient. This derived from Fuchs's immersion in a tradition that emphasised local aspects, such as the nature of singular points of an analytic function, and provided techniques for dealing with them, but was much less well equipped to handle global questions. But nonetheless, it was Poincaré, not Fuchs, who was rigorous and Poincaré who, through this insistence on rigour was able to reach the situation where the attention to the behaviour of the inverse of the quotient and the nature of its domain was to lead to the great discovery of the importance of non-Euclidean geometry.

The same story can be told with Poincaré's work on asymptotic series, presented in (Poincaré 1886). Astronomers had observed that certain power series expansions that are not known to be convergent, and may even be known to be divergent, can even so be truncated after a certain number of terms and give useful information. But it is a delicate journey from there to a theory of how this can be, and what operations (such as differentiation and integration) are permissible under what conditions. As with Poincaré's work on several other technical topics in the theory of analytic functions, there could be nothing without rigour.

Poincaré also had reluctant criticisms of rigour. Proofs can be too large, he argued in *L'Avenir*, and well-chosen terms, such as 'uniform convergence' would encapsulate progress and prevent rigorous proofs from becoming almost incomprehensibly too long. Likewise, calculation should be an irreducible minimum, and never blind. Such proofs, he suggested, while valid, could not be properly understood.

A more substantial objection was that proofs can be wrong in kind, as was the case, he suggested, in potential theory, where they do not mimic the actual processes involved. More-or-less intuitive proofs, he said in an analysis of his own scientific work in a memoir[2] written in 1901, are of the right sort to satisfy a physicist because they leave the mechanism of the phenomena apparent. More rigorous arguments for the existence of solutions depended on convergence arguments but this convergence was usually too slow, and

---

[2] Published as (Poincaré 1921a), see *Œuvres* 9, 2.

the approximations involved too complicated for such approaches to yield effective numerical procedures. The implication is not only that there was a better proof to be found that would speak to both the physicists and the mathematicians. Poincaré was also explicit, in (Poincaré 1890b), that the physicists' understanding was not good enough. He argued instead that one could not be content with the lack of a rigorous proof; analysis itself should be able to solve such problems. Any rigorous solution is, of course, a solution, and even if crude nonetheless teaches us something. But was it not needlessly pedantic to seek the rigorous solution of equations that had only been established by approximate methods and which rested on imprecise experimental foundations? His answer was 'no': how could one be sure that something less than a rigorous proof was not actually flawed; had one the right to say that something inadequate for mathematics was yet good enough for physics?—the line was impossible to draw. One could not, as a mathematician, settle for less, and in any case many of these equations had applications not only in physics but also in pure mathematics (for example, he observed, Riemann himself had based his magnificent theory of Abelian functions on his use of Dirichlet's principle).

A further objection to rigour that Poincaré held was that there are occasions when it is is not enough. He observed in his (1905a) that Hilbert had exposed the formal character of reasoning in geometry, and remarked that even if the same was done for arithmetic and analysis, mathematics could not be reduced to an empty form without mutilating it and the origin of the axioms would still have to be investigated, however conventional they were taken to be. In (Poincaré 1908a, 932), he remarked that logical correctness is not all.

> A lengthy calculation that has led to a striking result is not satisfying until we understand why at least the characteristic features of the result could have been predicted.

And because it is not order per se, but only unexpected order that has a value, the mechanical pursuit of mathematics would be worthless,

> "A machine can take hold of the bare facts, but the soul of the fact will always escape it".

So the problem for Poincaré was: How to proceed? Isolated facts had no appeal for him, but a class of facts held together by analogy brings us into the presence of a law, and as he continued in L'Avenir, in explicit agreement with Ernst's Mach's principle of the economy of thought,

> The importance of a fact is measured by the return it gives—that is, by the amount of thought it enables us to economise.

Poincaré argued that the elegance of a good proof reflects an underlying harmony that in turn introduces order and unity and "enables us to obtain a clear comprehension of the whole as well as its parts. But that is also precisely what causes it to give a large return" (Poincaré 1908a). The aesthetic response to mathematics was regarded by Poincaré as a sign of its efficacy, and this pair of ideas then shaped the rest of his address.

2 Poincaré on progress in mathematics and physics

Poincaré was not seduced by flashes of insight. He explicitly commented that these, although convincing at the time, can mislead. As he put it in his address to the Parisian Society of Psychologists in 1908 (see his 1908b), the unconscious provides points of departure for calculations that must be made consciously, but operates by chance. And one must be careful, for the unconscious presents these ideas with a feeling of certainty even when, on rational analysis, they prove to be worthless.

There was, however, an in-built activity of the mind that was capable of providing knowledge, and that was our ability to reason by recurrence, and this allows for the growth of knowledge. And, he implied in his (1902a), "Who doubts arithmetic?" (Perhaps no-one in 1900, when he made these remarks at the Paris ICM.)

Not many years later there were people who did indeed, if not doubt arithmetic at least deny it a fundamental status. In 1909 Poincaré, who was also fighting acrimonious battles with Russell, Couturat, and other logicians, responded to Zermelo's first attempt at an axiomatisation of set theory (as presented in papers written 1904 and 1908 and now collected in (Zermelo 2010)). Poincaré began his (Poincaré 1912a) by observing that an axiom system of any kind must be free of contradiction. If this could not be done by an appeal to some other system, as Hilbert had done with his axioms for various geometries by appealing to arithmetic, then the only hope is that the axioms be self-evident. This was the situation Zermelo was in, but Poincaré found Zermelo's axioms far from self-evident. In particular, he was unconvinced by Zermelo's use of the term 'Menge' to identify the type of collection about which we can reason. For Poincaré these would be sets with predicative definitions, so that each member has, as it were, its own entitlement to membership. For Zermelo, these were collections with a 'definite' membership criterion, but by not requiring definiteness to mean predicativity Poincaré felt that Zermelo had not been careful enough:

But even though he has closed his sheepfold carefully, I am not sure that he has not set the wolf to mind the sheep. (Poincaré 1912a, 87, 67)[3]

Predicative definitions permit clear checks on membership of a set and impose limits on the size of sets, or so Poincaré believed, which is why he rejected the well-ordering axiom; no set larger than the first uncountable set can be surveyed. This was Poincaré's second objection: set theorists spoke to him far too easily of very large sets. Since this was a consequence of their approach, he took it as evidence that their concept of a set was not self-evident, and accordingly rejected it. The nub of Poincaré's opinion was that what could not be understood by the human mind should not be talked about, however formally it could be expressed.

If neither naive intuition nor strict logic nor axiomatic set theory was the right basis, what could be? Poincaré invested considerable effort in deciding how to conduct his own research,as perhaps many a researcher does, but unusually he also spelled out explicitly for others the way in which he operated and which, he believed, it was most propitious to proceed. Among the topics he considered was a lifelong interest of his, number theory, and because his work in that subject is less familiar than are his achievements in some of the other fields he occupied I shall draw examples from it here.

Poincaré on progress in mathematics

Higher arithmetic is difficult, he explained in *L'Avenir*, and progress slow because there can be no appeal to continuity. Therefore the subject should be guided by the numerous analogies with algebra, and in his first works he argued that it can be (partially) unified by use of transformations. This was an approach he had taken in his first substantial number theory paper, (Poincaré 1881). Here he had begun by observing that Hermite had completely and elegantly solved the problem of finding canonical representatives for quadratic forms, and then offered an extension of Hermite's methods to forms of higher degree, specifically cubic forms in 3 or 4 variables.[4] Poincaré considered the effect of linear changes of variables on a given form. He noted that the effect of following one change of variables by another depended on the order in which the transformations were carried out, and then considered different types of linear transformation: they could be unitary (they have determinant 1), real (they have real coefficients) or integral (integer

---

[3] Page references of this kind refer to the French and English editions of the text where appropriate; I have used (Poincaré 2001).

[4] Hermite had looked for all transformations of a ternary quadratic form to itself in 1853 in papers in *JfM* 47, see his *Œuvres*, I, and had considered the group that maps an indefinite ternary quadratic form $x^2 + y^2 - z^2$ to itself.

coefficients). He called two forms F' and F'' equivalent if there is a third form F and transformations T' and T'' such that $F' = FT'$ and $F'' = FT''$, and said this equivalence is algebraic, real, or arithmetic if the corresponding transformations are, respectively, unitary, real, or integral. Because a linear transformation can be thought of as a change of coordinates, Poincaré set himself the task of classifying the possible transformations where T and $\Sigma = S\nu TS$ are regarded as equivalent.

He then showed how the groups of transformations that arise yield a classification of the forms that agrees with the geometrical classification of them as loci or surfaces, which German geometers such as Hesse, Clebsch, and Gordan had already presented. The first part of the paper ended with a table of the cubics in four variables that are indecomposable, do not reduce to forms in three variables, and have non-trivial self-transformations, and the second part of the paper, (Poincaré 1882), Poincaré turned to the real and integral theories, which give a finer classification of the forms.

Hermite's response to this paper was to urge Poincaré to make an explicit investigation of the reduced forms, because calculation can reveal what noone could otherwise see or predict. This was never to be Poincaré's way. Hadamard, much later, on the other hand observed of this work (Hadamard 1921, 168), that the problem of indecomposable forms "disappeared, in this sense, that an idea of rare simplicity gave the rule applicable to all problems of this sort at a stroke". What was left, he explained was a purely algebraic problem of reducing the form to a canonical type, and then a problem about the arithmetic group.

Poincaré's interest in Hermite's work on number theory explains an otherwise mysterious but famous incident in Poincaré's discovery of the riches of non-Euclidean geometry. In the summer of 1880, as he recalled in his lecture in 1908, he realised while walking by the sea-side, that the arithmetical transformations of ternary indefinite quadratic forms were identical with those of non-Euclidean geometry.[5] This not only helped to illuminate the reduction of these forms to canonical form, it was the key that opened the way to Poincaré's theory of Fuchsian groups.

Poincaré's belief that the group idea was central to many different problems in mathematics continued to animate his work in number theory. In 1887 he wrote a major paper on Fuchsian functions and arithmetic (Poincaré 1887). He was interested in the famous modular equation, but not, as Hermite had been because it bridged elliptic function theory and number theory, but because he wanted to understand why it existed at all.

He argued that the modular function $J(z)$ is invariant under the group $\Gamma = SL(2,\mathbb{Z})$, and that the transformation S given by $z \mapsto z/n$ is not in

---

[5] See (Poincaré 1908b), in (Poincaré 1908c, 52–53, 393).

this group, but the relationship between $J(z)$ and $J(z/n)$ is governed by the celebrated modular equation. To generalise this, he introduced the idea of commensurable groups. He said that two groups G_1 and G_2 are commensurable if their intersection H is a subgroup of finite index of both G_1 and G_2, and he noted that the groups $SL(2,\mathbb{Z})$ and $S\nu SL(2,\mathbb{Z})S$ are commensurable, so there is an equation between $J(z)$ and $J(z/n)$—the modular equation.

Next, Poincaré looked for commensurable Fuchsian groups. He observed that there are three groups of interest that map a given ternary form to itself, and the matrices involved have either real, rational, or integer coefficients, so he denoted the groups

$$\Gamma_{\text{rec}}, \quad \Gamma_{\mathcal{Q}}, \quad \Gamma_{\mathbb{Z}}.$$

Each of these groups gives rise to a corresponding Fuchsian group. When he used the group $\Gamma_{\mathbb{Z}}$ Poincaré called the corresponding Fuchsian group the principal group, $G_{\mathbb{Z}}$, and this led him to the generalised modular equations. He took a $\Gamma_{\mathcal{Q}}$, and $S \in \Gamma_{\mathcal{Q}}$ with rational, non-integral coefficients, and argued that the Fuchsian group corresponding to $\Gamma_{\mathcal{Q}}$ is not discontinuous, so an element S in this group gives rise to a Fuchsian transformation $s \notin G_{\mathbb{Z}}$. But the groups $G_{\mathbb{Z}}$ and $s\nu G_{\mathbb{Z}} s$ are commensurable, so a Fuchsian function for the principal group is algebraically related to its transform by s. The relation takes the form of a polynomial equation, and so the existence of a family of Fuchsian groups with arithmetic properties explains why there is, in particular, the modular equation. (It should be added that this generalisation is not trivial, the proofs involved difficult and unfamiliar extensions of familiar ideas such as summing over the elements of a Fuchsian group and not over the integers. Good analogies are seldom simple.)

In short, what Poincaré had done was not, in his view, just a generalisation. Rather, where Hermite had found a pre-existing phenomenon, the modular equation, Poincaré went looking for a generalisation, and found it via a group-theoretic analysis which explained why the modular equation existed.

Poincaré on progress in physics

Here again analogy and generalisation played fundamental roles. One brief example will illustrate the point. It is well-known that one of Poincaré's great innovations in the theory of celestial mechanics was the idea of perturbing the system studied, and seeking to see if solutions known to exist in a simple case survived the transition to more general cases that were, in some sense, nearby. In his work for King Oscar II of Sweden's prize competition (Poincaré 1890a) Poincaré supposed that a system of equations depending on a parameter μ was solvable when $\mu = 0$ and that among the solutions were some that were periodic.[6] For example, consider the three body problem with masses $\alpha_1, \alpha_2\mu$, and $\alpha_3\mu$, where μ i very small. When $\mu = 0$, the explicit solutions have the two small bodies orbiting the large one in Keplerian ellipses, and there is an infinity of periodic solutions.

Poincaré now asked under what conditions has one the right to conclude that there will still be periodic solutions for small values of μ, and he promised to show that when μ is sufficiently small the problem for each value of μ still has infinitely many periodic solutions, as he did also later in 1892 in his (1892), the *Les Méthodes nouvelles de la mécanique céleste*.

The work was very difficult, and he looked for a simplified version. Eventually, to understand the problem in the simplest analogous case, he considered the topic of closed geodesics on spheroids in his (1905b). A spheroid is a surface that differs only slightly (in ways Poincaré did not specify) on a parameter μ from a sphere, where, of course, the closed geodesics are the great circles. To derive the appropriate equations for geodesics on the spheroid, Poincaré imagined that as μ increased from zero each point on the sphere moved to its corresponding point on the spheroid, (corresponding points have parallel tangent planes). This gave him a good way to relate the maps of the sphere and the spheroid on a plane. He could now investigate when a closed geodesic on the sphere (a great circle) remains a geodesic on the spheroid.[7]

3 Poincaré on mathematics and physics

Most importantly, Poincaré argued at the ICM in 1897, (Poincaré 1897) that mathematics and physics are inseparable.[8]

> Mathematics has a triple purpose: it must provide an instrument for the study of nature; it has a philosophical purpose and, I would say, an aesthetic purpose.
>
> It must aid the philosopher to make our ideas of number, space, and time more profound.

Mathematics, he went on, is not a mere provider of formulae for physics. Indeed

> The first reason why the physicist cannot give up mathematics is: it provides him with the only language he can speak.

---

[6] For a rich historical account, (see Barrow-Green 1997).

[7] For an excellent account of Poincaré's paper, (see Anantharaman 2006/2010).

[8] To be precise, Poincaré did not attend this ICM because his mother had died on 17 July 1897.

On the other hand,

> The only natural object of mathematical thought is the integer [...] It is the external world that has imposed the continuum upon us, which we would have invented without doubt, but we have been forced to invent. Without it there would be no infinitesimal analysis, and all of mathematical science would reduce to arithmetic or to the theory of groups.

The remark about the continuum is particularly noteworthy because Poincaré had elsewhere spoken favourably of a different continuum introduced by Du Bois Reymond (see Poincaré's 1893) which contains infinitesimals.

A crucial test of any philosophy of science is how it deals with theory change, and Poincaré was very aware that theories change and die. What survived, he believed, were the relations in which the theory was expressed. The aim of mathematical physics he proclaimed as being "to reveal the hidden harmony of things..."—a harmony of *relations* between facts—by 'facts' he meant the results of accepted experiments and mathematical theorems. Confronted, as one is, with what he called the melancholy remains of failed theories, he argued that the equations in which the old theories were expressed are still true, and the relations they capture preserve their reality. But as for what is related, as Poincaré put it at the International Congress of Physics (Poincaré 1900):

> these are merely names of the images we substituted for the real objects which Nature will hide for ever from our eyes. The true relations between these real objects are the only reality we can attain, [and] When theories seem to contradict each other, it is likely that it is the images we have supplied which stand in contradiction.

On this occasion Poincaré gave the examples of billiard ball atoms and the fluids of Coulomb, once old-fashioned "yet here they are re-appearing under the name of electrons".

4 Poincaré on philosophy and conventionalism

All of which brings us to the place Poincaré assigned to conventions in mathematics and physics. It is important to recognise that these are of two kinds. What is called geometric conventionalism was expressed in his repeated argument that we cannot tell if space is Euclidean or non-Euclidean. This is because there are no logical grounds for distinguishing the claim that Space is Euclidean and light rays are curved from the claim that Space is non-Euclidean and light rays are straight. (Here light rays stand in for any physical embodiment of straight lines.) Accordingly, we make a choice on grounds of convenience—but this choice, as he explained in his article in the *Monist* for 1898, is made because of our inherent ability to construct a theory of space out of our innate appreciation of rigid bodies. Geometrical conventionalism is fundamental to our ability to have knowledge of the external world at all, knowledge that is acquired before we are capable of receiving formal education of any kind.

Other conventions arise in our construction of physical theories. These included, by Poincaré's time, Newton's laws of motion, and the law of conservation of energy. At the International Congress of Physics, 1900, but also on other occasions, Poincaré repeatedly stressed the conventional element in mechanics, such as Newton's laws of motion, the definition of force, and the conservation of energy, are not increasingly well confirmed experimental results, rather, they have been elevated to the status of conventions. They function as axioms, and they are not to be put in question by an unexpected result. On such occasions scientists looks for an unexpected, and possibly new, process at work; they do not doubt the basic principles. The principles are true by convention, and to deny them on such grounds is not to be a radical scientist but to cease to be a scientist altogether.

Poincaré was challenged at that Congress by the radical conventionalism of his former student Édouard le Roy, who argued that science was a mere game that produced rules for action but no actual knowledge and supposed scientific facts were the creation of the scientist. Poincaré replied (Poincaré 1902b) that science provided knowledge because it makes predictions that are, in the main, correct. Scientists, he argued, do not create scientific facts. They start with the brute facts, and all they create is the language in which they express these facts. There is a greater creative role, he allowed, when it comes to scientific laws that have been raised to the status of principles. But here he insisted that any substantial disagreement will be settled by appeal to convention: principles are neither true nor false but conventional and convenient (4). True to his conventionalism, Poincaré set great store by the ability to converse effectively. For him science was objective because it rested on communication between people. In 4 he returned to his hypothetical confrontation between Euclideans and non-Euclideans, and argued that if they have analogous senses and accept the same logic then it would be possible to translate their language into ours. And in every case where translation is possible, there is an invariant (which is what is being said in each language) and these invariants are laws which in turn are relations between crude facts, expressed differently in each language. As he put it in 6, "No discourse, no objectivity".

Science, he went on, speaks only of relations between sensations, and once the role of conventions is understood it is objective precisely because it is a system of relations. But, he insisted, science was not about objects in themselves. Indeed, to say that science cannot be objective because it can speak only of relations and never of things 'in themselves' or 'as they really are' is absurd. Nothing can reveal the true nature of things, and, in words surely chosen to hint at le Roy's theology, Poincaré added (see Poincaré 2001, 267, 347–348) or (Poincaré 1904a), that if some god did know the true state of things "he could not find the words to express it. Not only would we not be able to guess the answer, but if one gave it to us we would not be able to understand it".

Even so, the fundamental principles can be challenged, and in his lectures in St Louis in 1904 (Poincaré 1905c) Poincaré discussed how this challenge was seemingly underway. His paper (Poincaré 1904b) is remarkable because it marks the closest he ever came to producing a relativistic theory of electrodynamics. He proposed that "the laws of physical phenomena must be the same for a fixed observer as for an observer who has a uniform motion of translation relative to him" (see Poincaré 2001, 176, 294), and deduced from this that

> From all these results would arise an entirely new mechanics which would above all be characterised by the rule that no velocity could exceed the velocity of light. (see Poincaré 2001, 197, 314)

Poincaré summarised what had driven him to contemplate a new physics and a revision of the conventions of contemporary physics under several headings. The fundamental theory of thermodynamics had no theoretical foundations. Newton's third law (action and reaction are equal and opposite) was contradicted by the best existing explanation of the persistent failure to detect the Earth's motion through the ether, which was Lorentz's theory in which the ether affected the electron but not the other way round. Lorentz had also suggested that all matter might be electro-magnetic in nature, in which case it might indeed not obey Newton's laws, a view that did not attract Poincaré. The study of cathode rays, by then understood as the motion of high speed electrons, suggested that their mass was electromagnetic in nature and depended on their velocity—in which case Poincaré observed, all of Newton's physics collapsed. Finally, the principle of conservation of energy was challenged by Becquerel's discovery of spontaneous radioactivity, for which he had won the Nobel Prize in 1903.

Poincaré's cautious proposal was to loyally defend present principles and not give up everything at once, because new experiments might yet restore harmony. But, he admitted, if even the best established experiments are to be overthrown, it is not clear what was left of his philosophy of science (see Poincaré 2001, 207, 312).

> Have you not written, you might say if you wished to seek a quarrel with me—have you not written that the principles, though of experimental origin, are now unassailable by experiment because they have become conventions? And now you have just told us that the most recent conquests of experiment put these principles in danger.
>
> Well, formerly I was right and today I am not wrong.

But it might be that a new approach, comparable to that which took physics from a theory of central forces to a physics of principles, would be created, in which recognisable traits of the old view would still be visible. For example, thermodynamics could become based on the laws of chance, and the physical law would no longer be a differential equation but a statistical law. This view, which the 20th, and still more it seems the 21st century only confirm, was prescient. So too was his suggest that the new theory of dynamics valid for high speeds, would have Newtonian dynamics as a limiting case (see Poincaré 2001, 211, 314).

We get a tantalising glimpse of what the new physics Poincaré contemplated might have been if we consider one of his last papers, which comes from a lecture he gave in London on 4 May 1912, (Poincaré 1912b). He reported on the impact of "the principle of relativity, as conceived by Lorentz", and so had to confront the problem that he had previously put our knowledge of the geometry of space beyond revision, yet here it was seemingly being revised.

Poincaré continued to argue that our knowledge of space is constructed from our representation of the sensations that accompany certain movements in space. Our measurements of space and time depend on instruments, starting with our own bodies, and an element of convention enters when we talk of perfect instruments. But Poincaré now distinguished between actual observations and the laws of motion derived from them by differentiation. Observational values are changed by a change of coordinate axes; the differential equations are not. The principle of relativity applies, said Poincaré, to the equations, their invariance under the appropriate coordinate changes is assured because they are second-order differential equations (and rotating axes can be handled by passing to third-order equations). By considering how we would treat a small piece of the universe distant from the rest and visibly rotating with respect to the rest, he deduced that physical relativity incorporates the idea that widely separated worlds may be treated independently, and is therefore not a necessity of the intellect but an experimental truth holding within limits. Relativity in this sense "is no longer a simple convention. It is verifiable, and consequently it might not be verified". As such, it differs from relativity in the broader psychological sense that draws on our sense of time, and cannot, for example decide of two events, one on Earth and one on Sirius, which came first except by a convention.

After Lorentz, then, there are two principles that can serve to define space: the old one involving rigid bodies, and a new one to do with the transformations that do not alter our differential equations. They are not essentially different, said Poincaré, because both are statements about the objects around us, but the new one is an experimental truth. Geometry can once again be made immune to revision by experimenters by making physical relativity a convention concerning distant objects. Then, whereas our conventional knowledge of geometry was formerly rooted in the group of Euclidean isometries, it could now be rooted in the Lorentz group: the Lorentz group that preserves our equations, at the price of placing us in a four-dimensional space. So, Poincaré concluded:

> What shall be our position in view of these new conceptions? Shall we be obliged to modify our conclusions? Certainly not; we had adopted a convention because it seemed convenient and we had said that nothing could constrain us to abandon it. Today some physicists want to adopt a new convention. It is not that they are constrained to do so; they consider this new convention more convenient; that is all. And those who are not of this opinion can legitimately retain the old one in order not to disturb their old habits. I believe, just between us, that this is what they shall do for a long time to come. (Poincaré 1913, 109, 24)

The ultimate test, for Poincaré, remained a pragmatic one. Conventions can be challenged if the theory that they support becomes incoherent, as it may under the impact of new experimental results. When this happens the transition to a new theory may be messy and uncertain, but if it has to be made and the old theory abandoned the new theory will rest on its own principles, which will again function as conventions. We adopt a mathematised theory of physics that we find most convenient, not one that is forced upon us (because no theory is).

5 A Wittgensteinian comparison

It is interesting to see how much of Poincaré's views make him a sceptic in the manner of Wittgenstein, Kripke, and Kusch. In his essay *On Certainty*

(Wittgenstein 1969) Wittgenstein remarked that "Certainty is attainable" (see Wittgenstein 1969, 56), and that

> Endless doubting is valueless: If you tried to doubt everything you would not get as far as doubting anything. (Wittgenstein 1969, 115)

This rather pragmatic sense of certainty was rooted in a concept of self-evidence, which for the purposes of mathematics in particular the self-evidence of a mathematical axiom system, Wittgenstein put this way: the assertion of self-evidence implies "that we have already chosen a definite kind of employment for the proposition without realising it. The proposition is not a mathematical axiom if we do not employ it precisely *for this purpose*. The fact, that is, that here we do not make experiments, but accept the self-evidence, is enough to fix the employment" (in Wittgenstein 1964, III). When we are forced to claim without a proof, or even the possibility of a proof, that a system we are using is consistent, the risk is that we shall turn out to be wrong, and of a system that was inconsistent but had never yet been made to generate a contradiction, Wittgenstein made the 'good angel' defence:

> Well, what more do you want? One might say, I believe: a good angel will always be necessary, whatever you do. (Wittgenstein 1964, V)

The usual contrary position is that we know what we are doing, and we know what we mean when we talk. Kusch (see 2006; 2009) calls sceptics people who reject as incoherent explanations of meaning in terms of the mental states of people. Sceptics argue, he says, nothing precludes anyone from having failed to exclude some other meanings. Kripke's example is that '+' could obey rules for large numbers that someone used to adding only small ones could never have ruled out (such as $a+b = 5$ for all numbers $a, b > 100$). Typically, the response to the sceptical challenge is to take it on its own terms and attempt to refute it, but Kusch argues that the proper thing to do is to see that it is harmless after all and accept it, and to replace talk about mental states with talk about intersubjectivity.

Poincaré's position, I suggest, is close to that of the Kusch's sceptic. He agreed that we rely on the testimony of experts and on a shared communication with others; that we speak a shared family of languages, natural, scientific, mathematical which work because of a shared set of conventions, and we have ideas about what we would do if our statements conflict or communication failed. As he put it in (Poincaré 1902b), (see Poincaré 2001, 292, 345): "No discourse—no objectivity". In his controversy with Zermelo he made it clear that he would accept self-evident axioms, and rested his case on the lack of self-evidence in Zermelo's system. His dispute with le Roy and his imaginary discussions between Euclideans and non-Euclideans are far from the only occasions where Poincaré put his trust in the possibility of effective communication. None of this involves knowing about meanings or have particular mental states.

If talk about meaning proceeds from introspection ('I know what I mean by X') to a charitable interpretation of what everyone else is saying as being sufficiently like what one says oneself, then Wittgenstein's alternative says that 'By their deeds ye shall know them'. It is clear that Poincaré did not talk about meanings, and certainly not about mental states, which he disparaged as carriers of truth in his lecture to the psychologists (Poincaré 1908b). He was reluctant to speak about eternally established facts, and placed great weight on actions and usage (for example, in measuring). Not only did he not think a list of facts was anything like as good as a theory (one cannot make predictions from a list of facts, science cannot progress by generalisation and analogy from a mere list) he often openly doubted if today's facts would be accepted tomorrow. But was he a sceptic in the sense just described?

The usual charge levelled at sceptics is that they are relativists. This charge alleges, to quote from (Kusch 2009, 19) again, that

> people using different epistemic systems (consisting of epistemic standards) can 'faultlessly disagree' over the question whether a given belief is epistemically justified or not. Faultless disagreement in such scenario is possible because (1) beliefs can be justified only within epistemic systems; (2) there are, and have been, many radically different epistemic systems; and (3) it is impossible to demonstrate by rational argument that one's own epistemic system is superior to all or most of the others [...].

But it can surely be argued that conventionalism in physics, and even geometric conventionalism, are akin to a language game. Poincaré was evidently a relativist over the question of the geometry of space, because faultless disagreement between a Euclidean and a non-Euclidean is exactly what he said would happen, and he was a relativist again when it came to a choice between the Galilean and the Lorentz group in special relativity. The nub of these disagreement is the existence of two distinct epistemic systems.

He was a sceptic about physics, for he agreed that we rely on the testimony of experts and on a shared communication with others "No discourse—no objectivity" he said in 1902 (see Poincaré 1902b). He argued that we speak a shared family of languages, natural, scientific, mathematical which work because of a shared set of conventions, and we have ideas about what we would do if our statements conflict or communication failed. None of this involves knowing about meanings or have particular mental states. Conventionalism is surely much more akin to a language game, and if scepticism is criticised for implying relativism, and if it is relativism to permit faultless disagreement, Poincaré's geometric conventionalism is relativist.

But Poincaré was not a sceptic about pure mathematics. He believed that we *know* what reasoning by recurrence is in an almost Kantian fashion. But recall that for Poincaré mathematics and physics are inseparable, and his deepest commitment was to discovery in mathematics. Now, no serious philosophy of mathematics can ignore or mistreat the role of discovery: without it there would be no mathematics! As Poincaré said in the first volume of *Enseignement mathématique* (see Poincaré 1899), even "the next generation of leading mathematicians will need intuition, for if it is by logic that one proves, it is by intuition that one invents".

6 A philosophical study of proof

A few brief remarks on this vexed topic may help to make a useful contrast. There is, of course, a sophisticated language for discussing proof known to mathematical logicians. This overlaps the ways mathematicians think about proofs, but it is not the whole story. It is excellent at questions of logical independence among axioms, at questions about the relative strengths of proofs, and of course, about logic itself. What Poincaré's comments were directed towards is better seen as ways of doing mathematics, and they are in some ways closer to how mathematicians regard their own work when they take a step back from it.

Central to Poincaré's approach was an idea of what it is to understand a concept or a mathematical argument. For him, grasping a proof should not be (only) a matter of psychology or mental states. Whatever aesthetic pleasure there might be in this or that piece of mathematics, it was more important to be able to advance the subject, and the aesthetic sense was, he argued, connected to the way in which the right new idea was enabling. For him, mathematics was a practice, and one to be judged ultimately on practical grounds. So a position of "Rigour—good: lack of rigour—bad" could not be the whole story, and alongside technical correctness a proof played a key part of acquiring, displaying, and using one's understanding. The "right" proof, for him, was one that got to the deepest relationship between the concepts and so enabled new work to be done, perhaps by displaying a new and valid use of the terms it involves. Research should aim for general ideas capable of wider application (by analogy), and would show not only what is the case but why it is the case. And it is striking, by the way, how often in his mathematics Poincaré relied on the general context and how little weight he attached to examples, and how often he insisted that physics dealt in relations that would survive changing beliefs or practices about objects.[9]

Bibliography

Anantharaman, N. (2006/2010). On the existence of closed geodesics. In *The Scientific Legacy of Poincaré, Hist. Math.*, vol. 36, Charpentier, E., Ghys, E., & Lesne, A., eds., Providence: American Mathematical Society, 143–160, Original title: *L'Héritage scientifique de Poincaré*.

Barrow-Green, J. (1997). *Poincaré and the three body problem*. Providence, RI; London: American Mathematical Society and London Mathematical Society.

Gray, J. (2012a). *Henri Poincaré: A Scientific Biography*. Princeton: Princeton University Press.

Gray, J. (2012b). Poincaré replies to Hilbert: On the future of mathematics ca. 1908. *The Mathematical Intelligencer*, *34*(3), 15–29, doi:10.1007/s00283-012-9299-7.

Hadamard, J. (1921). L'œuvre mathématique de H. Poincaré. *Acta mathematica*, *38*, 203–287, in Hadamard, *Œuvres*, 4, 1921–2005 and Poincaré, *Œuvres*, 11, 152–242).

Kusch, M. (2006). *A sceptical guide to meaning and rules: defending Kripke's Wittgenstein*. Chesham: Acumen.

Kusch, M. (2009). Kripke's Wittgenstein, On Certainty, and Epistemic Relativism. In *The Later Wittgenstein on Language*, Whiting, D., ed., New York: Palgrave Macmillan, 213–230.

Poincaré, H. (1881). Sur les formes cubiques ternaires et quaternaires I. *Journal de l'École Polytechnique*, *50*, 190–253, *Œuvres* 5, 28–72.

Poincaré, H. (1882). Sur les formes cubiques ternaires et quaternaires II. *Journal de l'École Polytechnique*, *51*, 45–91, *Œuvres* 5, 293–334.

Poincaré, H. (1886). Sur les intégrales irrégulières des équations linéaires. *Acta*, *8*(1), 295–344, *Œuvres* 1, 290–332.

Poincaré, H. (1887). Les fonctions fuchsiennes et l'arithmétique. *Journal de Mathématiques Pures et Appliquées*, *3*, 405–464, *Œuvres* 2, 463–511.

Poincaré, H. (1890a). Sur le problème des trois corps et les équations de la dynamique. *Acta*, *13*, 1–270, *Œuvres* 7, 262–479.

---

[9] A much fuller account of all these issues appears in (Gray 2012a).

Poincaré, H. (1890b). Sur les équations aux dérivées partielles de la physique mathématique. *American Journal of Mathematics*, *12*, 211–294, *Œuvres* 9, 28–113.

Poincaré, H. (1892). *Les Méthodes nouvelles de la mécanique céleste*. Paris: Gauthier-Villars, 3 vols: vol. 2, 1893; vol. 3, 1899.

Poincaré, H. (1893). Le continu mathématique. *Revue de métaphysique et de morale*, *1*, 26–34, in *La Science et l'Hypothèse*.

Poincaré, H. (1897). Sur les rapports de l'analyse pure et de la physique mathématique. *Acta*, *21*, 331–341, in *Verhandlungen des ersten internationalen Mathematiker-Kongresses in Zürich*, F. Rudio (ed.) 81–90, and *La Valeur de la science*.

Poincaré, H. (1899). La logique et l'intuition dans la science mathématique et dans l'enseignement. *Enseignement mathématique*, *1*, 157–162, in *Oeuvres* 11, 129–133.

Poincaré, H. (1900). Sur les rapports de la physique expérimentale et de la physique mathématique. In *Rapports présentés au Congrès international de physique*, vol. 1, Guillaume, C.-E. & Poincaré, L., eds., Paris: Gauthier-Villars, 1–29, Rep. as "Les hypothèses en physique" and "Les théories de la physique moderne" in *La Science et l'Hypothèse*.

Poincaré, H. (1902a). Du rôle de l'intuition et de la logique en mathématiques. In *Comptes rendus du IIe Congrès international des mathématiciens*, Duporcq, E., ed., Paris: Gauthier-Villars, 115–130, in *La Valeur de la science*.

Poincaré, H. (1902b). Sur la valeur objective de la science. *Revue de métaphysique et de morale*, *10*, 263–293, rep. with modifications, as 'La science est-elle artificielle?' and 'La Science et la Réalité' in *La Valeur de la science*, 213–247 and 248–276.

Poincaré, H. (1904a). *La Valeur de la science*. Paris: Flammarion.

Poincaré, H. (1904b). L'état actuel et l'avenir de la physique mathématique. *Bulletin des sciences mathématiques*, *28*, 302–324, in *La Valeur de la science*.

Poincaré, H. (1905a). Les mathématiques et la logique. *Revue de métaphysique et de morale*, *13*, 815–835, modified in *Science et Méthode*.

Poincaré, H. (1905b). Sur les lignes géodésiques des surfaces convexes. *Trans. AMS*, *6*, 237–274, in *Oeuvres* 6, 38–84.

Poincaré, H. (1905c). The Principles of Mathematical Physics. In *Congress of Arts and Science, Universal Exposition St. Louis*, vol. 1, Rogers, H. J., ed., Boston: Houghton, Mifflin & Co., 604–622.

Poincaré, H. (1908a). L'avenir des mathématiques. *Revue générale des sciences pures et appliquées*, *19*, 930–939, also in *Atti del IV congresso internazionale dei matematici*, 1909, 167–182. Only partially in *Science et Méthode*, and in the English trl. "The Future of Mathematics" in *Science and Method*.

Poincaré, H. (1908b). L'invention mathématique. *Enseignement mathématique*, *10*, 357–371, in *Année psychologique*, 15, 1909, 445–459, and *Science et Méthode*.

Poincaré, H. (1908c). *Science et Méthode*. Paris: Flammarion.

Poincaré, H. (1912a). La logique de l'infini. *Scientia (Rivista di Scienza)*, *12*, 1–11, in *Dernières pensées*. English trl. 'Mathematics and logic'.

Poincaré, H. (1912b). L'espace et le temps. *Scientia (Rivista di Scienza)*, *12*, 159–170, in *Dernières pensées*.

Poincaré, H. (1913). *Dernières pensées*. Paris: Flammarion.

Poincaré, H. (1921a). Analyse des travaux scientifiques de Henri Poincaré faite par lui-même. *Acta*, *38*, 1–135.

Poincaré, H. (1921b). Lettres à L. Fuchs (1880, 1881). *Acta*, *38*, 175–184, *Œuvres*, 11 volumes, various editors, 19261955, 11, 13–25.

Poincaré, H. (1997). *Trois suppléments sur la découverte des fonctions fuchsiennes*. Berlin: Akademie-Verlag, edited by J.J. Gray and S. Walter.

Poincaré, H. (2001). *The Value of Science: Essential Writings of Henri Poincaré*. New York: Random House, Gould, S. J. (ed.).

Wittgenstein, L. (1964). *Remarks on the Foundations of Mathematics*. Oxford: Blackwell.

Wittgenstein, L. (1969). *On Certainty*. Oxford: Blackwell, G.E.M. Anscombe and G.H. von Wright (eds).

Zermelo, E. (2010). *Collected Works, Gesammelte Werke*. Berlin: Springer, Ebbinghaus, H.-D., Fraser, C. G. and Kanamori, A. (eds.).

Jeremy Gray
Faculty of Mathematics, Computing, and Technology,
The Open University,
Milton Keynes, MK7 6AA
United Kingdom
j.j.gray@open.ac.uk

Einstein and Bohr Meet Alice and Bob

JEFFREY BUB

ABSTRACT. The Bohr-Einstein debate was ultimately about the nature of quantum reality. Here I consider how the puzzling questions at issue have been transformed by the information-theoretic turn in quantum foundations, and what we have learned about the possible answers.

1 Correlations

The debate between Bohr and Einstein about the nature of quantum reality reached its high point in 1935 with the publication of the Einstein-Podolsky-Rosen argument for the incompleteness of quantum mechanics (Einstein et al. 1935) and Bohr's reply (Bohr 1935). Roughly thirty years later, John Bell's critique (Bell 1964) turned the EPR argument on its head and was seminal in the development of quantum information theory. The conceptual issues have been transformed by the associated information-theoretic turn in quantum foundations, and we can now see that the puzzling counter-intuitive features of quantum mechanics at the heart of the Bohr-Einstein debate have their source in the peculiar nonclassical correlations of quantum phenomena. In this paper, I discuss some aspects of this change in perspective.

To fix notation and terminology, consider the simple case of measurements of two binary-valued observables, $x \in \{0,1\}$ with outcomes $a \in \{0,1\}$, performed by Alice in a region **A**, and $y \in \{0,1\}$ with outcomes $b \in \{0,1\}$, performed by Bob in a separated region **B**. I shall refer to the x-values and a-values as Alice's inputs and outputs, respectively, and similarly for Bob with respect to the y-values and b-values. So the two Alice-inputs ($x = 0$ or $x = 1$) correspond to the two Alice-observables, and the two Bob-inputs ($y = 0$ or $y = 1$) correspond to the two Bob-observables, and each observable can take two values, 0 or 1.

Correlations are expressed by a correlation array of joint probabilties as in Table 1:

x\y	0		1	
0	$p(00\|00)$ $p(01\|00)$	$p(10\|00)$ $p(11\|00)$	$p(00\|10)$ $p(01\|10)$	$p(10\|10)$ $p(11\|10)$
1	$p(00\|01)$ $p(01\|01)$	$p(10\|01)$ $p(11\|01)$	$p(00\|11)$ $p(01\|11)$	$p(10\|11)$ $p(11\|11)$

Table 1. Correlation array

The probability $p(00|00)$ is to be read as $p(a = 0, b = 0|x = 0, y = 0)$, i.e., as a joint conditional probability, and the probability $p(01|10)$ is to be read as $p(a = 0, b = 1|x = 1, y = 0)$, etc. (I drop the commas for ease of reading; the first two slots in $p(--|--)$ before the conditionalization sign '|' represent the two possible outputs for Alice and Bob, respectively, and the second two slots after the conditionalization sign represent the two possible inputs for Alice and Bob, respectively.)

The sum of the probabilities in each square cell of the array in Table 1 is 1, since the sum is over all possible outcomes, given the two observables that are measured. The marginal probability of 0 for Alice or for Bob is obtained by adding the probabilities in the left column of each cell or the top row of each cell, respectively, and the marginal probability of 1 for Alice or for Bob by adding the probabilities in the right column of each cell or the bottom row of each cell, respectively. The measurement outcomes are said to be uncorrelated if the joint probability is expressible as a product of marginal or local probabilities for Alice and Bob; otherwise they are correlated.

Now consider all possible correlation arrays of the above form. They form an 8-dimensional regular polytope with 256 vertices and 1024 edges, where the vertices are the extremal deterministic arrays with probabilities 0 or 1 only, e.g., the array in Table 2:[1] A general correlation array is represented by a point inside this polytope, so the probabilities in the array can be expressed (in general, non-uniquely) as convex combinations of the 0, 1 probabilities in extremal correlation arrays (in a similar sense in which a probability can be represented as a point on a line between the points 0 and

---

[1] A polytope is the analogue of a polygon in many dimensions. A convex set is, roughly, a set such that from any point in the interior it is possible to see any point on the boundary. There are four possible arrangements of 0's and 1's that add to 1 in each square cell of the correlation array (i.e., one 1 and three 0's), and four cells, giving $4^4 = 256$ vertices. Each of the four pairs of inputs, 00, 01, 10, 11, is associated with two dimensions.

x	0		1	
y				
0	$p(00\|00)=1$	$p(10\|00)=0$	$p(00\|10)=0$	$p(10\|10)=0$
	$p(01\|00)=0$	$p(11\|00)=0$	$p(01\|10)=1$	$p(11\|10)=0$
1	$p(00\|01)=0$	$p(10\|01)=1$	$p(00\|11)=0$	$p(10\|11)=0$
	$p(01\|01)=0$	$p(11\|01)=0$	$p(01\|11)=0$	$p(11\|11)=1$

Table 2. Extremal signaling deterministic correlation array

1 because it can be expressed as a convex combination of the extremal end points).

The correlation array in Table 2 defines a set of correlations that allow signaling between Alice and Bob. Alice's output is the same as Bob's input. Similarly, Bob's output is the same as Alice's input. So an input by Alice or Bob is instantaneously revealed in a remote output. There are 240 signaling extremal deterministic correlation arrays in the total set of 256 extremal deterministic correlation arrays. The remaining 16 extremal deterministic correlation arrays are non-signaling.

The 'no signaling' condition can be formulated as follows: no information should be available in the marginal probabilities of outputs in region **A** about alternative choices made by Bob in region **B**, i.e., Alice, in region **A** should not be able to tell what Bob measured in region **B**, or whether Bob performed any measurement at all, by looking at the statistics of her measurement outcomes, and conversely. Formally:

(1) $$\sum_b p(a,b|x,y) \equiv p(a|x,y) = p(a|x), \text{ for all } y$$

(2) $$\sum_a p(a,b|x,y) \equiv p(b|x,y) = p(b|y), \text{ for all } x$$

Here $p(a,b|x,y)$ is the probability of obtaining the pair of outputs a,b for the pair of inputs x,y. The probability $p(a|x,y)$ is the marginal probability of obtaining the output a for x when Bob's input is y, and $p(b|x,y)$ is the marginal probability of obtaining the output b for y when Alice's input is x. The 'no signaling' condition requires Alice's marginal probability $p(a|x,y)$ to be independent of Bob's choice of input in region **B** (and independent of whether there was any input in region **B** at all), i.e., $p(a|x,y) = p(a|x)$, and similarly for Bob's marginal probability $p(b|x,y)$ with respect to Alice's inputs: $p(b|x,y) = p(b|y)$. Note that 'no signaling' is simply a constraint on the marginal probabilities, not a relativistic constraint *per se*. But if this constraint is violated, instantaneous (hence superluminal) signaling is possible.

The joint probabilities in the 16 non-signaling deterministic correlation arrays can all be expressed as products of marginal or local probabilities for Alice and Bob separately. For example, the deterministic correlation array in which the outputs are both 0 for all possible input combinations, as in Table 3, is a non-signaling array and the joint probabilities can be expressed as a product of local probabilities: a marginal Alice-probability of 1 for the output 0 given any input, and a marginal Bob-probability of 1 for the output 0 given any input.

x \ y	0		1	
0	$p(00\|00) = 1$ $p(01\|00) = 0$	$p(10\|00) = 0$ $p(11\|00) = 0$	$p(00\|10) = 1$ $p(01\|10) = 0$	$p(10\|10) = 0$ $p(11\|10) = 0$
1	$p(00\|01) = 1$ $p(01\|01) = 0$	$p(10\|01) = 0$ $p(11\|01) = 0$	$p(00\|11) = 1$ $p(01\|11) = 0$	$p(10\|11) = 0$ $p(11\|11) = 0$

Table 3. Extremal non-signaling deterministic correlation array

In the following, I shall refer to the probabilistic arrays as states, since the classical and quantum correlation arrays correspond to classical and quantum pure and mixed states.

2 Nonlocal boxes

Now suppose the correlations are as in Table 4. These correlations define a Popescu-Rohrlich (PR) box, a hypothetical device proposed by Popescu and Rohrlich (Popescu & Rohrlich 1994) to bring out the difference between classical, quantum, and superquantum non-signaling correlations.

x \ y	0		1	
0	$p(00\|00) = 1/2$ $p(01\|00) = 0$	$p(10\|00) = 0$ $p(11\|00) = 1/2$	$p(00\|10) = 1/2$ $p(01\|10) = 0$	$p(10\|10) = 0$ $p(11\|10) = 1/2$
1	$p(00\|01) = 1/2$ $p(01\|01) = 0$	$p(10\|01) = 0$ $p(11\|01) = 1/2$	$p(00\|11) = 0$ $p(01\|11) = 1/2$	$p(10\|11) = 1/2$ $p(11\|11) = 0$

Table 4. PR-box correlations

PR-box correlations can be defined as follows:

(3) $\quad a \oplus b = x \cdot y$

where $\oplus$ is addition mod 2, i.e.,

> same outputs (i.e., 00 or 11) if the inputs are 00 or 01 or 10
>
> different outputs (i.e., 01 or 10) if the inputs are 11

with the assumption that that the marginal probabilities are all 1/2 to ensure 'no signaling', so the outputs 00 and 11 are obtained with equal probability when the inputs are not both 1, and the outputs 01 and 10 are obtained with equal probability when the inputs are both 1.

A PR-box functions in such a way that if Alice inputs a 0 or a 1, her output is 0 or 1 with probability 1/2, irrespective of Bob's input, and irrespective of whether Bob inputs anything at all; similarly for Bob. The requirement is simply that whenever there are in fact two inputs, the inputs and outputs are correlated according to (3).

A PR-box can function only once, so to get the statistics for many pairs of inputs one has to use many PR-boxes. This avoids the problem of selecting the 'corresponding' input pairs for different inputs at various times, which would depend on the reference frame. In this respect, a PR-box mimics a quantum system: after a system has responded to a measurement (produced an output for an input), the system is no longer in the same quantum state, and one has to use many systems prepared in the same quantum state to exhibit the probabilities associated with a given quantum state.

The 16 vertices defined by the deterministic states form a convex polytope, the local polytope. The correlations represented by points in the local polytope have a common cause explanation, where the common causes can be represented geometrically by the vertices of a simplex, a polytope generated by $n + 1$ vertices that are not confined to any $(n - 1)$-dimensional subspace, e.g., a tetrahedron as opposed to a rectangle. The lattice of subspaces of a simplex (the lattice of vertices, edges, and faces) is a Boolean algebra, with a 1-1 correspondence between the vertices, corresponding to the atoms of the Boolean algebra, and the facets (the $(n - 1)$-dimensional faces), which correspond to the co-atoms. The 16-vertex simplex represents the correlation polytope of probabilistic states of a bipartite classical system with two binary-valued observables; the associated Boolean algebra represents the classical event structure. Probability distributions of these extremal states—mixed states—are represented by points in the interior of the simplex.

The local polytope is included in a non-signaling nonlocal polytope defined by the 16 vertices of the local polytope together with an additional 8 nonlocal vertices, one vertex representing the standard PR box as defined above, and the other seven vertices representing PR boxes obtained from the standard PR box by relabeling the x-inputs, and the a-outputs conditionally on the x-inputs, and the y-inputs, and the b-outputs conditionally on the y-inputs. For example, the correlations in Table 5 define a PR-box. Note that the 16 vertices of the local polytope can all be obtained from the vertex represented by Table 3 by similar local reversible operations.

x	0		1	
y				
0	$p(00\|00) = 0$	$p(10\|00) = 1/2$	$p(00\|10) = 0$	$p(10\|10) = 1/2$
	$p(01\|00) = 1/2$	$p(11\|00) = 0$	$p(01\|10) = 1/2$	$p(11\|10) = 0$
1	$p(00\|01) = 0$	$p(10\|01) = 1/2$	$p(00\|11) = 1/2$	$p(10\|11) = 0$
	$p(01\|01) = 1/2$	$p(11\|01) = 0$	$p(01\|11) = 0$	$p(11\|11) = 1/2$

Table 5. Locally transformed PR-box correlations (relative to Table 4)

3 Simulating a PR-Box

Suppose Alice and Bob are allowed certain resources. What is the optimal probability that they can perfectly simulate the correlations of a PR box?

In units where $a = \pm 1, b = \pm 1$,[2]

(4) $\quad \langle 00 \rangle = p(\text{outputs same}|00) - p(\text{outputs different}|00)$

so:

(5) $\quad\quad\quad p(\text{outputs same}|00) \;=\; \dfrac{1 + \langle 00 \rangle}{2}$

(6) $\quad\quad\quad p(\text{outputs different}|00) \;=\; \dfrac{1 - \langle 00 \rangle}{2}$

and similarly for input pairs 01, 10, 11.

It follows that the probability of successfully simulating a PR-box is given by:

$$p(\text{successful sim}) \;=\; \frac{1}{4}(p(\text{outputs same}|00) + p(\text{outputs different}|01) +$$

---

[2] It is convenient to change units here to relate the probability to the usual expression for the Clauser-Horne-Shimony-Holt correlation, where the expectation values are expressed in terms of ± 1 values for x and y (corresponding to the relevant observables). Note that 'outputs same' or 'outputs different' mean the same thing whatever the units, so the probabilities $p(\text{outputs same}|xy)$ and $p(\text{outputs different}|xy)$ take the same values whatever the units, but the expectation value $\langle xy \rangle$ depends on the units for x and y.

(7) $\qquad p(\text{outputs same}|10) + p(\text{outputs different}|11))$

(8) $\qquad\qquad = \frac{1}{2}(1 + \frac{K}{4}) = \frac{1}{2}(1 + E)$

where $K = \langle 00 \rangle + \langle 01 \rangle + \langle 10 \rangle - \langle 11 \rangle$ is the Clauser-Horne-Shimony-Holt (CHSH) correlation.

Bell's locality argument in the Clauser-Horne-Shimony-Holt version (Clauser et al. 1969) shows that if Alice and Bob are limited to classical resources, i.e., if they are required to reproduce the correlations on the basis of shared randomness or common causes established before they separate (after which no communication is allowed), then $|K_C| \leq 2$, i.e., $|E| \leq \frac{1}{2}$, so the optimal probability of successfully simulating a PR-box is $\frac{1}{2}(1+\frac{1}{2}) = \frac{3}{4}$.

If Alice and Bob are allowed to base their strategy on shared entangled states prepared before they separate, then the Tsirelson bound for quantum correlations requires that $|K_Q| \leq 2\sqrt{2}$, i.e., $|E| \leq \frac{1}{\sqrt{2}}$, so the optimal probability of successful simulation limited by quantum resources is $\frac{1}{2}(1 + \frac{1}{\sqrt{2}}) \approx .85$.

Clearly, relativistic causality does not rule out simulating a PR box with a probability greater than $\frac{1}{2}(1+\frac{1}{\sqrt{2}})$. As Popescu and Rohrlich observe, there are possible worlds described by superquantum theories that allow nonlocal boxes with non-signaling correlations stronger than quantum correlations, in the sense that $\frac{1}{\sqrt{2}} \leq E \leq 1$. The correlations of a PR box saturate the CHSH inequality ($E = 1$), and so represent a limiting case of non-signaling correlations.

I use the term 'nonlocal box' to refer to any non-signaling device with a probability array for which the probability of a successful simulation is greater than the classical value of 3/4. We do, in fact, live in a nonlocal box world: a pair of qubits in an entangled quantum state constitutes a nonlocal box for certain pairs of measurements.

For two binary-valued observables of a bipartite quantum system the correlations form a spherical convex set that is not a polytope, with extremal points between the 16-vertex local polytope and the 24-vertex non-signaling nonlocal polytope, which is itself included in the 256-vertex nonlocal polytope with 240 vertices that represent deterministic signaling states. The correlations of the local polytope have a common cause explanation, represented by a 16-vertex simplex, where the vertices of the simplex represent common causes or classical states.

A simplex has the rather special property that a mixed state, represented by a point in the interior of the simplex, can be expressed *uniquely* as a mixture (convex combination) of extremal or pure states, the vertices of the simplex. *No other convex set has this feature.* So in the class of non-signaling theories, classical theories are rather special. For all nonclassical (= non-simplex) theories, the decomposition of mixed states into pure states is not unique. For such theories, there can be no general cloning procedure capable of copying an arbitrary extremal state without violating the 'no signaling' condition, and similarly *there can be no measurement in the non-disturbing sense that one has in classical theories*, where it is in principle possible, via measurement, to extract sufficient information about an extremal state to produce a copy of the state without irreversibly changing the state. For a nonlocal box theory, *there is a necessary information loss on measurement*.

The quantum theory is a nonlocal box theory, i.e., it is a non-signaling, non-simplex theory with counter-intuitive probabilistic features like those of an extremal PR box. Hilbert space as a projective geometry (i.e., the subspace structure of Hilbert space) represents a non-Boolean event space, in which there are built-in, structural probabilistic constraints on correlations between events (associated with the angles between the rays representing extremal events)—just as in special relativity the geometry of Minkowski space-time represents spatio-temporal constraints on events. These are kinematic, i.e., pre-dynamic, objective probabilistic or information-theoretic constraints on events to which a quantum dynamics of matter and fields conforms, through its symmetries, just as the structure of Minkowski space-time imposes spatio-temporal kinematic constraints on events to which a relativistic dynamics conforms.

4 Why quantum mechanics?

The basic question underlying the Bohr-Einstein debate was the question of the completeness of quantum mechanics, which is essentially the question *why quantum mechanics rather than a classical theory*, i.e., a simplex theory? In effect, Einstein's assumption here was that there is something metaphysically privileged about a simplex theory (see the quotation at the end of this section). From the perspective of the previous analysis, we see that the more interesting question (first raised by Popescu and Rohrlich) is *why quantum mechanics rather than a superquantum theory*, i.e., a non-simplex theory that violates the Tsirelson bound: $\frac{1}{\sqrt{2}} \leq |E| \leq 1$.

The revolution in physics associated with relativity theory involves the discovery of a contingent fact that conflicts with what one might call a structural principle. The contingent fact is the discovery that there is no overtaking of light by light, as Hermann Bondi puts it (Bondi 1980). The structural principle is the relativity principle, roughly that velocity doesn't matter (Bondi): the laws of physics are the same in different reference frames moving at constant relative velocity. If velocity doesn't matter and there is no overtaking of light by light, then Newtonian space-time has to go. It was

Einstein's genius to see that the behavior of light could be reconciled with the relativity principle by replacing Newtonian space-time with Minkowski space-time.

The analogous contingent fact for the quantum revolution is the discovery of nonlocal entanglement; specifically, that there are correlations outside the classical simplex. This involves the extension of classical information theory to quantum information theory. We know that the simplex structure for probabilistic correlations should be extended to the quantum convex set. But what is the structural principle that *constrains* correlations to the quantum convex set?

There are various proposals in the literature for such a principle, e.g., the principle of *information causality* proposed by Pawłowski et al. (2009). Information causality is a generalization of the 'no signaling' principle. It can be interpreted as a principle characterizing system separability, or a limitation of what Bohr referred to as quantum 'wholeness'.

Information causality says that Bob's information gain about a data set of Alice (previously unknown to him), on the basis of his local resources (which may be correlated with Alice's local resources) and a single use by Alice of an information channel with classical capacity m, is bounded by the classical capacity of the channel. For $m = 0$, this is equivalent to 'no signaling'. Zukowski calls the principle 'causal information access'.[3] The proposal is that quantum mechanics optimizes causal information access.

Here is a simple way to see the significance of information causality as a constraint. If Bob has to guess the value of any designated one of N bits held by Alice, and Alice can send Bob just one bit of information, then Bob can do better exploiting quantum correlations than classical correlations (shared randomness). Information causality sets a limit on how much better he can do.

Suppose the probability of Bob guessing the k'th bit correctly is P_k. The binary entropy of P_k is defined as:

(9) $\quad h(P_k) = -P_k \log P_k - (1 - P_k) \log (1 - P_k)$

If Bob knows the value of the bit he has to guess, $P_k = 1$, so $h(P_k) = 0$. If Bob has no information about the bit he has to guess, $P_k = 1/2$, i.e., his guess is at chance, and $h(P_k) = 1$. So $h(P_k)$ varies between 0 and 1.

If Alice sends Bob one classical bit of information, information causality requires that Bob's information about the N unknown bits increases by at most one bit. If the bits in Alice's list are unbiased and independently distributed, Bob's information about an arbitrary bit $b = k$ in the list cannot

---

[3] Private communication.

increase by more than $1/N$ bits, i.e., for Bob's guess about an arbitrary bit in Alice's list, the binary entropy $h(P_k)$ is at most $1/N$ closer to 0 from the chance value 1: $h(P_k) \geq 1 - 1/N$.

So information causality is violated when $h(P_k) < 1 - 1/N$ or, taking $N = 2^n$, the condition for a violation of information causality is $h(P_k) < 1 - \frac{1}{2^n}$. Since it can be shown that $P_k = \frac{1}{2}(1 + E^n)$ (Pawłowski et al. 2009; Bub 2012), we have a violation of information causality when $h(\frac{1}{2}(1 + E^n)) < 1 - \frac{1}{2^n}$.

For classical correlations, $E = \frac{1}{2}$, so in the case $n = 1$ (i.e., Bob has to guess one of two bits), $P_k = \frac{3}{4}$, and:

$$h(P_k) \approx .81$$

For quantum correlations, $E = \frac{1}{\sqrt{2}}$, so for $n = 1$, $P_k = \frac{1}{2}(1 + \frac{1}{\sqrt{2}}) \approx .85$, and:

$$h(P_k) \approx .60$$

For PR-box correlations, $E = 1$, so for all n, $P_k = 1$, and:

$$h(P_k) = 0$$

It can be shown that if $E \leq E_T = \frac{1}{\sqrt{2}}$, information causality is satisfied (Pawłowski et al. 2009; Bub 2012), i.e.,

$$(10) \quad h(\frac{1}{2}(1 + E_T^n)) \geq 1 - \frac{1}{2^n}, \text{ for any } n$$

If $E > E_T$, information causality is violated.

If E is very close to the Tsirelson bound $E_T = \frac{1}{\sqrt{2}} \approx .707$, then n must be very large for a violation of information causality. For example, if $n = 10$ and $E = .708$, then $h(P_k) \approx .99938$. There is no violation of information causality because $.99938 > 1 - \frac{1}{1024} \approx .9990$. In fact, we require $n \geq 432$ for a violation of information causality (Bub 2012).

Quantum and classical theories satisfy information causality. For the correlations between the outcomes of two binary-valued observables for each subsystem of a bipartite system, Allcock et al. (2009) have been able to exclude all but a very small part of the superquantum region on the basis of information causality. For tripartite systems, however, Yang et al. (2012) have shown that information causality fails to exclude all superquantum correlations. So this principle can't be the whole story, but it represents a promising start in the search for a principle or principles characterizing quantum theory.

It is rather striking how closely Lorentz's reluctance to accept the theory of relativity is paralleled by Einstein's reluctance to accept the significance of the quantum revolution. In the 1916 edition of *The Theory of Electrons and its Applications to the Phenomena of Light and Radiant Heat* (Lorentz 1916, 229), Lorentz writes:

> I cannot speak here of the many highly interesting applications which Einstein has made of this principle [of relativity]. His results concerning electromagnetic and optical phenomena ...agree in the main with those which we have obtained in the preceding pages, the chief difference being that Einstein simply postulates what we have deduced, with some difficulty and not altogether satisfactorily, from the fundamental equations of the electromagnetic field. By doing so, he may certainly take credit for making us see in the negative result of experiments like those of Michelson, Rayleigh and Brace, not a fortuitous compensation of opposing effects, but the manifestation of a general and fundamental principle.
>
> Yet, I think, something may also be claimed in favour of the form in which I have presented the theory. I cannot but regard the aether, which can be the seat of an electromagnetic field with its energy and its vibrations, as endowed with a certain degree of substantiality, however different it may be from all ordinary matter. In this line of thought, it seems natural not to assume at starting that it can never make any difference whether a body moves through the aether or not, and to measure distances and lengths of time by means of rods and clocks having a fixed position relative to the aether.

Einstein's complaint against the quantum theory has a similar flavor (Einstein 1948). He writes:

> If one asks what, irrespective of quantum mechanics, is characteristic of the world of ideas of physics, one is first of all struck by the following: the concepts of physics relate to a real outside world, that is, ideas are established relating to things such as bodies, fields, etc., which claim a 'real existence' that is independent of the perceiving subject—ideas which, on the other hand, have been brought into as secure a relationship as possible with sense-data. It is further characteristic of these physical objects that they are thought of as arranged in a physical space-time continuum. An essential aspect of this arrangement of things

in physics is that they lay claim, at a certain time, to an existence independent of one another, provided that these objects 'are situated in different parts of space'. Unless on makes this kind of assumption about the independence of the existence (the 'being-thus' ['So-sein']) of objects which are far apart from one another in space—which stems in the first place from everyday thinking—physical thinking in the familiar sense would not be possible. It is also hard to see any way of formulating and testing the laws of physics unless one makes a clear distinction of this kind.

Just as a contingent fact about light requires us to drop the notion of Newtonian space-time, so the discovery that there are probabilistic correlations outside the classical simplex requires us to drop the notion that physical systems have an independent 'being-thus' (corresponding to the vertices of the classical simplex).

Acknowledgements

My research is supported by the Institute for Physical Science and Technology at the University of Maryland. This publication was made possible through the support of a grant from the John Templeton Foundation. The opinions expressed in this publication are those of the author and do not necessarily reflect the views of the John Templeton Foundation.

Bibliography

Allcock, J., Brunner, N., *et al.* (2009). Recovering part of the quantum boundary from information causality. *arxiv: quant-ph/0906.3464v3*.

Bell, J. S. (1964). On the Einstein-Podolsky-Rosen Paradox. *Physics*, *1*, 195–200, reprinted in John Stuart Bell, *Speakable and Unspeakable in Quantum Mechanics*, Cambridge University Press, Cambridge, 1989.

Bohr, N. (1935). Can quantum-mechanical description of physical reality be considered complete? *Physical Review*, *48*, 696–702, doi:10.1103/PhysRev.48.696.

Bondi, H. (1980). *Relativity and Common Sense*. New York: Dover Publications.

Bub, J. (2012). Why the tsirelson bound? In *Probability in Physics*, Ben-Menahem, Y. & Hemmo, M., eds., The Frontiers Collection, Berlin; Heidelberg: Springer, 167–185, doi:10.1007/978-3-642-21329-8_11.

Clauser, J., Horne, M., *et al.* (1969). Proposed experiment to test hidden variable theories. *Physical Review Letters*, *23*, 880–883, doi: 10.1103/PhysRevLett.23.880.

Einstein, A. (1948). Quantum mechanics and reality. *Dialectica*, *3-4*, 320–324, doi:10.1111/j.1746-8361.1948.tb00704.x.

Einstein, A., Podolsky, B., & Rosen, N. (1935). Can quantum-mechanical description of physical reality be considered complete? *Physical Review*, *47*, 777–780, doi:10.1103/PhysRev.47.777.

Lorentz, H. A. (1916). *The Theory of Electrons and its applications to the phenomena of light and radiant heat*. New York: Columbia University Press.

Pawłowski, M., Patarek, T., et al. (2009). A new physical principle: information causality. *Nature*, *461*, 1101–1104, doi:10.1038/nature08400.

Popescu, S. & Rohrlich, D. (1994). Quantum nonlocality as an axiom. *Foundations of Physics*, *24*(3), 379–385, doi:10.1007/BF02058098.

Yang, T. H., Cavalcanti, D., et al. (2012). Information-causality and extremal tripartite correlations. *New Journal of Physics*, *14*(1), 013061, doi:10.1088/1367-2630/14/1/013061.

Jeffrey Bub
Philosophy Department and Institute for Physical Science and Technology
University of Maryland
College Park, MD 20742
USA
jbub@umd.edu

The Utility of the Uncountable[1]

Justin Tatch Moore

ABSTRACT. In my lecture at the 2011 Congress on Logic, Methodology, and the Philosophy of Science in Nancy, France, I spoke on an additional axiom of set theory—the *Proper Forcing Axiom*—which has proved very successful in settling combinatorial problems concerning uncountable sets. Since I have already written a exposition on this subject (Tatch Moore 2011), I have decided to address a broader question in this article: *why study uncountability?*

In some circles within logic, there has been an ongoing campaign to stress the importance of countability in mathematics—and to marginalize the uncountable. While much of mathematics does concern objects which can be codified as hereditarily countable sets, this often does not reflect how mathematics is discovered or developed. More significantly, there are technical difficulties which can arise in mathematics—often quite unexpectedly—which are fundamentally uncountable in their character. The purpose of this article is survey some instances where uncountability has been useful in the discovery process, essential to the solution of a problem, or at least has offered a fruitful perspective. We will also examine settings in which restricting attention to countable objects artificially limits the perspective and gives an incomplete picture of the mathematical phenomenon under consideration.

In this article, we will take *countable mathematics* to mean the study of that which can be encoded in the hereditarily countable sets—the domain of discourse of second order arithmetic. For instance a separable complete metric space can be encoded as the completion of a countable metric space. Even Borel or suitably definable subsets of such a space have a countable description and as such lie within the scope of "countable mathematics". Nonseparable spaces or nonmeasurable subsets of $\mathbb{R}$ are typical examples of objects which are essentially uncountable in their nature.

---

[1] The research of the author was supported in part by NSF grant DMS–0757507.

None of the mathematics in this article is my own. I have generally tried to include references to the original works when it is reasonable to do so and otherwise provide a standard reference where the material can be found.

1 The theory of algebraically closed fields

One of the great ironies of logic is surely that the theory of algebraically closed fields of characteristic 0 is complete while Peano's Axioms for $\mathbb{N}$ are not only incomplete but cannot be completed in any intelligible way. Ostensibly, ACF_0 attempts to achieve more generality through abstraction than just to axiomatize the theory of the complex numbers. On the other hand, PA was formulated with the intention of axiomatizing a single model, namely $(\mathbb{N}, +, \cdot, 0, 1, <)$.

Equally remarkable is how natural it is to employ uncountability to prove the completeness of ACF_0—a statement which itself is purely arithmetical in nature. To illustrate this, I will sketch the argument presented in (Marker 2002). To be clear, this is not the original argument of Robinson (Robinson 1951), but it is an elegant illustration of how uncountability can play a role in proving an arithmetical statement.

The following are the two main ingredients:

VAUGHT'S TEST If T is a consistent theory in a countable language, T has no finite models, and any two models of T of cardinality $\aleph_1$ are isomorphic, then T is complete.

TRANSCENDENCE DEGREE (see, e.g., Hungerford 1980) If two algebraically closed fields have the same characteristic and transcendence degree, then they are isomorphic.

Vaught's Test has a very short proof using the Lowenheim-Skolem Theorem: if T does not decide ϕ, then there are consistent extensions T_0 and T_1 of T which include ϕ and $\neg \phi$ respectively and which have infinite models. By the Lowenheim-Skolem Theorem, T_0 and T_1 have models of cardinality $\aleph_1$. Such models are then isomorphic, contradicting that one satisfies ϕ and the other satisfies $\neg \phi$. Notice that the form of the Lowenheim-Skolem Theorem needed here is fundamentally *uncountable* in character.

The proof that ACF_0 is complete can now be finished as follows. By Vaught's Test, it is sufficient to show that any two algebraically closed fields of characteristic 0 and of cardinality $\aleph_1$ are isomorphic. This is true by observing that the transcendence degree of an uncountable algebraically closed field is equal to its cardinality. As with the Lowenheim-Skolem theorem, we fundamentally need here the notion of not only infinite but of uncountable transcendence degree.

2 Semigroup dynamics and Ramsey theory

Recall the following two theorems concerning partitions of $\mathbb{N}$:

VAN DER WAERDEN'S THEOREM (van der Waerden 1927) If $\mathbb{N} = \bigcup_{i<d} K_i$, then there is an $i < d$ such that K_i contains arbitrarily long arithmetic progressions.

HINDMAN'S THEOREM (Hindman 1974) If $\mathbb{N} = \bigcup_{i<d} K_i$, then there is an $i < d$ and an infinite $H \subseteq \mathbb{N}$ such that all finite sums of distinct elements of H are in K_i.

Both of these theorems were first proved by elementary means. Still, these elementary proofs are quite complex and the modern perspective is that the standard proofs of these statements go by way of *semigroup dynamics*. The basic idea is as follows. We begin with the discrete semigroup $(\mathbb{N}, +)$ and then form the Čech-Stone compactification $\beta\mathbb{N}$. The operation $+$ extends to a semigroup operation on $\beta\mathbb{N}$. This operation is moreover continuous in the left argument: $p \mapsto p + q$ is continuous for each q. The compactness of $\beta\mathbb{N}$ allows for the construction of algebraic objects which have powerful combinatorial consequence for $\mathbb{N}$.

For instance Glazer observed that the following lemma of Ellis implies that $\beta\mathbb{N}$ contains an idempotent (other than 0).

ELLIS'S LEMMA (Ellis 1958) If S is a left topological compact semigroup, then S contains an idempotent.

Galvin had already observed that such idempotents can be used to prove Hindman's Theorem. If $p + p = p$, then any element K of p contains a set H such that all finite distinct sums from H lie in K; the set H can be constructed by an easy recursive procedure (see Hindman & Strauss 1998) or (Todorčević 2010). Gowers later extended this argument in (Gowers 1992) to prove a stronger combinatorial statement which he then used to draw geometric conclusions about the Banach space c_0. Unlike Hindman's Theorem, there is currently no known elementary proof of Gowers's result.

The reader may also find Harrington's proof of the Halpern-Läuchli theorem interesting (see Todorchevich & Farah 1995). This proof utilizes both the Erdős-Rado theorem (i.e., the partition relation $\beth_d^+ \to (\aleph_1)_\omega^{d+1}$) and the method of forcing. While the other proofs of the Halpern-Läuchli theorem are more elementary, this proof offers those comfortable with forcing a more intuitive proof.

3 Serre's conjecture

Next we will turn to an example from group theory. The point here is not only to mention a very remarkable result, but to give an example of how "real mathematicians" are not satisfied with limiting themselves to second order arithmetic, even when this might seem to be a completely natural thing to do.

A *profinite* group is an inverse limit of a directed system of finite groups. These can be equivalently characterized as being those compact topological groups which are *totally disconnected*—they have no non-trivial connected subsets. Notice that, when separable, such groups have a countable description—the inverse system of groups which defines them is countable. Serre made the following conjecture after proving that it is true for pro-p groups (this also was asked by Mel'nikov in 7.37 of (Mazurov & Khukhro 2006)).

CONJECTURE *If G is a profinite group which is topologically finitely generated and H is a finite index subgroup of G, then H is open.*

So in particular, the subgroup structure of G already determines the topology of G; in any profinite group, the open subgroups form a neighborhood of the identity. If the requirement that G be topologically finitely generated is dropped, then it is easy to construct counterexamples.

EXAMPLE 1 *Let $G = 2^{\mathbb{N}}$, equipped with coordinatewise addition modulo 2. Let $\mathcal{U}$ be an ultrafilter on $\mathbb{N}$ and let H be the collection of all g in G such that $\{i \in \mathbb{N} : g(i) = 0\}$ is in $\mathcal{U}$. It is easily verified that H is a subgroup of index 2 and that H is not open unless $\mathcal{U}$ is a principal ultrafilter.*

On the other hand, it is not difficult to show using Pettis's Theorem (see Kechris 1995, 9.9) that if H is a subgroup of a Polish group G and H has finite index, then either H is open or else H fails to have the Property of Baire (a set has the Baire Property if it differs from an open set by a set of first category). In particular, Serre's conjecture is true even without the assumption that G is topologically finitely generated *if* we require that H is Borel or even analytic.

Thus Serre's conjecture becomes equivalent to asserting that H has additional regularity properties which it obtains just by virtue of the algebraic structure. While the analysis using Pettis's Theorem is presumably well known (and not at all difficult), a proof of Serre's Conjecture was only given very recently by Nikolov and Segal (Nikolov & Segal 2007). The proof itself is a *tour de force* in the theory of finite groups and brings closure to a long line of research on the subject (Hartley 1979), (Martinez & Zelmanov 1996), (Saxl & Wilson 1997), (Segal 2000). It should be remarked that this is related to another more general pursuit: understanding when algebraic constraints on functions imply topological constraints such as continuity. The study of *automatic continuity* dates back to Cauchy; (see Rosendal 2009) for a recent survey of work in this area.

4 The additivity of strong homology

If one wishes to have a theory of homology which extends to general topological spaces, this becomes a rather subtle matter. One such theory which was developed was that of *strong homology*. While the development is beyond the scope of this paper (the interested reader is referred to (Mardešić 2000) for a complete treatment) we will discuss an example of how a computation in strong homology reduces to a problem in uncountable combinatorics.

In (Milnor 1962), Milnor proposed the following natural axiom known as *additivity* that a homology theory might satisfy. It asserts for every family X_i ($i \in I$) of topological spaces, the natural inclusions of X_i into $\coprod_{i \in I} X_i$ induce an isomorphism of groups

$$\bigoplus_{i \in I} H_p(X_i) \simeq H_p(\coprod_{i \in I} X_i).$$

Now consider the following example due to Mardešić and Prasolov.

EXAMPLE 2 *(Mardešić & Prasolov 1988) For each $d > 0$, set $z_n = (2^{-n}, 0, 0, \ldots, 0)$ and define*

$$X_d = \bigcup_{n=0}^{\infty} \{x \in \mathbb{R}^{d+1} : |x - z_n| = 2^{-n}\}.$$

Thus X_d is a sequence of nested d dimensional spheres which converge to the origin. The space X_d is compact and its homology groups coincide with the Steenrod homology groups:

$$H_p(X_d) = \begin{cases} \mathbb{Z}^{\mathbb{N}} & \text{if } p = d \\ \mathbb{Z} & \text{if } p = 0 \\ 0 & \text{otherwise} \end{cases}$$

The additivity axiom would imply that

$$H_p(X_d \times \mathbb{N}) = \bigoplus_{n=0}^{\infty} H_p(X_d).$$

Mardešić and Prasolov have shown, however, that if strong homology is used then

$$H_i(X_d \times \mathbb{N}) = \begin{cases} \bigoplus_{i=0}^{\infty} \mathbb{Z}^{\mathbb{N}} & \text{if } p = d \\ \lim^{d-p} \mathbf{A} & \text{if } 0 < p < k \\ \lim^{d} \mathbf{A} \oplus (\bigoplus_{i=0}^{\infty} \mathbb{Z}) & \text{if } p = 0 \\ 0 & \text{if } p > d \end{cases}$$

Here $\mathbf{A}$ is the inverse system which is defined as follows. Set $D_f = \{(i,j) \in \mathbb{N}^2 : j < f(i)\}$ and define

$$A_f = \bigoplus_{(i,j) \in D_f} \mathbb{Z}.$$

If $f \leq g$ are in $\mathbb{N}^\mathbb{N}$, then $D_f \subseteq D_g$ and we have a natural restriction map $\varrho_{g,f} : A_g \to A_f$. The family A_f ($f \in \mathbb{N}^\mathbb{N}$), equipped with these restrictions, becomes an inverse system of abelian groups.

The derived limits $\lim^k \mathbf{A}$ are only understood in the single case mentioned below. They are in all cases apparently sensitive to set-theoretic assumptions. In particular, Mardešić and Prasolov have shown that if the Continuum Hypothesis is true, then $\lim^1 \mathbf{A} \neq 0$ and in particular that strong homology is not additive. Dow, Simon, and Vaughan have shown on the other hand, that the Proper Forcing Axiom implies that $\lim^1 \mathbf{A} = 0$.

Combinatorially, $\lim^1 \mathbf{A} = 0$ is equivalent to the following assertion (Mardešić & Prasolov 1988): if ϕ_f ($f \in \mathbb{N}^\mathbb{N}$) is a *coherent* family of functions with $dom(\phi_f) = D_f$, then there is a single $\Phi : \mathbb{N}^2 \to \mathbb{N}$ such that, for each $f \in \mathbb{N}^\mathbb{N}$, ϕ_f is, modulo a finite error, equal to the restriction $\Phi \upharpoonright D_f$. Here ϕ_f ($f \in \mathbb{N}^\mathbb{N}$) is *coherent* if for each $f, g \in \mathbb{N}^\mathbb{N}$ the set

$$\{(i,j) \in D_f \cap D_g : \phi_f(i,j) \neq \phi_g(i,j)\}$$

is finite.

So far, only $\lim^1 \mathbf{A} = 0$ has been examined in the set-theoretic literature. It appears to be a highly non-trivial problem to determine whether these groups can all be trivial in a single model of set theory.

5 Gaps and automorphisms of $\mathcal{P}(\mathbb{N})/\text{fin}$

The problem of whether $\lim^1 \mathbf{A} = 0$ discussed in the previous section is a special instance of a more general set-theoretic problem which frequently arises in applications of set theory: what types of gaps are present in quotients of $\mathcal{P}(\mathbb{N})$ and under what circumstances can they arise? What is interesting is that when questions arising outside of set theory are boiled down to a question concerning gaps, the gaps involved rarely if ever come with regularity restrictions. That is, these naturally arising questions are of an uncountable nature. Moreover, the development of the general theory of gaps has in turn guided a parallel theory of definable gaps.

Before proceeding, we will review some terminology. A *gap* in $\mathcal{P}(\mathbb{N})/\text{fin}$ is a pair $\mathcal{A}, \mathcal{B}$ of subsets of $\mathcal{P}(\mathbb{N})$ such that:

$A \cap B$ is finite whenever $A \in \mathcal{A}, B \in \mathcal{B}$, but there is no single $C \subseteq \mathbb{N}$ satisfying $C \cap B$ is finite for all $B \in \mathcal{B}$ and $A \setminus C$ is finite for all $A \in \mathcal{A}$.

Gaps in $\mathcal{P}(\mathbb{N})/\text{fin}$ where first studied by Hausdorff in (Hausdorff 1909). Todorcevic was the first to emphasize the Ramsey-theoretic nature of gaps and also stress their important role in applications. In (Todorčević 1989), he formulated a powerful graph-theoretic dichotomy known as the *Open Coloring Axiom* in order to study its influence on gaps:

OCA If G is a graph whose vertex set is a separable metric space and whose edge set is topologically open, then either G has a countable vertex coloring or else contains an uncountable clique.

(It is interesting to note that the formulation of OCA can be traced to problem of studying the isomorphism types of subsets of $\mathbb{R}$, something seemingly unrelated to gaps. Specifically, the definition of OCA was derived from similar statements considered by Abraham, Rubin, and Shelah in (Abraham et al. 1985) which in turn were derived from a result of Baumgartner (Baumgartner 1973).) Further information on gaps can be found in (Todorčević 1998).

Next we will turn to a problem whose solution involved the analysis of gaps.

PROBLEM If ϕ is an automorphism of the Boolean algebra $\mathcal{P}(\mathbb{N})/\text{fin}$, is there a function $f : \mathbb{N} \to \mathbb{N}$ which induces ϕ?

That is, is there an f such that $\phi([A]) = [B]$ if and only if the image of A under f and B differ by a finite set? If this is the case, we say that ϕ is a *trivial* automorphism. It is interesting to note here that while an automorphism of $\mathcal{P}(\mathbb{N})/\text{fin}$ is not *a priori* an object of second order arithmetic, a trivial automorphism is.

It turns out that the answer to the above problem is independent of ZFC. If one assumes the Continuum Hypothesis, then $\mathcal{P}(\mathbb{N})/\text{fin}$ is $\aleph_1$-saturated and there are $2^{2^{\aleph_0}}$ automorphisms of $\mathcal{P}(\mathbb{N})/\text{fin}$ (and so in particular not all are induced by a map from $\mathbb{N}$ to $\mathbb{N}$). On the other hand Shelah has shown that it is consistent with ZFC that all automorphisms of $\mathcal{P}(\mathbb{N})/\text{fin}$ are trivial (Shelah 1982). Later Shelah and Steprāns showed that PFA implies all automorphisms of $\mathcal{P}(\mathbb{N})/\text{fin}$ are trivial (Shelah & Steprāns 1988). Their proof was further simplified and carried out under the weaker assumption of OCA and MA by Veličković (Veličković 1993). The reader is referred to (Just & Krawczyk 1984), (Just 1992), and (Farah 2000) for subsequent work on this subject. More recently Philips-Weaver (Phillips & Weaver 2007) and Farah (Farah 2011) have adapted this method to solve a longstanding problem in the theory of operator algebras originating in (Brown *et al.* 1977).

What is also interesting about Shelah's solution of the automorphism problem was that it was later discovered that there is an *effective* analog of Shelah's theorem: any automorphism of $\mathcal{P}(\mathbb{N})/\text{fin}$ which has a Baire measurable lifting is trivial (Veličković 1986). It is important to note, however, that this effective theorem—which could be regarded as a result in second order number theory—was discovered by analyzing the combinatorics of Shelah's independence proof and the Shelah-Steprān's proof from PFA. Moreover, while proofs which utilize PFA often yield effective counterparts as corollaries, the converse is not true.

There are, in fact, other instances where solutions to effective versions of problems have been given while the original problem remains open and apparently intractable. The following are two examples.

PROBLEM (see Gruenhage & Tatch Moore 2007) Suppose that C is a compact convex subset of a locally convex topological vector space. If every closed subset of C is a G_δ set, is C necessarily metrizable?

PROBLEM (Arhangel′skiĭ & Malykhin 1996) (see Tatch Moore 2007) If G is a separable Fréchet group, must G be metrizable?

In the case of the first problem, Todorcevic has shown that the answer is positive if C is homeomorphic to a compact subset of the the Baire class one functions on a Polish space (Todorčević 1999) (this is a natural regularity assumption on C in this context). In the case of the second problem, it is not difficult to show that the problem reduces to the case in which G is countable. Todorcevic and Uzcágeti have shown that if G is a countable Fréchet group and the topology on G is analytic as a subset of the compact metric space $\mathcal{P}(G) \equiv 2^G$, then G is metrizable (Todorčević & Uzcátegui 2001). Consistent counterexamples to both problems are known (see Lopez-Abad & Todorčević 2011; Tatch Moore 2007, respectively).

6   The separable quotient problem

Next we turn to another instance in which restricting attention to objects of countable character does not give the full picture. One of the most basic questions about Banach spaces concerns the existence of Schauder bases in these spaces. The following question is often attributed to Banach himself, although it was only later that it was made explicit in print.

PROBLEM (see Pełczyński 1964) Does every infinite dimensional Banach space have an infinite dimensional quotient with a basis?

Nothing about this problem suggests "uncountability"—which in this context should be interpreted as nonseparability. Still, Johnson and Rosenthal were able to prove that this problem has a positive answer within the class of separable Banach spaces (Johnson & Rosenthal 1972). This reduced Banach's original problem to the following form, which is more prevalent in the literature today.

SEPARABLE QUOTIENT PROBLEM Does every infinite dimensional Banach space have an infinite dimensional separable quotient?

The reader is referred to (Mújica 1997) for a survey of this problem. I will note two more recent results in the positive direction.

THEOREM 3 *(Todorčević 2006) Assume PFA. Every Banach space of density $\aleph_1$ admits a nonseparable quotient with a basis.*

THEOREM 4 *(Argyros et al. 2008) If X is an infinite dimensional Banach space, then X^* has an infinite dimensional separable quotient.*

7 The determinacy of Gale-Stewart games

One of the most profound examples of how large sets can influence countable combinatorics is surely the determinacy of Gale-Stewart games. Recall that in a Gale-Stewart game, two players alternately play elements x_n of some set X, one element for each natural number. Both players have perfect information. Player I wins if the outcome $\langle x_n : n < \infty \rangle$ is in some prespecified set Γ; Player II wins otherwise. Such a game is *determined* if one of the two players has a winning strategy. The simplest theorem concerning the determinacy of Gale-Stewart games was already known to Gale and Stewart.

CLOSED DETERMINACY If $\Gamma \subseteq X^{\mathbb{N}}$ is closed, then the Gale-Stewart game specified by Γ is determined.

The proof is quite simple: Player I always plays to maintain that Player II does not have a winning strategy. Either this is impossible and Player II has a winning strategy from the beginning of the game or else Player I has arranged that at no point in the game did Player II have a winning strategy. The key point is that, in a closed game, if Player II wins a play of the game, she has already won at a finite stage of the game (i.e., all further plays are irrelevant).

The determinacy of Gale-Stewart games is of interest primarily because regularity properties of subsets of $\mathbb{R}$ and other Polish spaces can be recast in terms of the existence of winning strategies in games which are associated to these sets (see, e.g., Kechris 1995, 21). In general, Gale-Stewart games need not be determined; the Axiom of Choice can readily be used to construct games which are not determined. On the other hand, sets $\Gamma \subseteq \mathbb{N}^{\mathbb{N}}$ which are in some sense *regular* do tend to specify determined games.

BOREL DETERMINACY (Martin 1975) If $\Gamma \subseteq \mathbb{N}^{\mathbb{N}}$ is Borel, then Γ is determined.

What is remarkable is that all known proofs of determinacy ultimately rely on the determinacy of closed games: one reduces the determinacy of $\Gamma \subseteq \mathbb{N}^{\mathbb{N}}$ to the determinacy of some equivalent closed game $\Gamma^* \subseteq X^{\mathbb{N}}$. The set X underlying this "unraveled" game is typically much larger than $\mathbb{N}$. For instance, H. Friedman has shown that Borel Determinacy is not provable in ZFC without the powerset axiom (Friedman 1971). In fact any proof of Borel determinacy must use, in an essential way, $\aleph_1$ iterations of the power set operation.

Earlier, Martin had proved the determinacy of analytic games from the existence of a measurable cardinal (Martin 1969-1970). Harrington proved that the determinacy of analytic games is equivalent to the existence of $x^{\sharp}$ for each $x \subseteq \mathbb{N}$, thus demonstrating the necessity of large cardinals in Martin's proof (Harrington 1978). The determinacy of projective games was proved by Martin and Steel from the existence of infinitely many Woodin cardinals (Martin & Steel 1989)—an assumption which was shown by Woodin to be essentially optimal.

Notice that the determinacy of projective games is formalizable in *second order arithmetic* and concerns the properties of the hereditarily countable sets. Even the proof of the determinacy of Borel games, however, already makes essential use of transfinite iterates of the powerset operation. In the case of analytic determinacy, the proof moreover requires the use of large cardinals. The reader is referred to (Kanamori 2003) for further reading on determinacy and large cardinals, as well as an extensive bibliography on the subject. Some further information on the history of determinacy can be found in (Larson 2012).

8 Large cardinals, braids, and left self distributivity

Next we turn to an example where very large sets have proved useful both in establishing facts about finite algebraic structures and in improving the efficiency of algorithms for comparing braids. A binary system $(S, *)$ is called a *LD system* if it satisfies the *left self distributive law*:

$$a * (b * c) = (a * b) * (a * c)$$

Left self distributivity showed up independently in the literature in two very different contexts. On one hand, it came naturally out of attempts by Brieskorn, Joyce, Kauffman, and their students to develop invariants for studying the braid group (Dehornoy 2000). Roughly speaking, one colors the strands at the top of a braid using colors of a binary system $(S, *)$. The operation dictates how the strands change the colors of other strands in a diagram representing the braid. In order for this procedure to yield an invariant for braids, $(S, *)$ must be an LD system. The reader is referred to (Dehornoy 2000) for more details. Suffice it to say that this use of an LD system makes it desirable to understand *free* LD systems—those not satisfying any laws other than those which follow logically from the LD law.

In a separate branch of mathematics, LD systems were being studied for a completely different purpose. It was part of the folklore in set theory that the family $\mathcal{E}_\lambda$ of non-identity elementary embeddings from V_λ into itself formed an algebraic structure which moreover satisfied the left self distributive law (Dehornoy 2000). Such embeddings are known as *rank-to-rank embeddings*. Postulating the existence of a λ for which there is a non-identity elementary embedding from V_λ to V_λ is an example of a *large cardinal axiom*; in fact it is among the strongest of the large cardinal axioms (see Kanamori 2003). In particular, the existence of rank-to-rank embeddings cannot be proved within ZFC.

In (Laver 1992), Laver proved that if j is a rank-to-rank embedding, then the algebra $(\mathcal{A}, *)$ generated by j is free. This was in sharp contrast to the LD systems—such as a group equipped with conjugation—which had been employed previously in the study of braids. Then in (Laver 1993), Laver used the existence of a rank-to-rank embedding to prove that the word problem in LD systems is decidable. This in turn led to efficient new algorithms for comparing braids (Dehornoy 1997) (Laver 1996). Only later was Dehornoy able to remove the use of large cardinals from solution to the decision problem for LD systems (Dehornoy 1992).

Still, large cardinals played a remarkable and unique role in this development. Furthermore, there are questions concerning certain finite LD systems which so far have only been settled using large cardinal assumptions. An LD system is *cyclic* if it has a single generator a and there is a $p > 1$ such that the left associated power

$$a_{[p]} = ((a * a) \ldots * a) * a$$

equals a. Laver has shown that any cyclic LD system has 2^n elements for some n and is unique up to isomorphism. If for a given $n \in \mathbb{N}$ we define $*$ on $\{1, \ldots, 2^n\}$ by

$$a * 1 = \begin{cases} a + 1 & \text{if } a < 2^n \\ 1 & \text{if } a = 2^n \end{cases}$$

then there is a unique extension of $*$ a binary operation which is left self distributive. This LD system is the nth *Laver table* A_n. The following summarizes the important properties of the Laver tables:

$a * p = (a+1)_p$ if $a < 2^n$ and $2^n * a = a$.

if $m < n$, then the function $\pi : A_n \to A_m$ defined by $\pi(a) = b$ if $a \equiv b$ mod 2^m is a surjective homomorphism.

if $a \in A_n$, then there is a $p \leq n$ such that $a * b = a * b'$ if $b \equiv b'$ mod 2^p and $a * b < a * (b+1)$ if $1 \leq b < 2^p$.

In fact if $a \in A_n$ and 2^p is the period of row a, then $b \mapsto a * b$ defines a monomorphism of A_p into A_n (this is nothing more than the left self distributive law). Moreover, since each A_n is cyclic, all such monomorphisms arise in this way.

If we work within the category of one generator LD systems, then the Laver tables have an inverse limit A_∞. We now have the following result which is a consequence of work of Laver and Steel (see Dehornoy 2000).

THEOREM 5 *If there is a rank-to-rank elementary embedding, then A_∞ is free.*

It is not known whether this result can be proved in ZFC. On the other hand, it still is possible that one might be able to prove this theorem within Peano Arithmetic.

The freeness of A_∞ has many equivalents, even working over a weak base theory such as Primitive Recursive Arithmetic (Dougherty & Jech 1997). One equivalent is that for every p there is an n such that the period of row 1 in A_n is at least p. On the other hand, Dougherty has shown that the function $p \mapsto n$ which witnesses this cannot be primitive recursive (Dougherty 1993). Moreover he has shown that the least n for which the period of row 1 is at least 32 is $A_9(A_8(A_8(254)))$, where $A_k(n)$ is the kth level of the Ackermann function (Dougherty 1993).

In addition to the original sources mentioned above, further reading can be found in (Dehornoy 2000), which serves as a comprehensive source on this subject.

9   Concluding remarks

Of course there has been no attempt at being comprehensive in choosing the topics presented above; I do not even pretend to have taken a representative selection. The examples all appear to have a somewhat *ad hoc* character to them. There is some truth to this and in fact that is partly the point—it is very difficult to predict from the outset of one's study of a problem whether uncountability or some higher order of infinity is at all relevant. For instance, the conventional wisdom even among set theorists would be that uncountability *should not* be at all relevant to understanding the completeness of ACF_0, the Ramsey theory of the countably infinite, or the freeness of A_∞. The above discussion shows that even when it can be avoided, uncountability can still play an illuminating role in understanding the countable.

Additionally, the study of uncountability for its own sake sometimes leads to unexpected results about objects of a countable or even finite nature. Even if the use of uncountability ultimately turns out to be inessential, its role in the discovery process should not be ignored. This can be seen in Dehornoy and Laver's algorithm for the word problems for LD systems and braids. It can also be seen in Veličković's observation that Shelah's proof shows that if an automorphism of $\mathcal{P}(\mathbb{N})/\mathrm{fin}$ has a Baire measurable lifting, then it is induced by a function from $\mathbb{N}$ to $\mathbb{N}$.

Finally, there is the lesson illustrated in Serre's conjecture and the Separable Quotient Problem: mathematicians *do* care about arbitrary subsets of Polish spaces and sets of unrestricted cardinality. All too often logicians make assumptions about what "real mathematicians" care about, what they are interested in, and what their biases are, without spending enough time exploring real mathematics itself. Even if these biases are as prevalent as we've come to believe they are (something I doubt), the examples above (and many more) are compelling testimony as to why these biases are misinformed and unnecessarily restrictive.

Bibliography

Abraham, U., Rubin, M., & Shelah, S. (1985). On the consistency of some partition theorems for continuous colorings, and the structure of $\aleph_1$-dense real order types. *Annals of Pure and Applied Logic*, *29*(2), 123–206, doi:10.1016/0168-0072(84)90024-1.

Argyros, S., Dodos, P., & Kanellopoulos, V. (2008). Unconditional families in Banach spaces. *Mathematische Annalen*, *341*(1), 15–38, doi:10.1007/s00208-007-0179-y.

Arhangel'skiĭ, A. V. & Malykhin, V. I. (1996). Metrizability of topological groups (Russian). *Moscow University Mathematics Bulletin C/C Of Vestnik-Moskovskii Universitet Mathematika*, *51*(3), 13–16.

Baumgartner, J. E. (1973). All $\aleph_1$-dense sets of reals can be isomorphic. *Fundamenta Mathematicae*, *79*(2), 101–106.

Brown, L., Douglas, R. G., & Fillmore, P. A. (1977). Extensions of C^*-algebras and K-homology. *Annals of Mathematics*, *105*(2), 265–324.

Dehornoy, P. (1992). Preuve de la conjecture d'irréflexivité pour les structures distributives libres. *Comptes rendus de l'Académie des sciences. Série 1, Mathématique*, *314*(5), 333–336.

Dehornoy, P. (1997). A fast method for comparing braids. *Advances in Mathematics*, *125*(2), 200–235, doi:10.1006/aima.1997.1605.

Dehornoy, P. (2000). *Braids and Self-Distributivity, Progress in Mathematics*, vol. 192. Basel: Birkhäuser, doi:10.1007/978-3-0348-8442-6.

Dougherty, R. (1993). Critical points in an algebra of elementary embeddings. *Annals of Pure and Applied Logic*, *65*(3), 211–241, doi:10.1016/0168-0072(93)90012-3.

Dougherty, R. & Jech, T. (1997). Finite left-distributive algebras and embedding algebras. *Advances in Mathematics*, *130*(2), 201–241, doi: 10.1006/aima.1997.1655.

Ellis, R. (1958). Distal transformation groups. *Pacific Journal of Mathematics*, *8*(3), 401–405.

Farah, I. (2000). *Analytic Quotients: Theory of liftings for quotients over analytic ideals on the integers, Memoirs of the American Mathematical Society*, vol. 148. Providence, RI: American Mathematical Society.

Farah, I. (2011). All automorphisms of the Calkin algebra are inner. *Annals of Mathematics*, *173*(2), 619–661.

Friedman, H. M. (1971). Higher set theory and mathematical practice. *Annals of Mathematical Logic*, *2*(3), 325–357, doi:10.1016/0003-4843(71)90018-0.

Gowers, W. T. (1992). Lipschitz functions on classical spaces. *European Journal of Combinatorics*, *13*(3), 141–151, doi:10.1016/0195-6698(92)90020-Z.

Gruenhage, G. & Tatch Moore, J. (2007). Perfect compacta and basis problems in topology. In *Open Problems in Topology {II}*, Pearl, E., ed., Amsterdam: Elsevier, 151–159, doi:10.1016/B978-044452208-5/50016-8.

Harrington, L. (1978). Analytic determinacy and $0^{\sharp}$. *Journal of Symbolic Logic*, *43*(4), 685–693, doi:10.2307/2273508.

Hartley, B. (1979). Subgroups of finite index in profinite groups. *Mathematische Zeitschrift*, *168*(1), 71–76, doi:10.1007/BF01214436.

Hausdorff, F. (1909). Die Graduierung nach dem Endverlauf. In *Abhandlungen der mathematisch-physischen klasse der königlich sachsischen Gesellschaft der Wissenschaften*, vol. 31, 296–334.

Hindman, N. (1974). Finite sums from sequences within cells of a partition of N. *Journal of Combinatorial Theory, Series A*, *17*(1), 1–11, doi:10.1016/0097-3165(74)90023-5.

Hindman, N. & Strauss, D. (1998). *Algebra in the Stone-Čech compactification*, de Gruyter Expositions in Mathematics: Theory and applications, vol. 27. Berlin: de Gruyter.

Hungerford, T. W. (1980). *Algebra*, Graduate Texts in Mathematics, vol. 73. New York: Springer, reprint of the 1974 original.

Johnson, W. B. & Rosenthal, H. P. (1972). On ω^*-basic sequences and their applications to the study of Banach spaces. *Studia Mathematica*, *43*, 77–92.

Just, W. (1992). A weak version of AT from OCA. In *Set Theory of the Continuum*, Mathematical Sciences Research Institute Publications, vol. 26, Judah, H., Just, W., & Woodin, H., eds., New York: Springer, 281–291, doi:10.1007/978-1-4613-9754-0_17.

Just, W. & Krawczyk, A. (1984). On certain Boolean algebras $\mathscr{P}(\omega)/I$. *Transactions of the American Mathematical Society*, *285*(1), 411–429.

Kanamori, A. (2003). *The Higher Infinite: Large cardinals in set theory from their beginnings*. Springer Monographs in Mathematics, Berlin; Heidelberg: Springer, 2nd edn.

Kechris, A. S. (1995). *Classical Descriptive Set Theory*, Graduate Texts in Mathematics, vol. 156. New York: Springer.

Larson, P. B. (2012). A brief history of determinacy. In *Sets and Extensions in the Twentieth Century*, Handbook of the History of Logic, vol. 6, Dov M. Gabbay, A. K. & Woods, J., eds., Amsterdam: North-Holland, 457–507, doi: 10.1016/B978-0-444-51621-3.50006-2.

Laver, R. (1992). The left distributive law and the freeness of an algebra of elementary embeddings. *Advances in Mathematics*, *91*(2), 209–231.

Laver, R. (1993). A division algorithm for the free left distributive algebra. In *Logic Colloquium '90: ASL Summer Meeting in Helsinki*, Lecture Notes in Logic, vol. 2, Oikkonen, J. & Väänänen, J., eds., Berlin: Springer, 155–162.

Laver, R. (1996). Braid group actions on left distributive structures, and well orderings in the braid groups. *Journal of Pure and Applied Algebra*, *108*(1), 81–98, doi:10.1016/0022-4049(95)00147-6.

Lopez-Abad, J. & Todorčević, S. (2011). Generic Banach spaces and generic simplexes. *Journal of Functional Analysis*, *261*(2), 300–386, doi: 10.1016/j.jfa.2011.03.008.

Mardešić, S. (2000). *Strong Shape and Homology*. Springer Monographs in Mathematics, Berlin; Heidelberg: Springer.

Mardešić, S. & Prasolov, A. V. (1988). Strong homology is not additive. *Transactions of the American Mathematical Society*, *307*(2), 725–744, doi: 10.1090/S0002-9947-1988-0940224-7.

Marker, D. (2002). *Model Theory: An introduction, Graduate Texts in Mathematics*, vol. 217. New York: Springer.

Martin, D. A. (1969-1970). Measurable cardinals and analytic games. *Fundamenta Mathematicae*, *66*, 287–291.

Martin, D. A. (1975). Borel determinacy. *Annals of Mathematics*, *102*(2), 363–371.

Martin, D. A. & Steel, J. R. (1989). A proof of projective determinacy. *Journal of The American Mathematical Society*, *2*(1), 71–125.

Martinez, C. & Zelmanov, E. (1996). Products of powers in finite simple groups. *Israel Journal of Mathematics*, *96*(2), 469–479, doi:10.1007/BF02937318.

Mazurov, V. D. & Khukhro, E. I. (2006). *The Kourovka Notebook*. Novosibirsk: Russian Academy of Sciences Siberian Division Institute of Mathematics: Unsolved problems in group theory, Including archive of solved problems, 16th edn.

Milnor, J. (1962). On axiomatic homology theory. *Pacific Journal of Mathematics*, *12*, 337–341.

Mújica, J. (1997). Separable quotients of Banach spaces. *Revista Matemática de la Universidad Complutense de Madrid*, *10*(2), 299–330.

Nikolov, N. & Segal, D. (2007). On finitely generated profinite groups. I: Strong completeness and uniform bounds. *Annals of Mathematics*, *165*(1), 171–238.

Pełczyński, A. (1964). Some problems on bases in banach and frechet spaces. *Israel Journal of Mathematics*, *2*(2), 132–138, doi:10.1007/BF02759953.

Phillips, N. C. & Weaver, N. (2007). The Calkin algebra has outer automorphisms. *Duke Mathematical Journal*, *139*(1), 185–202.

Robinson, A. (1951). *On the Metamathematics of Algebra*. Amsterdam: North-Holland.

Rosendal, C. (2009). Automatic continuity of group homomorphisms. *Bulletin of Symbolic Logic*, *15*(2), 184–214.

Saxl, J. & Wilson, J. S. (1997). A note on powers in simple groups. In *Mathematical Proceedings of the Cambridge Philosophical Society*, vol. 122, 91–94.

Segal, D. (2000). Closed subgroups of profinite groups. *Proceedings of the London Mathematical Society*, *81*(1), 29–54, doi:null.

Shelah, S. (1982). *Proper Forcing, Lecture Notes in Mathematics*, vol. 940. Berlin; Heidelberg: Springer.

Shelah, S. & Steprāns, J. (1988). PFA implies all automorphisms are trivial. *Proceedings of the American Mathematical Society*, *104*(4), 1220–1225.

Tatch Moore, J. (2011). The proper forcing axiom. In *Proceedings of the International Congress of Mathematicians 2010 (ICM 2010)*, Hyderabad, India, 3–29, doi:10.1142/9789814324359_0038.

Tatch Moore, S., J.and Todorčević (2007). The metrization problem for Fréchet groups. In *Open Problems in Topology II*, Pearl, E., ed., Amsterdam; Oxford: Elsevier, 201–206.

Todorčević, S. (1989). *Partition Problems In Topology, Contemporary mathematics*, vol. 84. Providence, RI: American Mathematical Society.

Todorčević, S. (1998). Gaps in analytic quotients. *Fundamenta Mathematicae*, *156*(1), 85–97.

Todorčević, S. (1999). Compact subsets of the first Baire class. *Journal of The American Mathematical Society*, *12*(4), 1179–1212.

Todorčević, S. (2006). Biorthogonal systems and quotient spaces via baire category methods. *Mathematische Annalen*, *335*(3), 687–715, doi:10.1007/s00208-006-0762-7.

Todorčević, S. (2010). *Introduction to Ramsey Spaces, Annals of Mathematics Studies*, vol. 174. Princeton: Princeton University Press.

Todorčević, S. & Uzcátegui, C. (2001). Analytic topologies over countable sets. *Topology and its Applications*, *111*(3), 299–326, doi:10.1016/S0166-8641(99)00223-0.

Todorchevich, S. & Farah, I. (1995). *Some Applications of the Method of Forcing*. Moscow: Yenisei.

van der Waerden, B. L. (1927). Beweis einer Baudetschen Vermutung. *Nieuw Archief voor Wiskunde. Tweede Serie*, *15*, 212–216.

Veličković, B. (1986). Definable automorphisms of $\mathscr{P}(\omega)/\text{fin}$. *Proceedings of the American Mathematical Society*, *96*(1), 130–135.

Veličković, B. (1993). OCA and automorphisms of $\mathscr{P}(\omega)/\text{fin}$. *Topology and its Applications*, *49*(1), 1–13.

Justin Tatch Moore
Department of Mathematics
Cornell University
Ithaca, NY 14853–4201
USA
justin@math.cornell.edu

Unifying Functional Interpretations: Past and Future

Paulo Oliva

ABSTRACT. This article surveys work done in the last six years on the unification of various functional interpretations including Gödel's dialectica interpretation, its Diller-Nahm variant, Kreisel modified realizability, Stein's family of functional interpretations, functional interpretations "with truth", and bounded functional interpretations. Our goal in the present paper is twofold: (1) to look back and single out the main lessons learnt so far, and (2) to look forward and list several open questions and possible directions for further research.

1 Introduction

When studying and working with the two main functional interpretations, namely the *dialectica* (Avigad & Feferman 1998; Gödel 1958) and the modified realizability (Kreisel 1959) interpretations, one notices a striking similarity in the way the two interpretations behave. For instance, they both interpret ∀∃-statements in precisely the same way, and their soundness (also called adequacy) proofs follow very similar patterns. Yet, for all purpose these are two very different interpretations, validating different principles,[1] and having different properties.[2] Several questions naturally arise. What is the common structure behind these two functional interpretations? How are the different witnesses obtained from a given proof when applying different interpretations related to each other?

It was with these questions in mind that I set out (Oliva 2006) to develop a general framework to unify functional interpretations. This initial work was followed by several other articles (Ferreira & Oliva 2009; 2012; 2011; Gaspar & Oliva 2010; Hernest & Oliva 2008; Oliva 2007a;b; 2008; 2009;

---

[1] For instance, the dialectica interpretation validates the Markov principle whereas modified realizability does not. On the other hand, modified realizability validates full extensionality whereas the dialectica interpretation does not.

[2] For instance, realizability interpretations always have a so-called "truth" variant, whereas the dialectica interpretation does not.

2012) further refining or generalising the original idea. These were mainly done in collaboration with Gilda Ferreira, Jaime Gaspar and Mircea-Dan Hernest. What started as a small modification of the dialectica interpretation to also capture realizability and the Diller-Nahm variant (Diller & Nahm 1974) ended up as a very general hybrid functional interpretation of intuitionistic affine logic,[3] also capturing Stein's family of functional interpretations (Stein 1979), functional interpretations "with truth" (Gaspar & Oliva 2010), and bounded functional interpretations (Ferreira & Nunes 2006; Ferreira & Oliva 2005; 2007).

This article will survey the work mentioned above, singling out what I believe to be the key lessons learnt so far. These are summarised as follows. For details see the corresponding sections and the articles mentioned.

(2) Modified realizability can also alternatively be presented as a *relation* between potential witnesses and challenges, in a way very similar to the way the dialectica interpretation is presented. This is originally observed in (Oliva 2006) and is key to extending realizability to affine logic (Oliva 2007b).

(3) Most functional interpretations of *intuitionistic logic* can be factored via *affine logic*. More interestingly, all functional interpretations considered, when extended to affine logic, coincide in the pure fragment, where modalities are absent. This factorisation allows us to clearly see that the only difference between most of the functional interpretations is in the treatment of contraction, which in affine logic is captured by $!A$. Although this was originally done in the setting of *classical* affine logic (Hernest & Oliva 2008; Oliva 2007a;b; 2009), it turned out that *intuitionistic* affine logic is not only enough, but the unification becomes much simpler (Ferreira & Oliva 2009; 2011; Gaspar & Oliva 2010) (albeit at the cost of losing symmetry).

(4) When designing the unified functional interpretation of intuitionistic affine logic we were only expecting to be able to capture the classic interpretations such as the dialectica, modified realizability and Diller-Nahm. We were therefore surprised when we discovered (Gaspar & Oliva 2010) that even the truth variants of functional interpretations fit in the framework almost effortlessly. Which means that even proof interpretations with truth only differ from their "non-truth" variants in the treatment of $!A$, but coincide in the treatment of all other connectives.

---

[3] Intuitionistic linear logic plus the weakening rule.

(5) Because the bang (!) of affine logic is not canonical, one can then effectively combine all the functional interpretations mentioned above, including their truth variants, into single interpretations which we called *hybrid functional interpretations* (Hernest & Oliva 2008; Oliva 2012). This means, for instance, that in a single proof one can try to make use of both the dialectica interpretation in some parts of the proof and modified realizability in others, combining their strengths to maximum benefit.

We will conclude (6) by listing thirteen open questions which indicate possible interesting directions for further research.

Acknowledgement. Most of the work presented here has been done in collaboration with Gilda Ferreira, Jaime Gaspar and Mircea-Dan Hernest. I would also like to acknowledge previous work done in this direction on which the current work builds, such as those of Martin Stein (Stein 1979; 1980), Valéria de Paiva (de Paiva 1989a;b), Masaru Shirahata (Shirahata 2006) and Andreas Blass (Blass 1992). Finally, many thanks to Thomas Powell, Jules Hedges and Gilda Ferreira for several comments and corrections on an earlier version of this paper.

Notation. We use $X :\equiv A$ to say that X is defined by A. We use $A \equiv B$ to mean A and B are syntactically equal.

2 A different view on realizability

The first obvious difference between modified realizability (Kreisel 1959) and the dialectica interpretation (Gödel 1958) is that the first interprets formulas A as *unary* predicates $A_r(\boldsymbol{x})$, normally written as "$\boldsymbol{x}$ realizes A", whereas the dialectica interpretation associates to formulas A *binary* predicates $A_D(\boldsymbol{x};\boldsymbol{y})$. Here $\boldsymbol{x}$ and $\boldsymbol{y}$ denote tuples of variables $\boldsymbol{x} = x_1,\ldots,x_n$ and $\boldsymbol{y} = y_1,\ldots,y_m$, where the length of the tuple and the types of the variables depend on the logical structure of the formula A. The two formulas $A_r(\boldsymbol{x})$ and $A_D(\boldsymbol{x};\boldsymbol{y})$ are defined inductively as[4]

---

[4] We are using the abbreviation $A \Diamond_b B :\equiv (b = \mathsf{true} \to A) \wedge (b = \mathsf{false} \to B)$. We also use the same macro in the context of affine logic where it stands for $A \Diamond_b B :\equiv (!(b = \mathsf{true}) \multimap A) \otimes (!(b = \mathsf{false}) \multimap B)$.

$$(A \wedge B)_r(x, y) :\equiv A_r(x) \wedge B_r(y)$$
$$(A \vee B)_r(x, y, b) :\equiv A_r(x) \diamond_b B_r(y)$$
$$(A \to B)_r(f) :\equiv \forall x (A_r(x) \to B_r(fx))$$
$$(\exists z A)_r(x, a) :\equiv (A[a/z])_r(x)$$
$$(\forall z A)_r(f) :\equiv \forall z A_r(fz)$$

$$(A \wedge B)_\mathsf{D}(x, v; y, w) :\equiv A_\mathsf{D}(x; y) \wedge B_\mathsf{D}(v; w)$$
$$(A \vee B)_\mathsf{D}(x, v, b; y, w) :\equiv A_\mathsf{D}(x; y) \diamond_b B_\mathsf{D}(v; w)$$
$$(A \to B)_\mathsf{D}(f, g; x, w) :\equiv A_\mathsf{D}(x; gxw) \to B_\mathsf{D}(fx; w)$$
$$(\exists z A)_\mathsf{D}(x, a; y) :\equiv (A[a/z])_\mathsf{D}(x; y)$$
$$(\forall z A)_\mathsf{D}(f; y, a) :\equiv (A[a/z])_\mathsf{D}(fa; y).$$

with the base case $(A_\mathsf{at})_r(\epsilon) = (A_\mathsf{at})_\mathsf{D}(\epsilon; \epsilon) = A_\mathsf{at}$, for atomic formulas A_at, with ϵ denoting the empty tuple (henceforth omitted). Note that for tuples of variables $f = f_1, \ldots, f_n$ and x we write fx for the tuple of terms $f_1 x, \ldots, f_n x$. Using these predicates $A_r(x)$ and $A_\mathsf{D}(x; y)$ we can define two sets of "functionals"

$$A \mapsto \{x \mid A_r(x)\} \qquad A \mapsto \{x \mid \forall y A_\mathsf{D}(x; y)\}$$

which we will refer to as the "realizability witnesses" and the "dialectica witnesses". The two functional interpretations, modified realizability and dialectica, can be viewed as algorithms to turn an intuitionistic proof of A into concrete (e.g. higher-order programs) elements of these sets.

The work on unifying different functional interpretations (Oliva 2006) started with the observation that one can also view modified realizability as associating formulas with a *binary* predicate $A_\mathsf{rr}(x; y)$ (which I will call "relational realizability") between two tuples x and y in a way very similar to the dialectica interpretation, namely

(1)
$$(A \wedge B)_\mathsf{rr}(x, v; y, w) :\equiv A_\mathsf{rr}(x; y) \wedge B_\mathsf{rr}(v; w)$$
$$(A \vee B)_\mathsf{rr}(x, v, b; y, w) :\equiv A_\mathsf{rr}(x; y) \diamond_b B_\mathsf{rr}(v; w)$$
$$(A \to B)_\mathsf{rr}(f; x, w) :\equiv \forall y A_\mathsf{rr}(x; y) \to B_\mathsf{rr}(fx; w)$$
$$(\exists z A)_\mathsf{rr}(x, a; y) :\equiv (A[a/z])_\mathsf{rr}(x; y)$$
$$(\forall z A)_\mathsf{rr}(f; y, a) :\equiv (A[a/z])_\mathsf{rr}(fa; y).$$

It is easy to show by induction on the formula A that these two different definitions of realizability lead to the same interpretation as the following equivalence is intuitionistically provable:

$$A_r(x) \quad \Leftrightarrow \quad \forall y A_\mathsf{rr}(x; y).$$

The relational presentation of realizability, however, makes it absolutely clear that realizability only differs from the dialectica interpretation in the clause for implication $A \to B$. While the realizability interpretation does not attempt to witness the universal quantifier $\forall \boldsymbol{y}$ in the clause for $A \to B$, the dialectica interpretation witnesses such quantifier via the extra tuple of functionals $\boldsymbol{g}$.

The two main ideas behind the original *unifying functional interpretation* (Oliva 2006) are the introduction of a common notation $|A|_{\boldsymbol{y}}^{\boldsymbol{x}}$ for such binary predicates, and a parametrised interpretation of $A \to B$. That is achieved via an abstract formula constructor $\forall \boldsymbol{x} \prec \boldsymbol{a}\, A$ that takes a tuple of terms $\boldsymbol{a}$ and a formula A (with free variables $\boldsymbol{x}$) and produces a new formula where $\boldsymbol{x}$ are no longer free. A parametrised functional interpretation can then be given as

$$
\begin{aligned}
|A \wedge B|_{\boldsymbol{y},\boldsymbol{w}}^{\boldsymbol{x},\boldsymbol{v}} &:\equiv |A|_{\boldsymbol{y}}^{\boldsymbol{x}} \wedge |B|_{\boldsymbol{w}}^{\boldsymbol{v}} \\
|A \vee B|_{\boldsymbol{y},\boldsymbol{w}}^{\boldsymbol{x},\boldsymbol{v},b} &:\equiv |A|_{\boldsymbol{y}}^{\boldsymbol{x}} \Diamond_b |B|_{\boldsymbol{w}}^{\boldsymbol{v}} \\
(2) \quad |A \to B|_{\boldsymbol{x},\boldsymbol{w}}^{\boldsymbol{f},\boldsymbol{g}} &:\equiv \forall \boldsymbol{y} \prec \boldsymbol{gxw}\, |A|_{\boldsymbol{y}}^{\boldsymbol{x}} \to |B|_{\boldsymbol{w}}^{\boldsymbol{fx}} \\
|\exists z A|_{\boldsymbol{y}}^{\boldsymbol{x},a} &:\equiv |A[a/z]|_{\boldsymbol{y}}^{\boldsymbol{x}} \\
|\forall z A|_{\boldsymbol{y},a}^{\boldsymbol{f}} &:\equiv |A[a/z]|_{\boldsymbol{y}}^{\boldsymbol{fa}}.
\end{aligned}
$$

Subject to a few conditions (cf. Oliva 2006) on $\forall \boldsymbol{x} \prec \boldsymbol{a}\, A$, one can then prove a uniform soundness theorem for intuitionistic logic. When the formula constructor is instantiated one obtains the three main functional interpretations as follows:

$\forall \boldsymbol{x} \prec \boldsymbol{a}\, A$	**Functional interpretation**
$A[\boldsymbol{a}/\boldsymbol{x}]$	Gödel's dialectica interpretation
$\forall \boldsymbol{x} \in \boldsymbol{a}\, A$	Diller-Nahm interpretation
$\forall \boldsymbol{x} A$	Kreisel modified realizability

In order to show that each of these three interpretations is sound one only needs to check that they satisfy the required conditions mentioned above.

REMARK 1 (Stein family of interpretations) *Let $M \in \mathbb{N} \cup \{\infty\}$. Given a tuple of variables $\boldsymbol{x} = x_0, \ldots, x_n$ let us denote by $\boldsymbol{x}^{\geq M}$ the tuple containing only the elements of $\boldsymbol{x}$ with type level $\geq M$. Similarly we denote by $\boldsymbol{x}^{<M}$ the tuple containing only the elements of $\boldsymbol{x}$ with type level $< M$. Note that $\boldsymbol{x}^{<\infty} = \boldsymbol{x}$ and $\boldsymbol{x}^{<0}$ is the empty tuple. Stein's family of functional interpretations (Stein 1979) also fits in the above framework as we can take for each given M*

$$\forall \boldsymbol{x} \prec \boldsymbol{a}\, A :\equiv \forall \boldsymbol{x}^{<M} \forall \boldsymbol{x}^{\geq M} \in \boldsymbol{a}\, A$$

where $\boldsymbol{a}$ is a set indexed by the pure type M, i.e., $\boldsymbol{a}\colon M \to \rho$ for some type ρ. When $M = \infty$ this coincides with modified realizability, whereas with $M = 0$ this is a variant of the Diller-Nahm interpretation that allows for infinite (countable) sets, as $\boldsymbol{a}\colon \mathbb{N} \to \rho$ ($\mathbb{N}$ is the pure type having type level 0).

3 Factoring through affine logic

Reformulating realizability as a binary predicate as described in Section 2 was an important step towards showing that modified realizability and the dialectica interpretation have much more in common than previously imagined. The fact is that they only differ on their handling of witnesses coming from the premise of an implication. But that opens a new question: What is special about the premise of an implication that allows for these different interpretations to exist? A satisfactory answer to this question came from the analysis of functional interpretations via affine logic.

Intuitionistic affine logic (AL_i) is a refinement of intuitionistic logic (IL) where particular attention is paid to the contraction rule (Benton et al. 1993; Girard 1987). We call this a *refinement* because the connectives of intuitionistic logic can be recovered from a combination of those from affine logic. This is formally expressed via Girard's translations of intuitionistic logic into linear logic. The two most commonly used are[5]

$$
\begin{array}{rclrcl}
P^* &:\equiv& P & P^\circ &:\equiv& !P \\
(A \wedge B)^* &:\equiv& A^* \otimes B^* & (A \wedge B)^\circ &:\equiv& A^\circ \otimes B^\circ \\
(A \vee B)^* &:\equiv& !A^* \oplus !B^* & (A \vee B)^\circ &:\equiv& A^\circ \oplus B^\circ \\
(A \to B)^* &:\equiv& !A^* \multimap B^* & (A \to B)^\circ &:\equiv& !(A^\circ \multimap B^\circ) \\
(\forall x A)^* &:\equiv& \forall x A^* & (\forall x A)^\circ &:\equiv& !\forall x A^\circ \\
(\exists x A)^* &:\equiv& \exists x !A^*. & (\exists x A)^\circ &:\equiv& \exists x A^\circ.
\end{array}
$$

The translations are such that if A is provable in IL then both $!A^*$ and A° are provable in AL_i.

While working on (Oliva 2006), in the setting of intuitionistic logic, I came across de Paiva's (de Paiva 1989b) dialectica (and Diller-Nahm) interpretation of *affine logic*. It then occurred to me that one could use the new formulation of realizability discussed in Section 2 to extend the *realizabillity* interpretation from intuitionistic logic to affine logic. This was developed

---

[5]The usual clause for $(A \wedge B)^*$ is $(A \wedge B)^* :\equiv A^* \,\&\, B^*$. We can take $(A \wedge B)^* :\equiv A^* \otimes B^*$ instead because we are embedding intuitionistic logic into affine logic (linear logic with the weakening rule).

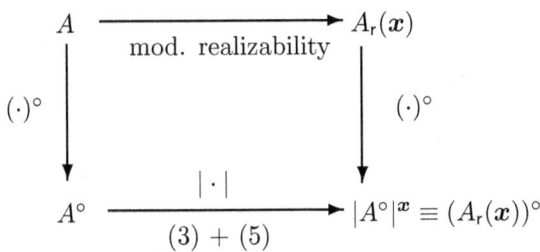

Figure 1. Factoring modified realizability

and presented in (Oliva 2007a;b). The starting point is the functional interpretation of pure affine logic (affine logic without the exponentials). As mentioned in the introduction, we consider the intuitionistic fragment of affine logic:

(3) $$\begin{aligned} |A \oplus B|_{y,w}^{x,v,z} &:\equiv |A|_y^x \Diamond_z |B|_w^v \\ |A \otimes B|_{y,w}^{x,v} &:\equiv |A|_y^x \otimes |B|_w^v \\ |A \multimap B|_{x,w}^{f,g} &:\equiv |A|_{gxw}^x \multimap |B|_w^{fy} \\ |\forall z A(z)|_{y,a}^{f} &:\equiv |A[a/z]|_y^{fa} \\ |\exists z A(z)|_y^{x,a} &:\equiv |A[a/z]|_y^x. \end{aligned}$$

What one notices is that the parameter constructor $\forall \boldsymbol{x} \prec \boldsymbol{a}\, A$ used to interpret $A \to B$ in (2) is in fact the interpretation of the affine logic modality $!A$. So we can extend the basic interpretation (3) to a parametrised interpretation of full intuitionistic affine logic as

(4) $\quad |!A|_a^x \;:\equiv\; !\forall y \prec a\, |A|_y^x.$

Via the translations $(\cdot)^\circ$ and $(\cdot)^*$ of IL into AL_i one can recover the interpretations of intuitionistic logic from those of intuitionistic affine logic as follows. For instance, consider the abbreviation $\forall \boldsymbol{x} \prec \boldsymbol{a}\, A :\equiv \forall \boldsymbol{x} A$, so that (4) simplifies to

(5) $\quad |!A|^x \;:\equiv\; !\forall y |A|_y^x.$

We call the resulting interpretation a *modified realizability* interpretation of affine logic because the diagram of Figure 1 commutes, i.e., given a formula A of intuitionistic logic we can either apply modified realizability directly and translate the result into liner logic, or alternatively, we can first translate

A into affine logic, and then apply the interpretation with $\forall x \prec a\, A := \forall x A$. Both paths result in the *same* formula. Note that we really mean syntactic equality, rather than logical equivalence.

Now, if instead of using the Girard translation A° we use instead the translation A^* we obtain a different diagram (Figure 2) which also commutes if we take in the upper arrow the *relational* realizability instead.

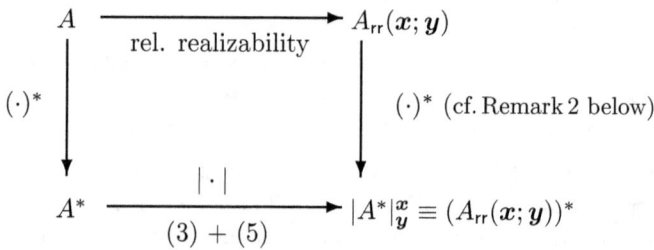

Figure 2. Factoring the relational variant of modified realizability

In other words, the two ways of presenting modified realizability arise from the two possible ways of translating intuitionistic logic into affine logic. In both cases the modified realizability interpretation of *affine logic* is fixed (the lower arrows of Figures 1 and 2). That illustrates how affine logic has a more fundamental nature, as it is able to capture precisely the inherent structure of realizability.

Just as we have factored the realizability interpretation through affine logic, we can also do the same for the dialectica interpretation by considering the abbreviation $\forall x \prec a\, A := A[a/x]$ leading to the interpretation of $!A$ as

(6) $\quad |!A|_y^x := !|A|_y^x.$

Again, we say that (6) is a dialectica interpretation of affine logic because it corresponds to the dialectica interpretation of intuitionistic logic as depicted in the commuting diagram of Figure 3.

Finally, a Diller-Nahm interpretation of affine logic is obtained by choosing the abbreviation

$$\forall x \prec a\, A := \forall x \in a\, A,$$

where a is a tuple of finite sets, and $x \in a$ denotes the usual set inclusion. For further details on the factorisation of the main functional interpretations via affine logic see (Ferreira & Oliva 2009; 2011).

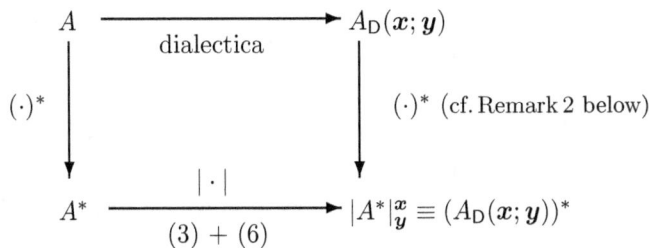

Figure 3. Factoring Gödel's dialectica interpretation

REMARK 2 *In the diagrams of Figures 2 and 3 we are taking a simplified form of the $(\cdot)^*$-translation, namely, one where the clauses for disjunction and existential quantifier are simply*

$$(A \vee B)^* :\equiv A^* \oplus B^*$$
$$(\exists x A)^* :\equiv \exists x A^*,$$

i.e., the bang is not used. The reason why we can work with this simpler translation of IL into AL_i is because we are considering AL_i extended with the following two principles

(7)
$$!A \oplus !B \multimap !(A \oplus B)$$
$$\exists x\, !A \multimap\, !\exists x A.$$

These principles are harmless because they are interpretable by the interpretation $|A|_y^x$ for any of the three choices of $\forall x \prec a\, A$ above. In general however, the combination of $|A|_y^x$ with the translation $(\cdot)^*$ will lead to interpretations of disjunction and existential quantifier as

(8)
$$|A \vee B|_{y,w}^{x,v,b} :\equiv \forall y \prec a\, |A|_y^x \lozenge_b \forall w \prec c\, |B|_w^v$$
$$|\exists z A|_c^{x,a} :\equiv \forall y \prec c\, |A[a/z]|_y^x.$$

This more general treatment is important for instance in the functional interpretation with truth as discussed in the following section.

4 Interpretations with Truth

The soundness of functional interpretations guarantees that from a proof of A a tuple of terms t can be extracted such that $|A|_y^t$. An important issue is that such a tuple t provides a witness to the statement $\exists x \forall y |A|_y^x$,

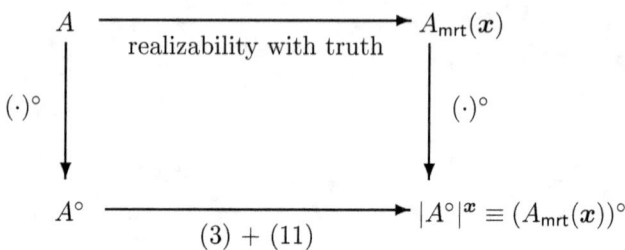

Figure 4. Factoring modified realizability with truth

but not necessarily a witness to the original theorem A. For realizability interpretations some variants have been developed so that a realiser for $\exists z A$ also contains a witness for z. These are the so-called q-*realizability* and *realizability with truth* (Grayson 1981; Kleene 1945; Troelstra 1998). In general what we would like is that

(9) $\quad \forall \boldsymbol{y} |A|_{\boldsymbol{y}}^{\boldsymbol{x}} \to A$

is derivable without the need for the characterisation principles[6] of the interpretation $|\cdot|$, because then we can extract actual witnesses from proofs of existential statements as follows

$$\vdash \exists z A(z) \stackrel{\text{soundness}}{\Rightarrow} \vdash |\exists z A(z)|_{\boldsymbol{y}}^{t,s} \stackrel{(3)}{\equiv} \vdash |A(s)|_{\boldsymbol{y}}^{t} \stackrel{(9)}{\Rightarrow} \vdash A(s).$$

In joint work with Jaime Gaspar (Gaspar & Oliva 2010) we have shown how interpretations with truth arise from a slight modification of the abstract interpretation of $!A$ from (4) to

(10) $\quad |!A|_{\boldsymbol{a}}^{\boldsymbol{x}} :\equiv \; !\forall \boldsymbol{y} \prec \boldsymbol{a}\, |A|_{\boldsymbol{y}}^{\boldsymbol{x}} \otimes \, !A.$

For instance, if we take the realizability abbreviation $\forall \boldsymbol{y} \prec \boldsymbol{a}\, A :\equiv \forall \boldsymbol{y} A$ in this case we obtain

(11) $\quad |!A|^{\boldsymbol{x}} :\equiv \; !\forall \boldsymbol{y} |A|_{\boldsymbol{y}}^{\boldsymbol{x}} \otimes \, !A.$

The composition of this affine logic interpretation with the translation $(\cdot)^{\circ}$ gives us precisely the *modified realizability with truth* (Kohlenbach 1992; 1998; 2008), as described in the diagram of Figure 4.

---

[6]The characterisation principles are the extra logical principles needed to show the equivalence between A and its interpretation $\exists \boldsymbol{x} \forall \boldsymbol{y} |A|_{\boldsymbol{y}}^{\boldsymbol{x}}$.

Consider then the q-variant of the relational realizability (1) where the clauses for disjunction and existential quantification are modified as

(12)
$$(A \vee B)_{\mathsf{qr}}(\boldsymbol{x}, \boldsymbol{v}, b;) :\equiv (\forall \boldsymbol{y} A_{\mathsf{qr}}(\boldsymbol{x}; \boldsymbol{y}) \wedge A) \Diamond_b (\forall \boldsymbol{w} B_{\mathsf{qr}}(\boldsymbol{v}; \boldsymbol{w}) \wedge B)$$
$$(\exists z A)_{\mathsf{qr}}(\boldsymbol{x}, a;) :\equiv \forall \boldsymbol{y}(A[a/z])_{\mathsf{qr}}(\boldsymbol{x}; \boldsymbol{y}) \wedge A[a/z].$$

The diagram of Figure 5 shows how such q-realizability corresponds to the $(\cdot)^*$ translation, making use in this particular case of the forgetful translation $(\cdot)^F$ of affine logic back into intuitionistic logic instead.[7]

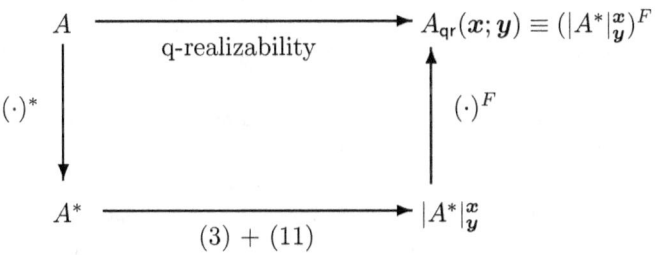

Figure 5. Factoring q-realizability

If one observes that the A° translation is affine logic equivalent to the "banged" A^* translation, i.e., $A^\circ \leftrightarrow \,! A^*$, one obtains the following interesting (apparently unobserved) correspondence between realizability with truth and q-realizability

$$A_{\mathsf{mrt}}(\boldsymbol{x}) \overset{\mathrm{IL}}{\Leftrightarrow} \forall \boldsymbol{y} A_{\mathsf{qr}}(\boldsymbol{x}; \boldsymbol{y}) \wedge A.$$

A great benefit of this analysis of truth interpretations via affine logic is that it gave us a handle to define truth variants of other functional interpretations. For instance, contrary to what was thought (Jörgensen 2004), we can immediately obtain a Diller-Nahm with truth instantiating (10) as

$$|\,! A|_a^x :\equiv \,! \forall \boldsymbol{y} \in a \,|A|_y^x \otimes \,! A.$$

---

[7] In this case a diagram similar to the ones considered before would not lead to a commuting diagram (not even if logical equivalence is taken instead of syntactic equality). The problem is that whereas A might contain existential quantifiers its interpretation $A_{\mathsf{qr}}(\boldsymbol{x}; \boldsymbol{y})$ does not. Hence, formulas which are duplicated in $|A^*|_{\boldsymbol{y}}^{\boldsymbol{x}}$ because of the ! in $\exists x \,! A$ are not duplicated in $(A_{\mathsf{qr}}(\boldsymbol{x}; \boldsymbol{y}))^*$ because the existential quantifiers have disappeared. One way to solve this is presented in (Gaspar & Oliva 2010), but uses logical equivalence. Here we present an alternative solution which is to use the forgetful translation that leads to a commuting diagram with syntactic equality instead. Obviously this is a weaker result than the previous four diagrams, as $(A^I)^* \equiv (A^*)^J$ implies $A^I \equiv ((A^*)^J)^F$ but not conversely.

For more details on the unification of functional interpretations with truth (see Gaspar & Oliva 2010).

REMARK 3 *It is essential here that one uses the full $(\cdot)^*$ translation, not the simplification of the previous section (cf. Remark 2) as the choice of interpretation (10) for $!A$, although sound for affine logic, it is not sound for the extra principles (7).*

5 Putting it all together

The analysis of different functional interpretations via *affine logic* not only provides a setting where the precise differences between the interpretations can be clearly seen, but surprisingly it also allows us to combine multiple interpretations when analysing a single proof. This follows because, as observed by Girard, the bang ($!A$) is not a canonical operator. One can add multiple instances $!'A, !''A, \ldots$ all with the same four rules without being able to show that any two are provably equivalent. This observation led us (Hernest & Oliva 2008; Oliva 2012) to consider a system of multi-modal affine logic with a different instance of $!A$ for each of the functional interpretations discussed above. For instance, we could add five different variants of $!A$ and interpret each as follows:

$$|!_k A|^x \; :\equiv \; !\forall y |A|^x_y \qquad \text{(Kreisel's modified realizability)}$$

$$|!_d A|^x_a \; :\equiv \; !\forall y \in a |A|^x_y \qquad \text{(Diller-Nahm interpretation)}$$

$$|!_g A|^x_y \; :\equiv \; !|A|^x_y \qquad \text{(Gödel's dialectica interpretation)}$$

$$|!_{kt} A|^x \; :\equiv \; !\forall y |A|^x_y \otimes \; !A \qquad \text{(Kreisel's modified realizability with truth)}$$

$$|!_{dt} A|^x_a \; :\equiv \; !\forall y \in a |A|^x_y \otimes \; !A \qquad \text{(Diller-Nahm interpretation with truth)}$$

This leads to what we have termed *hybrid functional interpretations*. If left completely unrelated, however, it would be difficult to make any practical use of this idea. We can observe, however, that there is a certain partial order between these different modalities, as for instance, a witness for $!_k A$ is clearly also a witness for $!_d A$. Therefore, we can add a rule that allows us to conclude $!_d A$ from $!_k A$, i.e.

$$\frac{\Gamma \vdash !_k A}{\Gamma \vdash !_d A}$$

In the diagram of Figure 5 we write $!_X$ above $!_Y$ if the interpretation of $!_X A$ implies the interpretation of $!_Y A$. As such, we could say that modified realizability with truth and Gödel's *dialectica* interpretation are the two "extreme" interpretations amongst these five. For more details on these hybrid functional interpretations (see Gaspar & Oliva 2010; Hernest & Oliva 2008; Oliva 2012).

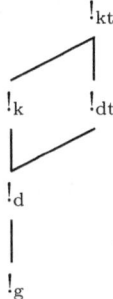

Figure 6. Ordering between different interpretations of $!A$

6 Directions for Further Work

Let us conclude by outlining a few possible directions for further work. These are either directly related to the unification of functional interpretation or to the actual nature and better understanding of functional interpretations themselves.

6.1 Functional interpretations with forcing

The combination of realizability with Cohen's notion of forcing was originally studied by Goodman (Goodman 1978) who showed it to be an effective way to prove conservation results that cannot apparently be shown by realizability alone. Goodman's work is related to the interpretations with truth (cf. Section 4) as forcing is used precisely to recover the truth property (9). Although Goodman presented a single combined interpretation, Beeson (Beeson 1979) showed that Goodman's interpretation can actually be seen as a simple composition of the Kleene number realizability based on Turing machines with oracles followed by an application of forcing. Recently, another variant of realizability, called *learning-based realizability* (Aschieri & Berardi 2010), has been developed providing an extension of realizability to classical arithmetic. Although different from Goodman's, the learning-based realizability has many similar features to Goodman's combination of realizability and forcing. For instance, the learning-based interpretation of formulas is described relative to a *memory*, which can be understood as a forcing condition approximating a non-computable oracle. Ineffective formulas (formulas without computable realisers) can be given an approximating realiser that works only when the memory has the correct information. The main result is that from a proof one can extract an agent that will be able to smartly build an approximation to the memory good enough to eventually produce a correct realiser. Finally, Alexander Miquel (Miquel 2011) has been working on extending Krivine's classical realizability with forcing, in the context of second-order arithmetic. This raises a few questions:

(**Q1**) What underlies the combination of realizability and forcing in general? Can forcing be combined with other functional interpretations, e.g., Diller-Nahm? What benefits could that bring?

(**Q2**) As with Goodman's interpretation, could the learning-based realizability be decomposed into a standard realizability interpretation followed by some variant of forcing?

6.2 Bounded-like interpretations

Very recently (van den Berg *et al.* 2012) variants of modified realizability and the *dialectica* interpretation have been proposed which apply to proofs in *nonstandard arithmetic*. The main feature of the interpretation is to extract from a proof of an existential statement a finite set of candidate witnesses (as in Herbrand's theorem), rather than a precise witness. The authors show that finite sets are the appropriate way to interpret existential *standard* quantifiers, while unrestricted existential quantifiers are interpreted uniformly (as in (Berger 2005) and (Krivine 2003)).

Also recently, so-called *bounded* variants of the dialectica and modified realizability interpretations (Ferreira & Nunes 2006; Ferreira & Oliva 2005; 2007) have been proposed which make use of the Howard/Bezem strong majorizability relation but in a more embedded way than Kohlenbach's monotone interpretation. The original motivation was to extend functional interpretations to deal with ineffective principles in analysis such as weak König's lemma even over weak fragments of analysis. The bounded modified realizability was then extended into a *confined* variant (Ferreira & Oliva 2010) which looks both for upper and lower bounds. There are striking similarities between the functional interpretation of non-standard arithmetic and the bounded and confined interpretations, as pointed out in (van den Berg *et al.* 2012). That raises the question:

(**Q3**) What is the common structure behind these *bounded-like* interpretations? In joint work with Gilda Ferreira (Ferreira & Oliva 2012) we have extended the unifying framework to deal with the bounded and confined interpretations, but unfortunately, this does not look to be general enough to include the non-standard arithmetic interpretation (van den Berg et al. 2012), as they make crucial use of a new form of functional application.

6.3 Type-free functional interpretations

We have so far only been discussing Kreisel's version of realizability known as *modified realizability*. The original realizability interpretation, however, due to Kleene (Kleene 1945), makes use of numbers (codes of Turing machines) as realizers, rather than functionals of higher type. The crucial difference is that not all codes n define a total function $\{n\}\colon \mathbb{N} \to \mathbb{N}$. As such, the realizability of an implication $A \to B$ was originally defined as

$$(A \to B)_{\mathsf{nr}}(n) \quad :\equiv \quad \forall k (A_{\mathsf{nr}}(k) \to \{n\}(k)\!\downarrow \wedge\, B_{\mathsf{nr}}(\{n\}(k))),$$

so $\{n\}$ only needs to be defined on k if k is indeed a realizer[8] for A. Let us refer to Kleene's original notion of realizability as *number realizability*. It is clear that a *relational* variant of number realizability also exists. For instance, the clause for implication would be:

$$(A \to B)_{\mathsf{rnr}}(n; k) \quad :\equiv \quad \forall m A_{\mathsf{rnr}}(k_0; m) \to \{n\}(k_0)\!\downarrow \wedge\, B_{\mathsf{rnr}}(\{n\}(k_0); k_1)$$

where k_0 and k_1 denote the first and second projections inverses of the standard coding $\mathbb{N} \times \mathbb{N} \to \mathbb{N}$. That raises the following questions:

(**Q4**) Is there a *number realizability* interpretation of affine logic? By that we mean an interpretation which works on numbers rather than functionals of finite type, and makes use of the fact that realizers might be partial. For instance, that might involve modifying the clause for $A \multimap B$ in (3) as

$$|A \multimap B|_k^n \quad :\equiv \quad |A|_{\{n_1\}(k)}^{k_0} \multimap |B|_{k_1}^{\{n_0\}(k_0)}.$$

But the question is when should we require that $\{n_0\}(k_0)$ and $\{n_1\}(k)$ be defined so as to obtain not only a sound interpretation but also possibly interpret new principles that are not interpreted by Kreisel's modified realizability? It seems none of the obvious choices work. But that of course does not rule out more comprehensive changes which could lead to a sound interpretation.

(**Q5**) Related to (**Q4**), can one in general show that every natural (e.g. modular) functional interpretation of intuitionistic logic can be extended

---

[8] To appreciate the difference between Kleene number realizability and Kreisel's modified realizability it is enough to point out that the former is sound for the Markov principle whereas the later isn't. In fact, Kreisel developed modified realizability (Kreisel 1959) precisely to show that the Markov principle is independent of intuitionistic arithmetic.

to an interpretation of intuitionistic affine logic? And, even if this is not the case, is it always possible to relate functional interpretations in a similar way to the one done in Section 5, perhaps using different parameters than the interpretation of $!A$?

(**Q6**) Is there a "number variant" of the other aforementioned interpretations? Beeson (Beeson 1978) has looked at the question for the dialectica interpretation, which he calls a *type-free* dialectica. Beeson points out that there cannot be one for the actual dialectica interpretation, as it requires decidability of quantifier-free formulas whereas statements of the form $\{n\}(k)\downarrow$ are not decidable in general. He then suggests a type-free variant of the Diller-Nahm interpretation as

$$|A \to B|_k^n \;:\equiv\; \{n_1\}(k)\downarrow \wedge (\forall i \in \{n_1\}(k)\, |A|_i^{k_0} \to \{n_0\}(k_0)\downarrow \wedge |B|_{k_1}^{\{n_0\}(k_0)}).$$

In other words, he requires the counter-example functions to be total,[9] whereas the witnessing functions might be partial. Could this be relaxed? Could this be translated to the setting of affine logic? Would this lead to extra principles that go beyond those interpreted by the typed Diller-Nahm interpretation?

6.4 Short games versus long games

The use of games between two players to model non-classical logics started with the work of Lorentzen (Felscher 2002) where formulas were put in correspondence with debates/dialogues so that those provable in intuitionistic logic corresponded to dialogues in which the first player had a winning "strategy". This idea was refined in the works of Blass (Blass 1992), Abramsky (Abramsky & Jagadeesan 1994) and several others, and led to complete semantics for fragments of linear logic.

The connections between games and the functional interpretations such as Gödel's dialectica have been there from the start (Scott 1968). In the final section 8 of (Blass 1992), Blass discusses at great length how one can view de Paiva's (de Paiva 1989b) categorical formulation of the Diller-Nahm interpretation of linear logic as arising from Blass' game semantics. Blass' suggestion is that the functional interpretation of linear logic arises by considering short two-move games combined according to his rules but including "Skolemisation" steps whenever it may be necessary to bring a long game into a two-move game.

---

[9] I confess to not have been able to completely verify the soundness of Beeson's interpretation. The problem seems to appear in the interpretation of the cut rule ($A \to B$ and $B \to C$ implies $A \to C$) as the "positive" witnesses for $A \to B$ need not be total, but that is used in building the "negative" witnesses for $A \to C$, which should be total (cf. Beeson 1978, middle of page 221).

(**Q7**) I feel that a better understanding of the differences between long games with concrete moves and the short games with higher-order moves is still lacking. Although Blass shows how one can think of the dialectica category as arising from his game semantics, it is well known that *dialectica-like* games are useful to interpret extra principles that go beyond the interpreted logic such as the Markov principle, independence of premise and the axiom of choice. Blass long games, however, capture precisely some fragments of the logic providing a sound and complete semantics.

(**Q8**) Related to (**Q7**), can functional interpretations be used to build fully abstract models? Another question that would provide guidance towards this is: How does the functional interpretation of the *propositional* fragment of linear logic relate to other models of linear logic such as proof nets, monoidal closed categories, coherent spaces and phase semantics?

(**Q9**) In the context of long games people have been able to fine tune the interpreted logic by restricting the kind of strategies one or both of the players is allowed to play (e.g., innocent (Hyland & Ong 2000), fair, history-free). Not much in this direction has been done in the setting of functional interpretations, whereby one could consider restrictions on the class of realisers in order to avoid interpreting certain principles. It seems hard, however, to think of any restrictions that would make the interpretation not sound with respect to the axiom of choice, for instance, as its realiser is the identity. But one could consider other restrictions such as linear functionals, functionals of certain complexity, etc.

(**Q10**) Using the nomenclature of game theory (Fudenberg & Tirole 1991), the long games considered by Blass and Abramsky are said to be in *extensive form*. Such games can be thought of as trees where each node in the tree is assigned one of the players and terminal nodes determine which player has won. Games in extensive form can be brought into a so-called *normal form*, a matrix specifying for each given pair of strategies for the two players which of the two wins the game if they follow these strategies. Games in normal form can also be thought of as two-move games. The two-move game arising from a functional interpretation is obviously not going to be the same as the normal form of the given strategic Blass/Abramsky game. Two questions arise: What is the relation between these two different two-move games that come for the same logical formula A? Moreover, could the functional interpretation way of constructing two-move games have any relevance to game theory?

6.5 Treading between linear and intuitionistic logic

We have seen that we can better understand and generalise an interpretation of intuitionistic logic by moving to the more general (and finer) setting of affine logic. There are, however, some interesting logics in between linear (no contraction) and intuitionistic (full contraction) logic. For instance, consider the following "intuitionistic" version of Lukasiewicz logic (LL_i) obtained by adding to affine intuitionistic logic the contraction schema

(13) $\quad A \multimap S_B A \otimes K_B A$

where $S_B A :\equiv B \multimap A$ and $K_B A :\equiv (A \multimap B) \multimap B$. Note that (13) clearly follows from $A \multimap A \otimes A$ since A implies over affine logic both $S_B A$ and $K_B A$. We can obtain "classical" Lukasiewicz logic (LL_c) by adding the double negation elimination $(A^\perp)^\perp \multimap A$. If we denote by CL = classical logic, IL = intuitionistic logic, AL_i = intuitionistic affine logic, and AL_c = classical affine logic, the relation between these six logics is shown in the diagram below, where an arrow from X to Y means that Y is an extension of X.

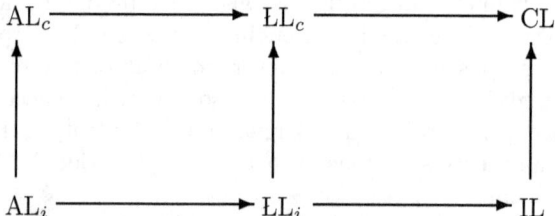

(**Q11**) Since LL_i is a fragment of IL, obviously any interpretation of IL also interprets LL_i. The question, however, is whether one can make use of the fact that only limited contraction is available in LL_i and hence restrict the kind of functionals needed for the interpretation. For instance, which kind of minimal fragment of the simply-typed lambda calculus would be sufficient to provide a *modified realizability* interpretation of LL_i? This is related to (**Q9**).

6.6 Endless possibilities?

The various functional interpretations discussed in Section 5 are only what one could call the "classic" interpretations. As has been discussed in this Section 6, several other new and fascinating functional interpretations have been discovered recently. Beyond those already mentioned one also has:

Kohlenbach's *monotone functional interpretations* (Kohlenbach 1996). These have been the cornerstone of the successful programme of *proof mining* (Kohlenbach 2008). It exploits a powerful combination of Gödel's original dialectica interpretation with Howard's (or Bezem's) majorizability relation (Bezem 1985; Howard 1973).

The *Copenhagen interpretation* (Blass & Gurevich 2008). A variant of the dialectica interpretation where essentially in the interpretation of $A \to B$ the negative witnessing functional is allowed to "give up" and not return a value. The original idea (apparently due to Martin Hyland) is that monads on types can quite often be lifted into an interpretation of (the comonad) !A. The Copenhagen interpretation carries this out for the monad $TX = X + 1$.

Krivine classical realizability (Krivine 2003). Realizability interpretation of classical second order arithmetic, recently extended to countable choice. Krivine's realizability can be viewed as a combination of negative translation with a simpler intuitionistic realizability interpretation (Oliva & Streicher 2008).

We close with some final questions:

(**Q12**) Is there a common structure behind all functional interpretations? What would be the appropriate way to define what functional interpretations are in general?

(**Q13**) Functional interpretations of classical logic have all been shown to arise from an interpretation of intuitionistic logic combined with a negative translation. Can one show that this is always the case?

Bibliography

Abramsky, S. & Jagadeesan, R. (1994). Games and full completeness for multiplicative linear logic. *Journal of Symbolic Logic*, *59*(2), 543–574, doi: 10.2307/2275407.

Aschieri, F. & Berardi, S. (2010). Interactive learning-based realizability for Heyting arithmetic with EM_1. *Logical Methods in Computer Science*, *6*(3), 1–22, doi:10.2168/LMCS-6(3:19)2010.

Avigad, J. & Feferman, S. (1998). Gödel's functional ("Dialectica") interpretation. In *Handbook of Proof Theory, Studies in Logic and the Foundations of Mathematics*, vol. 137, Buss, S. R., ed., Amsterdam: North Holland, 337–405.

Beeson, M. (1978). A type-free Gödel interpretation. *Journal of Symbolic Logic*, *43*(2), 213–227, doi:10.2307/2272820.

Beeson, M. (1979). Goodman's theorem and beyond. *Pacific Journal of Mathematics*, *84*(1), 1–16, doi:10.2140/pjm.1979.84.1.

Benton, P. N., Bierman, G. M., & de Paiva, V. C. V. (1993). A term calculus for intuionistic linear logic. In *Proceedings of Conference on Typed Lambda Calculi and Applications, Lecture Notes in Computer Science*, vol. 664, Bezem, M. & Groote, J. F., eds., Berlin; New York: Springer, 75–90.

Berger, U. (2005). Uniform Heyting arithmetic. *Annals of Pure and Applied Logic*, *133*(1–3), 125–148, doi:10.1016/j.apal.2004.10.006.

Bezem, M. (1985). Strongly majorizable functionals of finite type: a model for bar recursion containing discontinuous functionals. *Journal of Symbolic Logic*, *50*(3), 652–660, doi:10.2307/2274319.

Blass, A. (1992). A game semantics for linear logic. *Annals of Pure and Applied Logic*, *56*(1–3), 183–220, doi:10.1016/0168-0072(92)90073-9.

Blass, A. & Gurevich, Y. (2008). *Dialectica Interpretations: A Categorical Analysis*. Ph.D. thesis, ITU, Denmark.

de Paiva, V. C. V. (1989*a*). The Dialectica categories. In *Proc. of Categories in Computer Science and Logic, Boulder, CO, 1987, Contemporary Mathematics*, vol. 92, Gray, J. W. & Scedrov, A., eds., American Mathematical Society, 47–62.

de Paiva, V. C. V. (1989*b*). A Dialectica-like model of linear logic. In *Category Theory and Computer Science, Lecture Notes in Computer Science*, vol. 389, Pitt, D., Rydeheard, D., *et al.*, eds., Berlin; New York: Springer-Verlag, 341–356.

Diller, J. & Nahm, W. (1974). Eine Variante zur Dialectica-interpretation der Heyting Arithmetik endlicher Typen. *Archiv für mathematische Logik und Grundlagenforschung*, *16*(1–2), 49–66, doi:10.1007/BF02025118.

Felscher, W. (2002). Lorentzen's game semantics. In *Handbook of Philosophical Logic*, vol. 5, The Netherlands: Kluwer Academic Publisher, 2nd edn., 115–145.

Ferreira, F. & Nunes, A. (2006). Bounded modified realizability. *Journal of Symbolic Logic*, *71*(1), 329–346, doi:10.2178/jsl/1140641178.

Ferreira, F. & Oliva, P. (2005). Bounded functional interpretation. *Annals of Pure and Applied Logic*, *135*(1–3), 73–112, doi:10.1016/j.apal.2004.11.001.

Ferreira, F. & Oliva, P. (2007). Bounded functional interpretation in feasible analysis. *Annals of Pure and Applied Logic*, *145*(2), 115–129, doi:10.1016/j.apal.2006.07.002.

Ferreira, G. & Oliva, P. (2009). Functional interpretations of intuitionistic linear logic. In *Proceedings of CSL, Lecture Notes in Computer Science*, vol. 5771, Grädel, E. & Kahle, R., eds., Berlin; New York: Springer, 3–19.

Ferreira, G. & Oliva, P. (2010). Confined modified realizability. *Mathematical Logic Quarterly*, 56(1), 13–28, doi:10.1002/malq.200810029.

Ferreira, G. & Oliva, P. (2011). Functional interpretations of intuitionistic linear logic. *Logical Methods in Computer Science*, 7(1), paper 9, March, doi: 10.2168/LMCS-7(1:9)2011.

Ferreira, G. & Oliva, P. (2012). On bounded functional interpretations. *Annals of Pure and Applied Logic*, 163(8), 1030–1049, doi:10.1016/j.apal.2011.12.025.

Fudenberg, D. & Tirole, J. (1991). *Game Theory*. Cambridge, MA: MIT Press.

Gaspar, J. & Oliva, P. (2010). Proof interpretations with truth. *Mathematical Logic Quarterly*, 56(6), 591–610, doi:10.1002/malq.200910112.

Girard, J.-Y. (1987). Linear logic. *Theoretical Computer Science*, 50(1), 1–102, doi:10.1016/0304-3975(87)90045-4.

Gödel, K. (1958). Über eine bisher noch nicht benützte Erweiterung des finiten Standpunktes. *Dialectica*, 12(3–4), 280–287, doi:10.1111/j.1746-8361.1958.tb01464.x.

Goodman, N. (1978). Relativized realizability in intuitionistic arithmetic of all finite types. *Journal of Symbolic Logic*, 43(1), 23–45, doi:10.2307/2271946.

Grayson, R. J. (1981). Derived rules obtained by a model-theoretic approach to realisability. Handwritten notes from Münster University.

Hernest, M.-D. & Oliva, P. (2008). Hybrid functional interpretations. In *Proceedings of CiE, Lecture Notes in Computer Science*, vol. 5028, Berlin; New York: Springer, 251–260.

Howard, W. A. (1973). Hereditarily majorizable functionals of finite type. In *Metamathematical investigation of intuitionistic Arithmetic and Analysis, Lecture Notes in Mathematics*, vol. 344, Troelstra, A. S., ed., Berlin: Springer, 454–461.

Hyland, J. M. E. & Ong, C.-H. L. (2000). On full abstraction for PCF: I. models, observables and the full abstraction problem, II. dialogue games and innocent strategies, III. a fully abstract and universal game model. *Information and Computation*, 163, 285–408.

Jörgensen, K. F. (2004). Functional interpretation and the existence property. *Mathematical Logic Quarterly*, 50(6), 573–576, doi:10.1002/malq.200410004.

Kleene, S. C. (1945). On the interpretation of intuitionistic number theory. *Journal of Symbolic Logic*, *10*(4), 109–124, doi:10.2307/2269016.

Kohlenbach, U. (1992). Pointwise hereditary majorization and some application. *Archive for Mathematical Logic*, *31*, 227–241, doi:10.1007/BF01794980.

Kohlenbach, U. (1996). Analysing proofs in Analysis. In *Logic: from Foundations to Applications*, Hodges, W., Hyland, M., et al., eds., New York: Oxford University Press, 225–260.

Kohlenbach, U. (1998). Relative constructivity. *Journal of Symbolic Logic*, *63*, 1218–1238.

Kohlenbach, U. (2008). *Applied Proof Theory: Proof Interpretations and their Use in Mathematics*. Monographs in Mathematics, Berlin: Springer.

Kreisel, G. (1959). Interpretation of analysis by means of constructive functionals of finite types. In *Constructivity in Mathematics*, Heyting, A., ed., Amsterdam: North Holland, 101–128.

Krivine, J.-L. (2003). Dependent choice, 'quote' and the clock. *Theoretical Computer Science*, *308*(1-3), 259–276, doi:10.1016/S0304-3975(02)00776-4.

Miquel, A. (2011). Forcing as a program transformation. In *Logic In Computer Science (LICS'11)*, Berlin: Springer, 197–206.

Oliva, P. (2006). Unifying functional interpretations. *Notre Dame Journal of Formal Logic*, *47*(2), 263–290.

Oliva, P. (2007a). Computational interpretations of classical linear logic. In *Proceedings of WoLLIC'07, Lecture notes in computer science*, vol. 4576, Berlin; New York: Springer, 285–296.

Oliva, P. (2007b). Modified realizability interpretation of classical linear logic. In *Proc. of the Twenty Second Annual IEEE Symposium on Logic in Computer Science LICS'07*, IEEE Press.

Oliva, P. (2008). An analysis of Gödel's *dialectica* interpretation via linear logic. *Dialectica*, *62*(2), 269–290, doi:10.1111/j.1746-8361.2008.01135.x.

Oliva, P. (2009). Functional interpretations of linear and intuitionistic logic. *Information and Computation*, *208*(5), 565–577.

Oliva, P. (2012). Hybrid functional interpretations of linear and intuitionistic logic. *Journal of Logic and Computation*, *22*(2), 305–328, doi: 10.1093/logcom/exq007.

Oliva, P. & Streicher, T. (2008). On Krivine's realizability interpretation of classical second-order arithmetic. *Fundamenta Informaticae*, *84*(2), 207–220.

Scott, D. (1968). A game-theoretical interpretation of logical formulae. manuscript, Jahrbuch 1991 der Kurt-Gödel-Gesellschaft, Wien.

Shirahata, M. (2006). The Dialectica interpretation of first-order classical linear logic. *Theory and Applications of Categories*, *17*(4), 49–79.

Stein, M. (1979). Interpretationen der Heyting-Arithmetik endlicher Typen. *Archiv für mathematische Logik und Grundlagenforschung*, *19*, 175–189, doi: 10.1007/BF02011878.

Stein, M. (1980). Interpretations of Heyting's arithmetic – An analysis by means of a language with set symbols. *Annals of Mathematical Logic*, *19*, 1–31.

Troelstra, A. S. (1998). Realizability. In *Handbook of Proof Theory*, vol. 137, Buss, S. R., ed., Amsterdam: North Holland, 408–473.

van den Berg, B., Briseid, E., & Safarik, P. (2012). A functional interpretation for nonstandard arithmetic. Submitted for publication, URL `arXiv:1109.3103v2`.

Appendix: Proofs

Diagram 1. $|A^\circ|^{\boldsymbol{x}} \equiv (A_r(\boldsymbol{x}))^\circ$. First note that for any A its interpretation is $|A|^{\boldsymbol{x}}$, with empty challenge tuple.

$$
\begin{aligned}
|(A \to B)^\circ|^{\boldsymbol{f}} &\equiv |!(A^\circ \multimap B^\circ)|^{\boldsymbol{f}} \\
&\equiv !\forall \boldsymbol{x} |A^\circ \multimap B^\circ|^{\boldsymbol{f}}_{\boldsymbol{x}} \\
&\equiv !\forall \boldsymbol{x} (|A^\circ|^{\boldsymbol{x}} \multimap |B^\circ|^{\boldsymbol{f}\boldsymbol{x}}) \\
&\equiv !\forall \boldsymbol{x} ((A_r(\boldsymbol{x}))^\circ \multimap (B_r(\boldsymbol{f}\boldsymbol{x}))^\circ) \\
&\equiv (\forall \boldsymbol{x} (A_r(\boldsymbol{x}) \to B_r(\boldsymbol{f}\boldsymbol{x})))^\circ \\
&\equiv ((A \to B)_r(\boldsymbol{f}))^\circ.
\end{aligned}
$$

$$
\begin{aligned}
|(\forall z A)^\circ|^{\boldsymbol{f}} &\equiv |!\forall z A^\circ|^{\boldsymbol{f}} \\
&\equiv !\forall a |\forall z A^\circ|^{\boldsymbol{f}}_a \\
&\equiv !\forall a |A^\circ|^{\boldsymbol{f}a} \\
&\equiv !\forall a (A_r(\boldsymbol{f}a))^\circ \\
&\equiv (\forall a A_r(\boldsymbol{f}a))^\circ \\
&\equiv ((\forall z A)_r(\boldsymbol{f}))^\circ.
\end{aligned}
$$

Diagram 2. $|A^*|^{\boldsymbol{x}}_{\boldsymbol{y}} \equiv (A_{rr}(\boldsymbol{x}; \boldsymbol{y}))^*$. Note that we are using the simpler $(\cdot)^*$ translation where the clauses for $\lor$ and $\exists$ do not make use of !.

$$
\begin{aligned}
|(A \to B)^*|^{\boldsymbol{f}}_{\boldsymbol{x},\boldsymbol{w}} &\equiv |!A^* \multimap B^*|^{\boldsymbol{f}}_{\boldsymbol{x},\boldsymbol{w}} \\
&\equiv |!A^*|^{\boldsymbol{x}} \multimap |B^*|^{\boldsymbol{f}\boldsymbol{x}}_{\boldsymbol{w}} \\
&\equiv !\forall \boldsymbol{y} |A^*|^{\boldsymbol{x}}_{\boldsymbol{y}} \multimap |B^*|^{\boldsymbol{f}\boldsymbol{x}}_{\boldsymbol{w}} \\
&\equiv !\forall \boldsymbol{y} (A_{rr}(\boldsymbol{x}; \boldsymbol{y}))^* \multimap (B_{rr}(\boldsymbol{f}\boldsymbol{x}; \boldsymbol{w}))^* \\
&\equiv !(\forall \boldsymbol{y} A_{rr}(\boldsymbol{x}; \boldsymbol{y}))^* \multimap (B_{rr}(\boldsymbol{f}\boldsymbol{x}; \boldsymbol{w}))^* \\
&\equiv (\forall \boldsymbol{y} A_{rr}(\boldsymbol{x}; \boldsymbol{y}) \to B_{rr}(\boldsymbol{f}\boldsymbol{x}; \boldsymbol{w}))^* \\
&\equiv ((A \to B)_{rr}(\boldsymbol{f}; \boldsymbol{x}, \boldsymbol{w}))^*.
\end{aligned}
$$

Diagram 3. $|A^*|^{\boldsymbol{x}}_{\boldsymbol{y}} \equiv (A_{\mathsf{D}}(\boldsymbol{x}; \boldsymbol{y}))^*$. Trivial.

Diagram 4. $|A^\circ|^{\boldsymbol{x}} \equiv (A_{\mathsf{mrt}}(\boldsymbol{x}))^\circ$. First note that for any A° its interpretation is $|A^\circ|^{\boldsymbol{x}}$, with empty challenge tuple.

$$
\begin{aligned}
|(A \to B)^\circ|^{\boldsymbol{f}} &\equiv |\,!(A^\circ \multimap B^\circ)|^{\boldsymbol{f}} \\
&\equiv \,!\forall \boldsymbol{x}|A^\circ \multimap B^\circ|^{\boldsymbol{f}}_{\boldsymbol{x}} \otimes \,!(A^\circ \multimap B^\circ) \\
&\equiv \,!\forall \boldsymbol{x}(|A^\circ|^{\boldsymbol{x}} \multimap |B^\circ|^{\boldsymbol{f}\boldsymbol{x}}) \otimes \,!(A^\circ \multimap B^\circ) \\
&\equiv \,!\forall \boldsymbol{x}((A_{\mathsf{mrt}}(\boldsymbol{x}))^\circ \multimap (B_{\mathsf{mrt}}(\boldsymbol{f}\boldsymbol{x}))^\circ) \otimes \,!(A^\circ \multimap B^\circ) \\
&\equiv (\forall \boldsymbol{x}(A_{\mathsf{mrt}}(\boldsymbol{x}) \to B_{\mathsf{mrt}}(\boldsymbol{f}\boldsymbol{x})))^\circ \otimes \,!(A^\circ \multimap B^\circ) \\
&\equiv ((\forall \boldsymbol{x}(A_{\mathsf{mrt}}(\boldsymbol{x}) \to B_{\mathsf{mrt}}(\boldsymbol{f}\boldsymbol{x}))) \wedge (A \to B))^\circ \\
&\equiv ((A \to B)_{\mathsf{mrt}}(\boldsymbol{f}))^\circ.
\end{aligned}
$$

$$
\begin{aligned}
|(\forall z A)^\circ|^{\boldsymbol{f}} &\equiv |\,!\forall z A^\circ|^{\boldsymbol{f}} \\
&\equiv \,!\forall a|\forall z A^\circ|^{\boldsymbol{f}}_a \otimes \,!\forall z A^\circ \\
&\equiv \,!\forall a|A^\circ|^{\boldsymbol{f}a} \otimes \,!\forall z A^\circ \\
&\equiv \,!\forall a(A_{\mathsf{mrt}}(\boldsymbol{f}a))^\circ \otimes \,!\forall z A^\circ \\
&\equiv (\forall a A_{\mathsf{mrt}}(\boldsymbol{f}a))^\circ \otimes \,!\forall z A^\circ \\
&\equiv (\forall a A_{\mathsf{mrt}}(\boldsymbol{f}a) \wedge \forall z A)^\circ \\
&\equiv ((\forall z A)_{\mathsf{mrt}}(\boldsymbol{f}))^\circ.
\end{aligned}
$$

Diagram 5. $A_{\mathsf{qr}}(\boldsymbol{x};\boldsymbol{y}) \equiv (|A^*|^{\boldsymbol{x}}_{\boldsymbol{y}})^F$. Note that in this case we must use the full $(\cdot)^*$ translation.

$$
\begin{aligned}
(|(A \to B)^*|^{\boldsymbol{f}}_{\boldsymbol{x},\boldsymbol{w}})^F &\equiv (|\,!A^* \multimap B^*|^{\boldsymbol{f}}_{\boldsymbol{x},\boldsymbol{w}})^F \\
&\equiv (|\,!A^*|^{\boldsymbol{x}} \multimap |B^*|^{\boldsymbol{f}\boldsymbol{x}}_{\boldsymbol{w}})^F \\
&\equiv ((!\forall \boldsymbol{y}|A^*|^{\boldsymbol{x}}_{\boldsymbol{y}} \otimes \,!A^*) \multimap |B^*|^{\boldsymbol{f}\boldsymbol{x}}_{\boldsymbol{w}})^F \\
&\equiv ((!\forall \boldsymbol{y}(A_{\mathsf{qr}}(\boldsymbol{x};\boldsymbol{y}))^* \otimes \,!A^*) \multimap (B_{\mathsf{qr}}(\boldsymbol{f}\boldsymbol{x};\boldsymbol{w}))^*)^F \\
&\equiv (!(\forall \boldsymbol{y} A_{\mathsf{qr}}(\boldsymbol{x};\boldsymbol{y}) \wedge A)^* \multimap (B_{\mathsf{qr}}(\boldsymbol{f}\boldsymbol{x};\boldsymbol{w}))^*)^F \\
&\equiv \forall \boldsymbol{y} A_{\mathsf{qr}}(\boldsymbol{x};\boldsymbol{y}) \wedge A \to B_{\mathsf{qr}}(\boldsymbol{f}\boldsymbol{x};\boldsymbol{w}) \\
&\equiv (A \to B)_{\mathsf{qr}}(\boldsymbol{f};\boldsymbol{x},\boldsymbol{w}).
\end{aligned}
$$

$$
\begin{aligned}
(|(A \vee B)^*|^{\boldsymbol{x},\boldsymbol{v},b})^F &\equiv (|\,!A^* \oplus \,!B^*|^{\boldsymbol{x},\boldsymbol{v},b})^F \\
&\equiv (|\,!A^*|^{\boldsymbol{x}} \Diamond_b |\,!B^*|^{\boldsymbol{v}})^F \\
&\equiv ((!\forall \boldsymbol{y}|A^*|^{\boldsymbol{x}}_{\boldsymbol{y}} \otimes \,!A^*) \Diamond_b (!\forall \boldsymbol{w}|B^*|^{\boldsymbol{v}}_{\boldsymbol{w}} \otimes \,!B^*))^F \\
&\equiv ((!\forall \boldsymbol{y}(A_{\mathsf{qr}}(\boldsymbol{x};\boldsymbol{y}))^* \otimes \,!A^*) \Diamond_b (!\forall \boldsymbol{w}(B_{\mathsf{qr}}(\boldsymbol{v};\boldsymbol{w}))^* \otimes \,!B^*))^F \\
&\equiv (!(\forall \boldsymbol{y} A_{\mathsf{qr}}(\boldsymbol{x};\boldsymbol{y}) \wedge A)^* \Diamond_b \,!(\forall \boldsymbol{w} B_{\mathsf{qr}}(\boldsymbol{v};\boldsymbol{w}) \wedge B)^*)^F \\
&\equiv (\forall \boldsymbol{y} A_{\mathsf{qr}}(\boldsymbol{x};\boldsymbol{y}) \wedge A) \Diamond_b (\forall \boldsymbol{w} B_{\mathsf{qr}}(\boldsymbol{v};\boldsymbol{w}) \wedge B) \\
&\equiv (A \vee B)_{\mathsf{qr}}(\boldsymbol{x},\boldsymbol{v},b;).
\end{aligned}
$$

$$\begin{aligned}
(|(\exists z A)^*|^{\boldsymbol{x},a})^F &\equiv (|\exists z\,!\,A^*|^{\boldsymbol{x},a})^F \\
&\equiv (|\,!(A[a/z])^*|^{\boldsymbol{x}})^F \\
&\equiv (!\forall \boldsymbol{y}|(A[a/z])^*|^{\boldsymbol{x}}_{\boldsymbol{y}} \otimes \,!(A[a/z])^*)^F \\
&\stackrel{\text{(IH)}}{\equiv} (!\forall \boldsymbol{y}((A[a/z])_{\mathsf{qr}}(\boldsymbol{x};\boldsymbol{y}))^* \otimes \,!(A[a/z])^*)^F \\
&\equiv \forall \boldsymbol{y}(A[a/z])_{\mathsf{qr}}(\boldsymbol{x};\boldsymbol{y}) \wedge A[a/z] \\
&\equiv (\exists z A)_{\mathsf{qr}}(\boldsymbol{x},a;).
\end{aligned}$$

Paulo Oliva
Queen Mary University of London
School of Electronic Engineering and Computer Science
United Kingdom
p.oliva@qmul.ac.uk

Questions about Compositionality[1]

DAG WESTERSTÅHL

Compositionality is currently discussed mainly in computer science, linguistics, and the philosophy of language. In computer science, it is seen as a desirable design principle. But in linguistics and especially in philosophy it is an *issue*. Most theorists have strong opinions about it. Opinions, however, vary drastically: from the view that compositionality is trivial or empty, or that it is simply false for natural languages, to the idea that it plays an important role in explaining human linguistic competence. This situation is unsatisfactory, and may lead an outside observer to conclude that the debate is hopelessly confused.

I believe there is something in the charge of confusion, but that compositionality is nevertheless an idea that deserves serious consideration, for logical as well as natural languages. In this paper I try to illustrate why, without presupposing extensive background knowledge about the issue.[1]

1 Not a vague concept

Here is Jerry Fodor, a well-known philosopher, on compositionality:

> So not-negotiable is compositionality that I'm not even going to tell you what it is.
> ...
> Nobody knows exactly what compositionality demands, but everybody knows why its demands have to be satisfied. (Fodor 2001, 6)

---

[1] Thanks to Wilfrid Hodges for helpful remarks, and to Peter Pagin for valuable comments and many years of joint work on compositionality-related issues. Work on this paper was supported by a grant from the Swedish Research Council.

[1] There are by now handbook accounts and journal overviews of compositionality, and I will have to refer to these for many details. A good source is the recent *Handbook of Compositionality* (Hinzen et al. 2012), which in addition to several useful articles has a bibliography that covers most of what has been published in this area. The surveys (Pagin & Westerståhl 2010a;b) provide definitions, properties, and overviews of several arguments for and against compositionality.

And here is the voice of a renowned linguist, David Dowty:

> I believe that there is not and will not be—any time soon, if eve—a unique precise and "correct" definition of compositionality that all linguists and/or philosophers can agree upon ....
>
> ...
>
> I propose that we let the term NATURAL LANGUAGE COMPOSITIONALITY refer to *whatever strategies and principles we discover that natural languages actually do employ to derive the meanings of sentences, on the basis of whatever aspects of syntax and whatever additional information (if any) research shows that they do in fact depend on.* (Dowty 2007, 25,27)

Both quotes find compositionality 'non-negotiable', but despair of a definition, either because it would be too complicated, or because theorists would disagree about it. Dowty in effect gives up and suggests using the term in a way that makes natural languages compositional by definition.[2]

An immediate reaction is that this is simply wrong: there are completely precise notions of compositionality of which one can ask whether a natural language has them or not. Or rather, whether the language under a given syntactic and semantic analysis has them or not. And this is of course the catch: questions about compositionality are never completely empirical. They also depend on theory. On the other hand, so do most scientific questions. That doesn't mean they have no answers.

To begin, we should bear in mind the following:

> *Given* a language L with a 'reasonable' syntax that identifies *parts* of complex expressions, and *given* an assignment μ of semantic values ('meanings') to expressions, the question whether μ is compositional *is not vague.*
>
> Indeed, although there are a few distinct notions of compositionality, *each notion is precise* and allows a definite answer to the question.
>
> Moreover, these notions are general: they don't depend on *how* the syntax or semantics of L is specified.

These observations (to be made good below) should be ground for some optimism. Of course, the real work lies in specifying the syntax-semantics

---

[2] I am being slightly unfair to both Fodor and Dowty: Dowty has interesting things to say in that paper about concrete applications of compositionality, and compositionality is a cornerstone in Fodor's criticism of meaning theories such as the prototype theory. My point is just that they unnecessarily obscure the very idea of compositionality.

interface, an enterprise guided by considerations which are empirical as well as theoretical. Indeed, compositionality may be one such consideration. If so, we should avoid mystifying or trivializing it.

2   The guiding intuition

The motivation behind postulating compositionality has always been that it helps explain successful linguistic communication, in particular how speakers apparently effortlessly understand sentences never encountered before. Sentences have both structure and meaning, and the thought is that the meaning somehow can be read off the structure. If you know the meanings of the words, and the rules by which they are put together, and also the meaning building operations corresponding to those rules, then you can figure out the meaning of any correctly construed sentence.

This thought has long historical roots. Classically, meanings are taken to be mental objects: concepts or thoughts in the mind, or at least graspable by the mind.[3] For example, the word "horse" corresponds to the concept or idea HORSE, under which all and only horses fall. The word "every" has a different kind of meaning: it does not itself correspond to a 'clear and distinct idea in the mind', but when combined with e.g. "horse", it yields such an idea (exactly which is often less clear), which in turn can be be combined with, say, the concept RUN, to give the meaning that every horse runs.

To make this precise, you need some mathematics: a notion of structure, applicable to linguistic expressions, and possibly also to meanings. The pioneer is Frege, who applied the notion of a *function*: a concept word like "horse" stands for a function HORSE from objects to *truth values*, "everything" corresponds to a *second-level* function Φ which can take a function F like HORSE as an argument, yielding True whenever F yields True for every object.[4] Details aside, sentences express thoughts, which are structured objects, and the structure of the thought is *reflected* in the structure of the sentence.

Or is it the other way around? Consider a scenario based on a much simplified, but still useful, idea of linguistic communication: A wants to communicate a thought T to B. She finds a sentence S that means T, and utters it. B hears S, and reconstructs T from it. Discussions of compositionality usually focus on the *second* part of this transaction: from linguistic

---

[3]For an exposé of historical expressions of this idea about compositionality, which Hodges calls the *Aristotelian* version, (see Hodges 2012).

[4]So Φ(HORSE) says that everything is a horse; to say that every horse runs you can either use a conditional, Φ(HORSE $\to$ RUN), or let 'every' correspond to a *binary* second-level function.

items to meanings. Compositionality is invoked to make this step work, even if B has never heard S before. But it seems that the first part is equally important: A may never have uttered S before, so how does she find it, given T? A natural idea is that that step too is compositional.

So we may need compositionality in both directions. At a suitably abstract level, it is presumably the *same* notion in each case. Here I follow tradition and focus on the direction from syntax to meaning.[5]

While theories of syntax are subject to obvious empirical constraints, it is less clear what the data are for theories of meaning. Modern discussions of compositionality tend to circumvent this problem by making the notion more abstract. What seems to matter for compositionality, one may argue, is not what meanings *are* but the fact that the meanings of complex expressions are *determined* by the meanings of their parts (and the way these parts are syntactically combined). Put differently, replacing parts with the same meaning should not change the meaning of the whole. We arrive at the following two modern formulations of compositionality:

(PC-1) The meaning of a complex expression is determined by the meanings of its immediate parts and the mode of composition.

(PC-2) Appropriately replacing (not necessarily immediate) parts of a complex expression with synonymous expressions preserves meaning.

(As to the role of immediacy, see below.) Note that there is no longer any requirement that meanings be mental objects, or objects which themselves can have parts. Indeed, there is no requirement at all on meanings, except that a notion of *sameness of meaning* (synonymy) is available.

3 Structured expressions

To get started, we need a notion of syntax general enough to cover most common forms of grammar. In fact, very little is required: a notion of *structured expression* with identifiable *constituents*. I will consider two similar but distinct ways to proceed, both due to Wilfrid Hodges.[6]

---

[5] Bidirectional compositionality is discussed in (Pagin 2003), where it is observed that Frege's famous opening paragraph in (Frege 1923) seems to be about both directions. Fodor hints at similar ideas, using the term 'reverse compositionality', (e.g. in Fodor 2000). Pagin provides a detailed formal analysis of bidirectionality, in particular of how non-trivial synonyms such as 'brother' and 'male sibling' can be dealt with.

[6] There are other abstract theories of structured objects, notably (Aczel 1990), whose notion of a *replacement system* generalizes both set-theoretic and syntactic structure. It doesn't seem directly applicable to questions of compositionality, however.

3.1 Syntactic algebras

Systematic attempts to represent natural language syntax in algebraic terms go back at least to Montague (1974), where, conforming to linguistic practice, expressions are assigned primitive *categories*, in effect making syntactic algebras *many-sorted*. Hodges (2001) uses *partial* algebras instead, a simpler approach in the present context. Moreover, Hodges provides an abstract representation of the link between constituent structure and surface form. Thus, a *syntactic algebra* is a structure

$$\mathbf{E} = (E, \alpha^{\mathbf{E}})_{\alpha \in \Sigma}$$

where E is the set of *expressions* and each symbol α in the *signature* Σ denotes an n-ary partial function $\alpha^{\mathbf{E}}$ on E (for some $n \geq 0$), to be thought of as a grammar rule. Partiality, rather than category assignment, is used to restrict the domain of rules to appropriate arguments. *Atomic* expressions can be identified with 0-ary functions.

Expressions in E can be structurally ambiguous, and operations on expressions may suppress meaningful information, so on this picture the syntactic objects of semantic interest are not the expressions themselves but their *derivation histories* ('analysis trees'). These are immediately obtained as the terms in the *term algebra* corresponding to $\mathbf{E}$. The inductive definition of the set GT of well-formed *grammatical terms* (a subset of the set of all terms), respecting the partiality constraints, simultaneously yields a (surjective) homomorphism val from GT to $\mathbf{E}$. For example,

$$\alpha(a, \beta(b))$$

(a, b atoms) is grammatical iff $\alpha^{\mathbf{E}}$ is defined for the arguments $(val(a), val(\beta(b)))$, and then

$$val(\alpha(a, \beta(b))) = \alpha^{\mathbf{E}}(val(a), val(\beta(b))) = \alpha^{\mathbf{E}}(val(a), \beta^{\mathbf{E}}(val(b)))$$

If $val(t) = val(u)$ for complex terms $t \neq u$, the expression $val(t)$ may be structurally ambiguous. Lexical ambiguity can be dealt with by adding new atoms to the term algebra, e.g. $\overline{bank_1}$ and $\overline{bank_2}$, with $val(\overline{bank_1}) = val(\overline{bank_2}) = bank$.

We now get the constituent relation for free: it is simply the *subterm* relation. Moreover, syntactic categories can be recovered. For $X \subseteq GT$, define

(1) $t \sim_X u$ iff for all terms $s[t]$, $s[t] \in X \Leftrightarrow s[u] \in X$

($s[t]$ indicates that t is a *subterm occurrence* in s, and $s[u]$ is the result of replacing that occurrence by u.) Syntactic categories can then be construed as equivalence classes of $\sim_{GT}$; indeed, a familiar way of identifying categories is precisely in terms of preservation of grammaticality under replacement.

This format fits Montague Grammar, various forms of Categorial Grammar, not to mention the syntax of most logical languages. It also fits the idea of *direct compositionality* of Jacobson (2002) and Barker & Jacobson (2007). One aspect of 'directness' consists in restrictions on the functions in Σ (e.g. that only concatenation of strings is allowed), and hence on the mapping *val*. But the main point is that the semantics runs 'in tandem' with the syntax, which means that *val exists*. In grammars using notions of Movement and Logical Form (LF) (see Heim & Kratzer 1998, for a textbook example), there is no such mapping. Meanings are (usually compositionally) assigned to LFs, but the rules for constructing LFs have no semantic counterpart; in particular, there need be no homomorphic connection to surface form.[7]

3.2 Constituent structures

Hodges' recent notion of a *constituent structure* (see Hodges 2011; 2012) distills the bare essentials needed for talking about compositionality, and in particular for his notion of a semantics based on Frege's Context Principle (section 7 below). Formally, a constituent structure $(\mathbb{E}, \mathbb{F})$ is quite similar to a syntactic algebra: $\mathbb{E}$ is a set of objects called expressions, and $\mathbb{F}$ is a set of partial functions on $\mathbb{E}$. But the intuition is different: think of the elements of $\mathbb{F}$ as *frames* (which is what they are called), obtained from expressions by deleting some parts, leaving *argument places* that can be filled with other expressions, i.e. those in the domain of the frame. For example, from the sentence

(2) Henry knows some students.

you can get various frames, such as

(3) x knows some students

(4) x knows D students

(5) x knows Q

(6) Henry R some A

---

[7]The syntactic algebra format also applies, *mutatis mutandis*, to the currently popular idea of formulating grammar rules as applying to *triples* consisting of a string, a syntactic category, and a meaning; (see Kracht 2003; 2007, for a formal account). So the meaning assignment is built into the grammar rules, but in practice it can be teased apart, and one can usually go between the two formats in a straightforward way—(Pagin & Westerståhl 2010a, sec. 3.6) has more details.

By definition, $\mathbb{F}$ is closed under composition, substitution, and contains unit frame 1 (a total identity function on $\mathbb{E}$), but no empty frame (function with empty domain).[8] Thus, syntactic term algebras are a special case, with $\mathbb{E}$ as the set of grammatical terms, and $\mathbb{F}$ as the set of *polynomially definable* partial functions on $\mathbb{E}$, i.e., those definable precisely by leaving out subterm occurrences (replacing them with variables) of grammatical terms.

e is said to be a *(proper) constituent* of f iff ($e \neq f$ and) there is a frame F such that $f = F(\ldots, e, \ldots)$. In fact (using (NS) in note 8), F can be assumed to be 1-ary. The relation $\sim_X$ now becomes:

$e \sim_X f$ iff for each 1-ary $G \in \mathbb{F}$, $G(e) \in X$ iff $G(f) \in X$

Constituent structures start from a quite concrete idea of syntactic structure. But the formal requirements are minimal. For example, there is no guarantee that the 'proper constituent' relation is *transitive*. Transitivity follows if the relation is *wellfounded*, a natural enough assumption, but not part of the definition.[9]

Since there are normally several ways to turn a given expression into a frame, we will often have

(7) $F(e_1, \ldots, e_n) = G(f_1, \ldots, f_m)$

for distinct n, m, e_i, f_j, F, G. But if different expressions are inserted into the *same* frame, the idea of a frame seems to require that the results be different. Thus, call a frame F *rigid* iff

(8) $F(e_1, \ldots, e_n) = F(f_1, \ldots, f_n)$ implies $e_i = f_i$ for $1 \leq i \leq n$,

---

[8] More precisely, *Nonempty Composition* is the following:

(NC) If $F(x_1, \ldots, x_n), G(y_1, \ldots, y_m) \in \mathbb{F}$, and
$F(e_1, \ldots, e_{i-1}, G(f_1, \ldots, f_m), e_{i+1}, \ldots, e_n) \in \mathbb{E}$, then

$$F(x_1, \ldots, x_{i-1}, G(y_1, \ldots, y_m), x_{i+1}, \ldots, x_n) \in \mathbb{F}$$

And *Nonempty Substitution* is

(NS) If $F(e_1, \ldots, e_n) \in \mathbb{E}$, then $F(x_1, \ldots, x_{i-1}, e_i, x_{i+1}, \ldots, x_n) \in \mathbb{F}$.

[9] The definition allows the existence of (let us call them) *2-loops*: distinct expressions e, f and frames F, G such that $f = F(e)$ and $e = G(f)$. Then e is a proper constituent of f, which is a proper constituent of e, but no expression is a proper constituent of itself, so transitivity fails. Clearly, wellfoundedness precludes 2-loops (or n-loops for any n). It is not hard to show that the 'proper constituent' relation is transitive if and only if there are no 2-loops.

2-loops are in principle allowed in syntactic algebras $\mathbf{E} = (E, \alpha^{\mathbf{E}})_{\alpha \in \Sigma}$ as well (though they would never appear with standard grammar rules), but grammatical terms are always wellfounded.

i.e. if it is an *injective* function. This looks like another reasonable requirement on constituent structures (which is satisfied in the special case of term algebras), but again it is not needed in Hodges' account.

4 Meanings

Once the wellformed structured expressions have been identified, we can simply let a *semantics* be any assignment μ of values ('meanings') to these. The semantics is *partial* if the domain of μ is a proper subset of the set of expressions, otherwise *total*.

With the syntactic algebra approach, the structured expressions are, not the surface expressions but the grammatical terms in GT. For a constituent structure $(\mathbb{E}, \mathbb{F})$ on the other hand, the only candidates are the expressions in $\mathbb{E}$. Thus, even when (7) holds, we have one expression and hence at most one semantic value. This means that structural ambiguity is not accounted for within the frame picture; some kind of disambiguation must be supposed to have taken place already.[10] Indeed, this seems to be the main conceptual difference between the two approaches to constituency.

Each semantics μ has a corresponding *synonymy* relation:

(9) $s \equiv_\mu t$ iff $\mu(s) = \mu(t)$

Here the right-hand side means: $\mu(s)$ and $\mu(t)$ are both defined, and equal. (The letters 's','t' stand for terms in the term algebra, but exactly the same definition gives the relation $e \equiv_\mu f$ for expressions $e, f \in \mathbb{E}$.)

$\equiv_\mu$ is a partial equivalence relation. Conversely, every partial equivalence relation $\equiv$ on the set of structured expressions generates a corresponding *equivalence class semantics*: $\mu_\equiv(t) = [t]_\equiv = \{s : s \equiv t\}$ provided $[t]_\equiv \neq \emptyset$, undefined otherwise. One easily shows that the buck stops here: $\equiv_{\mu_\equiv} = \equiv$.

5 Compositionality

Now we get precise versions of (PC-1) and (PC-2), in each of the syntactic settings above.

Compositionality, functional version

(i) A semantics μ for GT, given by a syntactic algebra $(E, \alpha^{\mathbf{E}})_{\alpha \in \Sigma}$, is *compositional* iff for each $\alpha \in \Sigma$ there is an operation r_α such that whenever $\mu(\alpha(t_1, \ldots, t_n))$ is defined,

$$\mu(\alpha(t_1, \ldots, t_n)) = r_\alpha(\mu(t_1), \ldots, \mu(t_n))$$

---

[10]Compare Montague's notion of a *language* in (Montague 1974), which is a pair of a *disambiguated language* (essentially a free syntactic algebra) and an unspecified disambiguation relation.

(ii) A semantics μ for $\mathbb{E}$, relative to a constituent structure $(\mathbb{E}, \mathbb{F})$, is *compositional* iff for each $F \in \mathbb{F}$ there is an operation s_F such that whenever $\mu(F(e_1, \ldots, e_n))$ is defined,

$$\mu(F(e_1, \ldots, e_n)) = s_F(\mu(e_1), \ldots, \mu(e_n))$$

The idea is the same in both cases: the value of a complex expression is *determined* by the values of its parts and the mode of composition. In the term algebra, we look at the *immediate* constituents. This notion is not in general available in constituent structures, so we need a separate condition for each frame. Thus, if the situation in (7) obtains, we must have

$$s_F(\mu(e_1), \ldots, \mu(e_n)) = s_G(\mu(f_1), \ldots, \mu(f_m))$$

Note that both versions of (PC-1) require that the domain of μ is *closed under constituents*. This is not necessary for (PC-2):

Compositionality, substitution version

(i) A partial equivalence relation $\equiv$ on GT is *compositional* iff for each term $s[t_1, \ldots, t_n]$, if $t_i \equiv u_i$ for $1 \leq i \leq n$, and $s[t_1, \ldots, t_n], s[u_1, \ldots, u_n]$ are both in the domain of $\equiv$, then

$$s[t_1, \ldots, t_n] \equiv s[u_1, \ldots, u_n]$$

(ii) A partial equivalence relation $\equiv$ on $\mathbb{E}$ is *compositional* iff for each expression $F(e_1, \ldots, e_n)$, if $e_i \equiv f_i$ for $1 \leq i \leq n$, and $F(e_1, \ldots, e_n), F(f_1, \ldots, f_n)$ are both in the domain of $\equiv$, then

$$F(e_1, \ldots, e_n) \equiv F(f_1, \ldots, f_n)$$

In (i), $t_1, \ldots, t_n$ are *disjoint* subterm occurrences in the complex term: if two subterm occurrences of a term are not disjoint, one is a subterm of the other. Constituent structures can model expressions with overlapping constituents, which allows a simpler formulation, and makes the second claim of the next fact trivial. The first claim is also straightforward, but requires an argument by induction over the complexity of terms.

FACT 1 *If $dom(\mu)$ is closed under constituents then, in the syntactic algebra setting as well as in the constituent structure setting, μ is compositional iff $\equiv_\mu$ is compositional.*

This is satisfactory since it shows that, under some assumptions, there is just *one* notion of compositionality. Thus, for any grammar or syntactic theory that satisfies the minimal requirement of having a reasonable notion of constituency, and for any proposed assignment of meanings to its expressions, the question of whether this assignment is compositional or not *has a definite answer*. Moreover, the only way of showing that such an assignment is *not* compositional, is to exhibit a complex expression that changes its meaning when some of its constituents are replaced by synonymous ones (wrt the meaning assignment).

EXAMPLE 2 (adjective-noun combinations) We can use this to immediately lay to rest certain arguments against compositionality. The extension of some adjective-noun combinations is the intersection of the extension of the adjective and the extension of the noun, for example, *male cat* or *prime number*. But in other cases it is not; cf. *white wine* or *red hair*. This has been taken to show that the Adj N construction is not compositional.[11] But it shows nothing of the sort. The extension of *white wine* can still be *determined* by the extension of *white and* the extension of *wine, and* the Adj N construction, even if it is not always intersection. Nor does the example show that *white* means something else in *white wine* than it means in, say, *white paper*. That might be the case, or not, but it has nothing to do with (failure of) compositionality. To repeat, the *only* way to show that the Adj N construction is non-compositional (wrt extension) would be to find an expression $Adj_1 N_1$ and an adjective Adj_2 with the same extension as Adj_1 (or a noun N_2 with the same extension as N_1) such that $Adj_2 N_1$ (or $Adj_1 N_2$ or $Adj_2 N_2$) is well-formed and differs in extension from $Adj_1 N_1$. Such examples may, or may not, exist, but as far as I know none have been suggested.

This is not to say that there are no variant notions of compositionality. One weaker version requires only that the meaning of the *atomic* constituents (words) of a structured expression, and the structure itself, determines its meaning.[12] A precise formulation is obtained by restricting the t_i, u_i and the e_i, f_i in the substitution version of compositionality to atomic constituents, where, in a constituent structure, an expression is atomic iff it has no proper constituents. The usual criticism is that this is too weak to figure in any explanation of speaker competence, since the speaker would have to learn, as it were, not just the grammar rules in Σ, but each of the infinitely many syntactic structures that they generate. But note that this sort of criticism can be levelled at the whole constituent structure approach: in the function version there is one semantic operation for each frame.

---

[11] Arguments of this kind occur in the literature, but I refrain from giving references.

[12] Dowty (2007, 23) calls this variant—for reasons unclear to me—Frege's Principle. Larson & Segal (1995) call it 'compositionality', and use 'strong compositionality' for compositionality as defined here.

The syntactic algebra approach brings out the *generative* aspect of syntax, and thereby of a compositional semantics; the constituent structure approach doesn't, and isn't intended to. But *wellfounded* constituent structures recover the generative element: it is then possible to generate $\mathbb{F}$ from a set of primitive frames, and compositionality for the primitive frames implies full compositionality. (But if $(\mathbb{E}, \mathbb{F})$ is not wellfounded, there need not even be any primitive frames, or any atoms.)

That said, it should be noted that full compositionality is still a very weak requirement. The best way to see this is via the following observation.

FACT 3 *If a semantics μ is one-one, it is compositional.*

(This follows from Fact 1, since $\equiv_\mu$ is then the identity relation.) The observation should not come as a surprise, but it highlights the fact that the word "determine" in (PC-1) just means 'is a function of': it doesn't mean that one is 'able to figure out' the meaning of complex expressions from the meanings of their parts. For that, one must impose extra requirements, notably that the meaning operations are *computable* in some suitable sense.

No doubt the computability aspect is also part of the intuitive motivation for compositionality. Still, it makes sense to isolate a *core meaning* of 'compositionality', as in the above definitions. It is the requirement expressed by (PC-1) or similar formulations. In the literature, it has been called *local* compositionality, *strong* compositionality, *homomorphism* compositionality, but the idea is the same. True, it is a weak requirement. But weak is not the same as trivial or empty.

6 Triviality

Compositionality has been charged with triviality for both mathematical and philosophical reasons. In the former case, the idea is roughly that any semantics can be made compositional by some trivial manipulations. There is a sense in which this is true. It is just that this fact tells us next to nothing about the unmanipulated semantics. The philosophical charge is rather that compositionality adds nothing to an account of linguistic meaning. I will look at one typical example of each kind.[13]

6.1 Mathematical triviality: Zadrozny

Zadrozny (1994) shows that given any semantics μ one can find another semantics μ^* with the same domain such that (a) μ^* is compositional; (b)

---

[13]Part of the discussion in this section comes from (Westerståhl 1998) and (Pagin & Westerståhl 2010b), where several other examples are examined as well.

μ can be recovered from μ^*.[14] In fact, the semantics μ^* is one-one, so its compositionality is indeed trivial (Fact 3). But the claim that a semantics satisfying (a) and (b) exists is *itself* trivial: just let, for each $e \in dom(\mu)$,

$$\mu'(e) = (\mu(e), e)$$

Then μ' is compositional (since it is one-one), and μ is easily recovered from μ' ($\mu(e)$ is the first element of the pair $\mu'(e)$). Clearly, this says *nothing at all* about the original semantics μ.

A very different observation is that it has often happened that a proposed semantics μ has been replaced by a compositional semantics μ^*, precisely because μ turned out not to be compositional. Perhaps the first example is Frege's introduction of indirect Sinn and Bedeutung in order to be able to deal (compositionally) with attitude reports. A recent case is Hodges' compositional trump semantics for the Hintikka-Sandu Independence-Friendly Logic, (Hodges 1997). These semantics are not obtained by trivial manipulations but by a deeper analysis of meaning.

If there is anything in the charge of triviality for mathematical reasons it comes from the observation in Fact 3. When the analysis of meaning is so fine-grained that there are no non-trivial synonymies, compositionality is indeed trivial. To take an extreme example, if the sound of the words themselves, or the associations they conjure up in the mind of the speaker, are taken to be part of the meaning expressed, very few distinct expressions will mean the same. This is not a notion of meaning for which compositionality makes a difference. It doesn't follow that there aren't others for which it does.

6.2 Philosophical triviality: Horwich

In "Deflating compositionality" in (Horwich 2005), Paul Horwich accepts compositionality but gives it no role whatsoever in explaining the meaning of complex sentences. The idea is that the meanings of words (atoms) and the rules of syntax provide all the information needed:

(a) That x means DOGS BARK consists in x resulting from putting together words whose meanings are DOGS and BARK, in that order, into a schema whose meaning is NS V.

(b) "dogs" means DOGS, "bark" means BARK, and "ns v" means NS V.

(c) "dogs bark" results from putting "dogs" and "bark", in that order, into the schema "ns v".

---

[14] He also shows that with a non-wellfounded set theory as metatheory, the only composition operation required for μ^* is function application. This is more interesting, but irrelevant to the issue of the triviality of compositionality.

(d) Hence, "dogs bark" means DOGS BARK.

Horwich's conclusion is that compositionality holds as a direct consequence of what it is for a complex expression to have meaning.

I think the possible attraction of this argument comes from the fact that the example is so simple that any meaning explanation is bound to appear trivial. Looking closer, however, this impression dissolves.

First, one may wonder if the idea is that no other string of words can mean DOGS BARK, and similarly for other sentences. If so, we have trivial compositionality because of a one-one meaning assignment, as just discussed. But that is not the reason offered. Second, the reason this is unclear is that we are not told what the meanings of DOGS or BARK are, and even less about the operation of concatenating two such meanings. Is the notation used a shorthand for a semantic operation of combining the meaning of a bare plural with the meaning of an intransitive verb? Compositionality says that such an operation *exists*. But the order of explanation is the reverse: *after* we have specified such an operation (not done in (a)–(d)), we can *conclude* that compositionality holds.

Third, the example may look trivial but the compositionality claim still has content. It says that other sentences of the *same* form, for example "Cats meow", should be analyzed with the *same* semantic operation. If you find that obvious, you have an *argument* for compositionality!

Finally, the appearance of triviality fades with more complex sentences:

(10) Everyone knows someone.

It is easy to specify schemas generating (10). It is less trivial to specify corresponding semantic operations that yield the intended meaning (rather, one of the intended meanings) of (10), though, of course, nowadays every semanticist knows ways to do that. To say that the meaning of (10) is EVERYONE KNOWS SOMEONE is completely uninformative until the semantic operations are specified. To say that language requires such operations to exist is to *presuppose* compositionality. But then it looks like an essential trait of language, and anything but trivial. However, it seems more fruitful to regard it as an *hypothesis* about natural language meaning. After all, it is easy to make up non-compositional languages. So it is a substantial hypothesis, to which empirical evidence is relevant. It may look 'deflated' with examples like the one in (a), but it really isn't.

6.3 Triviality: conclusion

Even if there are various uninteresting ways to make a non-compositional semantics compositional, isn't it a significant fact that in so many cases, what looked like non-compositional linguistic constructions have been amenable to a compositional treatment? To evaluate the significance of this, one would have to look at the instances case by case, and there is no space for that here. But, hypothetically, suppose that in each case it was in fact possible to replace the non-compositional semantics by an *improved* semantics which was compositional. That would certainly count as evidence for the *truth* of the compositionality hypothesis. Or, suppose instead that some constructions would resist a compositional treatment. This need not mean we must give up compositionality altogether; it could still be that large *fragments* of natural languages are compositional.

In the second case, compositionality is surely not trivial: it would be false for some parts of language and true for others. What about the first case? For all we have said so far, it could still be that compositionality is trivially true, in the sense that on the ultimately best account of how language works, it plays no significant explanatory role. But we are not at that point yet. In the meantime, it still looks like an hypothesis worth exploring further.

Besides, once we have a well-defined framework in which to talk about compositionality, several related but distinct issues suggest themselves. We look at one in the next section.

7 Hodges and the Context Principle

Frege's second methodological maxim in the introduction to *Grundlagen der Aritmetik* famously reads:

> Nach der Bedeutung der Wörter muß im Satzzusammenhange, nicht in ihrer Vereinzelung gefragt werden. (Frege 1884)

Frege's application was that the meaning of *number words* is given by the sentences in which they occur, but the general idea seems to be:

(F) The meaning of an expression is the contribution it makes to the meanings of sentences in which it occurs.

Hodges observed that this is in fact a recipe for recovering expression meanings, up to synonymy, from sentence meanings.[15] Let a language L be given

---

[15] See (Hodges 2001; 2005). Hodges is one of those who have contributed most to our understanding of compositionality, so it is no accident that his name appears so often in this paper. Apart from contributions mentioned here, Hodges resolved the issue of the compositionality of Hintikka's Independence-Friendly (IF) Logic (Hintikka 1996; Hintikka & Sandu 2011), by providing it with a compositional semantics (Hodges 1997), but also showing (Cameron & Hodges 2001) that no semantics with sets of assignments as values (as for first-order logic) is compositional (see Galliani 2011, for a strengthened version of this). His compositional so-called trump semantics sparked off a surge of research

as a constituent structure $(\mathbb{E}, \mathbb{F})$ with a semantics μ, where $X = dom(\mu)$. For the next definition, recall sections 3.2 and 4.

DEFINITION 4 (fregean semantics) *For $e, f \in \mathbb{E}$, define*

$e \equiv_\mu^F f$ *iff*
$e \sim_X f$ *and for each 1-ary $G \in \mathbb{F}$, if $G(e) \in X$ then $G(e) \equiv_\mu G(f)$*

Note that $\equiv_\mu^F$ is a total equivalence relation on $\mathbb{E}$. Let $|e|_\mu$ be the equivalence class of e (alternatively, a chosen label for that class); this is called the *fregean semantics* for L.

LEMMA 5 (Hodges' Lifting Lemma) *Suppose $F(e_1, \ldots, e_n)$ is a constituent of some expression in X, and $e_i \equiv_\mu^F f_i$ for each i. Then*

(a) $F(f_1, \ldots, f_n) \in \mathbb{E}$
(b) $F(e_1, \ldots, e_n) \equiv_\mu^F F(f_1, \ldots, f_n)$

Proof. (outline) The fact that $\mathbb{F}$ is closed under substitution allows us to restrict attention to the case $n = 1$. The assumption about $F(e_1)$, and that $e_1 \sim_X f_1$, together with the fact that $\mathbb{F}$ is closed under composition, yields (a). (b) follows by a similar argument. $\square$

A crucial property of the set of sentences is that it is *cofinal*: every expression is a constituent of some sentence. So if we assume that $X = dom(\mu)$ is cofinal, the Lifting Lemma immediately shows that the fregean semantics is compositional. Thus (Fact 1), for each $F \in \mathbb{F}$ there is an operation h_F such that whenever $F(e_1, \ldots, e_n) \in \mathbb{E}$,

$$|F(e_1, \ldots, e_n)|_\mu = h_F(|e_1|_\mu, \ldots, |e_n|_\mu)$$

How does the fregean semantics relate to the original semantics μ? By Definition 4, we get (since the unit frame belongs to $\mathbb{F}$):

(11) If $e \in X$ and $e \equiv_\mu^F f$, then $e \equiv_\mu f$ (so $f \in X$).

That is, $\equiv_\mu^F$ *refines* $\equiv_\mu$: it may make more meaning distinctions than $\equiv_\mu$ does, but it will never declare synonymous two expressions in X that are not μ-synonymous. However, if μ is already compositional, and satisfies what Hodges calls the *Husserl property*,

---

on logics where notions of (in)dependence are treated explicitly, notably Dependence Logic (DL) (Väänänen 2007); (see also Kontinen et al. 2013). He has also contributed significantly to our knowledge of the *history* of the idea of compositionality, especially in Arabic medieval philosophy commenting on Aristotle, but also its modern history with Frege and Tarski. And he has applied his mathematical insights to careful discussion of various linguistic constructions.

(12) if $e \equiv_\mu f$, then $e \sim_X f$

(recall that $X = dom(\mu)$), then it follows that $\equiv_\mu^F$ coincides with $\equiv_\mu$ on X.[16] This in fact means that it is possible to choose a label $\nu(e)$ for each $|e|_\mu$ such that ν *extends* μ, i.e. for $e \in X$, $\nu(e) = \mu(e)$. In other words, under these circumstances, the meaning of sentences is unchanged, and the fregean semantics extends the given meaning assignment to (all) other expressions of L. If we in addition assume that the constituent structure of L is wellfounded (section 3.2), Hodges observes (the *Abstract Tarski Theorem*) that the fregean semantics can be presented as recursive definition, with base clauses for atomic expressions, and clauses for complex expressions of the special form

(13) $\nu(F(e_1,\ldots,e_n)) = h_F(\nu(e_1),\ldots,\nu(e_n))$

These abstract results already have interesting applications to formal languages: Hodges notes that they establish the existence of a Tarski-style truth definition for IF logic (see note 15) as well as for the (closely related) logic with branching quantifiers (i.e. branching of $\forall$ and $\exists$). What do they tell us about natural languages?

First of all that the Context Principle, in the form (F), is indeed viable. But there are some caveats. One is that the fregean semantics is only defined up to synonymy, so it tells us nothing about what suitable fregean values are. Here Hodges is optimistic: in practice it has turned out that natural ways of finding out when two expressions have different fregean values yield natural ways of choosing suitable labels for the equivalence classes. In any case, if our main interest is compositionality, synonymy is enough.

A seemingly more pressing issue is—again—triviality. One might think that, even if the sentence semantics μ is not one-one, it is fine-grained enough that for any two distinct expressions you can find a sentence such that replacing one by the other in it changes its meaning. If so, the fregean semantics is one-one outside X, and thus essentially trivial. But this is another instance of the fact that you need a substantial notion of synonymy for properties like compositionality to make a difference. In this case, not all nuances of meaning should be taken into account; perhaps sameness of truth conditions, or sameness of the expressed proposition (in some suitable sense), is enough. Moreover, as Hodges notes, it makes sense to restrict attention to fragments of languages, deliberately excluding certain construc-

---

[16]In more detail, we have the following fact, which is immediate from the definitions:
FACT 6 (Hodges) *The following are equivalent:*
 (a) $\equiv_\mu^F$ *coincides with* $\equiv_\mu$ *on* X.
 (b) *For all* $e, f \in X$ *and* $F \in \mathbb{F}$, $e \equiv_\mu f$ *and* $F(e) \in X$ *implies* $F(e) \equiv_\mu F(f)$.

tions. Rather than as a way to avoid complications, this can be seen as abstracting from some details of reality in order to bring out underlying uniformities, a common procedure in the natural sciences.

Still, what are we to make of the fact that the fregean semantics is *always*—provided $X = dom(\mu)$ is cofinal—compositional? Simply, I think, that this is a feature of the most natural way of recovering expression meanings from sentence meanings. It doesn't in itself have empirical content. But the properties of the fregean semantics tell us, to begin with, to direct our attention to the sentence semantics μ. For only when μ is well behaved, in particular, is itself compositional, will the fregean semantics *extend* μ. Only then is it related in a reasonable way to the semantics we started with. And μ shouldn't be compositional for trivial reasons, and it shouldn't make the fregean semantics trivial either.

Furthermore, the fregean semantics may clarify our reflection on intuitive notions of meaning, or rather synonymy. As Hodges says, we have to solve the equation

$$\equiv_\mu^F = \sim_X \cap \approx$$

where $\sim_X$ comes from syntax (provided identifying sentences is a syntactic matter) and $\approx$ is an intuitive synonymy relation. The relation $\equiv_\mu^F$ itself is a trivial solution, but finding more reasonable solutions involves real semantic work (Hodges gives several illustrations). These are the real lessons, it seems to me, from Frege's Context Principle.

8 Quotation: a counter-example?

I will not discuss here the many counter-examples to compositionality that have been proposed, and the compositional solutions that have been suggested. But I will look at one case, which is perhaps the clearest of them all: (pure) quotation, i.e. the ability to refer in the language to linguistic expressions (meaningful or not). In a perfectly clear, and in principle familiar, sense, quotation is not compositional. Let us make this a bit more precise.

A language L is, as above, identified with a constituent structure $(\mathbb{E}, \mathbb{F})$ with a distinguished cofinal set $X \subseteq \mathbb{E}$ of (declarative) sentences, and a semantics μ with domain X. We say that L is *interpreted* if each sentence is either true or false, and that μ *respects truth values* if whenever e and f differ in truth value, $\mu(e) \neq \mu(f)$.

I will further say that L *has quotation* if there is a unary frame $Q \in \mathbb{F}$ such that, intuitively, $Q(e)$ is a quote frame of e (e.g. e surrounded by quotation marks) when $e \in X$,[17] and L is able to express elementary syntactic properties of sentences. The details need not be specified, but the point is there are sentences in L, with $Q(e)$ as a constituent, which are true iff, say, e begins with the letter "a", or e consists of five words, etc. Then we have:

(NQ) *Suppose L is an interpreted language that has quotation and whose sentence semantics μ respects truth values. Then, either μ is one-one or it is not compositional.*

For suppose μ is not one-one, i.e. that there are distinct $e, f \in X$ such that $\mu(e) = \mu(f)$. Since they have distinct shapes, some true sentence s in X with $Q(e)$ as a constituent is sensitive to this difference: it becomes false when e is replaced by f. There is a frame $G \in \mathbb{F}$ such that $s = G(e)$. Since μ respects truth values, $\mu(G(e)) \neq \mu(G(f))$. So μ is not compositional. And so $\equiv_\mu^F$ does not coincide with $\equiv_\mu$ on sentences: we have $e \equiv_\mu f$ but $e \not\equiv_\mu^F f$. Indeed, as remarked in the preceding section, the fregean semantics becomes trivial.

This is essentially nothing but the familiar 'opacity' of quotation, but formulated in general terms which reveal the very minimal assumptions needed about L; for example, it doesn't rely on identifying meaning with reference. There are statements in the literature which appear to contradict (NQ), but on a closer look, they don't.[18]

What should we conclude? The strategy of weakening the synonymy $e \equiv_\mu f$ doesn't seem helpful, since respecting truth values looks like a minimum requirement. The remaining alternative is to simply leave out quotation from the language. That is certainly possible. On the other hand, quotation, in the pure form of having a means of referring to linguistic items, is such a natural mechanism with such a straightforward semantics. And if we admit this mechanism in the language, compositionality is lost.

But maybe not completely lost. Section 10 will sketch a generalization of compositionality that admits quotation, and certain other recalcitrant linguistic constructions as well. But first I need to say something about compositionality and *context*.

---

[17]It is enough to assume here that we can quote sentences. In general, of course, one wants to quote arbitrary expressions, perhaps even arbitrary sequences of atomic symbols.

[18]For example, Potts (2007) presents an elegant semantics for (not only pure) quotation, which he claims to be compositional. What he in effect does is to give a recursive truth definition whose clauses for complex expressions are not of the form (13) but rather

$$\nu(F(e_1, \ldots, e_n)) = h_F(\nu(e_1), \ldots, \nu(e_n), e_1, \ldots, e_n)$$

Thus, the expressions themselves, as well as their meanings, are arguments of the semantic operations. This is much weaker than (homomorphism) compositionality; (see also Pagin & Westerståhl 2010a, sec. 3.2).

9 Dependence on extra-linguistic context

Context dependence in natural languages is ubiquitous. The clearest case is *indexicals*. Normally one wants to assign a *meaning* to sentences like

(14) I am hungry.

But if this meaning is to have anything to do with *truth conditions*, you need to account for the fact that the truth of (14) varies with the context of utterance. There are basically two ways to proceed. Either you let the meaning assignment μ take expressions *and* contexts as arguments. Or you *curry*, that is, you introduce, in the words of Lewis (1980), 'constant and complex semantic values', values which themselves are functions from contexts to ordinary meanings.[19] On the curried approach the notion of compositionality as we have defined it applies. But on the first approach we have this extra argument, requiring a slight reformulation. How slight, and what are the relations between the two approaches? Abstractly, the situation is easy to describe.[20]

As before, the language L has a constituent structure $(\mathbb{E}, \mathbb{F})$ and a semantics μ, but now μ is a function from $\mathbb{E} \times C$ to some set Z of 'ordinary' meanings, where C is a set of *contexts*. For simplicity, I'll assume μ is total. Contexts can be any objects; typical cases are

$\mu(\forall x \varphi, f) = \mathrm{T}$ iff for all $a \in M$, $\mu(\varphi, f(a/x)) = \mathrm{T}$ (contexts as assignments)

$\mu(P\varphi, t) = \mathrm{T}$ iff for some $t' < t$, $\mu(\varphi, t') = \mathrm{T}$ (contexts as times)

$\mu(I, c) = speaker_c$ (contexts as utterance situations)

Currying, we get the 1-ary function

$\mu_{curr} \colon \mathbb{E} \longrightarrow [C \longrightarrow Z]$

($[X \longrightarrow Y]$ is the set of functions from X to Y), defined by

$\mu_{curr}(e)(c) = \mu(e, c)$

We know what compositionality of μ_{curr} amounts to. For μ, there are two slightly different natural notions (using the functional formulation):

---

[19] This is the *functional* version. On the *structured* version, meanings are structured objects, possibly with 'holes' that can be filled with e.g. contexts. Everything I say below holds, with slight alterations, for the structured approach as well.

[20] For a details, motivation, proofs, and discussion of the issues raised in this section, (see Pagin 2005; Westerståhl 2012). Note that the 'meanings' in Z can themselves be functions, say, from possible worlds to truth values.

Context-sensitive compositionality

(i) μ is *compositional* iff for each $F \in \mathbb{F}$ there is an operation s_F such that for all $c \in C$,

$$\mu(F(e_1, \ldots, e_n), c) = s_F(\mu(e_1, c), \ldots, \mu(e_n, c))$$

(ii) μ is *weakly compositional* iff for each $F \in \mathbb{F}$ there is an operation s_F such that for all $c \in C$,

$$\mu(F(e_1, \ldots, e_n), c) = s_F(\mu(e_1, c), \ldots, \mu(e_n, c), c)$$

So the only difference is that the context itself is allowed to be an argument of the semantic operations in the weak case. This is actually an important weakening, and allows as compositional several phenomena that are often considered pragmatic rather than semantic. Here is how these notions are related.

PROPOSITION 7 *(Contextual) compositionality of μ implies weak (contextual) compositionality of μ, which in turn implies (ordinary) compositionality of μ_{curr}, but none of these implications can in general be reversed.*

The first two examples above, with contexts as assignments and as times, respectively, are typical instances of semantics which are *not* (not even weakly) contextually compositional, but where the curried version *is* compositional. The first of these reflects the familiar fact that Tarski's truth definition for first-order logic is compositional if you take *sets of assignments* (not truth values) as semantic values. The third example, on the other hand, with contexts as utterance situations, you typically expect to belong to a (contextually) compositional semantics. The reason is that in the first two cases contexts are *shifted* in the right-hand side of the clause, but this is usually not thought to happen in the third case.

There is much to say about which notion applies to which kind of linguistic construction, but here the points to take home are these: (a) Compositionality makes perfect sense also when meaning is context-dependent (which is the rule rather than the exception in natural languages). (b) But there are (at least) three distinct notions involved, related as in Proposition 7, and in applications one needs to be aware of which one is at stake.

10 General compositionality

Once extra-linguistic context dependence is seen to be compatible with compositionality, there is no reason why linguistic context dependence shouldn't also be. Such dependence can be understood in different ways. One is dependence on other parts of *discourse*, as when an anaphoric pronoun refers back to something introduced earlier by a name or, as in (15), an indefinite description:

(15) A woman entered the room. Only Fred noticed her.

Here I am interested in dependence on *sentential* context, of the kind Frege talks about in the following well-known passage:

> If words are used in the ordinary way, what one intends to speak of is their reference. It can also happen, however, that one wishes to talk about the words themselves or their sense. This happens, for instance, when the words of another are quoted. One's own words then first designate words of the other speaker, and only the latter have their usual reference. We then have signs of signs. In writing, the words are in this case enclosed in quotation marks. Accordingly, a word standing between quotation marks must not be taken to have its ordinary reference. (Frege 1892, 58–9)

What Frege says here is that *the type of linguistic context can change the meaning*. Quotation is one such type, sometimes indicated by quotation marks, and in this context, words no longer refer to what they usually refer to, but to themselves. Attitude contexts is of another type (only hinted at in this passage but developed in other parts of (Frege 1892)); then we use the same words to "talk about ... their sense."

In the syntactic algebra framework (section 3.1), terms are construction trees, so you can identify the (linguistic) context of a term occurrence t *in* a sentence s (or any complex term with t as a subterm) with the unique *path* from the top node to t. Let a *context typing* be a partition of the set of such paths, with the property that the type of each daughter t_i of a node $\alpha(t_1, \ldots, t_n)$ is determined by the type of that node, α, and i. Then we can formulate compositionality with C as the set of context types just as we did weak compositionality for arbitrary C, but with the difference that the meaning of $\alpha(t_1, \ldots, t_n)$ at c is determined by α, c, and the meanings of the t_i at c_i, where c_i is the context type determined by c, α, and i.

This version doesn't easily extend to the constituent structure framework (section 3.2), but there is another formulation, equivalent to the one just sketched for syntactic algebras, but applying more generally.[21] In the constituent structure framework, the idea would be to let a semantics be a *set S* of mappings from $\mathbb{E}$ to meanings, together with a *selection function*

---

[21]This formulation is due to Peter Pagin. For full details of these notions of compositionality (in the syntactic algebra setting), their properties, and the application to quotation, (see Pagin & Westerståhl 2010c).

Ψ, telling which function $\mu_i \in S$ should be applied to e_i when μ applies to $F(e_1, \ldots, e_n)$. Thus, *compositionality of* (S, Ψ) is the property that for each $F \in \mathbb{F}$ and each $\mu \in S$ there is an operation $r_{\mu, F}$ such that when $F(e_1, \ldots, e_n) \in \mathbb{E}$,

$$\mu(F(e_1, \ldots, e_n)) = r_{\mu, F}(\mu_1(e_1), \ldots, \mu_n(e_n)),$$

where $\mu_i = \Psi(\mu, F, i)$. So there is no extra argument to the meaning assignment, but instead there may be more than one meaning assignment function. We call this *general compositionality*. (If S is a unit set we have the ordinary notion.)

The application to quotation is now straightforward: in the simplest version you just need two meaning assignment functions, a *default* function μ_d and a *quotation* function μ_q, and the quote frame Q (section 8) has the property that whatever function is applied to $Q(e)$, μ_q is applied to e. And of course, for all $e \in \mathbb{E}$, $\mu_q(e)$ is (the surface representation of) e itself.

The idea of a semantics that allows switching between different meaning assignments appears quite natural, not only for quotation but for certain other linguistic phenomena as well.[22] Frege had a similar idea for attitude contexts. Glüer & Pagin (2006; 2008; 2012) use such a semantics for the modal operators, to deal with rigidity phenomena without treating names or natural kind terms as rigid designators. The point here has just been to show that compositionality, in its general form, is still a viable issue for such semantics.

11 Summing up

The question about compositionality, given a semantics for a set of structured expressions, is not vague. I have illustrated how it is spelled out relative to two (closely related) abstract accounts of syntax, accounts that fit most current syntactic theories. I have also emphasized that the real work lies in the choice of the syntax/semantics interface. Compositionality can be a factor in this choice, provided it is thought to play a role in an account of how language works. Most semanticists believe that it does. There are dissenting voices, but I have not yet seen a convincing purely mathematical or purely philosophical argument that it is trivial or empty. Nor have I seen a proposed counter-example that is not amenable to a compositional treatment—two kinds of context dependence were given as illustrations. And even if counter-examples should exist, it seems beyond doubt

---

[22] According to Hodges (2012, 249), this is the notion of *taḥrīf* used by the eleventh-century Persian-Arabic writer Ibn Sīnā ('Avicenna'), among others. Hodges is skeptical of its usefulness in semantics (2012, 255–6), but at least the application to quotation (not discussed by him) seems very natural.

that efforts to insure compositionality have lead to exciting developments in semantics and in logic. That is one kind of evidence that compositionality is a good thing. Another is the cluster of related issues that the study of compositionality brings to light, such as the relation between word meaning and sentence meaning, or the Husserl property (section 7). I think we may conclude that in the present state of language research, it would be ill-advised to disregard the issue of compositionality.

Bibliography

Aczel, P. (1990). Replacement systems and the axiomatization of situation theory. In *Situation Theory and its Applications*, vol. 1, Cooper, R., Mukai, K., & Perry, J., eds., Stanford: CSLI Publications, 3–31.

Barker, C. & Jacobson, P. (Eds.) (2007). *Direct Compositionality*. Oxford: Oxford University Press.

Cameron, P. & Hodges, W. (2001). Some combinatorics of imperfect information. *Journal of Symbolic Logic*, 66(2), 673–684, doi:10.2307/2695036.

Dowty, D. (2007). Compositionality as an empirical problem. In *Direct Compositionality*, Barker, C. & Jacobson, P., eds., Oxford: Oxford University Press, 23–101.

Fodor, J. (2000). Reply to critics. *Mind & Language*, 15(2–3), 350–374, doi: 10.1111/1468-0017.00139.

Fodor, J. (2001). Language, thought, and compositionality. *Mind & Language*, 16(1), 1–15, doi:10.1111/1468-0017.00153.

Frege, G. (1884). *Grundlagen der Arithmetik,* translated by J. Austin as *Foundations of Arithmetic*. Oxford: Blackwell, 1950.

Frege, G. (1892). Über Sinn und Bedeutung. In *Translations from the Philosophical Writings of Gottlob Frege*, Geach, P. & Black, M., eds., Oxford: Blackwell, 56–78, 1952.

Frege, G. (1923). Gedankengefüge. In *Logical Investigations. Gottlob Frege*, Geach, P. & Stoothof, R. H., eds., Oxford: Blackwell, 55–78, 1977.

Galliani, P. (2011). Sensible semantics of imperfect information. In *Logic and Its Applications, Lecture notes in computer science*, vol. 6521, Banerjee, M. & Seth, A., eds., Berlin; Heidelberg: Springer, 79–89.

Glüer, K. & Pagin, P. (2006). Proper names and relational modality. *Linguistics and Philosophy*, 29(5), 507–535, doi:10.1007/s10988-006-9001-7.

Glüer, K. & Pagin, P. (2008). Relational modality. *Journal of Logic, Language and Information*, 17(3), 307–322, doi:10.1007/s10849-008-9059-4.

Glüer, K. & Pagin, P. (2012). General terms and relational modality. *Noûs*, *46*, 159–199, doi:10.1111/j.1468-0068.2010.00783.x.

Heim, I. & Kratzer, A. (1998). *Semantics in Generative Grammar*. Oxford: Blackwell.

Hintikka, J. (1996). *The Principles of Mathematics Revisited*. Cambridge: Cambridge University Press.

Hintikka, J. & Sandu, G. (2011). Game-theoretical semantics. In *Handbook of Logic and Language*, Benthem, J. v. & Meulen, A. t., eds., London: Elsevier, 2nd edn., 415–465.

Hinzen, W., Machery, E., & Werning, M. (Eds.) (2012). *The Oxford Handbook of Compositionality*. Oxford: Oxford University Press.

Hodges, W. (1997). Compositional semantics for a language of imperfect information. *Logic Journal of the IGPL*, *4*(5), 539–563, doi:10.1093/jigpal/5.4.539.

Hodges, W. (2001). Formal features of compositionality. *Journal of Logic, Language and Information*, *10*(1), 7–28, doi:10.1023/A:1026502210492.

Hodges, W. (2005). A context principle. In *Intensionality*, Kahle, R., ed., Wellesley, MA: Association for Symbolic Logic and A. K. Peters, 42–59.

Hodges, W. (2011). From sentence meanings to full semantics. In *Proof, Computation and Agency, Synthese Library*, vol. 352, van Benthem, J., Gupta, A., & Parikh, R., eds., Dordrecht: Springer, 261–276.

Hodges, W. (2012). Formalizing the relationship between meaning and syntax. In *The Oxford Handbook of Compositionality*, Hinzen, W., Machery, E., & Werning, M., eds., Oxford: Oxford University Press, 245–261.

Horwich, P. (2005). *Reflections on Meaning*. Oxford: Oxford University Press.

Jacobson, P. (2002). The (dis)organisation of the grammar: 25 years. *Linguistics and Philosophy*, *25*(5–6), 601–626, doi:10.1023/A:1020851413268.

Kontinen, J., Väänänen, J., & Westerståhl, D. (Eds.) (2013). *Dependence and Independence in Logic*, vol. 101. Special issue of *Studia Logica*.

Kracht, M. (2003). *The Mathematics of Language*. Berlin: Mouton de Gruyter.

Kracht, M. (2007). The emergence of syntactic structure. *Linguistics and Philosophy*, *30*(1), 47–95, doi:10.1007/s10988-006-9011-5.

Larson, R. & Segal, G. (1995). *Knowledge of Meaning. An Introduction to Semantic Theory*. Cambridge, MA: MIT Press.

Lewis, D. (1980). Index, context, and content. In *Philosophy and Grammar*, Kanger, S. & Öhman, S., eds., Dordrecht: D. Reidel, 79–100, reprinted in D. Lewis, *Papers in Philosophical Logic*, 1998, 21–44. Cambridge: Cambridge University Press.

Montague, R. (1974). Universal grammar. In *Formal Philosophy. Selected papers by Richard Montague*, New Haven: Yale University Press, 222–246, first published in 1970.

Pagin, P. (2003). Communication and strong compositionality. *Journal of Philosophical Logic*, *32*(3), 287–322, doi:10.1023/A:1024258529030.

Pagin, P. (2005). Compositionality and context. In *Contextualism in Philosophy*, Preyer, G. & Peter, G., eds., Oxford: Oxford University Press, 303–348.

Pagin, P. & Westerståhl, D. (2010a). Compositionality I. Definitions and variants. *Philosophy Compass*, *5*(3), 250–264, doi:10.1111/j.1747-9991.2009.00228.x.

Pagin, P. & Westerståhl, D. (2010b). Compositionality II. Arguments and problems. *Philosophy Compass*, *5*(3), 265–282, doi:10.1111/j.1747-9991.2009.00229.x.

Pagin, P. & Westerståhl, D. (2010c). Pure quotation and general compositionality. *Linguistics and Philosophy*, *33*(5), 381–415, doi:10.1007/s10988-011-9083-8.

Potts, C. (2007). The dimensions of quotation. In *Direct Compositionality*, Barker, C. & Jacobson, P., eds., Oxford: Oxford University Press, 405–431.

Väänänen, J. (2007). *Dependence Logic*. Cambridge: Cambridge University Press.

Westerståhl, D. (1998). On mathematical proofs of the vacuity of compositionality. *Linguistics and Philosophy*, *21*, 635–643, doi:10.1023/A:1005401829598.

Westerståhl, D. (2012). Compositionality in Kaplan style semantics. In *The Oxford Handbook of Compositionality*, Hinzen, W., Machery, E., & Werning, M., eds., Oxford: Oxford University Press, 192–219.

Zadrozny, W. (1994). From compositional to systematic semantics. *Linguistics and Philosophy*, *17*(4), 329–342, doi:10.1007/BF00985572.

Dag Westerståhl
Department of Philosophy
Stockholm University
Sweden
dag.westerstahl@philosophy.su.se

General Proof Theory[1]

KOSTA DOŠEN

By the end of the last century Saunders Mac Lane wrote:

> So "proof" is the central issue in mathematics. There ought to be a vibrant specialty of "proof theory". There is a subject with this title, started by David Hilbert in his attempt to employ finitistic methods to prove the correctness of classical mathematics. This was used essentially by Gödel in his famous incompleteness theorem, carried on further by Gerhard Gentzen with his cut elimination theorem. In 1957, at a famous conference in Ithaca, proof theory was recognized as one of the four pillars of mathematical logic (along with model theory, recursion theory and set theory). But the resulting proof theory is far too narrow to be an adequate pillar... (Mac Lane 1997, 152)

General proof theory—the term is due to Dag Prawitz—should lead to the proof theory Mac Lane was looking for. It addresses the philosophically-looking question "What is a proof?" by dealing with technical questions related to normal forms of proofs, and in particular with the question of identity criteria for proofs. It follows Gentzen more than Gödel, and in doing that it deals with the structure of proofs, as exhibited for example by the Curry-Howard correspondence, rather than with their strength measured by ordinals.

Much of general proof theory is the field of categorial proof theory, opened up by Joachim Lambek. Fundamental notions of category theory like the notion of adjoint functor, and very important structures like cartesian closed categories, came to be of central concern for logic in that field. William Lawvere's contribution here is decisive. Besides that, results of the kind categorists call coherence results provide a model theory for equality of proofs. For example, the coherence theorem for symmetric monoidal closed

---

[1] This short text is the introduction to the Symposium on General Proof Theory that was organized at the congress by its author. The speakers at this symposium were Dag Prawitz, William Lawvere and Philip Scott.

categories, which antedates considerably the advent of linear logic, is about equality of proofs in a fragment of this logic. Much of Philip Scott's work in categorial proof theory is inspired by linear logic. Mac Lane, who with Max Kelly proved this coherence theorem by relying on Gentzen's cut elimination, introduced the subject of coherence in category theory, through which logic finds new ties with geometry, topology and algebra.

Mac Lane wrote a doctoral thesis in proof theory under the supervision of Paul Bernays. In that thesis, contrary to the spirit of the proof theory that came to dominate the twentieth century, and in accordance with the spirit of general proof theory, he concentrated on the justification of the inference steps, rather than on the propositions that make the proofs. In general proof theory one looks for an algebra of proofs, and for that one should concentrate on the operations of this algebra, which come with the inference rules. The propositions that make the proofs are secondary. Usually, however, the propositions are in the forefront, and not the rules, which are noted only in the margins (and this marginal "bureaucracy" may even be found undesirable).

As an equational theory, the algebra of proofs involves the question of identity criteria for proofs, the central question of general proof theory. This is a question that may be found, at least implicitly, in David Hilbert's discarded 24th problem:

> Überhaupt, wenn man für einen Satz zwei Beweise hat, so muss man nicht eher ruhen, als bis man die beide aufeinander zurückgeführt hat oder genau erkannt hat, welche verschiedenen Voraussetzungen (und Hilfsmittel) bei den Beweisen benutzt werden:... [In general, if one has two proofs for a proposition, one must keep going until one has derived each of them from the other, or until one has discerned clearly enough what different conditions (and means) have been used in these proofs:...] (Thiele 2005, Section 10.4, 280–282)

Bibliography

Mac Lane, S. (1997). Despite physicists, proof is essential in mathematics. *Synthese*, *111*, 147–154.

Thiele, R. (2005). Hilbert and his twenty-four problems. In *Mathematics and the Historian's Craft: The Kenneth O. May Lectures*, Van Brummelen, G. & Kinyon, M., eds., New York: Springer, 243–295.

Kosta Došen
Faculty of Philosophy
University of Belgrade
Mathematical Institute, SANU
Belgrade
Serbia
kosta@mi.sanu.ac.rs

Uniformization in Automata Theory

ARNAUD CARAYOL & CHRISTOF LÖDING

ABSTRACT. We survey some classical results on uniformizations of automaton definable relations by automaton definable functions. We consider the case of automatic relations over finite and infinite words and trees as well as rational relations over finite and infinite words. We also provide some new results concerning the uniformization of automatic and rational relations over finite words by subsequential transducers. We show that it is undecidable whether a given rational relation can be uniformized by a subsequential transducer and provide a decision procedure for the case of automatic relations.

1 Introduction

A uniformization of a (binary) relation is a function that selects for each element in the domain of the relation a unique image that is in relation with this element. In other words, a uniformization of $R \subseteq X \times Y$ is a function $f_R : X \to Y$ with the same domain as R and whose graph is a subset of R. The origin of uniformization problems comes from set theory, where the complexity of a class of definable relations is related with the complexity of uniformizations for these relations (see Moschovakis (1980) for results of this kind).

The aim of this paper is to give an overview of some uniformization results in the setting where the relations and functions are defined by finite automata. In analogy to the problems studied in set theory, the uniformization question for a given class of relations defined by some automaton model is whether each such relation has a uniformization in the same class. A motivation for studying these kinds of questions in computer science arises when the relation describes a specification relating inputs to allowed outputs, for example for a program or a circuit, as in Church's synthesis problem Church (1962). A uniformization of such a specification can be seen as a concrete implementation conforming to the specification because it selects for each input exactly one allowed output. In this view, the uniformization question then asks whether each specification from a given class can implemented within the same class.

These uniformization questions have already been studied in the early times of automata theory. The first class considered is that of word relations defined by finite automata Elgot & Mezei (1965) also called rational relations. In (Kobayashi 1969, Theorem 3), it is first shown that any relation accepted by a finite automaton admits a uniformization also accepted by a finite automaton. Several alternative and simplified proofs were published Eilenberg (1974); Arnold & Latteux (1979); Choffrut & Grigorieff (1999). The proof of Arnold & Latteux (1979) builds on a decomposition theorem for rational functions from Elgot & Mezei (1965) and shows that uniformization can be realised by the composition of a sequential (i.e., input deterministic) transducer working from left to right followed by one working from right to left. Choffrut & Grigorieff (1999) contains the idea of the reduction to the length-preserving case that we use in this article as well as a generalisation to the infinite word case. The uniformization problem for relations on trees defined by finite automata was only considered more recently Kuske & Weidner (2011); Colcombet & Löding (2007) but can be traced back to Engelfriet (1978).

The decision problem corresponding to the uniformization question is to decide whether a given relation has a uniformization. We consider this question in a setting where the relation comes from one class $\mathbb{C}$ and we are looking for a uniformization in another class $\mathbb{C}'$ (which is more restrictive than $\mathbb{C}$ in our setting). An early decidability results in this spirit was provided by Büchi and Landweber in Büchi & Landweber (1969) for relations of infinite words given by synchronous two-tape Büchi automata and to be uniformized by synchronous deterministic sequential transducers, that is, deterministic transducers that produce one output symbol for each input symbol. Modern presentations of these results can be found, e.g., in Thomas (2008) and Löding (2011).

This decidability result has been studied for variations where the transducer defining the uniformization is allowed to skip a bounded number of output symbols in order to obtain a bounded look-ahead Hosch & Landweber (1972). In Holtmann *et al.* (2010) the condition is relaxed further by allowing an arbitrary finite number of skips. However, as shown in Holtmann *et al.* (2010), this case can be reduced to a bounded number of skips, where the bound depends on the size of the automaton defining the specification.

We study a similar setting in the case of finite words. A standard model for deterministically computing functions over finite words are subsequential transducers. These are basically deterministic finite automata that output a finite word on each transition. We show that it is decidable whether an automatic relation can be uniformized by a subsequential transducer. The setting is similar to the one mentioned above for infinite words studied in Holtmann et al. (2010). However, in the setting of finite words the number of required skips (where the transducer outputs the empty word) cannot be bounded because the input and the output might be of different length. This adds some new phenomena compared to the setting of Holtmann et al. (2010).

Furthermore, we show that it is undecidable whether a rational relation can be uniformized by a subsequential transducer.

The paper is structured as follows. In Section 2 we consider uniformization automatic relations over finite and infinite words and trees. Section 3 is about uniformization of rational relations over finite and infinite words. In Section 4 we present some new results on uniformization of rational and automatic relations over finite words by subsequential transducers.

We assume the reader to be familiar with basic notions from automata theory and only provide some definitions for fixing the terminology.

2 Automatic relations on words and trees

In this section, we consider the class of relations that are definable by synchronous finite automata, that is, automata that process the input and the output at the same time and at the same speed. Such relations are referred to as automatic Khoussainov & Nerode (1995); Blumensath & Grädel (2000). Automatic relations can be defined over finite words, infinite words, finite trees, and infinite trees. For these four classes, we study the question whether a given automatic relation always has an automatic uniformization. We only consider the case of binary relations; uniformization questions for automatic relations of higher arity can be reduced to the binary case (which is not the case for rational relations, see Section 3). To simplify notation, we usually assume that the alphabets for the two components are the same, which is not a restriction.

2.1 Finite words

We start by considering automatic relations over finite words. As usual, a finite word is finite sequence of letters over an alphabet Σ, where an alphabet is just a finite set of symbols. The set of all finite words over Σ is denoted by Σ^* and the empty word by ε. The length of a word w is denoted by $|w|$.

We use the standard model of finite automata on finite words. To fix the notation, a nondeterministic finite automaton (NFA) is of the form $\mathcal{A} = (Q, \Sigma, q_0, \Delta, F)$, where Q is a finite set of states, Σ is the input alphabet, $q_0 \in Q$ is the initial state, $\Delta \subseteq Q \times \Sigma \times Q$ is the transition relation, and $F \subseteq Q$ is the set of final (or accepting) states. A run of $\mathcal{A}$ on $w \in \Sigma^*$ is a sequence $p_0, \ldots, p_n$ of states such that $(p_i, a_i, p_{i+1}) \in \Delta$ for all $i \in$

$\{0, \ldots, n-1\}$, where $w = a_0 \cdots a_{n-1}$. We write $\mathcal{A} : p_0 \xrightarrow{w} p_n$ to indicate that there is a run of $\mathcal{A}$ on w from p_0 to p_n. An accepting run is a run that starts in q_0 and ends in a state from F. The set of words that labels an accepting run of $\mathcal{A}$ is denoted by $L(\mathcal{A})$ and is called the language accepted by $\mathcal{A}$. The class of languages that can be accepted by NFAs is called the class of regular languages. An NFA is deterministic (a DFA) if for each $q \in Q$ and $a \in \Sigma$ there is at most one $q' \in Q$ with $(q, a, q') \in \Delta$. In this case, we usually write the transition relation as a (partial) function $\delta : Q \times \Sigma \to Q$. We refer the reader who is not familiar with the basic results on regular languages and finite automata to Hopcroft & Ullman (1979).

One way to make finite automata process tuples of words (for defining a relation) is to use a product alphabet such that the automaton processes one letter from each word in the tuple in one step. This leads to the notion of automatic relations defined more formally below. To handle the case of words of different length, a dummy symbol $\square$ is used to pad shorter words.

Formally, for $w_1, w_2 \in \Sigma^*$ we define

$$w_1 \otimes w_2 = \begin{bmatrix} a'_{11} \\ a'_{21} \end{bmatrix} \cdots \begin{bmatrix} a'_{1n} \\ a'_{2n} \end{bmatrix} \in (\Sigma_\square^2)^*$$

where $\Sigma_\square = \Sigma \cup \{\square\}$, n is the maximal length of one of the words w_i, and a_{ij} is the jth letter of w_i if $j \leq |w_i|$ and $\square$ otherwise. A language $L \subseteq ((\Sigma \cup \{\square\})^2)^*$ defines a relation $R_L \subseteq (\Sigma^*)^2$ in the obvious way: $(w_1, w_2) \in R_L$ iff $w_1 \otimes w_2 \in L$. A relation $R \subseteq (\Sigma^*)^2 s$ is called automatic if $R = R_L$ for some regular language $L \subseteq ((\Sigma \cup \{\square\})^2)^*$. These definitions can easily be generalized to relations of higher arity, but as mentioned earlier, we restrict our considerations to the binary case.

EXAMPLE 1 *Consider the following relation of words over the alphabet* $\{0, 1\}$:

$$R = \{(1^n, 1^m 01^k 0^*) \mid n = m + k + 1\}.$$

This relation is accepted by the DFA depicted in Figure 1 and hence, is an automatic relation (the initial state is marked by an incoming arrow and the accepting states by a double circle).

As mentioned above, we study in the section the question whether (binary) automatic relations have automatic uniformizations. A simple technique to define a uniformization is to fix a total well-ordering (a total ordering without infinite decreasing chains) on the elements of the domain (finite words in this case), and then to select for each word u the smallest word v such that u and v are in relation. One such well-ordering is the length-lexicographic ordering, which first orders words by their length

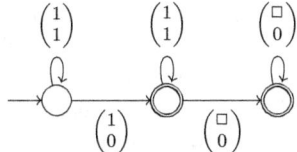

Figure 1. A finite automaton for the relation from Example 1.

and the words of the same length are ordered lexicographically according to some fixed order on the letters of the alphabet. We denote this ordering by $<_{\text{llex}}$. We observe that $<_{\text{llex}}$ is itself an automatic relation, and thus is a useful tool for uniformization.

For a relation $R \subseteq \Sigma^* \times \Sigma^*$ we define the length-lexicographic uniformization by

$$f_R(u) = \min_{<_{\text{llex}}} \{v \in \Sigma^* \mid (u,v) \in R\}$$

for all u in the domain of R. For the relation R from Example 1 we obtain $f_R(1^n) = 01^{n-1}$.

In general, one can always construct from the automaton for R a new automaton that checks for $u \otimes v$ whether $(u,v) \in R$ and at the same time verifies that there is no smaller word $v' <_{\text{llex}} v$ such that $(u,v') \in R$. This technique of selecting smallest representatives (according to some ordering) in a regular way goes back to Eilenberg's Cross-Section Theorem Eilenberg (1974). In the context of automatic relations, this technique is used with the ordering $<_{\text{llex}}$ in Khoussainov & Nerode (1995) to show that automatic equivalence relations have automatic cross-sections, which means that there is a regular set of representatives from the equivalence classes. The following theorem summarizes our considerations. We do not attribute this theorem to a specific paper because its proof is rather simple and it can easily be derived in many ways from other results. An explicit statement of the result can be found in Choffrut & Grigorieff (1999).

THEOREM 2 *For an automatic relation R over finite words, the length-lexicographic uniformization of R is also automatic. In particular, each automatic relation has an automatic uniformization.*

2.2 Finite trees

We now turn to the uniformization problem for automatic relations over finite trees. To define trees, we fix a ranked alphabet Σ, that is, each symbol in Σ has a rank (or arity), which is a natural number. A tree domain dom is a non-empty finite subset of $\mathbb{N}^*$ (the set of finite sequences of natural numbers) with the property that for each $x \in \mathbb{N}^*$ and $i \in \mathbb{N}$ with $x \cdot i \in dom$ we also have $x \in dom$ and $x \cdot j \in dom$ for each $j < i$. By using sequences of natural numbers, a tree domain is naturally equipped with a successor relation, namely $x \cdot i$ is a successor of x.

A (finite Σ-labeled) tree t is a mapping from a tree domain $dom(t)$ to the ranked alphabet Σ such that for each $x \in dom(t)$ the rank of $t(x)$ corresponds to the number of successors of x in $dom(t)$.

To define tree-automatic structures, we need a way to code tuples of finite trees, i.e., we need an operation $\otimes$ for finite trees similar to the one for words. For a tree $t : dom(t) \to \Sigma$ let $t^{\square} : \mathbb{N}^* \to \Sigma_{\square}$ be defined by $t^{\square}(u) = t(u)$ if $u \in dom(t)$, and $t^{\square}(u) = \square$ otherwise. For finite Σ-labeled trees t_1, t_2, we define the $\Sigma_{\square}^2$-labeled tree $t = t_1 \otimes t_2$ by $dom(t) = dom(t_1) \cup dom(t_2)$ and $t(u) = (t_1^{\square}(u), t_2^{\square}(u))$. When viewing words as unary trees, this definition corresponds to the operation $\otimes$ as defined for words. As in the case of words, a set T of finite $\Sigma_{\square}^2$-labeled trees defines the relation R_T by $(t_1, t_2) \in R_T$ iff $t_1 \otimes t_2 \in T$.

We are not going to use any details on automata on finite trees. For the purpose of this paper, it is enough to know that there is a model of finite automata on finite trees that defines the class of regular tree languages whose closure and algorithmic properties are comparable to the class of regular word languages. We refer the reader to Comon et al. (2007) or Löding (2012) for an introduction to automata on finite trees.

We call a (binary) relation R over the domain of finite Σ-labeled trees tree-automatic if $R = R_T$ for some regular language T of finite $\Sigma_{\square}^2$-labeled trees.

A crucial difference to the setting of finite words is that there is no tree-automatic total well-ordering over the set of all finite Σ-labeled trees. This fact can be deduced from Theorem 6 presented in Section 2.4. Hence, the technique used for the uniformization of automatic relations cannot be extended to finite trees. However, a result from Kuske & Weidner (2011) (based on Colcombet & Löding (2007)) shows that tree-automatic equivalence relations have regular cross-sections (that is, given a tree-automatic equivalence relation $\sim$, it is possible to construct a tree automaton accepting a language that contains exactly one tree from each equivalence class of $\sim$). We can use this result to obtain a tree-automatic uniformization of an arbitrary (binary) tree-automatic relation R as follows. We define an equivalence relation $\sim_R$ over the set of finite $\Sigma_{\square}^2$-labeled trees by $t_1 \otimes t_2 \sim_R t_3 \otimes t_4$ if $(t_1, t_2), (t_3, t_4) \in R$ and $t_1 = t_3$. Then $\sim_R$ is a tree-automatic equivalence relation and we obtain a regular tree language T of finite $\Sigma_{\square}^2$-labeled trees that contains exactly one representative from each equivalence class of $\sim_R$. The relation R_T defines a uniformization of R because for each possible first component t in the domain of R it contains exactly one pair (t, t').

The uniformization result for tree-automatic relations can also be deduced from (the proof of) a result in Engelfriet (1978) that shows that relations computed by nondeterministic top-down tree transducers can be uniformized by deterministic top-down tree transducers with regular look-ahead. The constructed transducer basically chooses at each node for a given input letter the least possible transition (according to some fixed ordering on the transitions) that admits a successful run on the remaining input. The regular look-ahead is used to check this property for the chosen transition. In our setting, the regular look-ahead could be simulated by nondeterministically guessing a transition and then verifying that all smaller transitions would not admit a successful computation on the remaining input.

THEOREM 3 *Every tree-automatic relation has a tree automatic uniformization.*

2.3 Infinite words

An infinite word α over an alphabet Σ corresponds to a function $\alpha : \mathbb{N} \to \Sigma$, which naturally defines an infinite ordered sequence of letters. We denote the set of infinite words (also called ω-words) over Σ by Σ^ω.

Automatic relations over infinite words, called ω-automatic relations, are defined in the same way as for finite words using Büchi automata instead of standard finite automata (the definition is even simpler because for infinite words no padding is required). A Büchi automaton is given in the same way as an NFA. It accepts an infinite word if there is a run on this word that starts in the initial state and infinitely often visits an accepting state. For an introduction to the theory of automata on infinite words, we refer the reader to Thomas (1997).

EXAMPLE 4 *Let Σ be some alphabet and consider the relation $R_\approx$ of ultimately equal words, that is,*

$$R_\approx := \{(\alpha, \beta) \mid \exists u, v \in \Sigma^*, \gamma \in \Sigma^\omega : |u| = |v| \text{ and } \alpha = u\gamma \text{ and } \beta = v\gamma\}.$$

This is an ω-automatic equivalence relation that is accepted by the Bchi automaton depicted in Figure 2 for $\Sigma = \{0, 1\}$.

The techniques for uniformization of relations over finite words and trees presented above are both based on the regular cross-section property for (tree-)automatic equivalence relations. For infinite words, a corresponding

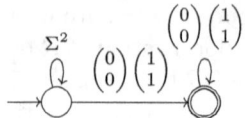

Figure 2. A Büchi automaton for the relation from Example 4.

result does not hold. As pointed out in Kuske & Lohrey (2006), the equivalence relation $R_\approx$ from Example 4 does not have the regular cross-section property (i.e., does admit a set of representatives that can be accepted by a Büchi automaton). This easily follows from the fact that each Büchi automaton that accepts infinitely many ω-words also accepts two different words that are ultimately equal.

However, uniformization of ω-automatic relations is still possible because selecting representatives from equivalence classes is a stronger requirement than selecting unique images for all elements in the domain of a relation (see also the reduction in Section 3.2). The uniformization result for ω-automatic relations was already obtained in Siefkes (1975). The technique that we present here is taken from Choffrut & Grigorieff (1999).

THEOREM 5 (Siefkes (1975); Choffrut & Grigorieff (1999)) *Every ω-automatic relation has an ω-automatic uniformization.*

Proof. The key idea for the construction is to take the accepting runs of an automaton for the relation as additional information for selecting an image. Given a Büchi automaton $\mathcal{A} = (Q, \Sigma^2, q_0, \Delta, F)$ recognizing $R \subseteq \Sigma^\omega \times \Sigma^\omega$, we consider an extended relation $R' \subseteq \Sigma^\omega \times \Sigma^\omega \times Q^\omega$ containing those tuples (α, β, ρ) such that $(\alpha, \beta) \in R$ and ρ is an accepting run of $\mathcal{A}$ on $\alpha \otimes \beta$.

Our aim is to define an ordering on the pairs (β, ρ) such that an automaton can select the smallest such pair for a given α. For ρ, we consider the unique sequence $i_1 < i_2 < \cdots$ of position at which ρ is in an accepting state, that is, with $\rho(i_j) \in F$.

This induces a factorization of the combined word $\beta \otimes \rho$ into finite segments $\beta_j \otimes \rho_j$ for $j \geq 1$ where each β_j is the segment of β from $i_{j-1} + 1$ to i_j and similarly for ρ (for the initial segment to be well defined, we set $i_0 = -1$). Note that the segments ρ_j all contain exactly one accepting state, namely at the last position.

Given another pair β' and ρ' with corresponding sequence $i'_1 < i'_2 < \cdots$, we obtain the factors $\beta'_j \otimes \rho'_j$. Now pick the first j such that $\beta_j \otimes \rho_j \neq \beta'_j \otimes \rho'_j$. We define $(\beta, \rho) < (\beta', \rho')$ if $(\beta_j, \rho_j) <_{\text{llex}} (\beta'_j, \rho'_j)$ for this j, where $<_{\text{llex}}$ refers to some fixed ordering on the set $\Sigma \times Q$.

One can verify that there is a Büchi automaton that accepts precisely those pairs $(\alpha, \beta) \in R$ such that there is ρ with $(\alpha, \beta, \rho) \in R'$ and for all $(\alpha, \beta', \rho') \in R'$ one has $(\beta, \rho) \leq (\beta', \rho')$.

It remains to verify that for each α in the domain of R such a minimal pair (β, ρ) exists. Such a minimal pair can be constructed for a given α by inductively defining a sequence of segments $\beta_j \otimes \rho_j$ as follows. The segment $\beta_j \otimes \rho_j$ is the $<_{\text{llex}}$-minimal segment with the property that ρ_j is in $(Q \backslash F)^* F$ and that the concatenation $\beta_0 \otimes \rho_0 \cdots \beta_j \otimes \rho_j$ can be extended to a sequence $\beta \otimes \rho$ such that $(\alpha, \beta, \rho) \in R'$. Taking the limit sequences $\beta_1 \beta_2 \cdots$ and $\rho_1 \rho_2 \cdots$ results in a pair with the desired properties. ∎

2.4 Infinite trees

The theory of automata on infinite words can be generalized to automata on infinite trees. For simplicity, we only consider Σ-labeled complete binary trees, which are mappings $t : \{0, 1\}^* \to \Sigma$. As for automata on finite trees, we do not detail the model here because it is not required for our considerations. We refer the reader to Thomas (1997) for the basics on this model.

The definitions for ω-tree-automatic structures are a straightforward generalization of the definitions for the other classes of automatic structures in this section. However, it is the only class of automatic structures that does not admit uniformization.

This result goes back to Gurevich & Shelah (1983) where it is shown that there is no choice function over the infinite binary tree that is definable in monadic second-order logic (MSO). In this setting, we view the (unlabeled) infinite binary tree as a structure $T_2 = (\{0, 1\}^*, S_0, S_1)$ with universe $\{0, 1\}^*$ and two successor relations S_0 and S_1 with the natural interpretation (S_0 corresponds to appending a 0 and S_1 to appending a 1).

As usual, monadic second-order logic is the extension of first-order logic by set quantifiers. An MSO definable choice function would be given by an MSO formula $\varphi(X, y)$ with one free set variable X and one free element variable y such that for each nonempty subset $U \subseteq \{0, 1\}^*$ there is exactly one element $u \in U$ such that $T_2 \models \varphi[U, u]$.

THEOREM 6 (Gurevich & Shelah (1983)) *There is no MSO-definable choice function over the infinite binary tree.*

While the proof given in Gurevich & Shelah (1983) uses set-theoretic tools, more recently a new proof only based on automata-theoretic methods has been given in Carayol & Löding (2007); Carayol et al. (2010). From this new proof, one can even derive a simple family of counter examples in the following sense. For $N \in \mathbb{N}$, define the set $U_N \subseteq \{0,1\}^*$ as

$$U_N := \{0,1\}^* (0^N 0^* 1)^N.$$

It turns out that this simple family of sets is sufficient to show that there is no MSO definable choice function over T_2.

THEOREM 7 (Carayol & Löding (2007); Carayol et al. (2010)) *For each MSO formula $\varphi(X,y)$ over the infinite binary tree, there exists N such that φ fails to choose a unique element from U_N.*

Using the tight connection between MSO and finite automata (see Thomas (1997)), one can rephrase the result from Gurevich & Shelah (1983) in automata theoretic terms. For this purpose, we define the relation $R_\in$ over infinite $\{0,1\}$-labeled trees as follows: $(t,t') \in R$ if

there is exactly one $u \in \{0,1\}^*$ with $t'(u) = 1$, and

$t(u) = 1$ for the unique $u \in \{0,1\}^*$ with $t'(u) = 1$.

Intuitively, the relation $R_\in$ corresponds to the element relation because the tree t represents a non-empty set U (via the 1-labeled nodes), t' corresponds to an element u (via the unique 1-labeled node), and the last condition states that $u \in U$. An automaton defining a uniformization of $R_\in$ could be turned into an MSO formula defining a choice function, which is not possible according to Theorem 6.

COROLLARY 8 *There are ω-tree-automatic relations that do not have an ω-tree-automatic uniformization. In particular, the relation $R_\in$ does not have an ω-tree-automatic uniformization.*

3 Rational word relations

In this section, we consider the uniformization of relations on finite and infinite words accepted by asynchronous automata. These relation are called rational relations (as for finite words, they are the rational subsets of the product monoid). Following Choffrut & Grigorieff (1999), we will see that these problems can be reduced to the automatic case using a standard decomposition theorem Eilenberg (1974). The generalization of rational (word) relations to trees is unclear and several notions of asynchronous tree automata models have been proposed (see for instance (Comon et al. 2007, Chapter 6) as well as the discussion in Raoult (1997)). The uniformization problem for these various classes goes beyond the scope of this article.

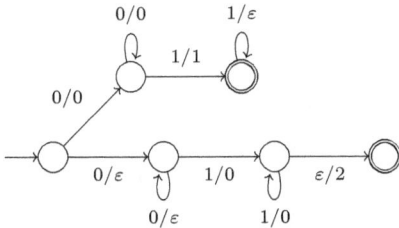

Figure 3. A transducer for the relation from Example 9.

3.1 Finite words

As in Section 2, we consider the case of binary relations where (most of the time) we assume the input and the output alphabet to be the same. The latter assumption is easily seen not to be a restriction. We comment on the case of higher arity at the end of this section.

In the case of finite words, a rational relation $R \subseteq (\Sigma^*)^2$ is accepted by a finite automaton which can read its two tapes (also called input and output tape in the binary setting) in an asynchronous fashion. Formally, such an automaton has its transitions labeled by elements of $(\Sigma \cup \{\varepsilon\})^2$. The language $L \subseteq ((\Sigma \cup \{\varepsilon\})^2)^*$ accepted by the automaton defines the relation $\{(\pi_1(w), \pi_2(w)) \mid w \in L\}$ where π_i is the morphism from $(\Sigma \cup \{\varepsilon\})^2$ to Σ^* corresponding to the projection over the i-th component.

EXAMPLE 9 *Consider the rational relational over the alphabet $\{0,1,2\}$ taken from (Sakarovitch 2009, Example 3.1,p. 687)*

$$R = \{(0^m 1^n, 0^m 1) \mid m, n \geq 1\} \cup \{(0^m 1^n, 0^n 2) \mid m, n \geq 1\}.$$

This relation is accepted by the automaton of Figure 3, where we use the standard notation x/y for pairs of input $x \in \Sigma \cup \{\varepsilon\}$ and output $y \in \Sigma \cup \{\varepsilon\}$ processed by a transition.

From the definitions, it is clear that automatic relations are special cases of rational relations. Furthermore, a word morphism which replaces every letter by a given word is an example of rational function. More generally, rational substitutions in which a letter is replaced by a word chosen from a regular language also define rational relations. As an example for such a rational substitution over the alphabet $\{0,1\}$, consider the mapping given by $0 \mapsto 0^+$ and $1 \mapsto 0^*10^*$. This defines a relation that relates a non empty word w to any word obtained by inserting 0s at arbitrary positions in w, which is easily seen to be rational. In general, using automata for the languages in the images of the substitution one can easily build an asynchronous automaton for the corresponding relation.

Contrarily to the case of automatic relations, the length-lexicographic uniformization of a rational relation is not in general a rational function. Consider for instance the relation R of Example 9. By ordering the alphabet in the natural way (i.e., $0 < 1 < 2$), the length-lexicographic uniformization of R is the function S defined for all $m, n \geq 1$, by $S(0^m 1^n) = 0^m 1$ if $m \leq n$ and $S(0^m 1^n) = 0^n 2$ otherwise. The function S is not a rational function as for instance the inverse image by S of the regular set $0^+ 1$ is the non-regular set $\{0^m 1^n \mid m \leq n\}$.

REMARK 10 *In Lombardy & Sakarovitch (2010), it is shown that the length-lexicographic uniformization of a rational relation in $\Sigma^* \times \Gamma^*$ with $|\Sigma| = 1$ is a rational function.*

Uniformization of rational relations is reduced to the automatic case using the following decomposition theorem which states that any rational relation whose domain does not contain the empty word can be expressed as the composition of an automatic relation followed by a rational substitution.

THEOREM 11 (Eilenberg (1974)) *For any rational relation R over an alphabet Σ whose domain does not contain the empty word, there exists an alphabet Γ, a length-preserving automatic relation $S \subseteq \Sigma^* \times \Gamma^*$ and a rational substitution $\rho \subseteq \Gamma^* \times \Sigma^*$ with $dom(\rho) = \Gamma^*$ such that:*

$$R = S \circ \rho := \{(u, v) \in \Sigma^* \times \Sigma^* \mid \exists w \in \Gamma^* : (u, w) \in S \text{ and } (w, v) \in \rho\}.$$

Proof. Let R be a rational relation over Σ whose domain does not contain the empty word. Consider an automaton $A = (Q, (\Sigma \cup \{\varepsilon\})^2, q_0, \Delta, F)$ accepting R. For all states p and $q \in Q$, we let $A_{p,q}$ denote the automata $(Q, (\Sigma \cup \{\varepsilon\})^2, p, \Delta, \{q\})$ where p is the new initial state and q the unique final state.

Let Γ be the alphabet $Q \times \Sigma \times Q$. Consider the length-preserving automatic relation $S \subseteq \Sigma^* \times \Gamma^*$ associating to a word $a_1 \cdots a_n$ with $n \geq 1$ any word $(q_0, a_1, q_1) \cdots (q_{n-1}, a_n, q_n)$ such that q_n belongs to F. The rational substitution ρ associates to (p, a, q) the image of the letter a by the relation accepted by $A_{p,q}$. It can be shown that $R = S \circ \rho$.

To guarantee that $dom(\rho) = \Gamma^*$, it is enough to ensure that for all $a \in \Gamma$, the language $\rho(a)$ is non empty. If it is not the case, we restrict ρ to the alphabet $\Xi = \{a \in \Gamma \mid \rho(a) \neq \emptyset\}$ and we restrict the image of S to Ξ^*. ∎

REMARK 12 *As a consequence of Theorem 11, we re-obtain the well-known result stating that all rational functions are unambiguous (i.e., accepted by* an automaton having at most one accepting run for each pair of the relation). Indeed it shows that any rational function is the composition of length-preserving automatic function and of a morphism. A length-preserving automatic function being accepted by a DFA labeled by $\Sigma \times \Sigma$ is an unambiguous rational function. The unambiguity is easily shown to be preserved in the composition with a morphism.

In conjunction with the uniformization result for automatic relations, we obtain the uniformization theorem for rational relations.

THEOREM 13 *Rational relations can be uniformized by rational functions.*

Proof. Let R be a rational relation. It is enough to consider the case where the domain of R does not contain the empty word.[1] By Theorem 11, there exists an alphabet Γ such that R can be expressed as $S \circ \rho$ where $S \subseteq \Sigma^* \times \Gamma^*$ is a length-preserving automatic relation and a rational substitution ρ with $dom(\rho) = \Gamma^*$ (and hence $dom(R) = dom(S)$).

By Theorem 2, S admits a length-preserving automatic uniformization T. Furthermore the rational substitution ρ is uniformized by a morphism φ such that for all $a \in \Gamma$, $\varphi(a) \in \rho(a)$. We have $T \circ \varphi \subseteq R$ and as $dom(\varphi) = \Gamma^*$, $dom(T \circ \varphi) = dom(T) = dom(S) = dom(R)$. ∎

Contrary to the case of automatic relations, the uniformization theorem cannot be extended to arity greater than 2. Consider for instance the following rational relation in $\{a,b\}^* \times \{a\}^* \times \{a\}^*$

$$\{(a^n b^m, a^n, a) \mid n, m \geq 0\} \cup \{(a^n b^m, a^m, \varepsilon) \mid m, n \geq 0\}$$

In (Choffrut & Grigorieff 1999, p. 7), it is shown by a pumping argument that this relation does admit any rational uniformization $f : \{a,b\}^* \times \{a\}^* \to \{a^*\}$. It is furthermore established that the class of rational relations in $\Gamma_1^* \times \cdots \times \Gamma_n^*$ for $n > 2$ enjoys the rational uniformization property if and only if all the Γ_i are unary alphabets. The converse implication follows from the fact that these relations which are essentially subsets of $\mathbb{N}^n$ are those relations definable in Presburger arithmetic. Their length-lexicographic uniformization is also definable by a Presburger formula and hence realizable by a rational relation.

Another notable difference with the automatic setting is that it is not known whether rational equivalence relations admit rational cross-sections Johnson (1986). Rational equivalence relations are more difficult to apprehend than their automatic counter-parts. For instance, it is undecidable whether a given rational relation is an equivalence relation Johnson (1986).

---

[1] The rational relation R can be decomposed into $R' \cup \{\varepsilon\} \times L$ where R' is a rational relation whose domain does not contain the empty word and L is a regular language.

3.2 Infinite words

Rational relations over infinite words are defined in the same way as for finite words using Büchi automata instead of standard finite automata. These relations are called ω-rational relations. For simplicity, we only consider rational relations whose domain only contains infinite words.

Using the same construction as in the finite word case, we can decompose an ω-rational relation into an ω-automatic relation followed by a rational substitution.

THEOREM 14 (Choffrut & Grigorieff (1999)) *For any ω-rational relation R over an alphabet Σ whose domain only contains infinite words, there exists an alphabet Γ, an ω-automatic relation $S \subseteq \Sigma^\omega \times \Gamma^\omega$ and a rational substitution $\rho \subseteq \Gamma^\omega \times \Sigma^\omega$ with $dom(\rho) = \Gamma^\omega$ such that $R = S \circ \rho$.*

As in the finite word case, the uniformization theorem for ω-rational relations follows from that of ω-automatic relations.

THEOREM 15 (Choffrut & Grigorieff (1999)) *ω-rational relations can be uniformized by ω-rational functions.*

4 Uniformization by sequential transducers

While in the previous section we considered the question whether relations from a given class have a uniformization within the same class, we now consider a setting in which the class of functions to choose the uniformization from is restricted.

As already mentioned, a uniformization of a relation can be viewed as a concrete implementation of a specification. The specification describes the admissible outputs for a given input, and the uniformization function selects one output for each input. In this section, we consider the setting where the output has to be constructed deterministically by a finite state device that reads the input letter by letter and can output finite words in each step. In the classical setting, this problem has been studied for infinite words, going back to a problem posed by Church Church (1962) that has been solved by Büchi and Landweber in Büchi & Landweber (1969) (see Thomas (2009) for a recent overview on this subject). For the setting of finite words, we use a standard transducer model, which basically can be seen as the subclass of asynchronous automata from Section 3 which are deterministic on their input.

A subsequential[2] transducer (ST) is of the form $\mathbf{T} = (S, \Sigma, \Gamma, s_0, \delta, F, f)$ where S is the finite set of states, Σ and Γ are the input and output alphabet,

---

[2]The prefix "sub" is added for transducers that can make a final output depending on the last state reached in a run, as opposed to sequential transducers that can only produce outputs on their transitions.

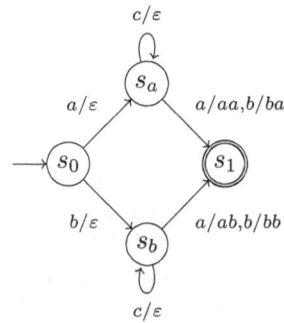

Figure 4. A subsequential transducer.

respectively, $s_0 \in S$ is the initial state, $\delta : S \times \Sigma \to S \times \Gamma^*$ is the transition function, $F \subseteq S$ is a set of final states, and $f : F \to \Gamma^*$ is the final output function. Such an ST behaves as a standard deterministic finite automaton but additionally produces a finite (possibly empty) output word in each transition. An input $u \in \Sigma^*$ is accepted if $s \in F$ for the state s reached after reading u. By $\boldsymbol{T}(u)$, we denote the output of $\boldsymbol{T}$ produced along the transitions while reading u (independent of whether u is accepted or not), and by $\boldsymbol{T}_f(u) := \boldsymbol{T}(u) \cdot f(s)$ the complete output including the final one at the last state s (if $s \notin F$, then $\boldsymbol{T}_f(u)$ is undefined). For a detailed introduction to this subject and functions definable by STs, we refer the reader to Berstel (1979).

EXAMPLE 16 *Figure 4 shows the transition graph of an ST $\boldsymbol{T}$ (where the notation $s \xrightarrow{a/u} s'$ denotes $\delta(s,a) = (s',u)$). We define the final output to be ε for the only final state s_1. Furthermore, we assume that the missing transitions lead to a rejecting sink state. The function defined by $\boldsymbol{T}$ has the domain $\{a,b\}c^*\{a,b\}$ with $\boldsymbol{T}(xc^*y) = \boldsymbol{T}_f(xc^*y) = yx$ for $x, y \in \{a,b\}$.*

While it is decidable whether a rational function can be defined by an ST (see (Berstel 1979, Theorem 6.2)), our first result shows uniformization of rational relations by STs is undecidable.

THEOREM 17 *It is undecidable whether a given rational relation has a uniformization by a subsequential transducer.*

Proof. We sketch a reduction from the halting problem for Turing machines (TM). Given such a TM M, we describe a rational relation R_M. The interesting cases are those pairs (u, v) in which the first component is of the

form
$$u = c_1\$c_2\$\cdots\$c_n\#^*X$$

where \$ and \# are special symbols in the alphabet, each c_i is a configuration of M (coded as a word in a standard way), c_1 is the initial configuration of M on the empty tape, c_n is a halting configuration of M, and $X \in \{A, B\}$ is a letter that determines how the word in the second component has to look like. If u is of this form, we say that it codes a configuration sequence.

If u is does not code a configuration sequence, then every word v with $|v| = |u|$ is allowed in the second component. If u codes a configuration sequence and is ending in A, then $(u, v) \in R_M$ if, and only if, $u = v$. If u codes a configuration sequence and is ending in B, then $(u, v) \in R_M$ if, and only if, v is of the form

$$v = c'_1\$c'_2\$\cdots\$c'_n\#^*B$$

such that c'_{i+1} is not the successor configuration of c_i for some $i \in \{1, \ldots, n-1\}$.

First note that this defines a rational relation. An asynchronous automaton can guess at the beginning whether u codes a configuration sequence, and whether it ends in A or B. If it ends in A, then the automaton synchronously checks whether $u = v$, and if it ends in B, then the automaton guesses a c_i and asynchronously verifies that c'_{i+1} is not the successor configuration of c_i. To check this it advances to the next configuration in the output and compares c'_{i+1} and c_i.

We claim that R_M can be uniformized by an ST if, and only if, M does not halt. If M does not halt, then the ST that simply reproduces the input is a uniformization of R_M for the following reason: The only case to verify is the one where u codes a configuration sequence and ends in B. But since M does not halt, the configuration sequence in u must contain two configurations c_i and c_{i+1} such that c_{i+1} is not the successor configuration of c_i. Since $c'_{i+1} = c_{i+1}$, the condition is satisfied.

Now assume that M does halt and that there is an ST **T** that uniformizes R_M. Let k be the maximal length of an output string on the transitions of **T**. Now consider an input u that codes the halting configuration sequence, followed by k times $\#$. Since the next input letter could be an A, **T** must have already reproduced the configuration sequence because it can produce at most k output letters on the last transition. But then the output does not satisfy the condition if the next input letter is B. ∎

The rational relation constructed in the reduction in the proof of Theorem 17 is not automatic, and in fact one can show that the problem becomes decidable when restricted to automatic relations.

THEOREM 18 *It is decidable whether a given automatic relation has a uniformization by a subsequential transducer.*

The proof of this result uses techniques similar to Holtmann et al. (2010) where a similar problem on infinite words is studied: Given an automatic relation R of infinite words, decide if there is a sequential transducer such that for each input α in the domain of R, the computed output β is such that $(\alpha, \beta) \in R$.

At first glance, it might seem that this problem is more general because it is studied in the setting of infinite words, and finite words can be coded by infinite words using a dummy letter that is appended to the finite word. However, in the setting studied in Holtmann et al. (2010), it is assumed that the sequential transducer produces an infinite output for each possible input. As a consequence, one can show that if there is a uniformization by a sequential transducer, then there is one of bounded delay, which means that the difference between the length of the processed input and the produced output is globally bounded.

In the setting of finite words that we study, this needs not to be true. Consider the following relation over the alphabet $\{a, b, c\}$, specified slightly informally using pairs of regular expressions, representing the Cartesian product of the respective languages:

$$(ac^*b, bc^*a) \cup (bc^*a, ac^*b) \cup (ac^*a, ac^*a) \cup (bc^*b, bc^*b) .$$

It is not difficult to verify that this is an automatic relation. It is uniformized by the ST from Example 16. This ST has arbitrarily long delays between input and output, and this cannot be avoided because the first letter of the output depends on the last letter of the input.

However, the key insight for the decidability proof is that if such a long delay is necessary, then the connection between the remaining input and output cannot be very complex. This intuition of not being very complex is captured by the notion of recognizable relation. A relation $R \subseteq \Sigma^* \times \Gamma^*$ is called recognizable if it is of the form

$$R = \bigcup_{i=1}^{n} (U_i \times V_i)$$

for regular sets $U_i \subseteq \Sigma^*$ and $V_i \subseteq \Gamma^*$. From the definition, it is obvious that the above example is a recognizable relation. It is easy to verify that a relation R is recognizable if, and only if, there is a finite automaton that reads two words u and v sequentially (e.g., as $u\$v$ separated by a unique marker \$), and accepts if $(u, v) \in R$. It directly follows that a recognizable relation can be uniformized by an ST that first scans the entire input and then outputs some word in the matching set of output words (as the one in Figure 4 does).

Before making the link between large delays in the output and recognizable relations, we need some notations. For the remainder of this section, fix an automatic relation $R \subseteq \Sigma^* \times \Gamma^*$ recognized by a DFA $\mathcal{A} = (Q, (\Sigma \cup \{\Box\}) \times (\Gamma \cup \{\Box\}), q_0, \delta, F)$. For simplicity, we assume that $dom(R) = \Sigma^*$. Since $dom(R)$ is a regular language and an ST can test membership in regular languages, this assumption is not a restriction.

For $u \in \Sigma^*$ and $q \in Q$, let

$$R_q^u := \{(ux, v) \mid \mathcal{A} : q \xrightarrow{(ux) \otimes v} F\}$$

be the set of pairs that are accepted from q and where the input component starts with u. For R_q^ε we write R_q, and for $R_{q_0}^u$ we write R^u.

Our aim is to show that if R can be uniformized by an ST, then there is a bound on the delay between the input and the output, or the remaining relation can be uniformized by a recognizable one. This bound will be chosen such that an input word of this length contains some idempotent factor w.r.t. to some monoid structure that we define in the following.

For each input word u, we are interested in the types of behavior of $\mathcal{A}$ that can be induced by the input u together with some output (of same or smaller length). Hence, for each $v \in \Gamma^*$ with $|u| \geq |v|$, we consider the function (also called state transformation) $\tau_{u,v} : Q \to Q$ defined by $\tau_{u,v}(q) = p$ if $\mathcal{A} : q \xrightarrow{u \otimes v} p$.

The profile P_u of a word u contains the set of possible state transformations for the different types of words in the second component (of same length, shorter, or empty). More formally, $P_u = (P_{u,=}, P_{u,<}, P_{u,\varepsilon})$ with

$$P_{u,=} := \{\tau_{u,v} \mid v \in \Gamma^* \text{ and } |v| = |u|\},$$

$$P_{u,<} := \{\tau_{u,v} \mid v \in \Gamma^+ \text{ and } |v| < |u|\},$$

$$P_{u,\varepsilon} := \{\tau_{u,v} \mid v = \varepsilon\} \text{ (this set only contains one function but for consistency of notation, we prefer this definition).}$$

It is not difficult to see that from the profiles of two words u and u' one can compute the profile of uu'. Hence, the set of profiles is naturally equipped with a concatenation operation and a neutral element (the profile of the empty word), and the mapping that assigns the profile to a word u is a morphism from Σ^* to the profile monoid. A word u is called idempotent if $P_u = P_{uu}$, that is, if the corresponding monoid element is idempotent.

A consequence of Ramsey's Theorem (see Diestel (2000)) is that there is some $K \in \mathbb{N}$ such that all words $u \in \Sigma^*$ with $|u| \geq K$ contain an idempotent factor. With this in mind, the following lemma is a technical statement formalizing the intuition between long output delays and recognizable relations.

LEMMA 19 *Let $q \in Q$ and $u, v \in \Sigma^+$ with v idempotent. If R_q^{uv} is uniformized by an ST $\boldsymbol{T}$ such that $|\boldsymbol{T}(uv^n)| \leq |u|$ for all $n \in \mathbb{N}$, then R_q^{uv} can be uniformized by a recognizable relation.*

Proof. Consider an arbitrary word $w \in \Sigma^*$. We show that there is $x \in \Gamma^*$ whose length only depends on u and v such that $(uvw, x) \in R_q^{uv}$. For fixed u and v, there are only finitely many such words x and thus, R_q^{uv} can be uniformized by a recognizable relation (which can be shown using the technique from the proof of Proposition 20 below).

Since $|\boldsymbol{T}(uv^n)| \leq |u|$ for each n, the length of $\boldsymbol{T}(uv^n w)$ is independent of n for large n. We can thus choose n such that $|\boldsymbol{T}(uv^n w)| \leq |uv^n|$. Let $y = \boldsymbol{T}(uv^n w)$. We now consider the factorization $y = y'y''$ such that $|y'| = |u|$. Since v is idempotent, $P_v = P_{v^n}$, and thus there is some $z \in \Gamma^*$ with $|z| \leq |v|$ such that $\tau_{v,z} = \tau_{v^n, y''}$. Letting $x = y'z$, we obtain that the pair (uvw, x) induces the same state transformation on $\mathcal{A}$ as $(uv^n w, y)$ and hence $(uvw, x) \in R_q^{uv}$. ∎

Lemma 19 basically shows us that we can focus on the construction of STs in which the output delay is bounded. Once the output delay has to be larger than this bound, the uniformization task is either impossible or very simple (reduced to a recognizable relation). The following proposition shows that we can decide in which cases uniformization by a recognizable relation is possible.

PROPOSITION 20 *It is decidable whether an automatic relation can be uniformized by a recognizable relation.*

Proof. Let $R \subseteq \Sigma^* \times \Gamma^*$ be an automatic relation and let

$$V := \{v \in \Gamma^* \mid \exists u \in \Sigma^* : (u,v) \in R \text{ and } \forall v' \in \Gamma^* : (u,v') \in R \to v \leq_{\text{llex}} v'\}$$

be the set of words that are the length-lexicographically least output for some input. It is easy to verify that V is a regular set (an automaton for V can be constructed from an automaton for R using the closure properties of finite automata). We claim that R can be uniformized by a recognizable relation if, and only if, V is finite (and since V is regular, finiteness of V can be decided).

If V is finite, then for each $v \in V$, let U_v consist of those words $u \in \Sigma^*$ such that v is the length-lexicographically minimal word with $(u,v) \in R$. Then U_v is regular and $\bigcup_{v \in V} U_v \times \{v\}$ is a recognizable uniformization of R.

If $\bigcup_{i=1}^{n}(U_i \times \{v_i\})$ is a uniformization of R by a recognizable relation (in fact, a recognizable function), then consider the set

$$W := \{w \in \Gamma^* \mid w \leq_{\text{llex}} v_i \text{ for some } i \in \{1, \ldots, n\}\}.$$

Since $\leq_{\text{llex}}$ is a well-ordering, the set W is finite. Thus, V must also be finite because $V \subseteq W$. ∎

We are now ready to describe the decision procedure. Similar to Holtmann et al. (2010), we consider a game between two players Input (In) and Output (Out). The game is played on a game graph such that In plays an input symbol and Out can react with a finite (possibly empty) sequence of output symbols. Player In wins the game if for the input sequence u that he has played so far, Out cannot extend her current output sequence such that the resulting pair is in R. Our goal is to construct the game graph such that an ST uniformizing R can be obtained from a winning strategy of Out.

The vertices of the game graph keep track of the current state of $\mathcal{A}$ on the combined part of the input and output, and possibly of the part of the input that is currently ahead. Making use of Lemma 19 (see the proof of Lemma 21 below), we can restrict the game to situations in which the input is ahead at most $2K$ steps (where K is such that words of length at least K contain an idempotent factor, see the description before Lemma 19). It turns out that we do not have to consider the case in which the output is ahead.

The game graph $G_{\mathcal{A}}^K$ consists of vertices for In and Out and a set of edges corresponding to the possible moves of the two players:

$V_{\text{In}} := \{(q, u) \in Q \times \Sigma^* \mid |u| \leq 2K\}$ is the set of vertices of player In.

$V_{\text{Out}} := V_{\text{In}} \times \{\text{Out}\}$ is the set of vertices of player Out.

From a vertex of In, the following moves are possible:

- $(q, u) \xrightarrow{a} (q, ua, \text{Out})$ if $|u| < 2K$ and $a \in \Sigma$

From a vertex of Out, the following moves are possible:

- $(q, u, \text{Out}) \xrightarrow{b} (q', u', \text{Out})$ for each $b \in \Gamma$ such that $u = au'$ for $a \in \Sigma$ and $q' = \delta(q, (a, b))$,

- $(q, u, \mathsf{Out}) \xrightarrow{\varepsilon} (q, u)$,

The initial vertex is (q_0, ε).

The winning condition should express the following property: at each point, In can extend the output such that the resulting pair of input and output is in R. For this purpose, we define a set B of bad vertices for player Out consisting of

1. all $(q, u) \in V_{\mathsf{Out}}$ such that $|u| < 2K$ and R_q^u is empty, and

2. all $(q, u) \in V_{\mathsf{Out}}$ such that $|u| = 2K$ and R_q^u cannot be uniformized by a recognizable relation.

Note that both conditions are decidable for a given vertex ((1) is emptiness of automatic relations and (2) is decidable by Proposition 20).

The objective of Out is to avoid the vertices in B. Games with an objective of this kind are usually referred to as safety games because the player has to stay within the safe region of the game graph.

A play is a maximal path in $G_{\mathcal{A}}^K$ starting in the initial vertex. Maximal means that the path is either infinite or it ends in a vertex without outgoing edges (a vertex (q, u) with $|u| = 2K$). Out wins a play if no vertex from B occurs. A strategy for Out is a function that defines for finite sequences of moves that end in a vertex of Out the next move to be taken by Out. Such a strategy is winning if Out wins all plays in which she makes her moves according to the strategy.

The following lemma reduces our question to the existence of winning strategies in $G_{\mathcal{A}}^K$.

LEMMA 21 *The relation R can be uniformized by an ST if, and only if, Out has a winning strategy in $G_{\mathcal{A}}^K$.*

Proof. Assume that Out has a winning strategy in $G_{\mathcal{A}}^K$. Since $G_{\mathcal{A}}^K$ is a safety game, the player who has a winning strategy also has a positional one, which means that the next move chosen by the strategy only depends on the current vertex (see Grädel et al. (2002)). Such a strategy can be represented by a function $\sigma : V_{\mathsf{Out}} \to \Gamma \cup \{\varepsilon\}$ (because the moves of Out are deterministically labeled by letters in Γ or by ε).

We now describe how to construct the ST **T** that uniformizes R. Consider a pair (q, u) such that $|u| = 2K$ and R_q^u can be uniformized by a recognizable relation. For each such pair we choose an ST $\mathbf{T}_q^u$ that uniformizes R_q^u. Then **T** consists of the (disjoint) union of the transducers $\mathbf{T}_q^u$ and a part that uses V_{In} as states. The initial state of $\mathbf{T}_q^u$ is identified with $(q, u) \in V_{\mathsf{In}}$. The transitions of **T** for some $(q, u) \in V_{\mathsf{In}}$ with $|u| < 2K$ is defined as follows.

Let $a \in \Sigma$. The strategy σ defines a unique finite sequence of moves of Out from (q, ua, Out). This sequence of moves corresponds to some finite word $w \in \Gamma^*$ and ends in a vertex (p, u'). We define $\delta_{\mathbf{T}}((q, u), a) := ((p, u'), w)$. Furthermore, we define the final output function f of $\mathbf{T}$ by $f(q, u) := v$ for some v such that $(u, v) \in R_q^u$, which exists since σ is a winning strategy and thus avoids all vertices in B. It is not difficult to verify that $\mathbf{T}$ indeed defines a uniformization of R.

For the other direction, assume that R is uniformized by some ST $\mathbf{T}$. A winning strategy for Out basically simulates $\mathbf{T}$ on the inputs played by In. However, it might happen that the output delay in $\mathbf{T}$ is larger than $2K$ or that for some input sequences the output sequence produced by $\mathbf{T}$ might be longer than the input sequence. These cases are not captured by the game graph.

To describe the strategy, we split the sequence of moves by In (simply referred to as input sequence) into blocks $u_i \in \Sigma^*$ of length K. So the current input sequence is always of the form $u_1 \cdots u_n u$ with $|u_i| = K$ and $|u| < K$. The strategy produces its output moves that are different from the ε-move in blocks of K. For the input $u_1 \cdots u_n u$, it will have produced output moves $v_1 \cdots v_{n-1}$ with $|v_i| = K$. This means that the corresponding vertex in the game graph is $(q_{n-1}, u_n u)$ with $q_{n-1} = \delta(q_0, (u_1 \cdots u_{n-1}) \otimes (v_1 \cdots v_{n-1}))$.

The output moves producing v_n are played once a vertex $(q_{n-1}, u_n u_{n+1}, \mathsf{Out})$ is reached (with $|u_n u_{n+1}| = 2K$). To define v_n, the ST $\mathbf{T}$ is not simulated on the original input sequence $u_1 \cdots u_n u_{n+1}$ but on a modification $u'_1 \cdots u'_n u'_{n+1}$ that is obtained by repeating some idempotent factors as follows: We let $u'_1 = u_1$. Now assume that u'_i is defined for all $1 \leq i \leq n$. Let $u_{n+1} = xyz$ with $y \neq \varepsilon$ idempotent. Then $u'_{n+1} = xy^m z$ for some m such that $|\mathbf{T}(u'_1 \cdots u'_{n+1})| \geq |u'_1 \cdots u'_n|$. If such an m does not exist, then Lemma 19 implies that $R_{q_{n-1}}^{u_n u_{n+1}}$ can be uniformized by a recognizable relation. Then Out can move to $(q_{n-1}, u_n u_{n+1})$ and wins.

Given this definition of the u'_i, we define v_n as follows: Let $v'_1 \cdots v'_n$ be the initial part of $\mathbf{T}(u'_1 \cdots u'_{n+1})$ such that $|v'_i| = |u'_i|$. Since u'_n is obtained from u_n by repeating an idempotent factor, the profiles of u_n and u'_n are the same and thus, there is some v_n such that $u_n \otimes v_n$ induces the same state transformation on $\mathcal{A}$ as $u'_n \otimes v'_n$. We pick such a v_n, and σ makes K moves from $(q_{n-1}, u_n u_{n+1}, \mathsf{Out})$ according to the letters in v_n, leading to some $(q_n, u_{n+1}, \mathsf{Out})$, and then takes the ε-move to (q_n, u_{n+1}).

To show that this defines a winning strategy for Out, it suffices to show that (q_n, u_{n+1}) is not in B. Consider $\mathbf{T}_f(u'_1 \cdots u'_{n+1})$, which is of the form $v'_1 \cdots v'_n v'$. Since $\mathbf{T}$ uniformizes R, we know that $(u'_1 \cdots u'_{n+1}, v'_1 \cdots v'_n v') \in R$. Because u'_{n+1} is obtained from u_{n+1} by repeating an idempotent factor, there is some v such that $u_{n+1} \otimes v$ induces the same state transformation on $\mathcal{A}$ as $u'_{n+1} \otimes v'$. In combination with the choice of the v_i, we obtain that $u'_1 \cdots u'_{n+1} \otimes v'_1 \cdots v'_n v'$ and $u_1 \cdots u_{n+1} \otimes v_1 \cdots v_n v$ induce the same state transformation in $\mathcal{A}$ and therefore, $(u_1 \cdots u_{n+1}, v_1 \cdots v_n v) \in R$ and $(u_{n+1}, v) \in R_{q_n}^{u_{n+1}}$. This shows that (q_n, u_{n+1}) is not in B. ∎

Theorem 18 follows from Lemma 21 and the fact that a winning strategy for Out can effectively be computed in $G_{\mathcal{A}}^K$ (see Grädel et al. (2002)).

5 Conclusion

In this paper, we have given an overview of uniformization results for relations defined by various automaton models. Automatic relations can be uniformized by automatic functions for relations defined over finite words and trees, as well as for relations over infinite words. For infinite trees, the uniformization fails, for example for the element relation.

For rational relations, uniformization can be shown by a reduction to automatic relations using a composition theorem. This technique works for finite as well as infinite words.

Concerning the uniformization of relations over finite words by subsequential transducers, we have presented a decidability result for automatic relations and an undecidability result for rational relations. It turns out that compared to the case of automatic relations over infinite words, some new phenomena arise because of the possible length difference in the input and output.

One direction for future research is the uniformization of tree relations beyond automatic relations. For example, there is no canonical adaption of the notion of rational relations. However, there are various models of transducers for defining relations over finite trees (see Comon et al. (2007) and Raoult (1997)), for which uniformization questions can be studied.

Bibliography

Arnold, A. & Latteux, M. (1979). A new proof of two theorems about rational transductions. *Theoretical Computer Science*, 8(2), 261–263, doi:10.1016/0304-3975(79)90049-5.

Berstel, J. (1979). *Transductions and Context-Free Languages*. Stuttgart: Tebuner, electronic edition available via the homepage of the author.

Blumensath, A. & Grädel, E. (2000). Automatic structures. In *Proceedings of the 15th IEEE Symposium on Logic in Computer Science, LICS 2000*, IEEE Computer Society Press, 51–62.

Büchi, J. R. & Landweber, L. H. (1969). Solving sequential conditions by finite-state strategies. *Transactions of the American Mathematical Society*, *138*, 295–311.

Carayol, A. & Löding, C. (2007). MSO on the infinite binary tree: Choice and order. In *Proceedings of the 16th Annual Conference of the European Association for Computer Science Logic, CSL 2007, Lecture Notes in Computer Science*, vol. 4646, Springer, 161–176.

Carayol, A., Löding, C., et al. (2010). Choice functions and well-orderings over the infinite binary tree. *Central European Journal of Mathematics*, *8*(4), 662–682, doi:10.2478/s11533-010-0046-z.

Choffrut, C. & Grigorieff, S. (1999). Uniformization of rational relations. In *Jewels are Forever, Contributions on Theoretical Computer Science in Honor of Arto Salomaa*, Springer, 59–71.

Church, A. (1962). Logic, arithmetic and automata. In *Proceedings of the International Congress of Mathematicians*, 23–35.

Colcombet, T. & Löding, C. (2007). Transforming structures by set interpretations. *Logical Methods in Computer Science*, *3*(2), 1–36, doi:10.2168/LMCS-3(2:4)2007.

Comon, H., Dauchet, M., et al. (2007). *Tree Automata Techniques and Applications*. Available on http://tata.gforge.inria.fr/, last Release: October 12, 2007.

Diestel, R. (2000). *Graph Theory*. New York: Springer, 2nd edn.

Eilenberg, S. (1974). *Automata, Languages and Machines*, vol. A. New York: Academic Press.

Elgot, C. C. & Mezei, J. E. (1965). On relations defined by generalized finite automata. *IBM Journal of Research and Development*, *9*(1), 47–68, doi: 10.1147/rd.91.0047.

Engelfriet, J. (1978). On tree transducers for partial functions. *Information Processing Letters*, *7*(4), 170–172.

Grädel, E., Thomas, W., & Wilke, T. (Eds.) (2002). *Automata, Logics, and Infinite Games, Lecture Notes in Computer Science*, vol. 2500. Berlin; New York: Springer.

Gurevich, Y. & Shelah, S. (1983). Rabin's uniformization problem. *Journal of Symbolic Logic*, *48*(4), 1105–1119, doi:10.2307/2273673.

Holtmann, M., Kaiser, L., & Thomas, W. (2010). Degrees of lookahead in regular infinite games. In *Foundations of Software Science and Computational Structures, Lecture Notes in Computer Science*, vol. 6014, Springer, 252–266.

Hopcroft, J. E. & Ullman, J. D. (1979). *Introduction to Automata Theory, Languages, and Computation.* Reading, MA: Addison Wesley.

Hosch, F. A. & Landweber, L. H. (1972). Finite delay solutions for sequential conditions. In *ICALP*, 45–60.

Johnson, J. H. (1986). Rational equivalence relations. *Theoretical Computer Science*, 47(3), 39–60, doi:10.1016/0304-3975(86)90132-5.

Khoussainov, B. & Nerode, A. (1995). Automatic presentations of structures. In *Logical and Computational Complexity. Selected Papers. Logic and Computational Complexity, International Workshop LCC '94, Indianapolis, Indiana, USA, 13-16 October 1994, Lecture Notes in Computer Science*, vol. 960, Berlin; Heidelberg: Springer, 367–392.

Kobayashi, K. (1969). Classification of formal languages by functional binary transductions. *Information and Control*, 15(1), 95–109, doi:10.1016/S0019-9958(69)90651-2.

Kuske, D. & Lohrey, M. (2006). First-order and counting theories of *omega*-automatic structures. In *Foundations of Software Science and Computation Structures, 9th International Conference, FOSSACS 2006, Proceedings, Lecture Notes in Computer Science*, vol. 3921, Springer, 322–336.

Kuske, D. & Weidner, T. (2011). Size and computation of injective tree automatic presentations. In *Mathematical Foundations of Computer Science 2011 – 36th International Symposium, MFCS 2011, Proceedings, Lecture Notes in Computer Science*, vol. 6907, Springer, 424–435.

Löding, C. (2011). Infinite games and automata theory. In *Lectures in Game Theory for Computer Scientists*, Apt, K. R. & Grädel, E., eds., Cambridge: Cambridge University Press.

Löding, C. (2012). Basics on tree automata. In *Modern Applications of Automata Theory*, D'Souza, D. & Shankar, P., eds., Singapore: World Scientific.

Lombardy, S. & Sakarovitch, J. (2010). Radix cross-sections for length morphisms. In *LATIN 2010: Theoretical Informatics, 9th Latin American Symposium, Oaxaca, Mexico, April 19-23, 2010. Proceedings*, 184–195.

Moschovakis, Y. N. (1980). *Descriptive Set Theory, Studies in Logic and the Foundations of Mathematics*, vol. 100. Amsterdam; New York; Oxford: North-Holland Publishing Company.

Raoult, J.-C. (1997). Rational tree relations. *Bulletin of the Belgian Mathematical Society*, 4(1), 149–176.

Sakarovitch, J. (2009). *Elements of Automata Theory.* Cambridge; New York: Cambridge University Press.

Siefkes, D. (1975). The recursive sets in certain monadic second order fragments of arithmetic. *Archiv für mathematische Logik und Grundlagenforschung*, *17*(1–2), 71–80, doi:10.1007/BF02280817.

Thomas, W. (1997). Languages, automata, and logic. In *Handbook of Formal Language Theory*, vol. III, Rozenberg, G. & Salomaa, A., eds., Berlin; New York: Springer, 389–455.

Thomas, W. (2008). Church's problem and a tour through automata theory. In *Pillars of Computer Science, Essays Dedicated to Boris (Boaz) Trakhtenbrot on the Occasion of His 85th Birthday, Lecture Notes in Computer Science*, vol. 4800, Berlin; Heidelberg: Springer, 635–655.

Thomas, W. (2009). Facets of synthesis: Revisiting Church's problem. In *Proceedings of the 12th International Conference on Foundations of Software Science and Computational Structures, FOSSACS 2009, Lecture Notes in Computer Science*, vol. 5504, Springer, 1–14.

Arnaud Carayol
Laboratoire d'Informatique Gaspard Monge,
Université Paris-Est & CNRS
France
arnaud.carayol@univ-mlv.fr

Christof Löding
RWTH Aachen, Informatik 7
Aachen
Germany
loeding@informatik.rwth-aachen.de

On the Church-Turing Thesis and Relative Recursion

YIANNIS N. MOSCHOVAKIS

1 Introduction

The *Church-Turing Thesis* is the claim that for every function $f : \mathbb{N}^n \to \mathbb{N}$ on the natural numbers $\mathbb{N} = \{0, 1, \ldots\}$,

CT: *f is computable* $\iff$ *f can be computed by a Turing machine*.

It implies that for every relation R on the natural numbers or (via an effective coding) on the set Λ^* of *strings* from some finite alphabet Λ,

     R is decidable $\iff$ *its characteristic function is Turing computable*.

CT was postulated (in different but equivalent forms) by (Church 1935; 1936b;a) and (Turing 1936), who applied it immediately to prove the *undecidability of provability in first order logic*. This solved the classical *Entscheidungsproblem*—or showed that it was "unsolvable", depending on your point of view; and similar invocations of CT have been used to establish some of the most important applications of logic to mathematics and computer science in the 20th century, including the unsolvability of Hilbert's 10th problem by Matiyasevich (after Davis, Putnam and Robinson), the construction of finitely generated, finitely presented groups whose word problem is unsolvable (Novikov and Boone), etc.

There was some initial scepticism (cf. (Moschovakis 1968)), but there is no doubt that the Church-Turing Thesis is almost universally accepted today, so much so that it is usually invoked with no explicit mention. At the same time, foundational questions about its epistemological status and *what exactly it means* continue to generate considerable discussion: is it a "logical", a "mathematical" or an "empirical truth"? And more significantly: can it be *proved* from simple, plausible assumptions?[1]

It is not my purpose here to give another proof of CT or a critical analysis of the arguments that have been given for it—a daunting task, given the vast material published on the problem. My aim is to introduce and discuss the *Thesis on Relative Recursion* RRT, a principle which is related to but very different from CT, not only in content but in kind; and to show that CT can be reduced to the conjunction of RRT and a nearly universally accepted view of *what the natural numbers are*. The move shifts the discussion about the meaning and truth of CT to the corresponding questions about RRT, which (I think) are substantially simpler.

The brief, last paragraph 8.1 summarizes what I believe is achieved in this article.

1.1 About algorithms

It is often assumed that CT is equivalent to

CT^*: $f : \mathbb{N}^n \to \mathbb{N}$ *is computable by an algorithm*
$\iff f$ *can be computed by a Turing machine.*

This is, in fact, how (Church 1936*b*) formulates CT, although the earlier (Church 1935) and (Turing 1936) do not mention algorithms.

We have direct intuitions about *algorithms* which can be brought to bear in discussing CT, as it is natural to assume that

(1) *if some algorithm computes a function f, then f is computable.*

However, it can be (and has) been argued that we also have independent, direct intuitions about functions on the natural numbers which are *computable by finite means* in Turing's words. In other words, it may well be that CT and CT^* are both true but they do not have the same meaning, and so arguments in favor of one of them do not necessarily apply to the other. For example: arguments for CT often depend on assumptions about the (natural) *primitives of computation*, while arguments for CT^* are naturally grounded on explications of *what algorithms are*, which is a difficult (and controversial) subject.

I take (1) to be obviously true, if it is understood correctly, and so algorithms will come up naturally and often in the sequel. However: I take CT to be the "official" version of the Church-Turing Thesis, and I will refrain

---

[1] Cf. (Gandy 1980; 1995), (Sieg 2002), (Kripke 2000) (which is a recording), (Dershowitz & Gurevich 2008) and the large bibliographies in these papers.

from discussing in any serious way the difficult problem of explicating the notion of algorithm; this is not what this article is about.[2]

2 Preliminary remarks,

on three issues which bear on our understanding of the Church-Turing Thesis.

2.1 The primitives of computation

(Turing 1936) starts with

> the "computable" numbers may be described briefly as the real numbers whose expressions as a decimal are calculable by finite means,

and then he argues (mostly) in the last Section 9 that his "computing machines" capture the natural notion of "calculability". His reasoning is driven by the following statement in the first paragraph of Section 9:

> The real question at issue is "What are the possible processes which can be carried out in computing a [real] number?"

The elementary operations that Turing machines can do involve manipulating tapes with symbols on them, and they are most directly understood as operations on (finite) *strings of symbols*. Turing argues that all *immediate* string operations (which intuitively can be effected in one step) can be simulated in finitely many steps by those basic, Turing machine operations. His arguments "are bound to be, fundamentally, appeals to intuition", he says, "and for this reason rather unsatisfactory mathematically". They are, however, very persuasive and were pivotal in securing the quick acceptance of **CT**. Gandy's seminal 1980 calls Turing's analysis an "outline of a proof" of[3]

---

[2] I think that the problem of giving a rigorous, mathematical definition of *algorithms* is very important for the foundations of the theory of computation and has not received the attention it deserves. For my own ideas about it, see (Moschovakis 1998), and for alternative proposals cf. (Gurevich 2000), (Dershowitz & Gurevich 2008) and (Tucker & Zucker 2000).

[3] (Gandy 1980) gives an alternative understanding of **CT** which limits computability by arbitrary *mechanical devices*, and then gives an explication of what these are and bases on it a proof of **CT**. Physics enters the picture and **CT** becomes an *empirical proposition*, burdened by the usual problems: what if, in some distant future, someone builds a *Higgs boson machine* which can use God as an oracle and get answers to arbitrary mathematical questions, perhaps by doing subtle experiments? (Kripke 2000) argues (in a more serious vein) that the empirical understanding of **CT** is problematic and impossible to settle without a great deal more knowledge about physical laws than physicists have today. He concludes that **CT** is most coherently understood as a *mathematical proposition* and

Theorem T. What can be calculated by an abstract human being working in a routine way is [Turing] computable.

2.2 Symbolic computation

Built into this picture of the "mindless clerk" scribbling away is the principle that *all computation is symbolic*, which is why the *primitives of computation* are assumed to be operations on strings of symbols. This is a plausible (and popular) slogan which, however, merits some discussion. We will return to it further down.

2.3 Input and output

(Church 1936b) formulates CT in the form CT* above, at least by the words he uses:

> ...every function, an algorithm for the calculation of the values of which exists, is effectively calculable [$\sim$ Turing computable].

He then goes on to explain that for a function $F(n)$ of one positive integer,

> an algorithm consists in a method by which, given any positive integer n, a sequence of expressions (in some notation) $E_{n1}, E_{n2}, \ldots, E_{nr_n}$ can be obtained; ...[and in the end] the fact that the algorithm has terminated becomes effectively known and the value of $F(n)$ is effectively calculable. ...If this interpretation or some similar one is not allowed, it is difficult to see how the notion of an algorithm can be given any exact meaning at all.

So Church's rather restricted understanding of "algorithms" invokes again this "all computation is symbolic" principle, albeit somewhat more loosely than Turing's. But I want to stop here on this innocent

> given any positive integer n ...

Exactly how is a positive integer n *given*? Perhaps Church has in mind the rather complex *numeral* for n of the untyped λ-calculus or, more likely, the term $S^n(0)$ in Herbrand-Gödel-Kleene systems of equations, the unary representation of n. But why not use binary notation, as is routinely done today? Or, for that matter, why not "give n" to the algorithm by coding the value $F(n)$ in the syntactic expression E_{n1}, rendering all further computation redundant?

---

outlines a proof of it, elaborating on arguments which are (at least implicit) in (Turing 1936) and (Church 1936b). I will not go into this reasoning here, but I also understand CT as a mathematical proposition on more basic grounds: *it refers essentially to the natural numbers*, and so its truth or falsity depends on what they are.

The joke is old and worn out, but it makes the point: if we are to understand all computation as symbolic, then in addition to identifying the *string primitives* which our (human or mechanical) computing machine can call directly, we must also specify an *input function* which turns a "given n" into a string and also an *output function* which decodes the required value from the contents of the tape when the machine stops; and to prove CT, we must argue that these input and output functions are "effective"—in fact "immediately effective"—without benefit of the Church-Turing Thesis.[4]

3 Mathematical algorithms

To understand better the connection between algorithms and computability expressed by (1), we discuss briefly two well known, classical algorithms and a simple recursive process which is a variation on many popular themes.

3.1 The Euclidean algorithm (before 300 BC)

For $a, b \in \mathbb{N} = \{0, 1, \ldots\}$, $a, b \neq 0$,

$\gcd(a, b) =$ the largest number which divides both a and b.

It is easy to check that if for $a, b \neq 0$, $\mathrm{rem}(a, b)$ is the *remainder* of a by b, the unique number r such that for some $q \in \mathbb{N}$

(2) $\quad a = qb + r$ and $0 \leq r < b$,

then

(3) $\quad \gcd(a, b) =$ if $(\mathrm{rem}(a, b) = 0)$ then b else $\gcd(b, \mathrm{rem}(a, b))$.

This is the basic mathematical fact about the *greatest common divisor* function, it defines it implicitly, and it expresses a *recursive algorithm* for computing it using iterated division:[5]

---

[4] (Turing 1936) avoids these problems by using machines with no input which compute (the decimal expansions of) real numbers and reading the (infinite) output from the digits "emitted" during the computation; but he would surely need to face up to them to "investigate computable functions of an integral variable" in much the same way, as he says he can do.

[5] Euclid defines first the so-called *subtractive Euclidean algorithm* which uses *anthypheresis*

$$A(x, y) = (\max(x, y) - \min(x, y), \min(x, y)),$$

an operation on unordered pairs of numbers which was very important in Greek mathematics. The relevant recursive equation now is

$$\gcd(x, y) = \text{if } (x = y) \text{ then } x \text{ else } \gcd(A(x, y)),$$

if $\text{rem}(a,b) = 0$, give output b,
otherwise replace (a,b) by $(b, \text{rem}(a,b))$ and repeat.

The important mathematical facts about the Euclidean algorithm are the following:

(a) *Correctness*: for all non-zero $a, b \in \mathbb{N}$, the Euclidean terminates and yields $\gcd(a,b)$.

(b) *Primitives*: The Euclidean is an algorithm on $\mathbb{N}$ *from* rem and $=_0$ (equality with 0).

(c) *Complexity*: if $\text{calls}(a,b)$ is the number of *calls to* rem that the Euclidean makes to compute $\gcd(a,b)$, then

$$\text{calls}(a,b) \leq 2\log_2(b) \quad (\text{for } a \geq b \geq 2).$$

We might be tempted to define $\text{calls}(a,b)$ as the *number of divisions* the algorithm makes to compute $\gcd(a,b)$, the basic *division algorithm* being the most obvious way to get at the remainder. But there is nothing in the specification of the Euclidean (by the recursive equation (3)) which tells us how $\text{rem}(x,y)$ must be obtained whenever it is needed: there might, in fact, be some *fast division algorithm* which is more efficient than the usual division process (as *fast multiplication* is more efficient than the elementary school algorithm for multiplication), or even some very clever, still unknown method which gets $\text{rem}(x,y)$ very quickly without also computing the quotient of x by y. None of that matters to the Euclidean which simply needs $\text{rem}(x,y)$ for various pairs x, y in the process of computing $\gcd(a,b)$; this is why we say that the Euclidean is an algorithm *from* rem *and* $=_0$ rather than "from division and $=_0$".

There is a large number of extensions and generalizations of the Euclidean algorithm, from which we mention here just two, also known to the Greeks before Euclid's time.

The Euclidean on positive real inputs

The division equation (2) holds for positive real numbers $a, b \in \mathbb{R}^+ = \{x \in \mathbb{R} \mid x > 0\}$ and determines a unique *integer quotient* $q = \text{iq}(a,b) \in \mathbb{N}$ and remainder $\text{rem}(a,b) \in \mathbb{R}$. It follows that the basic recursive equation (3) makes sense when $a, b \in \mathbb{R}^+$ and expresses (as before) an algorithm on pairs

---

and so the subtractive Euclidean is an algorithm from = and A. Later on he switches to our version of the Euclidean without much comment, most likely thinking that the subtractive version simply "implements" division by "iterated subtraction".

Perhaps more importantly, Euclid defines his algorithms using *iteration* (like modern *while programs*) rather than recursive equations. The relation between *iterative* and *recursive algorithms* is important but subtle and I will not discuss it in this article, cf. (Moschovakis 1998).

from $\mathbb{R}^+$, except that in this case the computation may go on forever, so that this algorithm computes a *partial function*[6] on $\mathbb{R}^+ \times \mathbb{R}^+$ with values in $\mathbb{N}$; moreover

(4)   the Euclidean terminates on $a, b \in \mathbb{R}^+$

$$\iff a \text{ and } b \text{ are commensurable, i.e., } \frac{a}{b} \text{ is a rational number,}$$

an important fact known to the Greeks and their basic method for proving *non-commensurability*.

The continuous fraction algorithm

The output of the Euclidean on positive reals, when it converges, is not especially interesting. More basic is the variation of the Euclidean which applies (3) again on pairs of positive real numbers but outputs the (finite or infinite) sequence $q_0, q_1, \ldots$ of the quotients produced during the computation, i.e., the *continuous fraction representation* of the quotient $\frac{x}{y}$. This is an algorithm on $\mathbb{R}^+$ from $\mathrm{rem}, =_0, \mathrm{iq}$ on $\mathbb{R}^+$ and $0, S, \mathrm{Pd}, =_0$ on $\mathbb{N}$, another of the fundamental algorithms of Greek mathematics with important applications in number theory.

3.2   The Sturm algorithm (1829)

This computes *the number of real roots* of a polynomial

(5) $p(x) = a_0 + a_1 x + \cdots + a_n x^n$

of degree $\leq n$ with real coefficients in a real interval (b, c). It operates on tuples $(a_0, \ldots, a_n, b, c)$ of real numbers and its primitives are the field operations $0, 1, +, -, \cdot, \div$ and the ordering $\leq$ of $\mathbb{R}$. A simple elaboration of it decides the relation

$$R(a_0, \ldots, a_n) \iff (\exists x \in \mathbb{R})[a_0 + a_1 x + \cdots + a_n x^n = 0].$$

The main subroutine of the Sturm algorithm is a version of the Euclidean, applied to the space of real polynomials of degree $\leq n$ and with a (critical) twist, in which the remainder $r(x)$ at each step is replaced by $-r(x)$. It is an important algorithm and the main algebraic fact extended and used by (Tarski 1951) in his famous proof of the *decidability of the first order theory of* $\mathbb{R}$ *as an ordered field*.

---

[6] A partial function $f : X \rightharpoonup Y$ is an ordinary function $f : X_0 \to Y$ on some arbitrary $X_0 \subseteq X$, the *domain of convergence* of f. As usual, $f(x) \downarrow \iff x \in X_0$, $f(x) \uparrow \iff x \notin X_0$, and for $f_1, f_2 : X \rightharpoonup Y$,

$$f_1(x) = f_2(x) \iff [f_1(x)\uparrow \ \& \ f_2(x)\uparrow] \vee f_1(x) = f_2(x).$$

3.3 The color of leaves

A (finite, rooted, binary, colored) *tree* is a tuple

$$\mathbf{T} = (T, \mathrm{root}, l, r, \mathrm{Leaf}, \mathrm{Red}),$$

where T is a finite set; root $\in T$; Leaf and Red are unary relations on T; and

$$l, r : T \setminus \{x \in T \mid \mathrm{Leaf}(x)\} \rightarrowtail T \setminus \{\mathrm{root}\}$$

are injections with disjoint images whose union exhausts $T \setminus \{\mathrm{root}\}$. A *path from x_0 to x_n* in $\mathbf{T}$ is any sequence $(x_0, \ldots, x_n)$ such that for each $i < n$, $x_{i+1} \in \{l(x_i), r(x_i)\}$. It follows easily that every *node* $x \in T$ is the endpoint of a unique path from root, and that every *maximal path* ends at a leaf. Set[7]

(6) $\quad R(x) \iff$ every leaf below x is red
$\qquad \iff$ (for every path $(x, x_1, \ldots, x_n))[\mathrm{Leaf}(x_n) \implies \mathrm{Red}(x_n)]$.

The basic mathematical fact about this relation is the equivalence

(7) $\quad R(x) \iff$ if $\mathrm{Leaf}(x)$ then $\mathrm{Red}(x)$ else $[R(l(x)) \ \& \ R(r(x))]$;

and as with the Euclidean, it expresses a recursive algorithm for deciding $R(x)$:

> if $\mathrm{Leaf}(x)$, output the truth value of $\mathrm{Red}(x)$,
> otherwise decide $R(l(x))$ and $R(r(x))$ using the same procedure
> $\qquad$ and output the Boolean product $R(l(x)) \ \& \ R(r(x))$.

This is an abstract version of many standard, recursive (*divide-and-conquer*) algorithms, including the *merge-sort*.[8] It operates on the set T and its primitives are those of the structure $\mathbf{T}$, i.e., root, l, r, Leaf and Red.

---

[7]This is a simplified version of the example in the basic article (Tiuryn 1989), which separates non-deterministic from deterministic recursive computability and also full recursion from *tail recursion*, i.e., iteration.

[8]The merge-sort orders (alphabetizes, sorts) finite sequences from a set L with respect to a given ordering of L, and is asymptotically optimal for the number of comparisons it needs to do the job. Its optimality among sorting algorithms (of the appropriate kind) is probably the only lower bound result proved in every introductory course in Computer Science, and so a discussion of it can be found in any standard, introductory text. See (Moschovakis 1998) for a discussion of its significance for the foundations of the theory of algorithms.

This *leaf-color algorithm* can also be applied when T is infinite. In this case, the computation described terminates only on *the well founded part of T*

$$\mathrm{WF}(\mathbf{T}) = \{x \in T \mid (\exists n)[\text{every path from } x \text{ has length } \leq n]\}$$

and decides correctly the relation R on $\mathrm{WF}(\mathbf{T})$.[9]

In a second variation we consider *arbitrary, well founded binary trees*, i.e., structures of the form

$$\mathbf{T} = (T, \mathrm{Roots}, l, r, \mathrm{Leaf}, \mathrm{Red})$$

where Leaf and Red are as above; Roots $\subseteq T$ is a non-empty set of *roots*;

$$l, r : T \setminus \{x \in T \mid \mathrm{Leaf}(x)\} \rightarrowtail T \setminus \mathrm{Roots}$$

are injections with disjoint images; every $x \in T$ occurs on some path which starts with a root; and there are no infinite paths. Now $\mathrm{WF}(\mathbf{T}) = T$ and the algorithm terminates for every x and decides whether all the (finitely many) leaves below x are red.

4 Computing on an arbitrary set from specified primitives

The algorithms in Section 3 are very different from those envisioned by Church or expressed by Turing machines. Some of their important features are:

(I) *Arbitrary universe*: They operate on sets other than $\mathbb{N}$ or the strings from some finite alphabet: the real numbers for the variations of the Euclidean and the Sturm and an arbitrary finite or infinite set T for the leaf-color algorithm and its variations.

In particular, the computations defined by them are not "symbolic".

(II) *No input function*: They operate *directly* on their arguments, i.e., there is no intermediary of an input or an output function.

(III) *Use of arbitrary primitives*: They can use (call) specified primitives (constants, functions and relations) on their domain of application which need not be (and often are not) intuitively effectively computable. For example, the Sturm uses the inequality relation on $\mathbb{R}$ which is not decidable in any meaningful sense, and the second variation of the leaf-color algorithm operates on an arbitrary set T, for which it does not make sense to ask whether the primitives Roots, $l.r. \ldots$ are effective—they are just *given*.

---

[9] By König's Lemma, $x \in \mathrm{WF}(\mathbf{T})$ exactly when no *infinite path* starts from x.

One might argue that the specifications we gave in (3) and (7) are not precise or at least not complete, and they do not meet today's standard of rigor unless they are complemented by directions for how to "implement" them. This is one of many legitimate issues which make the foundational problem of explicating the notion of algorithm subtle (and even controversial). But this is not our issue here: what we will do in the next section is to extract from the robust intuitions behind these classical algorithms a precise notion of *computability from arbitrary primitives* for functions and relations on arbitrary sets. This is a natural and useful notion, and we will show in Section 8 that it is closely—and usefully—related to the kind of computability on the natural numbers that the Church-Turing Thesis is about.

5 Recursion in an arbitrary partial structure

We outline here very briefly the basic definitions of recursion in first order structures, for the sake of completeness. A full exposition of the (much richer) *recursion in a many-sorted, functional structure* is given in (Moschovakis 1989), and a summary of a mildly restricted case is included in (van den Dries & Moschovakis 2004).

It is convenient to think of a relation $R \subseteq A^n$ as a function $R : A^n \to \{T, F\}$ and of an element $c \in A$ as a nullary function with value c. This makes it natural to also allow *partial relations* $R : A^n \rightharpoonup \{T, F\}$, and so to deal uniformly with (partial) functions, relations and objects.

With these conventions, a *vocabulary* (signature) is a tuple

$$\Phi = (\phi_0, \ldots, \phi_k)$$

of function symbols, together with two functions, arity and sort, which assign to each ϕ_i its *arity*, the number of arguments that it expects, and its *sort*, the kind of values it takes, **ind** or **boole**; and a (partial) Φ-*structure* is a tuple

$$\mathcal{A} = (A, \Phi) = (A, \phi_0^{\mathcal{A}}, \ldots, \phi_k^{\mathcal{A}})$$

where each $\phi_i^{\mathcal{A}}$ is a *partial* constant, relation or function on the universe A of the appropriate arity and sort, e.g., if arity$(\phi_i) = n$ and sort$(\phi_i) = $ **ind**, then

$$\phi_i^{\mathcal{A}} : A^n \rightharpoonup A.$$

The $\mathcal{A}$-*terms* (with parameters and conditionals) are defined by the recursion

$$E :\equiv T \mid F \mid x \mid \mathsf{v}_j \mid \phi_i(E_1, \ldots, E_n) \mid \text{if } E_0 \text{ then } E_1 \text{ else } E_2$$

where x is any member of A; $\{v_0, v_1, \ldots\}$ is a fixed sequence of variables of sort ind; arity$(\phi_i) = n$ and $E_1, \ldots, E_n$ are of sort ind; and in the conditional, E_0 is of sort boole and sort$(E_1) = $ sort(E_2). The definition also assigns to each term its *parameters* (the members of A which occur in it), its *variables*, and in the obvious way, its *sort*. The sort of the conditional construct is sort(E_1) $(= $ sort$(E_2))$.

A term is *closed* if it has no variables and a *pure Φ-term* if it has no parameters. We use the customary notation for *substitution*: if $E(\mathsf{u}_1, \ldots, \mathsf{u}_n)$ is a term in which the distinct variables $\mathsf{u}_1, \ldots, \mathsf{u}_n$ may occur and $u_1, \ldots, u_n \in A$, then $E(u_1, \ldots, u_n)$ is the result of replacing in E each u_i by u_i.

The denotations of closed $\mathcal{A}$-terms are defined as one might expect:

$$\mathrm{den}(\mathrm{T}) = \mathrm{T}, \quad \mathrm{den}(\mathrm{F}) = \mathrm{F}, \quad \mathrm{den}(x) = x,$$
$$\mathrm{den}(\phi_i(E_1, \ldots, E_n)) = \phi_i^{\mathcal{A}}(\mathrm{den}(E_1), \ldots, \mathrm{den}(E_n)),$$
$$\mathrm{den}(\text{if } E_0 \text{ then } E_1 \text{ else } E_2) = \text{if } \mathrm{den}(E_0) \text{ then } \mathrm{den}(E_1) \text{ else } \mathrm{den}(E_2).$$

We write $\mathrm{den}(\mathcal{A}, E)$ if it is important to specify the structure in which the denotation is computed, and we note that $\mathrm{den}(\mathcal{A}, E)$ need not always converge, because we have allowed partial functions in Φ.

Recursive (McCarthy) programs

An n-ary (deterministic) *recursive Φ-program*[10] E is a syntactic expression

(8) $\quad E \equiv E_0(\vec{\mathsf{x}}, \vec{\mathsf{p}})$ where $\{\mathsf{p}_1(\vec{\mathsf{u}}_1) = E_1(\vec{\mathsf{u}}_1, \vec{\mathsf{p}}), \ldots, \mathsf{p}_k(\vec{\mathsf{u}}_k) = E_k(\vec{\mathsf{u}}_k, \vec{\mathsf{p}})\}$

where the following conditions hold:

(RP1) $\vec{\mathsf{p}} \equiv \mathsf{p}_1, \ldots, \mathsf{p}_k$ is a sequence of distinct function and relation symbols which do not occur in Φ. These are the *recursive variables* of E.

(RP2) For $i = 0, \ldots, k$, the *part* $E_i(\vec{\mathsf{u}}_i, \vec{\mathsf{p}})$ of E is a pure term (no parameters) in the vocabulary $\Phi \cup \{\mathsf{p}_1, \ldots, \mathsf{p}_k\}$ whose variables are in the list $\vec{\mathsf{u}}_i$ (where by convention, $\vec{\mathsf{u}}_0 \equiv \vec{\mathsf{x}} \equiv \mathsf{x}_1, \ldots, \mathsf{x}_n$).

(RP3) For $i = 1, \ldots, k$, sort$(E_i) = $ sort(p_i).

---

[10] Deterministic recursive programs were introduced by (McCarthy 1963), who used them to develop clean foundations for call-by-value computability from arbitrary, specified primitives. Especially significant was McCarthy's explicit identification of the *conditional* (branching) as an essential ingredient of computation: he used it to give an elegant characterization of the general recursive functions on $\mathbb{N}$ which avoids the non-determinism inherent in the *Herbrand-Gödel-Kleene* systems of (Kleene 1952).

The sort of E is the sort of its *head term* $E_0(\vec{x}, \vec{p})$; the *free occurrences of variables* of E are the occurrences of $x_1, \ldots, x_n$ in its head; and its *bound occurrences* of variables are those in the lists $\vec{u}_i$ in its *body* and all occurrences of $p_1, \ldots, p_k$.

For example,

(9) $\quad E \equiv \mathsf{p}(\mathsf{x}, 0) \text{ where } \{\mathsf{p}(\mathsf{x},\mathsf{y}) = \text{if } (\phi(\mathsf{x},\mathsf{y}) = 0) \text{ then } \mathsf{y} \text{ else } \mathsf{p}(\mathsf{x}, S(\mathsf{y}))\}$

is a program in the vocabulary $\{0, S, \phi, =_0\}$ of sort the sort of p, with x free in its first occurrence and bound in its occurrence in the body and y and p bound in all their occurrences.

The body of a recursive program E specifies a system of mutually recursive equations in the partial function variables $p_1, \ldots, p_k$. The denotation of a recursive program E in a Φ-structure $\mathcal{A}$ is obtained, intuitively, by "solving" this system and then substituting the solutions into the *head* E_0 of E.

More precisely, the parts of E can be evaluated in *expansions*

$$(\mathcal{A}, p_1, \ldots, p_k) = (A, \Phi, p_1, \ldots, p_k)$$

of a Φ-structure $\mathcal{A}$ by arbitrary partial functions of the correct arity and sort, and they define the following *system of recursive equations* on A:

$$\begin{aligned} p_0(\vec{x}) &= \text{den}((\mathcal{A}, p_1, \ldots, p_k), E_0(\vec{x}, \vec{p})), \\ p_1(\vec{u}_1) &= \text{den}((\mathcal{A}, p_1, \ldots, p_k), E_1(\vec{u}_1, \vec{p})), \\ &\vdots \\ p_k(\vec{u}_k) &= \text{den}((\mathcal{A}, p_1, \ldots, p_k), E_k(\vec{u}_k, \vec{p})). \end{aligned}$$

The expressions on the right of these equations define partial functions which are *monotone* and *continuous* in their partial function arguments, and so by a standard set theoretic construction the system has a *least tuple of solutions*

$$\overline{p}_0, \overline{p}_1, \ldots, \overline{p}_k;$$

the *denotation of E in $\mathcal{A}$* is then the partial function of the appropriate sort defined by the head,[11]

$$f_E^{\mathcal{A}} = \overline{p}_0 : A^n \rightharpoonup A \text{ if sort}(E_0) = \mathbf{ind} \text{ and}$$

---

[11] In many cases, $E_0(\vec{x}, \vec{p}) \equiv p_1(\vec{x})$, so that $f_E = \overline{p}_1$, i.e., the function computed by E is simply the first of the mutual fixed points of the system determined by the body of E.

$$f_E^{\mathcal{A}} = \bar{p}_0 : A^n \rightharpoonup \{\mathrm{T}, \mathrm{F}\} \text{ if sort}(E_0) = \mathtt{boole}.$$

Finally, a partial function or relation is $\mathcal{A}$-*recursive* or *recursive from* the primitives Φ of $\mathcal{A}$ if it is computed in $\mathcal{A}$ by some deterministic recursive program, and we set

rec$(\mathcal{A})$ =**rec**(A, Φ)
=the set of all partial functions and relations which are recursive in $\mathcal{A}$.

If S and Pd are the *successor* and *predecessor* functions on $\mathbb{N}$, then

(10) $f \in \mathbf{rec}(\mathbb{N}, 0, S, =) \iff f \in \mathbf{rec}(\mathbb{N}, 0, S, \mathrm{Pd}, =_0)$
$\iff f$ is Turing computable.

This was one of the first results about Turing computability, albeit somewhat differently formulated, and it is often used to infer that a certain f is Turing computabile by giving a recursive equation or system which computes it.[12]

6 The Relative Recursion Thesis

It is basically trivial that *all algorithms in Section 3 compute (partial) functions or relations which are recursive from the relevant primitives*: just turn the given recursive equation into a recursive program by "formalizing" it and adding a trivial head, cf. Footnote 11.[13] There is also a rich theory of *recursion on arbitrary structures* which covers most "intuitive" definitions of algorithms and claims of computability on abstract sets from specified primitives. The examples in Sections 4 and 5 do not provide sufficient evidence that all "algorithms" from specified primitives can be expressed by

---

[12]For example, if S is the successor on $\mathbb{N}$ and ϕ is total, then the program in (9) computes in $(\mathbb{N}, 0, S, \phi, =_0)$ the *minimalization* of ϕ,

$$f_E(x) = \mu y[\phi(x, y) = 0] = \text{ the least } y \text{ such that } \phi(x, y) = 0.$$

This is the key idea in Kleene's proof that the class of Turing computable functions is closed under the minimalization operator, perhaps the earliest important connection of fixed point recursion with Turing computability.

[13]The continuous fraction algorithm operates on both reals and natural numbers and it outputs finite or infinite sequences of numbers. In the approach we are taking here, it is best viewed as an algorithm of the structure $(\mathbb{R}^+, \mathbb{N}, \mathrm{iq}, \mathrm{rem}, =_0^{\mathbb{R}}, 0^{\mathbb{N}}, S, \mathrm{Pd}, =_0^{\mathbb{N}})$ with two universes, which computes the partial function $q : \mathbb{R}^+ \times \mathbb{R}^+ \times \mathbb{N} \rightharpoonup \mathbb{N}$, where $q_n = q(x, y, n)$ is the n'th term in the continuous fraction expansion of $\frac{x}{y}$, defined for all n when x and y are not commensurable. The reduction of many-sorted recursion to recursion on one sort (other than $\mathtt{boole}$) is routine, and so is putting down a recursive program which expresses the continuous fraction algorithm.

recursive programs, of course, but this is not the issue here; so I will leap immediately, Church-style,[14] to the strongest claim:

The Relative Recursion Thesis

For every function $f : A^n \to A$ or relation $f : A^n \to \{T, F\}$ on a set A and any set of primitives Φ on A,

RRT: f is computable from Φ

$\Longleftrightarrow$ f is recursive in the structure $\mathcal{A} = (A, \Phi)$.

As with CT, the "easy" direction ($\Leftarrow$) of RRT can be proved by *implementing* recursive programs using *oracles* to represent the primitives; one needs to appeal only to simple and non-controversial properties of algorithms from primitives on an arbitrary set and to assume that some basic operations on finite sequences are intuitively effective, much as Turing does for the easy direction of CT. A proof of the non-trivial direction ($\Rightarrow$) would require showing that all computation from primitives can be reduced to *calling* (composition), *branching* and *mutual recursion*. This is possibly a simpler task than what is needed to prove CT, but I do not see now how to go about it.[15]

7 Logical notions and propositions

(Tarski 1986) gave a famous explication of *logical notions*, by "applying" to logic Felix Klein's classical *Erlangen Program* for classifying geometries. We give here a (very) abbreviated and somewhat simplified version of Tarski's definitions with the aim to show that the Relative Recursion Thesis RRT is a logical proposition, of a very different *kind* than CT.

The simple type structure over a set A

For every non-empty set A, set

$$T_0(A) = A, \qquad T_{n+1}(A) = \mathcal{P}(T_n(A)) = \text{ the set of all subsets of } T_n(A),$$

where A is viewed as a set of *individuals* (atoms) with no internal set structure and every member of *the type $T_n(A)$* is "tagged" with n, so that every object in these sets belongs to exactly one $T_n(A)$, n being its *type*. We set

$$x \in_n y \iff x \in y \in T_{n+1}, \quad T^*(A) = \bigcup_n T_n(A), \quad \mathbf{T}^*(A) = (T^*(A), \{\in_n\}_n).$$

---

[14] I have heard it said that Church claimed his version of CT as soon as Kleene proved that the *predecessor function* on $\mathbb{N}$ is λ-definable. Kleene took a little longer to believe it.

[15] The best arguments I know which support RRT come from the analysis of the notion of algorithm in (Moschovakis 1998).

This is the *simple type structure above A*.

We can use standard, set-theoretic constructions to identify in $T^*(A)$ much more complex objects than typed pure sets (of sets of sets ... of members of A). For example, using the Kuratowski pair,

$$x, y \in T_n(A) \implies (x,y) = \{\{x\}, \{x,y\}\} \in T_{n+2}(A),$$

which gives us the Cartesian product

$$x, y \in T_{n+1}(A) \implies x \times y = \{(u,v) \mid u \in x \ \& \ v \in t\} \in T_{n+3}(A)$$

of two sets in the same type; and iterating the process as usual, we get k-fold products, relations and (partial and total) functions of any arity, etc. Moreover, the embedding $x \mapsto \{x\}$ injects each $T_n(A)$ into $T_{n+1}(A)$, and its iterates give us simple embeddings

$$j_n^{n+k} : T_n(A) \rightarrowtail T_{n+k}(A)$$

which "code" each $T_n(A)$ into every larger type. The upshot is that we can think of any set

$$X \subseteq T_{k_1}(A) \times \cdots \times T_{k_n}(A)$$

as a member of $T_l(A)$ for any sufficiently large l and operate on these sets by the standard set operations, union, intersection, etc.[16] For example, we can code truth and falsity in $T_1(A)$ by setting

$$\mathrm{T} = A, \ \mathrm{F} = \emptyset,$$

and for any n, we can think of the relations

$$\in_n = \{(x,y) \mid x \in_n y\}, \quad =_n = \{(x,y) \mid x = y \in T_n(A)\}$$

as members of $T_l(A)$ for some l, which is not very hard to compute in this case. More significantly, for what we aim to do, fix a number n and a vocabulary $\Phi = (\phi_0, \ldots, \phi_k)$ and set

(11) $\mathrm{Rec}_n(A, \Phi) = \left\{(f, \Phi) \mid f : A^n \rightharpoonup A \ \& \ f \in \mathbf{rec}(A, \Phi)\right\}$

$\cup \left\{(f, \Phi) \mid f : A^n \rightharpoonup \{\mathrm{T}, \mathrm{F}\} \ \& \ f \in \mathbf{rec}(A, \Phi)\right\},$

where $\Phi = (\varphi_0, \ldots, \varphi_k)$ stands for any tuple of partial functions on A which have the arities and sorts specified by Φ so that $(A, \varphi_0, \ldots, \varphi_k)$ is a Φ-structure; we can locate (a code of) this set in $T_l(A)$, for every sufficiently large l (as determined by Φ and n).

---

[16]This appeal to codings can be avoided, of course, by adopting a modern, richer definition of the type structure, with product and function types. We will not do enough in this brief note to justify the additional machinery, however, and I thought it best to stick with the simpler, classical definition.

Logical notions

Every permutation $\pi : A \rightarrowtail\!\!\!\!\!\rightarrow A$ extends naturally to a permutation $\pi^* : T_n(A) \rightarrowtail\!\!\!\!\!\rightarrow T_n(A)$ by the recursion

$$\pi^*(x) = \pi^*[x] = \{\pi^*(y) \mid y \in x\} \quad (x \in T_{n+1}(A)),$$

and so to $T^*(A)$; and then, easily, π^* is an *automorphism* of the type structure $\mathbf{T}^*(A)$, i.e., it is a bijection of $T^*(A)$ with itself such that

$$x \in_n y \iff \pi^*(x) \in_n \pi^*(y) \quad (x, y \in T^*(A)).$$

Following (Tarski 1986), a set $X \in T^*(A)$ is *logical above* A if it is fixed by every such automorphism, i.e.,

for every permutation $\pi : A \rightarrowtail\!\!\!\!\!\rightarrow A$, $\pi^*(X) = X$.

The motivation comes from a basic feature of definability: *if $X \in T^*(A)$ is definable* (without parameters) *in a reasonable language $\mathcal{L}$, then X is fixed by every automorphism of $\mathbf{T}^*(A)$*—and this applies not only to the natural formal language of type theory, but to every reasonable, precisely formulated language which is naturally interpreted in $\mathbf{T}^*(A)$, including languages with second and higher order quantifiers, infinitary connectives, etc. Tarski replaces the elusive search for a characterization of "definability in some reasonable language" by a rigorous, semantic criterion which should be satisfied by all definable objects. We can then give rigorous proofs of *logicality* and *non-logicality*:

THEOREM 1 *For each n and every vocabulary Φ, the set $\mathrm{Rec}_n(A, \Phi)$ in (11) is logical above A.*

Outline of proof.

For any $g : A^m \rightharpoonup A$ and any permutation $\pi : A \rightarrowtail\!\!\!\!\!\rightarrow A$, let $g^\pi = \pi^*(g) : A^m \rightharpoonup A$ and check that

(12) $\quad g^\pi(x_1, \ldots, x_m) = \pi g(\pi^{-1} x_1, \ldots, \pi^{-1} x_m) \quad (x_1, \ldots, x_m \in A).$

By a simple exercise in *fixed point recursion*, for any $f, \varphi_0, \ldots, \varphi_m$,

(13) $\quad f$ is recursive in $(A, \varphi_0, \ldots, \varphi_k)$
$$\iff f^\pi \text{ is recursive in } (A, \varphi_0^\pi, \ldots, \varphi_k^\pi);$$

this implies that the function part of $\mathrm{Rec}_n(A)$ is fixed by π^*, and the corresponding argument about relations finishes the proof. ∎

More interesting is the next result about (intuitively understood) computability from arbitrary primitives. We label it a "Claim" rather than a theorem, because we will appeal in its proof to some assumptions about "computability from primitives", which we will not (and cannot) prove without a precise definition.

For each n and each vocabulary Φ, let

(14) $\operatorname{Comp}_n(A, \Phi) = \{(f, \Phi) \mid f : A^n \rightharpoonup A \ \& \ f \text{ is computable from } \Phi)\}$
$\cup \{(f, \Phi) \mid f : A^n \rightharpoonup \{\mathrm{T}, \mathrm{F}\} \ \& \ f \text{ is computable from } \Phi\},$

where $\overline{\Phi}$ is related to Φ as in the formulation of (11) above.

CLAIM 2 *For each n and every vocabulary Φ, the set $\operatorname{Comp}_n(A, \Phi)$ in (14) is logical above A.*

Outline of proof.

The key step—and where the intuitions about computability from primitives come in—is the following

Lemma. If some process computes $f : A^n \rightharpoonup A$ from the primitives $\varphi_0, \ldots, \varphi_k$ on A, then for every permutation $\pi : A \rightarrowtail A$, the same process computes f^π from $\varphi_0^\pi, \ldots, \varphi_k^\pi$.

This yields

(15) f is computable from $\varphi_0, \ldots, \varphi_k$
$\iff f^\pi$ is computable from $\varphi_0^\pi, \ldots, \varphi_k^\pi,$

from which the proof of the Claim can be completed as in Theorem 1.

Proof of the Lemma. Suppose α is some kind of process which computes a partial function $f : A^n \rightharpoonup A$ from $\varphi_0, \ldots, \varphi_k$. Our basic intuition is that for any $\vec{y} = (y_1, \ldots, y_n) \in A^n$, there is a "computation" of α which derives the value $f(\vec{y})$; that in the course of this computation, α may request from "the oracle" representing any φ_i any particular value $\varphi_i(u_1, \ldots, u_m)$ for $u_1, \ldots, u_m$ which it has already computed from $\vec{y}$; and that if the oracles cooperate and respond to all requests, then the computation of $f(\vec{y})$ is completed in a finite number of steps. This much is probably non-controversial, and certainly true of all precisely defined "processes" (i.e., algorithms) from primitives like those in Section 3, with reasonable, precise notions of "computation". We also assume that

the primitives $\varphi_0, \ldots, \varphi_k$ are the only non-logical operations used by α, which is the most important part of our understanding of "computation from $\varphi_0, \ldots, \varphi_k$". It insures that α does not have access to any "hidden primitives" other than $\varphi_0, \ldots, \varphi_k$ and is again true of all standard algorithms. This supports the claim that

> if we replace the input $\vec{y}$ by $\pi(\vec{y}) = (\pi y_1, \ldots, \pi y_n)$ and also replace every request in the computation for $\varphi_i(u_1, \ldots, u_m)$ by a request for $\varphi_i^\pi(\pi u_1, \ldots, \pi u_m)$, we get a computation of $\pi f(\vec{y})$ from $\varphi_0^\pi, \ldots, \varphi_k^\pi$; and if we apply this construction to $\vec{y} = \pi^{-1}(\vec{x})$, then the output is $\pi f(\pi^{-1}(\vec{x})) = f^\pi(\vec{x})$,

which then implies the Lemma. ∎

More—or less—could be put into this "proof" of Claim 2, which depends fundamentally on "appeals to intuition" and "for this reason [is] rather unsatisfactory mathematically", to use Turing's words. Or we might just *assume* Claim 2 as flowing naturally from the

> *Basic intuition*: Each value $f(\vec{x})$ of a partial function $f : A^n \rightharpoonup A$ computable from specified primitives, depends in some uniform way only on finitely many values of those primitives—and on nothing else.

We can now claim the basic result of this section:

CLAIM 3 *The Relative Recursion Thesis* RRT *is a logical proposition.*

Proof. Without defining *logical propositions* in general, we just assume the following which is, I think, quite plausible: if for every m, $X_m(A)$ and $Y_m(A)$ are logical objects over A, then the identity $X_m(A) = Y_m(A)$ is logical over A and the universal closure

$$(\forall m)(\forall A \neq \emptyset)[X_m(A) = Y_m(A)]$$

is a logical proposition. By Theorem 1 and Claim 2, this is exactly the form of

$$\text{RRT} \iff (\forall \Phi)(\forall n)(\forall A \neq \emptyset)[\text{Comp}_n(A, \Phi) = \text{Rec}_n(A, \Phi)]$$

once we enumerate all pairs (n, Φ) of numbers and vocabularies. ∎

8 The punchline

The Relative Recursion Thesis restricts computability on arbitrary sets from arbitrary, specified primitives and does not say anything directly about (absolute) computability or recursion on the natural numbers. However:

CLAIM 4 *A function $f : \mathbb{N}^n \to \mathbb{N}$ or relation $f : \mathbb{N}^n \to \{T, F\}$ is computable if and only if f is computable on $\mathbb{N}$ from $0, S, =$*

— because $(\mathbb{N}, 0, S, =)$ is just what the natural numbers are. This is also a Claim rather than a Theorem, because it is grounded on an assumption about the nature of natural numbers—*what they are*—which cannot be proved any more than CT or RRT can be proved.

Very briefly, about a problem which has been discussed as extensively as any other in the Philosophy of Mathematics since the 1870s, there are two basic facts about the natural numbers, both due to Frege and (mostly) Dedekind:

(i) $(\mathbb{N}, 0, S, =)$ is a *Peano system*, i.e., $0 \in \mathbb{N}$, the successor function is a bijection of $\mathbb{N}$ with its non-0 elements, and the *Induction Axiom* holds, i.e., for every $X \subseteq \mathbb{N}$,[17]

$$\Big(0 \in X \ \& \ (\forall x \in X)[S(x) \in X]\Big) \implies X = \mathbb{N}.$$

(ii) *Dedekind's Theorem*: Any two Peano systems are (uniquely) isomorphic.

These lead to what I will call

The Standard View. *The natural numbers are a Peano system—and that is all they are.*

There are many well-known and much discussed problems with the claim that the numbers are *just* a Peano system, some of them stemming from the fact that there is no natural way to "select" a particular (privileged) one, cf. (Benacerraf 1965). At the same time, there are also many responses to this problem, e.g., structuralist or modal approaches, which get around the problem in various (sometimes very sophisticated) ways. I do think, however, that the common, starting point for all philosophical views about the natural numbers are (i) and (ii). This is what I am trying to convey by saying *"and that is all the numbers are"*: the idea is that if we derive some results about the numbers using only the fact that they are a Peano system, then these results will find a natural expression in any coherent approach to the foundations of number theory. This should apply to Claim 4, which can then be used in the proof of the following

THEOREM 5 *The Relative Recursion Thesis and the Standard View about numbers imply the Church-Turing Thesis, i.e.,*

$$\text{RRT} + \text{the Standard View} \implies \text{CT}.$$

---

[17] The equality relation $=$ is not usually included in the definition of a Peano system, but it is implicit in the definition of *isomorphism*.

Proof. For the non-trivial direction of CT, suppose $f : \mathbb{N}^n \to \mathbb{N}$ is computable; f is then computable from $0, S, =$ by the Standard View; and so it is recursive in $(\mathbb{N}, 0, S, =)$ by RRT; and so it is Turing computable by (10). ∎

8.1 Concluding remarks

Theorem 5 suggests that the meaning and truth value of the Church-Turing Thesis do not depend on any deep properties of numbers or any assumptions about "the primitives of computation" or whether "all computation is symbolic"; these are now replaced by a well understood view of "what the numbers are". It does not prove CT, because RRT is not immediate: its meaning and justification require identifying a basis for the *logical primitives of computation on an arbitrary set from arbitrary functions and relations*. They should be composition, branching and recursion, of course, but it is not obvious to me how to prove this beyond a reasonable doubt.

Bibliography

Benacerraf, P. (1965). What numbers could not be. *Philosophical Review*, *74*, 47–73, reprinted in Philosophy of Mathematics, Selected readings, eds. Benacerraf, P. and Putnam, H., Cambridge University Press, 1983.

Church, A. (1935). An unsolvable problem in elementary number theory. *Bulletin of the American Mathematical Society*, *41*, 332–333, this is an abstract of Church (1936b).

Church, A. (1936a). A note on the Entscheidungsproblem. *Journal of Symbolic Logic*, *1*(1), 40–41, doi:10.2307/2269326.

Church, A. (1936b). An unsolvable problem in elementary number theory. *American Journal of Mathematics*, *58*(2), 345–363, an abstract of this paper was published in Church (1935).

Dershowitz, N. & Gurevich, Y. (2008). A natural axiomatization of computability and proof of Church's Thesis. *The Bulletin of Symbolic Logic*, *14*, 299–350, doi:10.2178/bsl/1231081370.

Gandy, R. (1980). Church's Thesis and principles for mechanisms. In *The Kleene Symposium*, Barwise, J., Keisler, H. J., & Kunen, K., eds., Amsterdam; New York: North Holland Publishing, 123–148.

Gandy, R. (1995). Church's Thesis and principles for mechanisms. In *The Kleene Symposium*, Barwise, J., Keisler, H. J., & Kunen, K., eds., North Holland Publishing Co., 123–148.

Gurevich, Y. (2000). Sequential abstract state machines capture sequential algorithms. *ACM Transactions on computational logic*, *1*, 77–111.

Kleene, S. C. (1952). *Introduction to Metamathematics*. New York: Van Nostrand; North Holland.

Kripke, S. A. (2000). From the Church-Turing Thesis to the First-Order Algorithm Theorem. In *Proceedings of the 15th Annual IEEE Symposium on Logic in Computer Science*, LICS '00, Washington, DC, USA: IEEE Computer Society, URL http://dl.acm.org/citation.cfm?id=788022.789011, the reference is to an abstract. A video of a talk by Saul Kripke at *The 21st International Workshop on the History and Philosophy of Science* with the same title is posted at www.youtube.com/watch?v=D9SP5wj882w, and this is my only knowledge of this article.

McCarthy, J. (1963). A basis for a mathematical theory of computation. In *Computer Programming and Formal Systems*, Braffort, P. & Herschberg, D., eds., Amsterdam: North-Holland, 33–70.

Moschovakis, Y. N. (1968). Review of four papers on Church's Thesis. *Journal of Symbolic Logic*, *33*, 471–472.

Moschovakis, Y. N. (1989). The formal language of recursion. *Journal of Symbolic Logic*, *54*(4), 1216–1252, doi:10.1017/S0022481200041086.

Moschovakis, Y. N. (1998). On founding the theory of algorithms. In *Truth in Mathematics*, Dales, H. G. & Oliveri, G., eds., Oxford: Clarendon Press, 71–104, posted on www.math.ucla.edu/~ynm/ papers/foundalg.pdf.

Sieg, W. (2002). Calculations by man and machine: mathematical presentation. In *In the Scope of Logic, Methodology and Philosophy of Science*, Gärdenfors, P., Woleński, J., & Kijania-Placek, K., eds., Dordrecht; Boston: Kluwer Academic Publishers, 247–262.

Tarski, A. (1951). A decision method for elementary algebra and geometry. RAND Corporation report. Prepared for publication with the assistance of J.C.C. McKinsey.

Tarski, A. (1986). What are logical notions? *History and Philosophy of Logic*, *7*, 143–154, doi:10.1080/01445348608837096, edited by John Corcoran.

Tiuryn, J. (1989). A simplified proof of DDL < DL. *Information and Computation*, *82*, 1–12.

Tucker, J. & Zucker, J. (2000). Computable functions and semicomputable sets on many-sorted algebras. In *Handbook of Logic in Computer Science*, vol. 5, Abramsky, S., Gabbay, D., & Maibaum, T., eds., New York: Oxford University Press, 317–523.

Turing, A. M. (1936). On computable numbers with an application to the Entscheidungsproblem. *Proceedings of the London Mathematical Society*, *42*, 230–265, *A correction*, ibid. volume 43 (1937), pp. 544–546.

van den Dries, L. & Moschovakis, Y. N. (2004). Is the Euclidean algorithm optimal among its peers? *Bulletin of Symbolic Logic, 10*(3), 390–418.

Yiannis N. Moschovakis
Department of Mathematics, University of California, Los Angeles
USA
`ynm@math.ucla.edu`
Department of Mathematics, University of Athens
Greece

Explanatory and Non-Explanatory Demonstrations

Carlo Cellucci

1 Premise

It is commonly held that a basic aim of natural science is to explain natural facts.

Thus Popper states that "a problem of pure science" always "consists in the task of finding an explanation, the explanation of a fact or of a phenomenon or of a remarkable regularity or of a remarkable exception from a rule" (Popper 1996, 76).

Conversely, several people believe that there is no explanation of mathematical facts, in particular, there is no objective distinction between explanatory and non-explanatory demonstrations.

Thus Resnik and Kushner state that, from the fact that several proofs "leave many of our why-questions unanswered", we "derive the mistaken idea that there is an objective distinction between explanatory and non-explanatory proofs" (Resnik & Kushner 1987, 154). But "the notion of explanatory proof is not viable" (Resnik & Kushner 1987, 156). In fact, "mathematicians rarely describe themselves as explaining" (Resnik & Kushner 1987, 151).

This, however, contrasts with the view of many mathematicians.

Thus Atiyah states: "I remember one theorem that I proved, and yet I really couldn't see why it was true", but "five or six years later I understood why it had to be true. Then", using "quite different techniques", I "got an entirely different proof" that made "quite clear why it had to be true" (Atiyah 1988, I, 305).

Similarly, Auslander states that, while "a proof is supposed to explain the result", it "must be admitted that not all proofs meet this standard" and this has "often led to the development of new, more understandable, proofs" (Auslander 2008, 66).

Notice that the question whether there are explanations of mathematical facts is not to be confused with the question whether there are explanations of non-mathematical facts—physical, biological, psychological, economical—using mathematics. Explanation of mathematical facts is explanation in mathematics, explanation of non-mathematical facts using mathematics is explanation with mathematics.

This paper is about explanation in mathematics, and specifically about the question whether there exists an objective distinction between explanatory and non-explanatory demonstrations. I have already dealt with this question elsewhere (Cellucci 2008a), but this paper somewhat modifies and expands the approach set forth there.

That explanation in mathematics concerns the question whether there exists an objective distinction between explanatory and non-explanatory demonstrations, contrasts with the opinion of several people.

Thus Sierpinska states that "the quest for explanation in mathematics cannot be a quest for proof, but it may be an attempt to find a rationale of a choice of axioms, definitions, methods of construction of a theory" (Sierpinska 1994, 76). This is based on the assumption that 'demonstration' is synonymous with 'axiomatic demonstration', so it is "a means for ascertaining the truth of a theorem" (Sierpinska 1994, 19).

But this is only one concept of demonstration. As it will be argued in this paper, there is also another concept of demonstration, according to which demonstration is a means for discovering solutions to problems. Therefore, the assumption that 'demonstration' is synonymous with 'axiomatic demonstration' is unjustified.

2 Explanation in science

In the last century the most widespread view about explanation in natural science has been the deductive view, which was stated by Popper in 1934 under the name 'causal explanation', but is often credited to (Hempel & Oppenheim 1948).

According to Popper, "to give a causal explanation of an event means to deduce a statement which describes it, using as premises of the deduction one or more universal laws, together with certain singular statements, the initial conditions" (Popper 1959, 59). Thus the premises of the deduction consist of universal statements which have "the character of natural laws", and singular statements "which apply to the specific event in question" and are called "initial conditions" (Popper 1959, 60). The "initial conditions describe what is usually called the 'cause' of the event in question" (Popper 1959, 60).

For example, suppose a certain thread breaks and the question is to explain why it breaks. An explanation might be that, whenever a thread is loaded with a weight exceeding its tensile strength, it will break (universal law), and that the thread has a tensile strength of 1 lb. and was loaded with a weight of 2 lbs. (initial conditions). From these premises one can deduce that the thread will break. Then the fact that the thread has a tensile strength of 1 lb. and was loaded with a weight of 2 lbs. "was the 'cause' of its breaking" (Popper 1959, 59).

The deductive view can also be extended to mathematics. Suppose the question is to explain why $1+3+5+7 = 16 = 4^2$. An explanation might be that, for all $n, 1+3+5+\ldots+(2n-1) = n^2$ (universal law), and $n = 4$ (initial conditions). From these premises one can deduce that $1+3+5+7 = 16 = 4^2$.

But the deductive view is inadequate because, even when the universal laws and the initial conditions are assumed to be true, such view does not necessarily provide an explanation of the event described by the statement deduced from them. A simple example is the following, which is attributed to Bromberger although "in fact it appears in none" of his "published papers" (Bromberger 1992, 8).

Suppose a flagpole casts a shadow of 30 meters across the ground, and the question is to explain why the shadow is 30 meters long.

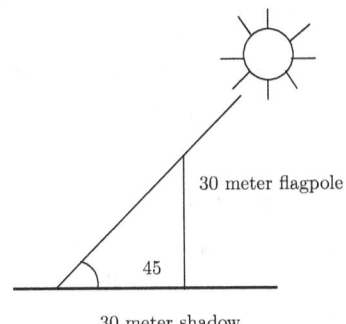

An explanation might be that light travels in straight lines and the laws of geometry hold (universal laws), and the angle of elevation of the sun is 45° and the flagpole is 30 meters high (initial conditions). From these premises one can deduce that the shadow is 30 meters long. Thus the universal laws and the initial conditions explain why the shadow is 30 meters long.

But suppose we swap the event to be explained, that the shadow is 30 meters long, with the initial condition, that the flagpole is 30 meters high. That is, suppose the flagpole is 30 meters high, and the question is to explain why it is 30 meters high.

An explanation might be that light travels in straight lines and the laws of geometry hold (universal laws), and the angle of elevation of the sun is 45° and the shadow is 30 meters long (initial conditions). From these premises one can deduce that the flagpole is 30 meters high. Thus the universal laws and the initial conditions explain why the flagpole is 30 meters high.

But it seems odd to consider this an explanation. The genuine explanation might be that the flagpole has been built 30 meters high so that it could be visible from all over the area. This has nothing to do with the length of the shadow that it casts.

From this example it is clear why the deductive view is inadequate. It implies that there is a symmetrical relation between the fact to be explained and the initial conditions. On the basis of it, when the angle of elevation of the sun is 45°, not only the flagpole being 30 meters high explains why the shadow is 30 meters long, but also the shadow being 30 meters long explains why the flagpole is 30 meters high. But this is not the case, thus the relation between the fact to be explained and the initial conditions is asymmetrical.

From the example it is also clear that Popper's calling the deductive view 'causal explanation' is improper. According to the deductive view, the shadow being 30 meters long would count as the causal explanation of the flagpole being 30 meters high, whereas this is not its cause. Therefore 'deductive view' is a better name.

3 Aristotle on explanation

It is surprising that, in the last century, the deductive view has been the most widespread view about explanation in natural science. For already Aristotle made it quite clear that such view is inadequate.

Aristotle distinguishes between two kinds of knowing, 'knowing that' and 'knowing why', for he states that "knowing that [*to oti*] is different from knowing why [*to dioti*]" (Aristotle, *Posterior Analytics,* A 13, 78 a 22). He also states that "to have knowledge" of a thing is "to have a demonstration of it" (Aristotle, *Posterior Analytics,,* B 3, 90 b 9–10). Therefore, his distinction between two kinds of knowing parallels a distinction between two kinds of demonstration, 'demonstration that' and 'demonstration why'. Indeed, he states that there are "differences between a demonstration that [*tou oti*] and a demonstration why [*tou dioti*]" (Aristotle, *Posterior Analyt-*

ics, A 13, 78 b 33–34). In a 'demonstration that' "the cause is not stated" (Aristotle, *Posterior Analytics,* A 13, 78 b 14–15). Thus a 'demonstration that' is non-explanatory. Only in a 'demonstration why' the cause is stated, hence only that kind of demonstration is explanatory.

Aristotle illustrates his distinction between 'demonstration that' and 'demonstration why' by a number of examples. One of them consists of the following demonstrations: (A) Something does not twinkle if and only if it is near; the planets do not twinkle; therefore, the planets are near; (B) Something does not twinkle if and only if it is near; the planets are near; therefore, the planets do not twinkle. Now, (A) is a 'demonstration that' since "it is not because the planets do not twinkle that they are near" (Aristotle, *Posterior Analytics,* A 13, 78 a 37–38). Conversely, (B) is a 'demonstration why' since "it is because" the planets "are near that they do not twinkle" (Aristotle, *Posterior Analytics,* A 13, 78 a, 38). Thus demonstration (A) is non-explanatory, while demonstration (B) is explanatory.

Aristotle's example shows that the relation between the fact to be explained and the initial conditions is asymmetrical, thus the deductive view of explanation is inadequate. It also shows that, for Aristotle, there exists an objective distinction between explanatory and non-explanatory demonstrations, since only the former show the cause.

There is, however, a problem with Aristotle's distinction. According to Aristotle, a demonstration is a deduction that proceeds "from premises which are true, and prime, and immediate, and better known than, and prior to, and causes of the conclusion" (Aristotle, *Posterior Analytics,* A 2, 71 b 22). But in a 'demonstration that' the cause is not stated. Thus, properly speaking, a 'demonstration that' is not a demonstration at all. For the same reason, 'knowing that' is not knowing at all.

The difficulty can be solved by conjecturing that, for Aristotle, there are two different senses of 'knowing', a weaker sense, expressed by 'knowing that', and a stronger sense, expressed by 'knowing why'. Correspondingly, there are two different senses of 'demonstration', a weaker sense, expressed by 'demonstration that', and a stronger sense, expressed by 'demonstration why'.

In this perspective, the distinction between explanatory and non-explanatory demonstrations parallels a distinction between knowing something through its explanation, or 'knowing why', and knowing something not through its explanation, or 'knowing that'.

4 Descartes on explanation

Like Aristotle, also Descartes holds that there exists an objective distinction between explanatory and non-explanatory demonstrations. But Descartes draws the distinction somewhat differently from Aristotle.

According to Descartes, although the axiomatic method or synthesis "demonstrates the conclusion clearly", such method is not satisfying "nor appeases the minds of those who are eager to learn, since it does not show how the thing in question was discovered" (Descartes 1996, VII, 156). Conversely, the analytic method or "analysis shows the true way by which a thing was discovered methodically" (Descartes 1996, VII, 155). In such a method, "it is the causes which are demonstrated by the effects", and "the causes from which I deduce" the effects "serve not so much to demonstrate them as to explain them" (Descartes 1996, VI, 76). Therefore, "there is a great difference between" merely "demonstrating and explaining" (Descartes 1996, II, 198).

Thus, according to Descartes, explanatory demonstrations are those which show how the thing was discovered and are based on the analytic method. Non-explanatory demonstrations are those which do not show how the thing was discovered and are based on the axiomatic method.

5 Axiomatic and analytic demonstration

That the axiomatic and the analytic method, to which Descartes refers, are essentially different, will be clear from the following description.

1. The axiomatic method is the method according to which, to demonstrate a statement, one starts from some given primitive premises, which are supposed to be true in some sense of 'true'—for example, consistent—and deduces the statement from them.
   This yields the axiomatic notion of demonstration, according to which a demonstration consists in a deduction of a statement from given prime premises which are supposed to be true, in some sense of 'true'. The goal of axiomatic demonstration is to provide justification for a statement. Thus axiomatic demonstration has a validation role, it is meant to establish the certainty of a statement.

2. The analytic method is the method according to which, to solve a problem, one looks for some hypothesis that is a sufficient condition for the solution of the problem. The hypothesis is obtained from the problem, and possibly other data, by some non-deductive rule and must be plausible, that is, compatible with the existing knowledge. But the hypothesis is in turn a problem that must be solved, and will be solved in the same way. That is, one looks for another hypothesis

that is a sufficient condition for the solution of the problem posed by the previous hypothesis. The new hypothesis is obtained from the previous hypothesis, and possibly other data, by some non-deductive rule and must be plausible. And so on, *ad infinitum*. Therefore, the solution of a problem is a potentially infinite process.

This yields the analytic notion of demonstration, according to which a demonstration consists in a non-deductive derivation of a hypothesis from a problem and possibly other data, where the hypothesis is a sufficient condition for the solution of the problem and is plausible; then, in a non-deductive derivation of a new hypothesis from the previous hypothesis, considered in turn as a problem, and possibly other data; and so on, *ad infinitum*.

The goal of analytic demonstration is to discover hypotheses that are sufficient conditions for the solution of a problem and are plausible. Thus analytic demonstration has a heuristic role. (For more on the analytic notion of demonstration, see (Cellucci 2008b).)

6 An example of analytic demonstration

A prototype of analytic demonstration is Hippocrates of Chios' solution of the problem of the quadrature of certain lunes, for example: Show that, if PQR is a right isosceles triangle and PRQ, PTR are semicircles on PQ, PR, respectively, then the lune $PTRU$ is equal to the right isosceles triangle PRS.

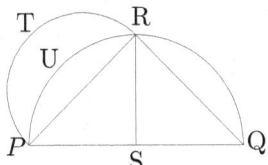

To solve this problem, Hippocrates of Chios formulates the hypothesis: Circles are as the squares on their diameters. The hypothesis is a sufficient condition for solving the problem. For, by the Pythagorean theorem, the square on PQ is twice the square on PR. Then, by the hypothesis, the semicircle on PQ, namely PRQ, is twice the semicircle on PR, namely PTR, and hence the quarter of circle PRS is equal to the semicircle PTR. Subtracting the same circular segment, PUR, from both the quarter of circle PRS and the semicircle PTR, we obtain the triangle PRS and the lune $PTRU$, respectively. Therefore, the lune $PTRU$ is equal to the triangle

PRS. This solves the problem. But the hypothesis is in turn a problem that must be solved, and it will be solved later on, presumably by Eudoxus. And so on, *ad infinitum*.

7 Static and dynamic approach to explanatory demonstration

Aristotle's distinction between 'demonstration that' and 'demonstration why', and Descartes' distinction between 'axiomatic demonstration' and 'analytic demonstration', suggest two different approaches to explanatory demonstration. Aristotle's distinction suggests a static approach, Descartes' distinction a dynamic approach.

According to the static approach, a demonstration of P is explanatory if it is an axiomatic demonstration that gives an answer to the question: Why is it the case that P?

According to the dynamic approach, a demonstration of P is explanatory if it is an analytic demonstration that gives an answer to the question: How can one arrive at P?

This means that, according to the static approach, a demonstration of P is explanatory if it is an axiomatic demonstration that shows the ground of the validity of P. On the other hand, according to the dynamic approach, a demonstration of P is explanatory if it is an analytic demonstration that reveals the way to the discovery of P, and specifically reveals to the researcher how to find a solution to the problem P, and to the audience how the solution to the problem P was found.

However, the above statement of the static approach needs to be refined. For, as van Fraassen points out, "being an explanation is essentially relative" because "what is requested, by means of the interrogative 'Why is it the case that P?', differs from context to context" (van Fraassen 1980, 156). In fact, "it is a use of science to satisfy certain of our desires", and "the exact content of the desire, and the evaluation of how well it is satisfied, varies from context to context" (van Fraassen 1980, 156).

For example, let P be the Pythagorean theorem and suppose we ask: Why is it the case that P? By this question we might desire to know, for example, 1) why it is the case that P for right-angled triangles and not for acute-angled or obtuse-angled triangles. Or 2) why it is the case that P in Euclidean space but not in certain non-Euclidean spaces. A demonstration of P might satisfy our desire 1) but not 2), and viceversa.

Since being an explanation is essentially relative, the above statement of the static approach must be modified as follows. According to the static approach, a demonstration of P is explanatory with respect to a certain context if it is an axiomatic demonstration that gives an answer to the question: Why is the case that P with respect to that context?

8 Descartes on published demonstrations

Descartes' distinction between 'axiomatic demonstration' and 'analytic demonstration' also suggests that there is a difference between how mathematical results are discovered and how they are presented in publications.

Indeed, after observing that "it was synthesis alone that the ancient geometers usually employed in their writings", Descartes states that "this was not because they were utterly ignorant of analysis", but rather because "they had such a high regard for it that they kept it to themselves like a sacred mystery" (Descartes 1996, VII, 156). They did so "with a kind of pernicious cunning" because, "as notoriously many inventors are known to have done where their own discoveries were concerned, they have perhaps feared" that their method, "just because it was so easy and simple, would be depreciated if it were divulged" (Descartes 1996, X, 376).

What is interesting in this statement is not so much the somewhat disputable claim that ancient geometers concealed their ways of discovery with a kind of pernicious cunning. It is rather the observation that many discoverers do not present their discoveries as they were made, that is, by the analytic method, but in a completely different way, specifically, by the axiomatic method.

This comes about because, as Davis and Hersh state, in mathematical papers "certain 'heuristic' reasonings" are "deemed 'inessential' or 'irrelevant' for purposes of publication" (Davis & Hersh 1986, 66). Or perhaps because discoverers are not fully aware of the processes by which the discovery came about, or feel uneasy to reveal that such processes were not rigorously deductive.

This shows a limitation of the present literature on explanation in mathematics, which uses demonstrations occurring in textbooks or papers as examples of explanatory demonstrations (for a survey, see, for example, (Mancosu 2008)). Such demonstrations do not reveal the way to their discovery, therefore they are useless as illustrations of explanatory demonstrations in the dynamic approach. This depends on the fact that the literature in question does not distinguish between the static and the dynamic approach.

9 Diagrams

In order to see what an explanatory demonstration looks like in the dynamic approach, it is useful to consider diagrammatic demonstrations, that is, demonstrations essentially based on diagrams.

The current literature on the subject tries to make diagrams fit into the axiomatic method. Thus Barwise and Etchemendy state that "diagrams and other forms of visual representation can be essential and legitimate components in valid deductive reasoning" (Barwise & Etchemendy 1996, 12). But this contrasts with the fact that, as Hilbert points out, in the axiomatic method "a theorem is only demonstrated when the demonstration is completely independent of the figure" (Hilbert 2004, 75).

Rather, diagrams naturally fit into the analytic method. To solve a problem, one draws a diagram and looks for some hypothesis that is a sufficient condition for the solution of the problem. The hypothesis is obtained from the problem, the diagram and possibly other data, by some non-deductive rule and must be plausible, that is, compatible with the existing knowledge. But the hypothesis is in turn a problem that must be solved, and will be solved in the same way. And so on, *ad infinitum*.

Considering diagrammatic demonstrations is useful because they more easily reveal the way to discovery. This is apparent from the earliest mathematical records, which show the notion of demonstration at its very formation.

For example, let us consider the demonstration that the area of a triangle is half of the base times the height, which is implicit in Problem 51 of the Rhind papyrus. The problem is: "What is the area of a triangle of 10 khet on the height of it and 4 khet on the base of it?" (1 khet = 52.5 meters), and the answer is: "Take 1/2 of 4, namely, 2, in order to get its rectangle. Multiply 10 times 2; this is its area" (Clagett 1999, 163).

The key to the demonstration is the observation that one should take 1/2 of 4 in order to get its rectangle. This suggests the hypothesis that a triangle is half the size of a rectangle with the same base and the same height.

Such hypothesis can be gathered from the diagram:

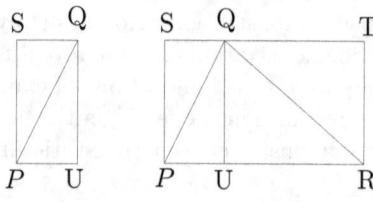

The diagram shows that, if the triangle PQU is right-angled, then PQU is half the size of the rectangle $PSQU$. Now, the height QU divides a triangle PQR into two right-angled triangles, PQU and RQU. Then PQU is half the size of the rectangle $PSQU$, and RQU is half the size of the rectangle $RTQU$. Therefore, the triangle PQR is half the size of the rectangle $PSTR$.

This is the key to the discovery, because it suggests the hypothesis that a triangle is half the size of a rectangle with the same base and the same height. The hypothesis is obtained from a single figure, thus by induction from a single case. From the hypothesis it follows that the area of a triangle is half of the base times the height.

10 Illusoriness of the justification role of axiomatic demonstration

It has been said above that the goal of axiomatic demonstration is to provide justification for a statement, thus axiomatic demonstration has a validation role, it is meant to establish the certainty of a statement.

This, in fact, is the received view. For example, Gowers states that an axiomatic demonstration is "an argument that puts a statement beyond all possible doubt" (Gowers 2002, 36). This "does make mathematics unique" (Gowers 2002, 40).

Similarly, Bass states that "the characteristic that distinguishes mathematics from all other sciences is the nature of mathematical knowledge and its certification by means of mathematical proof", that is, proof based on "the deductive axiomatic method", which makes mathematics "the only science that thus pretends to claims of absolute certainty" (Bass 2003, 769).

But this is unjustified. A statement established by an axiomatic demonstration cannot be more certain than the axioms on which the demonstration depends, and, by Gödel's second incompleteness theorem, no absolutely certain justification can be given for the axioms of any of the basic theories of mathematics. Then it is illusory to think that an axiomatic demonstration can establish the certainty of a statement, and mathematics is a science that can pretend to claims of absolute certainty.

At the origin of the belief that an axiomatic demonstration can establish the certainty of a statement there is foundationalism, the view that mathematics is a house with an absolutely secure foundation—the axioms. But mathematics is not that. Rather, as van Benthem says, it is "a planetary system of different theories entering into various relationships", so "damage one, and the system will continue, maybe with some debris orbiting here and there" (van Benthem 2008, 38).

11 The rhetorical role of axiomatic demonstration

If axiomatic demonstration cannot establish the certainty of a statement, what is its goal? It is rather to persuade the audience that the statement must be accepted. Then, as several people have argued, axiomatic demonstration can be said to have a rhetorical role.

Thus Hardy states that "proofs are what Littlewood and I call 'gas', rhetorical flourishes designed to affect psychology, pictures on the board in the lecture, devices to stimulate the imagination of pupils" (Hardy 1929, 18).

Davis and Hersh state that in mathematics "continual and essential use is made of rhetorical modes of argument and persuasion" (Davis & Hersh 1986, 58).

Kitcher states that in mathematics "a rhetorical function is served by the presentation of the proof", since "a proof presentation that is effective for one audience" can "be useless for others", and "'gas' is necessary even in professional mathematics" (Kitcher 1991, 7).

Resnik states that, to the question why the "proof of Euclid's theorem induces us to believe that there are infinitely many primes", we must answer that this occurs "because we have been prepared through our mathematical education" to "follow its reasoning" (Resnik 1992, 15). But we might give the same answer if mathematics were "an elaborate mythology passed from generation to generation by the priesthood of mathematics" (Resnik 1992, 16).

Because of the negative connotations often associated to rhetoric, one may dislike saying that axiomatic demonstration has a rhetorical role. Alternatively, one may say that it has a didactic role. But the substance is the same, both expressions referring to the capability of the demonstration to persuade the audience that a statement must be accepted.

That axiomatic demonstration has a rhetorical role has an implication for the static approach. If the goal of axiomatic demonstration is to persuade the audience that a statement must be accepted, this means that, according to the static approach, a demonstration of P that is explanatory with respect to a certain context is an axiomatic demonstration that is capable of persuading the audience that P must be accepted in that context—by giving an answer to the question: Why is the case that P with respect to that context?

That the goal of axiomatic demonstration is to persuade the audience that a statement must be accepted, clarifies why, as it has been said above, demonstrations occurring in textbooks or papers are useless as illustrations of explanatory demonstrations in the dynamic approach. Such demonstrations are intended to persuade the audience that the statement must be accepted, rather than to reveal how the demonstration was discovered.

Persuasion, however, plays some role also in the dynamic approach. The persuasion that a result is plausible often motivates the researcher to search for hypotheses toward a solution. This is an aspect of the general question of the role of emotion in mathematics.

12 Functions of explanatory demonstrations

That explanatory demonstrations in the static approach are intended to persuade the audience that a statement must be accepted, shows that such demonstrations have a social function. In fact, they can be addressed to a variety of audiences, from students to fellow researchers.

As Davis and Hersh say, mathematics "is a form of social interaction where 'proof' is a complex of the formal and the informal, of calculations and casual comments, of convincing argument and appeals to the imagination and the intuition" (Davis & Hersh 1986, 73).

The social function of explanatory demonstrations in the static approach is different from the creative function of explanatory demonstrations in the dynamic approach. What is essential to the latter is the capability of the demonstration to serve as a means of extending mathematical knowledge, by suggesting a hypothesis which is the key to the discovery of a solution to a problem.

Such is the hypothesis that a triangle is half the size of a rectangle with the same base and the same height, which was the key to the discovery, by an unknown Egyptian mathematician, that the area of a triangle is half of the base times the height.

In view of the different functions of explanatory demonstrations in the static and the dynamic approach, a distinction between these two approaches is quite natural. In particular, "the role of proof in class isn't the same as in research" (Hersh 1997, 59).

Then the alternative is not, as Lord Rayleigh says, between demonstrations which "command assent" and demonstrations which "woo and charm the intellect", evoking "delight and an overpowering desire to say 'Amen, Amen'" (Huntley 1970, 6). The alternative is rather between demonstrations which have a social function and demonstrations which have a creative function.

13 Relevance to mathematical practice

The existence of an objective distinction between explanatory and non-explanatory demonstrations is important only if it is relevant to mathematical practice. Now, the distinction as made above is actually relevant to mathematical practice, but the reason for its relevance is different in the static and the dynamic approach.

In the static approach, the distinction is relevant to mathematical practice, because explanatory demonstrations persuade the audience that a statement must be accepted. Thus they are essential to the acceptance of mathematics.

In the dynamic approach, the distinction is relevant to mathematical practice, because explanatory demonstrations extend mathematical knowledge, by suggesting a hypothesis which is the key to the discovery of a solution to a problem. Thus they are essential to the growth of mathematics.

In the dynamic approach, however, explanatory demonstrations are relevant to mathematical practice also in another respect. The growth of mathematics is often viewed as being cumulative. Mathematical discoveries are considered to be mere additions or increments to the growing stockpile of mathematical results.

Thus Devlin states that "mathematical knowledge is cumulative" (Devlin 1990, 33). This depends on the fact that "mathematics consists in making deductions from axioms" (Devlin 1990, 34).

This viewpoint, however, is appropriate only as concerns the addition of corollaries of known theorems, not as concerns the addition of really innovative results, that is, results which involve the introduction of really new ideas. The latter kind of addition reacts back upon some traditional parts of mathematics. By providing a new perspective on previously familiar objects, it affects the way in which those traditional parts are built up, leading to reconstruct them on a new basis. Since this involves making substantial changes to them, mathematical knowledge cannot be said to be cumulative.

The addition of really innovative results shows a basic feature of analytic demonstrations. The hypotheses introduced to solve a problem may establish new connections between the problem and the existing knowledge, which may lead to look at the latter in a new perspective and may also involve making changes to it. In the dynamic approach, explanatory demonstrations are relevant to mathematical practice also in this respect.

14 Global and local view of explanation

The static and the dynamic approach to explanatory demonstration involve a global and a local view of explanation, respectively. This depends on the fact that the axiomatic method has a global character, the analytic method a local one.

Indeed, in the axiomatic method, on which explanatory demonstration is based in the static approach, axioms serve to demonstrate, and hence to explain, all mathematical facts of a given theory. The explanation of all such facts depends on the same axioms, and hence is global. This motivates Sierpinska's claim, quoted in the Premise, that the quest for explanation in mathematics cannot be a quest for demonstration, but it may be an attempt to find a rationale of a choice of axioms, definitions, methods of construction of a theory. Indeed, in the static approach, the explanation stands not so much in the demonstration in itself as rather in the theory as a whole, hence in the rationale of the choice of axioms, definition, methods of construction.

On the other hand, in the analytic method, on which explanatory demonstration is based in the dynamic approach, the hypotheses for the solution of a problem are not general principles, good for all problems, but are aimed at a specific problem. In this sense, they are local rather than global. Not being global, they need not belong to the same field as the problem, but may belong to any field. For this reason, unlike the axiomatic method, which is incompatible with Gödel's first incompleteness theorem, the analytic method is compatible with it (see Cellucci 2008b). Moreover, being aimed at a specific problem, the hypotheses for the solution of a problem can take care of the peculiarities of the problem, and hence can spawn an explanation tailor-made for the problem. Thus, in the dynamic approach, the explanation can account for the peculiarity of the problem.

15 Explanation and understanding

A concept which has been traditionally considered to be strictly related to explanation is understanding. Indeed, explanation has been often viewed as a precondition for understanding. Thus Plato states: "As for the man who is unable to do so, wouldn't you say that, to the extent that he is unable to give an explanation of it, to himself or to anyone else, he has no understanding of it?" (Plato, *Republic*, VII, 534 b, 4–6).

Genuine understanding can come only from a genuine explanation. In particular, this can be said of demonstration. For, if a demonstration is explanatory, this founds the understanding of the demonstration on a deeper basis.

Once again, we must distinguish between demonstrations which are explanatory in the static approach and demonstrations which are explanatory in the dynamic approach.

In the static approach, that the audience is able to follow each step of a demonstration does not mean that they fully understand the demonstration. They may not see the idea of the demonstration, and hence may not get the deeper context.

Thus Poincaré says: "Does understanding the demonstration of a theorem consist in examining each of the syllogisms of which it is composed in succession, and being convinced that it is correct and conforms to the rules of the game?" (Poincaré 1914, 118) For some people, yes, but "not for the majority" since, as long as the syllogisms "appear to them engendered by caprice, and not by an intelligence constantly conscious of the end to be attained, they do not think they have understood" (Poincaré 1914, 118). Even when the logician will have cut up "each demonstration into a very great number of elementary operations" and will have "ascertained that each is correct", he will not "have grasped the real meaning of the demonstration" (Poincaré 1958, 21–22). For he will not have seized "that I know not what which makes the unity of the demonstration" (Poincaré 1958, 22).

Conversely, when a demonstration is explanatory in the static approach, the audience is able not only to follow each step of the demonstration, but also to see the idea of the demonstration.

On the other hand, in the dynamic approach, that the researcher or the audience are able to see the idea underlying the discovery of a demonstration, does not mean that they are able to see the full train of inferences which leads to the discovery.

Conversely, when a demonstration is explanatory in the dynamic approach, the researcher and the audience are able to see not only the idea underlying the discovery of the demonstration, but also the full train of inferences which leads to the discovery.

16 Understanding and intuition

That the audience is able to follow the full train of inferences which leads to the discovery of a demonstration, may give them the illusion that they could have discovered the demonstration themselves.

In fact, Poincaré states: "As I repeat an argument I have learned", it seems to me "that I could have discovered it. This is often only an illusion; but even then", I "rediscover it myself as I repeat it" (Poincaré 1914, 50).

Indeed, seeing the full train of inferences which leads to the discovery of a demonstration is all that is necessary to discovery. This means that discovery is an entirely rational process, which requires no appeal to intuition.

This contrasts with the widespread belief that discovery is only a matter of intuition, and hence cannot be accounted for rationally. What is rational is only demonstration, whose use is to check the suggestions of intuition.

Thus Wilder states that "it is the mathematical intuition that makes mathematics. Without it, we would have nothing to prove" since "the mathematical theorem comes from the intuition", and demonstration is only "a testing process that we apply to these suggestions of our intuition" (Wilder 1944, 318).

This view is inadequate because, on the one hand, the analytic method is a fully rational discovery procedure, and, on the other hand, by Gödel's second incompleteness theorem, demonstration is unable to test all the suggestions of our intuition.

If seeing the full train of inferences which leads to the discovery of a demonstration is all that is necessary to discovery, then discovery is not a matter of intuition but of logic—of course, non-deductive logic, because the analytic method essentially involves non-deductive rules.

Bibliography

Atiyah, M. (1988). *Collected Works*. Oxford: Oxford University Press.

Auslander, J. (2008). On the roles of proof in mathematics. In *Proof and other Dilemmas: Mathematics and Philosophy*, B. Gold & R. A. Simon, eds., Washington: The Mathematical Association of America, 62–77.

Barwise, J. & Etchemendy, J. (1996). Visual information and valid reasoning. In *Logical Reasoning with Diagrams*, G. Allwein & J. Barwise, eds., Oxford: Oxford University Press, 3–25.

Bass, H. (2003). The Carnegie initiative on the doctorate: The case of mathematics. *Notices of the American Mathematical Society*, *50*, 767–776.

Bromberger, S. (1992). *On What We Know We Don't Know*. Chicago: The University of Chicago Press.

Cellucci, C. (2008a). The nature of mathematical explanation. *Studies in History and Philosophy of Science. Part A*, *39*, 202–210.

Cellucci, C. (2008b). Why proof? What is a proof? In *Deduction, Computation, Experiment. Exploring the Effectiveness of Proof*, R. Lupacchini & G. Corsi, eds., Berlin: Springer, 1–27.

Clagett, M. (1999). *Ancient Egyptian Science, Vol. 3: Ancient Egyptian Mathematics*. Philadelphia: American Philosophical Society.

Davis, P. & Hersh, R. (1986). *Descartes' Dream. The World According to Mathematics*. Brighton: The Harvester Press.

Descartes, R. (1996). *Œuvres*. Paris: Vrin.

Devlin, K. (1990). *Mathematics: The New Golden Age*. London: Penguin Books.

Gowers, T. (2002). *Mathematics. A Very Short Introduction*. Oxford: Oxford University Press.

Hardy, G. (1929). Mathematical proof. *Mind, 38*, 1–25.

Hempel, C. & Oppenheim, P. (1948). Studies in the logic of explanation. *Philosophy of Science, 15* (2), 135–175.

Hersh, R. (1997). *What Is Mathematic, Really?* Oxford: Oxford University Press.

Hilbert, D. (2004). Die Grundlagen der Geometrie. In *David Hilbert's Lectures on the Foundations of Geometry (1891–1902)*, M. Hallett & U. Majer, eds., Berlin: Springer, 72–81.

Huntley, H. (1970). *The Divine Proportion. A Study of Mathematical Beauty.* Mineola: Dover.

Kitcher, P. (1991). Persuasion. In *Persuading Science. The Art of Scientific Rhetoric*, M. Pera & W. Shea, eds., Sagamore Beach: Science History Publications, 3–27.

Mancosu, P. (2008). Mathematical explanation: Why it matters. In *The Philosophy of Mathematical Practice*, P. Mancosu, ed., Oxford: Oxford University Press, 134–150.

Poincaré, H. (1914). *Science and Method.* London: Nelson.

Poincaré, H. (1958). *The Value of Science.* Mineola: Dover.

Popper, K. (1959). *The Logic of Scientific Discovery.* London: Hutchinson.

Popper, K. (1996). *In Search of a Better World.* London: Routledge.

Resnik, M. (1992). Proofs as a source of truth. In *Proofs and Knowledge in Mathematics*, M. Detlefsen, ed., London: Routledge, 6–32.

Resnik, M. & Kushner, D. (1987). Explanation, independence and realism in mathematics. *British Journal for the Philosophy of Science, 38*, 141–158.

Sierpinska, A. (1994). *Understanding in Mathematics.* London: The Falmer Press.

van Benthem, J. (2008). Interview. In *Philosophy of Mathematics: 5 Questions*, V. Hendricks & H. Leitgeb, eds., New York: Automatic Press/VIP, 29–43.

van Fraassen, B. (1980). *The Scientific Image.* Oxford: Oxford University Press.

Wilder, R. (1944). The nature of mathematical proof. *The American Mathematical Monthly, 51*, 309–323.

Carlo Cellucci
Sapienza University of Rome
Dept. of Philosophy
Italy
`carlo.cellucci@uniroma1.it`

X-Ray Data and Empirical Content

PAUL HUMPHREYS

With the demise of traditional empiricism, the project of reconciling scientific realism with an acceptable scientific empiricism becomes pressing. This paper will address two questions relevant to that project. The first is: Is it possible to make a reliable and justified inference from the data to the properties that generate the data? The second is: Can the use of techniques to correct 'raw data' be an epistemologically useful practice? The answer to both of these questions is 'Yes'. The first affirmative answer supports a form of scientific realism. The second affirmation suggests that a simple foundationalist attitude towards data is untenable. I shall discuss these questions within the context of a certain kind of causal-computational instrument, X-ray computed tomography scanners. These can provide us with a better insight into the answers to our questions than can purely causal instruments, such as optical telescopes, because the use of models and inferences within them is explicit, reasonably well understood, and in some cases can be assessed on a priori grounds. An additional benefit is that we can see how much more complex is scientific empiricism than traditional empiricism while allowing the former to act as a justifiable basis for scientific knowledge.

By 'scientific realism' I mean a selective position that asserts the existence of specified entities for which there is substantial scientific evidence but denies that all entitites referred to in successful scientific theories exist. A realist position in which all non-mathematical terms that occur in successful scientific theories, models, or simulations refer to entities in the natural world is indefensible. Idealizations, quantitities introduced for computational convenience, stylized facts, the subject matter of toy models, and many other things occur in scientific representations, often with an explicit denial by their users that they are committed to the existence of anything that corresponds to such conveniences. Perfectly smooth continuous surfaces on macroscopic objects, artificial viscosity, that the complementary cumulative probability distribution of real returns in financial markets satisfies a power law, and artificial insects in a model of social cooperation are, respectively, examples of the four types of unreal components listed above. Regarding empiricism, an acceptable scientific empiricism is one that accords a special place to data with empirical content that is gathered and processed according to approved scientific practice but that does not uniformly give priority to data that are accessible to human sense perception. To so restrict oneself is to deny that instruments can in some cases provide epistemically superior data.

If one is a realist, at least some data have empirical content that is generated by something distinct from the data themselves and moving from the data to the generating source requires an inference. It is these inferences at which traditional empiricists balk. Logical constructions from data reports are acceptable, inferences to unobservables are not. From the realist perspective, in conditions where well established knowledge of the processes that generated the data is available, that attitude is unnecessarily risk-averse.

An epistemological principle, which we can call the Data Principle, is frequently used in traditional empiricism: *The closer we stay to the raw data, the more likely those data are to be reliable sources of evidence.* Here is a typical presentation of that view: '[...] we must observe [the world] neutrally and dispassionately, and any attempt on our part to mould or interfere with the process of receiving this information can only lead to distortion and arbitrary imagining' (Lacey 1995, 226). Although the Data Principle had sound reasons for adoption by traditional empiricists on the grounds that any theoretical interpretation of data undermined their objectivity, and that any inference from raw data reduced the certainty associated with that data, an empiricism suitable for science should reject the principle. It is not just, as Suppes (1962) argued, that the judicious use of models is necessary to connect theory with data, although that is certainly true. It is that raw data coming from certain kinds of scientific instruments must be corrected and understood to improve the representational accuracy of these instruments. The better we are able to correct the raw data, the better the representations from those instruments will be.

1 X-ray Computed Tomography

I take as my running example X-ray computed tomography (CT) imaging devices. Although other imaging modalities such as positron emission tomography, magnetic resonance imaging, and single positron emission tomography use related computational techniques and many of the conclusions reached here generalize to those instruments, the physics underlying those devices is different from that occurring in CT scanners and this requires subtle but important modifications to the methods here described.

The mathematical methods behind CT image construction are complex and probably unfamiliar to many philosophers. For one of the few philosophical discussions see Israel-Jost (2011). A good introduction to the scientific and mathematical aspects of CT is Kak & Slaney (1988). I shall provide what I hope is a helpful outline of the main techniques but I stress that the current generation of scanners contains many more construction and correction algorithms than I can describe here. It is also worth keeping in mind that similar mathematical techniques are applicable in other sciences such as astronomy and geophysics in some of which, unlike medical imaging, a direct check on the accuracy of the images is not possible. Thus, although this direct accessibility to the target objects is epistemologically helpful, it is not essential in most cases to drawing the conclusions reached here.

One aspect of CT scanners that is relevant to selective scientific realism concerns the physical processes used by these instruments. X-rays were first systematically studied in 1895 and in the many years since, we have learned an enormous amount about their properties. To suggest that we do not know that they exist or that we should remain agnostic about their existence is to allow evidential concerns about genuinely problematical cases, such as the possible existence of dark energy, to affect our attitude towards entities that are of the same kind as well-understood entities such as visible light. Perhaps some of what we claim to know about X-rays is incorrect but it is sufficiently unlikely that most of what we claim to know about X-rays will turn out to be false that to reject realism in this realm is to engage in a level of scepticism that is unwarranted in scientific pursuits. We can accommodate those worries to a certain extent by adopting an ontology of properties rather than objects. That is, we take X-rays to consist in a cluster of property instances. Then claims concerning what X-rays 'really are' can be fallible in the sense that a few of those properties may turn out to be absent when X-rays are generated while acknowledging that there exist well established properties such as diffractibility, energies in the region 120 eV to 120 keV, being associated with an increased risk of cancer, and so on.[1] That is the realist side of our position with respect to features of the instrument that are present regardless of which target the instrument is applied. The selective side of the realism enters when we consider the image of a specific target.

The reason for having to be selective about our scientific realism is the presence of artifacts of the instrument; features of the output image that are

---

[1] The exact classification of X-rays is in any case a fluid matter. The older criterion that distinguished X-rays from γ rays on the basis of wavelengths has largely been supplemented by the criterion that γ rays originate in the atomic nucleus whereas X-rays have an electronic origin.

the result of the physical or computational processes that are used in image generation and that do not represent features of the target. Not everything displayed in the outputs of CT scanners corresponds to something real in this sense: the representational content of at least one part of the image does not correspond to the real structure of the associated part of the object being imaged even if all parts of the image have their causal origins in the target.

Data from medical imaging devices are, as with other types of instruments, about specific individuals. Although in some cases the primary purpose is to make an inference about the individual being imaged, in other cases, especially in academic settings, inductive inferences to a wider class of subjects are the goal. Yet even in cases where a specific individual is the object of study and one might have worries that this specificity rules out the use as scientific, the concerns are misplaced. Galileo's telescopic observations of the Moon were no less scientific than were his observations of bodies moving on inclined planes. The broader point is that theories may need to be general but scientific data do not.

The basic set-up for X-ray computed tomography consists of:

(a) a target object, located within a fixed coordinate frame that I call the *target frame*. Target frame positions are represented by Cartesian coordinates (x, y). In practical applications the target object will usually be a biological entity with an internal structure that differentially attenuates X-rays that pass through it. For example, bone attenuates the X-rays to a greater degree than does muscle. Physical calibration of the instrument is implemented using specially designed targets with known attenuation coefficients such as water and computational calibration is carried out on phantoms – artificial figures – such as the Shepp-Logan figure.[2]

(b) M sources of X-rays and an array of M detectors. The sources, regularly spaced Δr apart, are attached to a frame that rotates around the target within a two dimensional plane. Attached to this frame is a row of detectors also spaced Δr apart, each of which is directly opposite one of the sources. This rotating frame defines the *detector frame* and is represented by the polar coordinates (r, θ). To keep things as simple as

---

[2] Calibration using phantoms has a similar motivation to those at work in random number generators – the properties of mathematically generated random numbers are known exactly, while those generated from physical processes are not. Analytic results are available for projections of the ellipses used in the Shepp-Logan phantom.

possible, I consider the situation in which the beams of X-rays emitted by the sources are parallel.[3]

After each firing of the X-ray sources, the detector frame is rotated around the target by θ degrees. There will be $N = \pi/\theta$ values of θ between 0 and π. Values between π and 2π in the planar case are unnecessary because of symmetry. An important part of the computations involved in CT concerns transformations between the target and the detector frames.

2 Empirical content of data

Let $f(x,y)$ represent the value of some spatially distributed quantity. The central task is to calculate the values of $f(x,y)$ at various points within the target frame using values of detector counts. Before proceeding we need to disambiguate the use of 'empirical content'. We can attribute empirical content to data considered as concrete entities that occur as elements of causal processes and also to data considered as representational entities. The latter can take a number of different forms including perceptual entities, linguistic entities, graphical images, and so on. (See Humphreys (2000); Kulvicki (2010); Vorms (2012) for discussions of different representational formats.) In what follows context will determine when notation picks out a concrete datum and when it denotes a representational datum. Next, define a datum as either a) any entity of the form $R(a_1, a_2, ..., a_n, s)$, where R is a (possibly compound) concrete relation or property and s is the spatiotemporal location of the property instance or b) a representational correlate $R(a_1, a_2, \ldots, a_n, s)$ such as a predicate instance or a set of spatial relations between pixels forming an image. A common use of 'empirical content' asserts that a concrete datum has empirical content just in case it can be accessed by whatever means the empiricist tradition in question considers permissible. I leave these means deliberately unspecific in order to cover different versions of traditional empiricism including sense-data empiricism, sub-species of logical empiricism, and constructive empiricism. More liberal is the position taken here that a concrete datum does not have to be accessible by the unaided human senses but it must be detectable by some reliable scientific instrument that is currently available. The empirical content of a datum can then be identified with the property instance, a position that is argued in detail in Humphreys (2004, chap. 2). This definition has the

---

[3]Fan beams require greater mathematical sophistication but do not introduce any essentially new philosophical issues. Continuous 3-dimensional tomography uses a spiral path for the detector frame for which there are important additional complications that will not be considered here. Even if the specific instruments described here are no longer used in practice the conclusions drawn about the compatibility of selective scientific realism and scientific empiricism remain.

consequence that data can lose their empirical content. One example would be an electron microscope image stored in an obsolete medium for which readers no longer exist. Although one could take the view that the data retain the disposition to be accessed, this would require abandoning the position that what is important for science is what can be observationally accessed in practice.

These degrees of strictness cover a division within empiricist traditions. One branch is opposed to rationalist methods of inquiry and so the approved means of access explicitly preclude the exclusive use of non-perceptual intuitions, a priori reasoning, and similar methods to arrive at knowledge. The other branch of empiricism is opposed to realism; the usual foil is scientific realism so that the use of scientific theories and instruments that, respectively, refer to or detect unobservables, are impermissible under traditional empiricism. All varieties of empiricism should fall into the first branch if the requirement that our knowledge has empirical content is correctly formulated but our goal is to avoid the second branch. A priori methods do play a role in empiricism in that the domain of empirical content is closed under computable transformations, where 'computable' means 'computable by an existing device within time constraints set by the nature of the problem'.

Data sets can be adjusted, a process that includes throwing away data and inserting data. The topic of whether to keep or throw away anomalous data is an important one but it has been discussed elsewhere and I shall not address it here. Data can be added to a data set either by providing additional raw data, by generating artificial data, or by transforming raw data. Transforming data involves taking existing data and performing physical or formal transformations on them. In the case of CT instruments, this can involve coordinate transformations, interpolations between existing data values, the use of other mathematical operations on data, and applying physical operations to data.

Physical and formal transformations do not always preserve reference to reality even when they preserve empirical content. Consider the representation of the scalar value of the velocity of a macroscopic object, v. This is a datum with empirical content, we know how to measure it, and it represents something real. Now consider the variable $v^{17.34}$. We know how to measure what it represents – measure v and apply the appropriate mathematical transformation. So it also has empirical content. But the variable $v^{17.34}$ does not correspond to anything real according to the current state of scientific knowledge. A specific value of $v^{17.34}$ can, by accident, correspond to something real, such as the exact value on another occasion of v but that does not entail that there is a property corresponding to $v^{17.34}$. Compare this with length l. If we take a particular value for the length, say 2 feet, and square the numerical value, we have another value of length, 4. If we take the numerical variable l^2, that will represent area, which is clearly a different property from length. Although empirical content is preserved by mathematical transformations, they do not always preserve the property of being observable under traditional criteria. If l represents the length of an object that is in the domain of the observable, then l^{-100} will not be an observable. Many empiricist traditions have held that any representation that is explicitly definable using only observable terms will itself count as an observable term, perhaps because it would be eliminable via the definition, but that reasoning falls prey to the example just given.

Now consider a physical transformation of the datum v by means of an interaction with some physical system. Under some such transformations, the empirical content will be lost altogether in the sense that the interaction will convert the datum into a property instance that cannot be detected by current technology. In other cases the empirical content will change as the interaction changes one physical property into another. This is why identifying the empirical content of a datum with its source is not to be recommended, since data from the same source can be physically transformed in different ways.

Which functions of data are taken to represent instances of real properties cannot be read off the representations. The selection is made by the users of the representation not, as social constructivists would have it, on the basis of a social consensus, but on the basis of evidence that is often subject matter specific.

The overall moral is then that when asking whether a datum subject to transformations has empirical content, we must pay careful attention both to the type of transformation (physical or mathematical) and to the type of data (concrete or representational).

3   Basic CT computations

Returning to the general case of CT procedures, $f(x,y)$ represents the attenuation coefficients for X-rays at points (x,y) in a two dimensional plane through the three dimensional object. f allows us to calculate the amount by which a given X-ray beam is attenuated over a distance Δy where Δy is the dimension of one pixel in a discretization of the target space. Although the degrees of attenuation are themselves a function of the densities of various types of tissue and these densities in turn correspond to the presence or absence of various anatomical features, for simplicity I shall deal only with the function f and its values. The underlying point is, nevertheless, important; what is 'observed' is a function of the spatial distribution of attenuation coefficients of the target and an interpretation or inference is needed to move from this spatial distribution of coefficients to the existence of biological features. Standard visual images consist in a spatial distribution of reflectances in the optical range; here I am generalizing this concept to allow an image to consist in the spatial distribution of any physical quantity. I retain the spatial aspect of images in part to maintain a connection with the idea that these are imaging devices producing a visual output but the values of the function f could be displayed as a matrix of numerical values or in some other representation.

Call a raw datum a datum to which no transformations external to the instrument have been applied – that is, the datum is part of the instrument's direct output. What is counted as a raw datum thus depends upon where the boundary of the instrument is drawn. In the present case, a raw datum consists of an X-ray count at one detector. For a given value of θ the elements of the raw data set are the counts received at each of the M detectors at positions $r_1, r_2,...,r_n$. From these and knowledge of the initial intensities of the X-rays at the source it is easy to calculate the total attentuation of the X-ray beam along each of the M parallel rays. The central task is to reconstruct the values of $f(x,y)$ at specific points along those rays. The most common method of reconstruction uses *filtered backprojection*.

In broad outline, filtered backprojection contains these steps:

Step 1 Calculate the total attenuation along a given ray between a source and its detector by integrating the values of f along that ray

Step 2 Convolve these spatial projections with a filter (a weighting function) to compensate for geometrical frame changes

Step 3 Fourier transform these convolutions into the frequency domain to facilitate computations

Step 4 Compute the convolutions in the frequency domain

Step 5 Inverse Fourier transform the results back to the spatial domain

Step 6 Compute the inverse Radon transforms in the spatial domain to arrive at values of $f(x,y)$ at the desired points (x,y) within the target frame.

We see here that the 'observed image'—that of a human hip, for example – is an entity that is constructed from classically unobservable data using explicitly considered transformations. Whereas sceptical doubts about unobservables are appropriate in cases where these transformations are unknown, in the current case a refusal to distinguish between real but traditionally unobservable entities and processes and non-existent and traditionally unobservable entities and processes to construct corrections to the image is epistemically counter-productive.

Before discussing the details of these steps, some general issues concerning the switch from continuous to discrete models is necessary. In the idealized case where we have a continuous set of parallel beams and a continuous set of detectors, we would have the entire profile of projections along r. In the actual case, we have to deal with these constraints:

1. The measurements consists in a discrete set of M equally spaced point values along r and this gives rise to the need for discrete models, with the accompanying need for approximations.

2. The discrete data set gives rise to an underdetermination problem that must be addressed.

3. Realism requires us to identify artifacts which can arise either from physical or from computational sources. To avoid physically generated artifacts, correction models to supplement the basic model are required because of noise and systematic bias in the measurement. Correction models to compensate for the kind of computational errors discussed below are also required.

This transition between the idealized methods that use continuous mathematics and the discrete models used to represent the physical situation requires attention. In the present case, the spatially continuous function f is rendered discrete both by the use of pixels and by a truncation method such as a finite series approximation to the values of f. These finite approximations are important because they can often make an inverse inference problem ill-posed. An ill-posed problem, in cases where a unique solution exists, is one in which small changes in the data can result in large changes in the solution values. In such cases the discrete approximations must be examined in order to avoid an originally well-posed problem becoming ill-posed so that small errors in the data lead to large errors in the solution. This mathematical pitfall is a major difference between causal-computational and purely causal instruments.

We can now consider each of the six steps outlined above.

Step 1 Consider the detector frame when it is oriented at an angle θ to the target frame. Each ray can be represented mathematically by the line parameterized by r and θ: $L_\theta(r) = \{(x,y) : r = x\cos(\theta) + y\sin(\theta)\}$ where r is the radial coordinate. The total attenuation along the line L is given by $\int_L f(x,y) dL$. This represents projected values of $f(x,y)$ along the ray orthogonal to the r axis of the detector frame when it is oriented at angle θ

to the target frame. As a reminder, the empirical content of the raw data are the X-ray counts; from those one can calculate the total attenuation value along a given ray. To calculate this value using the target frame coordinates we have:

(1) $$(1) P_\theta(r) = \int_{y=-\infty}^{\infty} \int_{x=-\infty}^{\infty} f(x,y)\delta(x\cos(\theta) + y\sin(\theta) - r)dxdy$$

which is the Radon transform of f over L. $P_\theta(r)$ is the X-ray intensity at the detector located at r in the detector frame. The coordinate transformations used on the right hand side of (1) do not alter the empirical content of the raw data; they are simply different representations of the same data set.

Step 2 Given $P_\theta(r)$, it would seem that if we performed an inverse Radon transformation for every value of θ, we could reconstruct an image of the object. There is one important caveat to note here. Because all of the N values of θ are involved in the inverse transform, a reconstructed value of f is a function of the data for all N values of $P_\theta(r)$, that is of the entire data set. There is thus a holistic aspect to the reconstructed data and this has the consequence that defects in the data at one point in the target can be transferred to many other points in the reconstructed image. Once we move to the discrete versions of the backprojection method that must be used in practice, simple backprojections produce a star effect around point sources due to the finite number of projections and blurring of the image occurs. A generalization of this problem results from the fact that the backprojection method regularly spaces values along the radial lines centered on the origin within the detector frame, and within that frame the data become sparser as we move towards the periphery of the object. Because of this the final image, which is represented within the Cartesian target frame, becomes less well defined, a feature known as 1/r blurring, which affects all regions of the image.

To correct for these problems, before we back-project the projection must be convolved with a weighting function (called a 'filter') to give $P_\theta(r) * h(r)$ This operation constitutes the filtered part of the filtered backprojection method.

The convolution $f * h$ of a function f with a shift-invariant operation h is defined by

$$z(t) = \int_{-\infty}^{\infty} f(t')h(t-t')dt'$$

In contrast to the coordinate transformations, the convolutions do not result in the same local empirical content. Rather, they redistribute the empirical content over the detector frame in a way that corrects for the different geometry of that frame. Globally, but not locally, the empirical content is the same and the convolution corrects for distortions introduced by the data transformations. We have now seen the extent to which data must be corrected in these instruments and in each case the corrections are epistemically advantageous rather than harmful.[4]

Steps 3 and 4 We now desire to infer from the values of $P_\theta(r)$ the values of $f(x,y)$ along a plane lying within the object. The key to doing this is the Fourier Slice Theorem which asserts:

The Fourier transform of a projection of a function $f(x,y)$, taken at an angle θ to the target frame, gives a slice of the two dimensional transform $F(u,v)$ that subtends an angle θ with the u axis in the frequency domain.

That is, if we take the calculated projections $P_\theta(x)$ in (1) above onto the spatial detector plane and Fourier transform them into the frequency domain, this gives the Fourier transform within the frequency domain along a plane passing through the target object as long as that plane also passes through the origin of the spatial X-Y coordinates. Thus, given the data values at the detectors for a given value of θ, by performing a Fourier transform on those values, we can obtain the values of f along a plane in the object.

This has the additional advantage that because convolutions are computationally intensive but a convolution in the spatial domain Fourier transforms to a multiplication in the frequency domain, it is a significant advantage to carry out these computations in the Fourier domain. At this point we are far from the raw data.

Steps 5 and 6 These steps require less comment than the previous steps but in the next section I shall discuss some of the modeling steps that are required.

---

[4]h is often the ramp filter which weights each value of the Fourier transform by v, the frequency in the Fourier domain. But there is a trade-off involved because this weighting increases the high frequency noise and to increase the signal to noise ratio the ramp filter is itself multiplied by a window function such as the Shepp-Logan, the Hamming, or the mode-p Butterworth. The backprojection is carried out in the spatial domain whereas the filter is added in the frequency domain and has different formal properties in each. For example, the Ram-Lak filter in the spatial domain becomes the ramp filter in the frequency domain. In the spatial domain the Ram-Lak filter is given by $h(r) = 1/2\tau^2[sinc(r/\tau)] - 1/4\tau^2[sinc^2(r/2\tau)]$ The Hamming filter does better for improving the signal to noise ratio but worse on reducing the star effect.

4 Inverse inferences and underdetermination

The process of inferring the values of f at specific points from the total attenuation values is what is called an inverse inference problem.[5] Underdetermination problems are inescapable for empiricist positions and inverse inferences are no exception. We can carry out Radon transforms only up to a constant of integration, but there is a much more dramatic underdetermination problem that illustrates how the use of discrete data sets requires choices in practice that are not required in the continuous case. The result is this (Herman 1980, 283–286):

Let $Rad[(f(r,\theta)]$ represent the Radon transform of the function f at (r, θ). Let M, K be positive integers and for $1 \leq k \leq K$, let (r_k, θ_k) be distinct points in the interior of the target. Let l_k be arbitrary real numbers. Then there is a continuous function g such that for $1 \leq k \leq K$, $g(r_k, \theta_k) = l_k$ and for $0 \leq m \leq M - 1$ and for all l, $Rad[g(l, m\pi/M)] = 0$.

Thus given any function f that truly represents the target property, there exists a function $f + g$ such that f and $f + g$ have the same line integrals along each of the M X-ray beams but f and $f + g$ differ by an arbitrarily large amount at each of an arbitraily large finite number of points in the image plane. The proof of this result rests on the availability of an oscillatory function such that its Radon transform is invariant under scaling and translation and at each of the M line integrals has value 0. Philosophers will recognize this result as a proof in this specific application of the abstract point that any function is underdetermined by a finite data set. In practice this underdetermination does not prevent construction of a near-correct image. The reason is that by setting the number of detectors M at a number beyond the minimum required by the Nyquist sampling theorem, it is unlikely that the target object will exhibit periodicities that replicate those that underlie g. In the case of a one-dimensional function, the Nyquist sampling theorem requires that in order to avoid aliasing – the appearance of artifacts in the image – values of the function must be sampled at least twice during each cycle of the highest frequency contained in the spectrum of the continuous function (see Buzug 2008, 135).

5 Artifacts and noise

An artifact is a systematic discrepancy between the real attenuation values and the values inferred from the measurements taken at the CT detectors. (For a detailed assessment of artifacts, see Humphreys 2013). Artifacts have a number of different sources, both physical and computational, and cor-

---

[5] Inverse inferences are a rich source of inductive knowledge, largely ignored by philosophers of science. I shall discuss the inductive aspects of inverse inferences in a future paper.

rections need to be made to the data to eliminate or reduce these artifacts. The need to correct for artifacts does not distinguish causal instruments from causal-computational instruments because lenses are physically corrected for chromatic aberration by using compound lenses. There are many possible sources of error that require correction but here I shall discuss one representative case, beam hardening. This occurs when the mean energy of the X-rays increases as they pass through matter because the lower energy X-rays are absorbed at a higher rate. This means that the effective linear attenuation coefficient of tissue decreases with distance from the X-ray source because the attenuation coefficient is a function of X-ray energy. The overall effect of higher energy X-rays is an increased incidence of Compton scattering and an accompanying loss of contrast in the image. Correction methods for this are applied before the image reconstruction takes place.

We can use this example of beam hardening to illustrate how our access to real, rather than artifactual, features changes with time. Algebraic and statistical reconstruction methods are superior to filtered backprojection in treating some cases of beam hardening but were not used for many years because of the excessive computational loads they require. With advances in technology, they are becoming feasible in practice and with them an increased ability to distinguish real features from artifacts. Data based knowledge is not founded upon a static epistemic basis as is traditional empiricism.

In addition to artifacts, noise is a factor in almost all instruments of this type. At a minimum there is a tradeoff between the achieved sharpness of the constructed image and keeping the signal to noise ratio at an acceptable level. For example, the correction factors used to reduce $1/r$ blurring introduce increased noise because the filter increases high frequency components in the Fourier domain. In optical instruments there has always been a tradeoff between different types of optical aberration and it is impossible to simultaneously optimize these errors for all wavelengths of light or over the entire surface of most lenses for monochromatic light. Here we have a similar situation but for computational corrections. The fact that raw data are not ideal should not prevent us from correcting them.

Suppose that the true values of a quantity y are given by a probability distribution $f(y)$ and that the noise distribution is given by $g(z)$. If we assume that the noise is stochastically independent of the value of f, then the probability that a given value of f will be shifted an amount $z - x$ to the value z by the noise contribution g is $f(x)dx.g(z-x)dz$. Then, integrating over all the values of x that give rise to $f(x)$ gives $\int_{-\infty}^{\infty} f(x)g(z-x)dx$

$= f * g = h(z).$[6] $h(z)$ is the value of the output of the instrument and will ordinarily have a greater spread of values than f and be biased if g is biased. If f is a delta function, so that there is no stochasticity in the underlying physical variable, then the convolution places a copy of the noise distribution at each value of that variable.

Whereas the usual representation of error in linear models is with an additive function, the convolution is a general representation of how system values and noise combine. More important for us is the inverse of convolution, deconvolution. If we have the output distribution h and have a good model of the noise function g, then in certain cases we can recover the system distribution f. The important point is that in some cases we have knowledge of the distribution of noise in particular parts of the instrument and since we have access to the output distribution, we can compute the system distribution. Many of the philosophical discussions of noise operate under the assumption that all we have available is the pattern of data that constitutes the output from an instrument. This assumption is false in some cases. Because we have constructed these instruments, we are in a different epistemic relation with respect to noise originating with them compared to noise originating from a naturally occuring system such as a galaxy. Instruments are not experiments but the two share the feature that strict controls can be placed on independent variables and on some sources of noise.

This is one more reason why knowledge of how the imaging device works is crucially important in improving its accuracy. In the early seventeenth century, Leewenhoek and Galileo could construct optical microscopes and telescopes, respectively, using trial and error to produce a reasonable image while having nothing close to a correct theory of how they worked. In contrast, it would be impossible to construct a CT scanner without a considerable amount of knowledge of applied mathematics.[7]

---

[6]The convolution does not add the noise to the true value, nor is the convolution the composition of two point functions The notation here is also important: z and x are representations of the same physical variable but will in many cases have different values for that variable.

[7]Thanks are due to audiences at the Knowing and Understanding Through Computer Simulations conference, Paris; the 14th Congress on Logic, Methodology, and Philosophy of Science, Nancy; the Plurality of Numerical Methods in Computer Simulations and their Philosophical Analysis conference, Paris; the Computer Simulations and the Changing Face of Scientific Experimentation, Stuttgart; and the Models and Simulations 5 conference, Helsinki for helpful comments on talks related to this topic. I am especially indebted to George Gillies of the Department of Mechanical and Aerospace Engineering at UVA for correcting errors in the penultimate draft.

Bibliography

Buzug, T. (2008). *Computed Tomography: From Photon Statistics to Modern Cone-Beam CT.* Berlin: Springer.

Herman, G. T. (1980). *Image Reconstruction from Projections: The Fundamentals of Computerized Tomography.* New York: Academic Press.

Humphreys, P. (2000). Scientific knowledge. In *Handbook of Epistemology*, Niiniluoto, I., Sintonen, M., & Woleński, J., eds., Dordrecht: Kluwer Academic Publishers, 549–569.

Humphreys, P. (2004). *Extending Ourselves.* New York: Oxford University Press.

Humphreys, P. (2013). What are data about? In *Computer Simulations and the Changing Face of Scientific Experimentation*, Durán, J. & Arnold, E., eds., Cambridge: Cambridge Scholars Publishing, 12–28.

Israel-Jost, V. (2011). *The impact of modern imaging techniques on the concept of observation: A philosophical analysis.* unpublished, Université de Paris 1 Panthéon-Sorbonne – IHPST.

Kak, A. & Slaney, M. (1988). *Principles of Computerized Tomographic Imaging.* New York: IEEE Press.

Kulvicki, J. (2010). Knowing with images: Medium and message. *Philosophy of Science*, 77, 295–313.

Lacey, A. (1995). Empiricism. In *The Oxford Companion to Philosophy*, Honderich, T., ed., Oxford: Oxford University Press, 226–229.

Suppes, P. (1962). Models of data. In *Logic Methodology, and Philosophy of Science: Proceedings of the 1960 Conference*, Nagel, E., Suppes, P., & Tarski, A., eds., Amsterdam: North-Holland, 252–261.

Vorms, M. (2012). Formats of representation in scientific theorizing. In *Models, Simulations, and Representations*, Humphreys, P. & Imbert, C., eds., chap. 13, London: Routledge.

Paul Humphreys
Corcoran Department of Philosophy
University of Virginia
Charlottesville VA 22904-4780
USA
pwh2a@virginia.edu

A Priori Principles of Reason

WOLFGANG SPOHN

ABSTRACT. Basically, A is a reason for B, if A speaks in favor of B, or makes B more plausible, or is positively relevant to B. And basically, a doxastic feature is a priori, if all rational doxastic states have that feature. The paper will unfold these notions. And it will argue, partially in a deductive way, for a series of a priori principles about the structure of reasons, starting with a basic empiricist principle (capturing the gist of the positivists' verifiability principle) and ending up with a weak principle of causality (which thus turns out to be a priori, after all).

1 Introduction

As my title indicates, I would like to present various a priori principles of reason: a basic empiricist principle, as I would like to call it, some coherence principles, principles about the connection between truth and reason, etc. They are familiar, indeed venerable. What my paper will add are precise explications of those principles and rigorous relations between them. Just in order to make you curious, I will at last derive a weak principle of causality from a principle characteristic of pragmatic truth. This connection sounds surprising, and in view of the recent persistent silence on the principle of causality this result is certainly alerting. Let me work up to those principles and relations.

These announcements indicate that this paper will be a *tour de force* through large and heavy philosophical terrain. It will be clear that nearly every paragraph refers to a lot of literature and could—and should!—be more thoroughly argued. However, the required brevity has its benefits, too. In this succinct presentation, the intended line of reasoning should stand out much more clearly, and even though I cannot argue for each point, it will be clear which points would need to be argued for. Thus, the paper will also produce an argumentative map of its philosophical terrain. This is, I hope, valuable by itself.

This paper is essentially an attempt at informally summarizing chapter 6 and sections 17.2–4 of Spohn (2012). Hence, a side benefit to the reader will be that he or she gets relatively easily accessible information about what goes in the last chapter of this book without having to fight through it in its entirety.

2 Reasons and apriority

We have first to focus on the two basic notions in my title: apriority and reasons. Let me take up the latter first. The word "reason" has certainly various uses. We might say, though, that reason—without a determiner—is the capacity to reason, i.e., to have, give, and accept reasons. So, the principles to be presented will in fact be about reasons—more precisely, about theoretical reasons or reasons to believe, not about practical reasons or reasons to act.

What are reasons to believe? I find, our basic notion is that an assumption or a belief, i.e., a belief type or belief content, i.e., a proposition, A is a reason for another assumption, belief, or proposition, B, if A supports or confirms B, if A speaks in favor of B—we have many words for the same thing—, that is, if A strengthens the degree of belief in B, or if B is more credible given A than given non-A, in short: if A is positively relevant to B.[1]

If we want to make this idea precise, we obviously have to refer to degrees of belief. In fact, we have to refer to conditional degrees of belief. Let them be represented by some belief function β. Then the basic notion of reasons is this:

Definition 1: A is a reason for B w.r.t. the belief function β if and only if $\beta(B|A) > \beta(B|\bar{A})$. I call this the *positive relevance notion* of a reason.

Of course, all rigorous theorems depend on a precise specification of that belief function β. Various proposals might work. The first idea is that β is a probability measure. My subsequent considerations work best, I find, when β is interpreted as a ranking function; cf. (Spohn 2012, chap. 6). Perhaps there are further alternatives. In this paper I will remain informal. We need not go into the large issue of how to formally represent belief and degrees of belief. The important message is that the notion of a reason presupposes some workable account of conditional degrees of belief, and that every suitable account unfolds into a theory of reasoning via Definition 1.

The positive relevance notion of a reason is entirely subjective, i.e., relative to some belief function β, which characterizes the belief state of some subject. Most philosophers are not satisfied thereby; they strive for a more

---

[1] The literature abounds in more or less vague notions of a reason, of a deductive, or a computational, or a causal, or in some way inductive kind. In Spohn (2001) I have argued the positive relevance notion to be the basic one.

objective notion of a reason. This is a most delicate issue. There is intersubjective pressure and agreement; relative to my belief function your reasons may appear unreasonable, and if so, I will criticize you and you might agree. Often, only true reasons count as reasons and false reasons at best count as would-be-reasons. This is one objectifying move, which I don't take here. So, keep in mind that I will use the reason relation in a non-factive sense; false assumptions may also be and have reasons. There are further objectifying moves (see Spohn 2012, chap. 15). However, I will not pursue this issue here. In any case, I am convinced that all more objective notions of a reason build—indeed must build—on the basic subjective positive relevance notion of a reason. This basic notion will do for the rest of my paper.

Other notions of a reason may come to your mind. The most salient one certainly is that of a deductive reason: A is a deductive reason for B if A deductively entails B. However, according to the afore-mentioned main interpretations of the belief function β, this entails that a consistent A is positively relevant to, and thus a reason for, a non-tautological B. Hence, the positive relevance notion encompasses the deductive notion. And it allows also for non-deductive or inductive reasons, as it obviously must do.

The positive relevance notion has a simple, but important consequence. Belief change or revision usually proceeds by conditionalization; the posterior degrees of belief we move to are the prior ones conditional on the given evidence. This means that belief change or revision is driven precisely by reasons in the positive relevance sense. Evidence provides reasons, and only those propositions unaffected by the evidence keep their prior degrees of belief.

In fact, we may distinguish here a weaker and a stronger sense of revisability. The cause of revising the attitude towards a certain proposition may lie in any other proposition whatsoever one gets informed about, or it may more specifically lie in an evidential or experiential proposition. Let me state this a bit more precisely:

Definition 2: A proposition B is *weakly revisable* relative to a belief function β if $\beta(B|A) \neq \beta(B|\bar{A})$ for some proposition A; otherwise, B is *strongly unrevisable* relative to β. And B is *strongly revisable* relative to β if $\beta(B|A) \neq \beta(B|\bar{A})$ for some experiential proposition A; otherwise, B is *weakly unrevisable* relative to β.

Partially, this definition is still indeterminate because I have not said what experiential propositions are. We will have to return to this later on.

Are there any strongly unrevisable propositions? Yes, certainly. According to any belief function logical truths must receive maximal certainty or the maximal degree of belief (which must therefore exist), and they are strongly unrevisable. In fact, we may easily prove that a proposition A is strongly unrevisable relative to β if and only if A or $\bar{A}$ is maximally certain in β. This has the important consequence that maximally certain propositions and in particular logical truths have no reasons in the positive relevance sense—and are no reasons for other propositions, since positive relevance is always symmetric. This is not to be criticized. It only means that this positive relevance notion is not made for mathematical reasoning and that maximal certainties are inductively barren.

Do the weakly unrevisable propositions extend beyond the strongly unrevisable ones? This is something we have to carefully discuss. First, however, we should attend to our second central notion: that of apriority.

Traditionally, a proposition, belief, or judgment is a priori if it is independent of all experience. This is ambiguous. It may mean that a belief is a priori if it is maintained given any experience whatsoever. Then I call it *unrevisably a priori*. Or it may mean that a belief is a priori if it is held given no experience whatsoever or prior to any experience. Then I call it *defeasibly a priori*.

The traditional notion[2] is too restrictive in another way. It is not only a belief that may be a priori; any feature of a doxastic state may be a priori. So, my preferred explication is the following:

Definition 3: A doxastic feature is *unrevisably a priori* if and only if each rational doxastic state has it. And a doxastic feature is *defeasibly a priori* if and only if each initial rational doxastic state has it.

Since holding a certain belief is also a feature of a doxastic state, this definition generalizes the traditional notion in its ambiguity.

Of course, it is still obscure what initial rational doxastic states are. Only by explaining that initiality will the notion of defeasible apriority be filled with substance. I believe the demand can be met (see Spohn 2012, sec. 17.1). However, I shall not pursue this issue here, since defeasible apriority is only of secondary relevance in my paper.

An important consequence of my generalization is this: All normative principles of epistemic rationality, whatever they are, are unrevisably a priori in my sense, since they are supposed to hold for all rational epistemic states. Of course, we argue about what those principles are; the normative issues are by no means settled. Still, these are arguments about the a priori constitution of our mind.

---

[2] One may well say that the traditional notion is only unrevisable apriority; at least this is the notion Kant continuously pondered about. However, defeasible apriority has historic precedent as well; in any case, a priori probabilities were always taken to be defeasible.

In particular, the formal shape of the belief function β is unrevisably a priori. There are various rational justifications of the axioms of subjective probability; they thus attempt to show that those axioms are unrevisably a priori. Similarly, there are various rational justifications of the axioms of ranking theory. And any alternative proposal for the form of rational degrees of belief must come up with a corresponding justification.

Usually, these examples for unrevisable apriority are not discussed under that heading. Rather, contemporary discussion predominantly focuses on the conceptual as a source of the a priori. This seems still to be an unfortunate heritage of logical empiricism and its strict denial of synthetic principles a priori. In the meantime, we have learned to distinguish analyticity and apriority. But, somehow, it is still only conceptual considerations that are seen to lie at the bottom of both, analyticity and apriority. Even with conceptual apriority, though, matters are more complicated. There are not only analytic conceptual truths and, with Kripke, contingent a priori conceptual truths. There also are defeasibly a priori claims of a purely conceptual nature, for instance reduction sentences for dispositional predicates. However, I will not argue the point here (cf. Spohn 2012, sec. 13.3 and 17.1).

Rather, I would like to discuss further a priori principles, neither of a conceptual nature, nor merely about the formal shape of rational doxastic states, but having some substantial content. It is certainly not the least of my intentions to thereby revive Kant's wider conception of the a priori as conditions of the possibility of experience, though I shall continue proceeding in quite un-Kantian ways.

3 The Basic Empiricist Principle and some consequences

In order to work up to the principles I have in mind we have first to take a look at the propositions which are possibly grasped by our doxastic states. They form an algebra that is closed under Boolean operations. I shall consider the universal algebra of all propositions whatsoever, even though it is unintelligibly large and possibly threatened by paradox, just as the universal set. However, let us not bother about such points. We could instead consider more intelligible, restricted subalgebras of that universal algebra. Again, though, there is no place for such subtleties.

This universal algebra first contains unrevisably a priori propositions, e.g., those which are logically or analytically true. Unrevisably a priori propositions are strongly unrevisable in the sense defined and hence maximally certain.

There is a second class of exceptional propositions, namely possible contents of consciousness, as they may be called. They, too, can only have extreme degrees of belief. Which contents of consciousness one actually has, is obviously contingent. But if such a content is given to one, one is maximally certain of it; and if another such content is not given to one, one is maximally certain not to have it. This is a traditional view that seemed obvious for a long time; only recently philosophers have become more cautious about it. However, it is correct, I think, and indeed derivable from an adequate explication of the nature of contents of consciousness (cf. Spohn 2012, sec. 16.4).

Let us give a label to all the other propositions:

Definition 4: A proposition is *empirical* if and only if it is neither unrevisably a priori true or false nor a possible content of consciousness.

These are traditional distinctions to be found, e.g., in the old empiricists.[3]

Having stated the epistemic status of the exceptional propositions, the issue I am now interested in is: What is the epistemic status of those empirical propositions? So far, we can only say that different doxastic states may take different attitudes towards them, since they are not unrevisably a priori; so much is true by definition.

However, more interesting is whether one and the same subject should be able to change her attitude towards empirical propositions. This may be taken to require that rational belief functions be such that empirical propositions are weakly revisable relative to those belief functions. This is not true by definition, but almost, as it were. It means that belief functions have to be regular, i.e., that only unrevisably a priori propositions and possible contents of consciousness are maximally certain and that all empirical propositions are less than maximally certain.

However, the ability to change one's degree of belief in empirical propositions may be given a stronger reading; we may require that empirical propositions are even strongly revisable, i.e., that their epistemic status can be changed through experiential propositions—where I still owe an explanation of the latter, even though we have an intuitive grasp of them. This is indeed my first principle:

The Basic Empiricist Principle: For each rational belief function β and each empirical proposition A, A is strongly revisable relative to β.

Why should we accept this principle as an unrevisably a priori rationality postulate? The way I have introduced it shows that it is only a slight strengthening of what is true by definition; so it looks convincing, at the

---

[3]See, e.g., Hume (1748), who, in the first paragraph of section IV, distinguishes between relations of ideas that are "intuitively or demonstratively certain" and matters of fact that "are not ascertained in the same manner."

least. However, I have no deeper justification; basic principles must start somewhere.

The grand label I have chosen suggests, though, that most philosophers and most scientists have taken it for granted for centuries. If we put to one side the two exceptional cases, unrevisably a priori propositions (which comprise all of mathematics) and contents of consciousness, the principle says that the entire rest should be under the control of evidence, where evidential control means here finding reasons among experiential propositions. A great many of affirmations of this principle could be cited. It can only be criticized if one interprets control too strongly, say, as verifiability of falsifiability—as the logical positivists have done—or through an inadequate account of confirmation. The principle may also be called a basic principle of learnability; our mind must be open to learn about *all* empirical matters.

The Basic Empiricist Principle still looks weak. It has, however, some significant consequences which I will call the Special and the General Coherence Principle. Very roughly, they say that all our empirical beliefs must cohere in the sense of being tightly connected by reason relations. This is vague, and we have first to work up to their intended precise formulation. Only afterwards I can sketch their derivation from the Basic Empiricist Principle.

As a first step, recall my observation that non-empirical propositions, i.e., unrevisably a priori propositions and contents of consciousness, have no reasons, since they are maximally certain. In analogy to my introduction of the Basic Empiricist Principle we may hence postulate the reversal, i.e., that all empirical propositions do have reasons and, by symmetry, are reasons for other empirical propositions. However, this is entirely trivial; empirical propositions, being weakly revisable, always have or are deductive reasons (in the non-factive sense explained above).

So, the idea should rather be that each empirical proposition has at least one inductive, i.e., non-deductive reason. However, even this is entirely trivial; it is provably satisfied for each belief function taking at least three different degrees of belief (cf. Spohn 2012, 530, assertion 17.5).

Hence, the coherence produced by reasons must receive a stronger reading. For this purpose I would like to give a bit more structure to the universal algebra of propositions. That is, I want to take this algebra to be generated by variables. Formally, a variable is simply a function from the underlying space of possibilities or possible worlds into some set of values. The atomic propositions associated with a variable then state that the variable takes a specific value or some value within a specific set of possible values. For instance, a variable may represent the velocity of a certain particle at a certain time and thus map the space of possible worlds into the space of three-dimensional vectors, i.e., each world to the velocity the particle has at that time in this world. Another variable may represent the temperature in Nancy at noon of July 20, 2011 taking values between -273 °C and, say 1000 °C. These variables generate atomic propositions, for instance the proposition that this particle moves at that time with 10-20 meters per second into eastern direction or the proposition that it is 25 °C in Nancy at noon of July 20, 2011. And so on. In this way, each variable produces a set of atomic propositions, and all other propositions are Boolean combinations of those atomic propositions. Thus, we may conceive of the universal algebra of propositions as being generated by the universal manifold of variables.

Now, the idea is this: We saw that each empirical proposition trivially has some deductive or inductive reasons. Hence, the way to be more restrictive is to require that each atomic proposition about a given variable has a reason which is not about that variable and which must then be an inductive reason, since variables are assumed to be logically independent. This requirement is not trivially satisfied.

In order to state it more precisely, let U be the manifold of all *empirical* variables which generates all empirical propositions. We presently need not look at variables generating unrevisably a priori propositions or contents of consciousness and may restrict attention to the empirical variables in U. Moreover, for any subset $V \subseteq U$ let us call A to be a V-proposition if A is only about, or generated by, the variables in V. Then we have:

The Special Coherence Principle: For each rational belief function β, each empirical variable $X \in U$, and each empirical X-proposition A there is a $U - \{X\}$-proposition that is a reason for A relative to β.

This principle certainly has the same empiricist credentials as the Basic Empiricist Principle, and we shall see in a moment how it derives from the latter. The Special Coherence Principle may even appear to be a semantic principle. If we weaken the verifiability theory of meaning to a confirmability theory, as it were, then, it seems, the Special Coherence Principle must hold in order for each atomic proposition to be meaningful.

However, I would like to stay away from that semantic perspective. One reason is that, as far as I see, all attempts at a verifiability or confirmability theory of meaning have stayed programmatic.[4] Another reason is that I am about to plausibly generalize the Special Coherence Principle to another principle that has no semantic appearance whatsoever.

---

[4]This is a strong claim. However, a close look at the relevant literature, for instance (Brandom 1994), would reveal that it is justified.

For, what is so special about the partition $\{X, U - \{X\}\}$ to which the Special Coherence Principle refers? Nothing. It looks just as convincing if it is stated in terms of any binary partition:

The General Coherence Principle: For each rational belief function β, each non-empty proper subset V of U, there is a V-proposition A and a $U - V$-proposition B such that A is a reason for B relative to β.

The general principle is much stronger than the special principle. We may arrange all the empirical variables in U in a huge graph, where the nodes represent the variables and the edges or vertices between the nodes represent the dependencies between the variables according to the belief function β. Then the special principle says that each node is connected to at least one other node, whereas the general principle says that the entire graph is connected, i.e., there is a path from each node to each other node. This may be properly called coherence.

If one is prone to grand labels, one may say that the General Coherence Principle affirms something like the unity of science or the unity of our world picture. No part of science or our world picture can be completely isolated from the other parts; reason relations directly or indirectly connect each part with each other part.

I claimed that the two coherence principles follow from the Basic Empiricist Principle. How do they do so? At least the proof of the Special Coherence Principle seems quite straightforward: Let A be an empirical proposition about the single variable X. Because of the Basic Empiricist Principle A is at least weakly revisable and hence less than maximally certain according to the given belief function. How can A be strongly revisable as well? Suppose the variable X would be independent from all other empirical variables, then no information about those other variables could change the degree of belief in the X-proposition A. But all experiential variables generating the experiential propositions are among those other variables. Hence, experience could not change the degree of belief in A, and A would not be strongly revisable. End of proof?

Not really. Note that I have still not explained what experiential variables and propositions are; so far I did not need to say this. However, this causes the argument I just gave to have a gap. X may itself be an experiential variable and A an experiential proposition. And the experience may change the degree of belief directly and not through the mediation of reasons. Moreover, A may not be a reason for propositions about other variables. In this strange case A would be an exception to the Special Coherence Principle.

In order to close that gap we have to scrutinize what those experiential propositions might be. This seems to be a hopeless task. Haven't the logical empiricists despaired of characterizing observation sentences? The claim of the so-called theory-ladenness of observation language is still around, and many philosophers have given up this distinction. In the old phenomenalist spirit one might say that our sense impressions provide the experiential base. But aren't they contents of consciousness, and didn't I say that they are excluded from the circle of reasons?

I don't think that the situation is so desperate. We certainly must acknowledge propositions of the form: it appears to s at t, or to me now, as if A. Here "appear" is a sense-neutral expression which stands for "look", "sound", etc. And it must be taken in the comparative or the phenomenal reading, not in the epistemic reading, in which it would mean something like "I now tend to believe A"; for these distinctions see (Chisholm 1957, chap. 4). In that epistemic reading anything can appear to me; it can even appear to me as if the continuum hypothesis were true. But this is not so in the non-epistemic comparative or phenomenal reading. It is still quite indeterminate then what we might substitute for A, for which propositions A it makes sense to say "it now appears to me as if A". However, there is no need to resolve the indeterminacy. Certainly, though, the meaningful substitution for A is heavily restricted.

The next important point is that propositions of the form "it now appears to me as if A" are not contents of consciousness, but empirical propositions. This point would require a longer argument; Spohn (cf. Spohn 2012, sec. 16.3). But the gist of the matter is that by saying "it now appears to me as if A" I am already subsuming my sense impressions under public concepts involved in the proposition A, and then all kinds of things may go wrong, and uncertainty creeps in. I may even be in error when I say: this now appears red to me! Therefore, such propositions are not contents of consciousness. The latter are in a way ineffable, expressible by "it now appears *thus* to me" accompanied, as it were, by an inner pointing. The step from there to the proposition "it now appears to me as if A" is the step from consciousness into the circle of reasons.

We can indeed be more specific about the latter. I just said that "it now appears to me as if A" makes sense for not so many propositions A. But if it makes sense, there is a close epistemic relation to the proposition A itself. That relation is stated in what I like to call:

The Schein-Sein-Principle: It is unrevisably a priori that, given that the subject s attends at time t to a certain external situation and given normal conditions, the proposition that it appears to s at t as if A is a reason (for s' at t') for the proposition A, and vice versa. This holds even if $s = s'$ and $t = t'$.

Again, this principle would require a long and careful argument (see Spohn 2012, sec. 16.3). Basically, I think it is a conceptual truth about appearances or secondary qualities, which are a special case of dispositions. Then the Schein-Sein-Principle looks like a reduction sentence for a disposition qualified by normal conditions. Take the following instantiation: Given normal conditions, the assumption that something looks red to me is a reason, for me just as for you, to believe that it is red, and vice versa. This sounds most plausible, indeed. The qualification by normal conditions is certainly in need of clarification. However, there is again no place for going into details.

The point why I am explaining all this should be obvious, though. We may either take experiential propositions to be of the form "it now appears to me as if A". Or we may take those A themselves to be experiential propositions. Either way is fine, and we need not decide. However, either way it cannot be that experiential propositions are so isolated as to refute the Special Coherence Principle. This is prevented precisely by the Schein-Sein-Principle.

My argument for the General Coherence Principle is in the same spirit, but involves some further solvable complications which I cannot now explain (see, however Spohn 1999). Let me simply summarize my findings so far:

Theorem 1 The Basic Empiricist Principle and the Schein-Sein-Principle entail the Special and the General Coherence Principle.

4 Reasons and truth

In the second main part of my paper I would like to proceed to a second family of principles. So far I have discussed the a priori structure of reasons by itself, how propositions must be minimally connected by reasons in order to allow any learning from experience. However, these connections, our almost obsessive search for reasons is no idle play; they seem to serve a purpose, and the purpose obviously is to find out about the truth. In short, we should somehow account for the truth-conduciveness of reasons, and so far I have not said anything about it.

This is an extremely vexed topic, and I have to steer fairly directly to the results I would like to present. The topic is vexed also because it is not clear which notion of truth is involved here. One may say that there is only one notion of truth, the correspondence notion or its deflationary descendents. BonJour (1985, chap. 8) argues that this is the only interesting notion to apply here, and his metajustification attempts to show that stable coherent belief is likely to be true. However, I find his argument blatantly circular, as it relies on the inference to the best explanation and thus on our inductive practices the truth-conduciveness of which needs to be shown in the first place. (I do not see this as an objection; I think, any argument is bound to be circular at this point.)

No, correspondence truth is truth from the third person perspective, and it seems to inevitably open the skeptical gap. The ornithologist can study the extent to which migrant birds succeed in finding their home, which is known to the ornithologist. Likewise, God, who knows all truths, can tell the extent to which humans find the truth with all their activities.

However, this is not a perspective we can ultimately take. We cannot leave our first person perspective. In judging our fellow humans we imitate the third person perspective; and all our claims are phrased within the third-person perspective. However, the first person perspective is not the individual subjective one, it is that of the cognitive enterprise of the entire, not actually parochial, but counterfactually eternal mankind. We have to think through our issue from that perspective. BonJour (1985, 158) disagrees and thinks the issue is trivialized if the issue is only how coherent belief is conducive to truth in the coherentist sense. But it is very unclear what the trivial argument might be; I think there is none. In any case, let us think through the matter from the first person perspective.

Within the first person perspective we might say that reasons induce beliefs, and to believe something is to believe it to be true. Therefore, reasons bring me closer to the truth; this is what I have to think and say. So much is indeed trivial. However, this triviality does not exhaust the first person perspective. For, even if I think that my present beliefs are true, I know well enough that they might turn out false; if they are really true, they must survive all further learning. So, what truth is within the first person perspective shows up only in a dynamic setting.

In fact, within this perspective truth is Peircean pragmatic truth or Putnamian internal truth. In this sense, a belief is true if it is maintained in the limit of inquiry, after complete experience and fully considered judgment that can be reached only counterfactually. A belief must be true then, simply because there is no experience and no consideration or reason left which could show it to be wrong.

For this notion of truth we must first claim that, rationally, each truth is believable, not in the static sense that there is some doxastic state in which it is believed—this is trivial—, but in the dynamic sense that each rational doxastic state must be able to come to believe it.[5] That is, a rational doxastic state must be open to reach this limit of inquiry, and each true belief must come to be believed on the way to that limit. Therefore we

---

[5]Therefore I am not worried by the knowability paradox of Fitch (1963); (cf. Spohn 2012, 542).

must secondly claim that for each truth there is a true reason. Let me spell out these ideas a bit more precisely.

Let U again be the manifold of all empirical variables generating all empirical propositions, among them maximal propositions or entire possible worlds. One of those worlds must be the actual one; let it be denoted by @. Then, a proposition A is true iff @ $\in A$; however, this is so far only a formal characterization of internal truth. It is also important to conceive of @ not as a rigid designator for the actual world; it is non-rigid or variable. For, we do not know which possible universe we live in, and the feasibility of reasons must not hold accidentally, only in the one actual universe, but in the actual world, whatever it might turn out to be.

Now, what should it mean that each truth is believable? As before, let me restrict attention to empirical truths about single variables in order not to trivialize the possibility of reason finding. Since such a truth is empirical and hence a posteriori, it need not be believed. Of course, it can be believed; but this is not the intended sense of believability. The intent rather is that each doxastic state should be able to come to believe that truth, i.e., that there are possible experiences and, hence, revisions of that state that result in believing this truth. This is still not specific enough, though. There should not only be some possible experiences and revisions with that effect. It must be possible to actually make the required experiences and revisions in the actual world @. This is the intended sense of "-able" in taking truth to be believable. This is summarized in:

The Basic Belief-Truth Connection: Let $X \in U$ be an empirical variable and A an empirical X-proposition with @ $\in A$. Then for any rational belief function β there exists a sequence of experiences available in @ such that β changes through those experiences into a belief function β' in which A is believed.

This is much stronger than the Basic Empiricist Principle that requires only the revisability of empirical propositions through some possible experiences. The present principle rather requires that true empirical propositions must be revisable through actually possible experiences so as to be believed.

A direct consequence is, again with the help of the Schein-Sein-Principle:

The Basic Reasons-Truth Connection: Let X and A be as before such that @ $\in A$. Then for each rational belief function β there is a $U - \{X\}$-proposition B such that @ $\in B$ and B is a reason for A relative to β.

For, if there were no true reason at all for A, there could not exist actual experiences moving us to believe A.

All in all, we have a nice square of entailments:

Theorem 2 Given the Schein-Sein Principle, the Basic Belief-Truth Connection entails both, the Basic Empiricist Principle and the Basic Reason-Truth Connection, each of which in turn entails the Special Coherence Principle.

Why should we accept the new principles? Well, they appear highly convincing, I think. But, again, I have no deeper justification; basic principles must start somewhere. There are two ways of looking at the two new principles, both of which are apt.

The first way is to take them as conceptual truths about truth, truth in the intended internal or pragmatic sense; this is how I have introduced them. Of course, they do not define this notion, but they provide at least a minimal characterization. Truth in that sense must be accessible to experience and reason, and the principles specify some minimal sense in which it is so accessible.

Indeed, they well fit the many ways in which Putnam characterizes internal truth; see in particular the papers collected in Putnam (1983). One way is his claim that the ideal theory must be true. If a proposition could not get believed after ever so many actually possible experiences that are all part of the ideal theory, then this proposition could not belong to the ideal theory and thus be true. And if a proposition finds no true reason, no support in any part of the ideal theory, it can again not belong to the ideal theory.

The second way to take the principles about the connection of belief and reason to truth is as substantial principles constraining rational belief functions. Indeed, I think that they should be taken both ways. The more we advance our account of epistemic rationality, the better we understand internal truth; and reversely, grasping internal truth helps us furthering our account of epistemic rationality. I do not claim that this entanglement is inescapable. But I presently see no better way of explaining the truth-conduciveness of reasons than this postulational approach. In want of alternatives the aim can only be to search for stronger principles, to work out their consequences, and to see whether they stand our critical normative examination. In virtue of the specificity of our formulations this is indeed a constructive program.

In fact, we have not at all exhausted the resources of our dynamic approach. So far, we have only stated that each truth must have a true reason. This allows for the possibility, however, that, given further evidence, that reason is no longer a reason for that truth. It also allows for the possibility that further evidence is taken to overwhelmingly speak against that truth.

The dynamic core idea of the believability of truth was certainly intended to exclude such scenarios. This opens a space of subtly different stronger principles. There is no place for more detailed discussion. Let me only give you the flavor of my favorite version.

The idea is that there should not only be a true reason B for the truth A, the reason B should also be stable in a suitable sense.

Definition 5: Let us call B an *ultimately @-stable* reason for A if and only if there is some true condition C (i.e., $@ \in C$) such that B is a reason for A given any true proposition stronger than C.

This ensures that B remains to be a reason for A in the course of inquiry and even in its limit. The believability of the truth A seems to be secured only if it has some reasons which are stable in this sense. This leads to

The Stable Reason-Truth Connection: Let X and B be as before with $@ \in A$. Then for any rational belief function β there is a $U - \{X\}$-proposition B that is an ultimately @-stable reason for A relative to β.

This principle is obviously stronger than the Basic-Truth Connection. In this spirit, we might constructively propose further principles, prove their relations, and thus a rich normative discussion might evolve (for more details see Spohn 2012, sec. 17.3). And to recall, this is a discussion about a priori principles of reason, grounding not in conceptual considerations, but in the normative structure of our ability to grasp the world through experience and reason.

5 Reasons and causes

As a sort of appendix let me add a final line of thought. So far, my considerations were confined to pure epistemology; I only spoke about reasons and their structure. I am convinced, however, that the principles discussed so far have immediate implications for the structure of causation. Of course, this presupposes an analysis or theory of causation, which I cannot unfold here; see, however, (Spohn 2012, chap. 14) or already (Spohn 1983). But a few sentences suffice for at least stating those implications.

First, I am deeply convinced that Hume is basically right: Causation needs to be explicated in a subjective way, relative to an observer, i.e., a doxastic state or belief function. This is deterringly counter-intuitive, but we need not be stuck with such a subjectivistic analysis. Rather, I believe that only on this basis we can develop an adequate objective understanding of causation (cf. Spohn 2012, chap. 15).

Secondly, I am prepared to fully defend the following analysis of causation, according to which causes are simply a special kind of conditional reasons:

Analysis of Direct Causation: The atomic X-proposition A is a direct cause of the atomic Y-proposition B in the world w relative to the belief function β if and only if $w \in A \cap B$, i.e., both A and B obtain in w, A (or X) precedes B (or Y), and A is a reason for B relative to β conditional on the entire past of B in w except A itself.

And then I would continue defending the analysis that causation *simpliciter* is the transitive closure of direct causation. In any case, if causes really are a special kind of conditional reasons, it is no surprise that the structure of reasons has implications for the structure of causes.

For instance, we may state

The Very Weak Principle of Causality: For each empirical variable $X \in U$ and each X-proposition A, A has some direct cause or direct effect in some world w relative to any rational belief function β.

And then we might prove that this principle to be equivalent with the Special Coherence Principle.

Or we may state

The Unity of the Causal Nexus: Each empirical variable in U is causally connected with each other empirical variable in U, i.e., there is an undirected path from one to the other variable in the universal causal graph.

And then we might prove that this unity is equivalent with the General Coherence Principle.

Presumably, we would like to know about the causal structure of the actual world and not only of some possible worlds. Here, our Reason-Truth Connections do help. We may state

The Weak Principle of Causality: Let X and A be as before with $@ \in A$. Then A has some direct cause or some direct effect in the actual world $@$ relative to any rational belief function β.

And we can prove that this Weak Principle is equivalent to the Stable Reason-Truth Connection given the assumption that each direct cause immediately temporally precedes its direct effect (cf. Spohn 2012, 553, assertion 17.25).

All this shows that our principles have considerable bite. However, the Weak Principle is still not the classic principle of causality stating that each atomic fact has a cause. I am not sure whether it can be established in an a priori manner; in any case, I have so far no idea how the epistemic principles might be appropriately strengthened. Still, I am satisfied that we are able to at least derive the Weak Principle of Causality from a priori principles about the connection between truth and reason.

Bibliography

BonJour, L. (1985). *The Structure of Empirical Knowledge*. Cambridge, MA: Harvard University Press.

Brandom, R. (1994). *Making It Explicit*. Cambridge, MA: Harvard University Press.

Chisholm, R. M. (1957). *Perceiving. A Philosophical Study*. Ithaca: Cornell University Press.

Fitch, F. B. (1963). A logical analysis of some value concepts. *Journal of Symbolic Logic*, *28*, 135–142, doi:10.2307/2271594.

Hume, D. (1748). *An Inquiry Concerning Human Understanding*.

Putnam, H. (1983). Realism and Reason. In *Philosophical Papers*, vol. 3, Cambridge: Cambridge University Press.

Spohn, W. (1983). Deterministic and probabilistic reasons and causes. *Erkenntnis*, *19*, 371–396, doi:10.1007/978-94-015-7676-5_20.

Spohn, W. (1999). Two coherence principles. *Erkenntnis*, *50*, 155–175, doi: 10.1023/A:1005561502394.

Spohn, W. (2001). Vier Begründungsbegriffe. In *Erkenntnistheorie. Positionen zwischen Tradition und Gegenwart*, Grundmann, T., ed., Paderborn: Mentis, 33–52.

Spohn, W. (2012). *The Laws of Belief. Ranking Theory and its Philosophical Applications*. Oxford: Oxford University Press.

Wolfgang Spohn
Department of Philosophy
University of Konstanz
Germany
wolfgang.spohn@uni-konstanz.de

Scientific Integrity
in a Politicized World[1]

HEATHER E. DOUGLAS

ABSTRACT. That politics has an influence on science is unavoidable. Political winds shape the amount and emphasis for research funding. Political discussions determine the ethical boundaries for research. When is a political influence a politicization of science? In this paper, I begin by defining scientific integrity, so that it can be both identified when present and defended when threatened. By delving into the roles for values in science (both acceptable and unacceptable), this paper presents a clear, albeit narrow, view of scientific integrity, and shows how common forms of politicization violate scientific integrity. I also argue that defending scientific integrity is not sufficient to prevent all politicization of science—it removes only the most egregious abuses. To address the full range of politicization concerns, we need to consider both the community of science and the reasons why we pursue science.

1 Introduction

When is a political influence on science a politicization of science? Politicization carries with it the connotation of serious trouble for science, of a dangerous corruption of science's nature and goals.[1] But not every political influence necessitates corruption. For example, the political forces that demanded the scientific community set up guidelines and oversight for human experimentation did not corrupt science (as much as scientists grumble about the resulting Institutional Review Boards). The political attention of Congressional Committees, coverage of the horrors of Nazi doctors, the

---

[1] I would like to thank the organizers of the 2011 LMPSS conference in Nancy for inviting me to give a talk which turned into this paper, for the helpful comments of a blind reviewer, and as ever, the incisive editing of Ted Richards, who always makes the work better.

[1] I will use "politicize" in this paper to mean problematic corruption, as opposed to political influence, which may not be problematic.

scandals at US hospitals, and the egregious Tuskegee experiments created external political pressure on scientists, at the same time as scientific leaders (e.g., Dr. H. K. Beecher) inside the scientific community pressed for reform (MacKay 1995). Without the external political pressure, it is doubtful that clear regulations and enforcement mechanisms for the ethical guidance of human subject research would exist.

Or consider the shaping of research funding in any modern state. In conversation with scientists, the state sets funding priorities for research, utilizing initiatives, areas of focus, and increased funding for some projects (with decreased funding for others). It is perfectly within the purview of the state to shape the scientific agenda in this way, creating financial incentives for the kind of work that looks to be most promising of some public or societal benefit. While the extent to which such efforts are successful is debatable, it seems a stretch to call this a politicization of science.

On the other hand, there are clear cases where political forces do politicize science. For the state to silence scientists with whom it disagrees, either through forced imprisonment (as in the Soviet era Lysenko case) or forced editing and gagging (as was charged under the Bush Administration with respect to climate change) is clearly politicization (Union of Concerned Scientists 2004). For the state to ensure, through funding structures or harsher political means, that only predetermined results be produced (as opposed to focusing efforts on a particular topic and being open to whatever results are produced), is squarely in the realm of politicization.

In order to sort acceptable political influences from the politicization of science, we need to have some sense of what we want to defend from political forces. In this paper, I will identify one central thing to defend as scientific integrity. As of late, the term "scientific integrity" has been used as an overly broad slogan encompassing everything good in research ethics. If scientific integrity is to have a distinctive meaning above and beyond just "integrity" (as in moral uprightness), we need a narrower view. In this paper, I provide a more precise and narrow account, where scientific integrity consists of proper reasoning processes and handling of evidence essential to doing science. Scientific integrity here consists of a respect for the underlying empirical basis of science, and it is this scientists are often most concerned to protect against transgressions, whether those transgressions arise from external pressures (e.g., politicization) or internal violations (e.g., fabrication of data to further one's scientific career).

In elucidating the nature of scientific integrity, I will describe it as the adequate individual behavior and reasoning necessary to protect what we value about science. Once defined precisely, I will show how it can be defended. But the precise definition comes at some cost, narrowing what falls under the purview of scientific integrity. On my account, not everything about the responsible conduct of research (RCR) has to do with maintaining scientific integrity. The aspects of RCR that do not fall within scientific integrity *per se* arise from two additional bases for the moral responsibilities of scientists: 1) the proper functioning of the scientific community (which is ultimately essential to the production of reliable knowledge), and 2) the (legitimate) demands of the larger society for ethically and socially acceptable behavior from scientists (Douglas 2013*a*). These two additional bases for responsible research are no less weighty than scientific integrity. A supportive and critical epistemic community is crucial for enabling individual scientists to be able to produce knowledge as reliable as they do. The demands of fostering that community—such as mentoring students and post-docs properly, doing timely and thorough peer reviews of each other's work, generating the forums which allow for critical discourse, etc.—are essential. Equally as important are the responsibilities scientists have to the broader society, which generate such demands as the ethical treatment of human and animal subjects (see also, De Winter & Kosolosky 2013). Narrowing the scope of scientific integrity is not meant to narrow the scope of scientists' responsibilities.

What do we gain with a narrow definition of scientific integrity? Such a definition will allow us to see how political pressures can threaten scientific integrity, either by putting pressure on scientists to violate integrity or by violating the integrity of scientific work directly (e.g., by changing scientific claims without consulting the scientists that produced the original work). Politicization of science consists in distorting the nature of science for one's political purposes. Once we have defined scientific integrity, it will be clear that damaging scientific integrity is certainly one way to politicize science. However, we will also find that science can be politicized in worrisome ways without threatening scientific integrity *per se*. Politicization concerns, we will see, are larger in scope than the defense of scientific integrity.

2 The challenge of defining scientific integrity

As noted above, views of scientific integrity today are frequently overly broad. Within the scientific community, scientific integrity is often equated with all concerns over RCR. For example, *Integrity in Scientific Research*, published in 2002 by the U.S. Institute of Medicine and the National Research Council, is subtitled "Creating an Environment that Promotes Responsible Conduct of Research" (IoM/NRC 2002). The report finds its motivation in cases of scientific fraud, and focuses on finding remedies for and ways to discourage "research misconduct", centered on fabrication, falsification, and plagiarism, but also including issues concerning treatment of humans and animals, authorship, mentoring, peer review, collegiality, and conflicts of interest (IoM/NRC 2002, 34–40). It breaks new ground in focusing on the institutional context of misconduct and showing concern for educating developing scientists about misconduct, but its definition of scientific integrity itself lacks cohesiveness (an aspect of integrity) (Steneck 2006, 55). Scientific integrity, equated with all moral concerns, generates a laundry list of responsibilities. What one is trying to protect when protecting integrity becomes diffuse.

Other attempts at definition run into other problems. For example, the recent Singapore Statement on Research Integrity struggles with both circularity and long lists. It signals the importance of integrity in the preamble, which declares "the value and benefits of research are vitally dependent on the integrity of research" (Singapore Statement 2010). But in the resulting list of fourteen different responsibilities, the first is integrity, explicated like this: "Researchers should take responsibility for the trustworthiness of their research" (Singapore Statement 2010). There is no further elucidation of what integrity is, nor how the value and benefits of research rest on it. Integrity, i.e., taking responsibility for the trustworthiness of research, is one of fourteen responsibilities that must be met (including adherence to regulations, keeping good research records, performing peer review properly, and reporting irresponsible research practices to the proper authorities) in order to protect the integrity of research. This opaque circularity obscures what scientific integrity is all about, or how we can construe it as an integral whole that we want to protect from political forces.

Discussions of scientific integrity in the political realm have similar problems. While the Obama Administration recently raised concern for protecting scientific integrity, definitions of what was to be protected are frustratingly unclear. One gets the sense that integrity is crucial for trustworthiness, and in order to have integrity, one needs to be trustworthy. While the policies being pursued under the efforts are laudable (e.g., whistleblower protection and freedom of scientists to speak to the press), the policies appear to be guided more by examples of past problems and concerns than a coherent understanding of scientific integrity (Holdren 2010; Thomas 2012).

Finally, a recent attempt at a precise definition by De Winter and Kosolosky has a different problem (De Winter & Kosolosky 2013). They define research integrity in terms of deceptiveness, and deceptiveness in terms of saying something false, or saying something from which others could legitimately derive false implications. But this definition demands too much of scientists, as scientists who are honestly mistaken (which would include many famous scientists historically) would then be found to be failing to have scientific integrity.

We need a different approach. I will start my discussion by first examining what we are trying to protect when we defend scientific integrity. To answer this question I begin with why we value science. From there, I argue that we can see which aspects of scientific practice and reasoning are essential for its proper functioning, for the achievement of what we value. We can also see how science can be properly ethically constrained by the society in which it functions. Finally, we can see when the constraints generated by the larger political context damage science by undermining the reason we value science, i.e., we can see when they politicize science.

3 The value of science and values in science

Why do we do science? Why is science such an important activity that politicization is a worry? Whether one is interested in science for a capacity to intervene in the world or for the pure joy of understanding, both those interests rest on the ability of science to produce reliable empirical knowledge. It is this ability that is at the heart of why we value science. Science manages this production of reliable empirical knowledge by being an iterative, ampliative process of developing explanations (including explanatory theories and models), using those explanations to produce further predictions/implications, and testing those predictions empirically (Douglas 2009a). In light of the evidence produced by such tests, the explanations are refined, altered, or utilized further.

This iterative and ampliative process produces an ever-developing body of empirical knowledge, but one that is also endemically uncertain. We can never be completely sure our explanations or theories are correct, because we might encounter evidence as yet ungathered which will fundamentally challenge current views, including views on what it means to gather reliable evidence (e.g., what a method can and cannot accomplish). But this ability of new evidence and experience to overturn currently held belief makes science both exciting for the practitioner and robust for the user. Because any particular part of science can be held open to challenge, we can have *prima facie* confidence that it is the best we can currently do and that science provides our most reliable empirical knowledge available. It is *because* science is uncertain that it is robust; because it is empirical that it is reliable.

If this value of science is to be protected, evidence must be able to challenge currently held views. This requirement creates certain demands for the structure of how other values (whether ethical, social, political, or cognitive) can play a role in science. Depending on where one is in the scientific process, values have different legitimate roles they can play, with legitimacy determined by the need to protect the value *of* science.

Consider the following two roles values can play in our reasoning: *direct* and *indirect*. In the *direct role*, values are a reason in themselves for our decisions (Douglas 2009b, chap. 5). We use them to assess our options and tell us which we should choose. For example, if I select a particular food because I value its health benefits, the value is playing a direct role in my choice. In the *indirect role*, values instead serve to assess the sufficiency of evidence for our choices. We use values here to assess whether we think the uncertainties concerning our choices are acceptable, by assessing the consequences of error rather than by assessing the choices themselves. For example, if I do not accept the claim that I need yearly mammograms between the ages of 40 and 50 because I do not think the currently available evidence is strong enough to support the claim, particularly given the known risk of cancer generated by the radiation needed to do the mammogram and the value I place on avoiding that risk, that is an indirect role for values in the judgment. If the evidence became stronger, I would reevaluate and change my mind accordingly. The value serves only to assess the acceptability of uncertainty, staying in the indirect role. If values served in the direct role (if, for example, I avoided x-rays at all costs) no amount of evidence of the benefits of mammograms would be sufficient to persuade me to get one.

An *indirect* role for all kinds of values (political, social, ethical, cognitive) is needed and acceptable throughout the scientific process. Science is thus a value-saturated process. The *direct* role, on the other hand, must be excluded at certain, crucial points, but is allowable at others. For example, a direct role for values is perfectly acceptable when a scientist is deciding which research projects to pursue. Perhaps the scientist has a personal interest in a particular species, or cares deeply about the geology of a particular location, or has a strong fascination in a particular chemical process. The value the scientist holds (whether ethically, socially, or cognitively based) is fine to drive the scientist to work in that area, to direct the scientist's attention and choices.

At other points, a direct role for value judgments would be deeply problematic and unacceptable. Consider the direct role in the following case: a scientist, in studying a particular ecosystem, really wants the ecosystem to show particular signs of health. Although not detecting these signs empirically, the desire for them to be there is so strong, the scientist begins to manufacture their presence, either by fudging the data or deluding him/herself into thinking they are there. Here, the same values so laudable in the choice of research project are damaging to the conduct of the research, undermining (indeed demolishing) the value of the science produced. If the values here serve as reasons in themselves for the decisions of the scientist, in the production, collection, characterization, and interpretation of the evidence, then the very value of the scientific enterprise is grievously damaged. We should have no confidence in the empirical basis of the scientist's claims, for there is no actual empirical basis. The values have replaced it, serving where evidential considerations should. It is for this reason that values, at the heart of the scientific process, should only serve an indirect role, i.e., of helping to assess whether the gathered evidence is strong enough for a claim. Without this crucial constraint on the roles values can play, the value of science is lost.

Such a constraint is crucial at other points in the scientific process as well. For example, the decision of which methodologies to employ for a particular project requires careful utilization of values. Values in a direct role can legitimately keep certain methodological options off the table. Because of our ethical values, we demand that scientists respect human autonomy and that human subjects for research projects be informed volunteers who freely agree to participate. We could surely learn many things if we relaxed this restriction, e.g., keeping people in controlled environments while they grew from infants into adulthood to examine the effects of the environment on development. But such experiments would be ethically abhorrent, and so the value of knowledge pales next to our ethical valuation of the methods that would be required. We find other ways, perhaps less methodologically robust, to study the impact of the environment on human development.

But the utilization of values in a direct role for methodological choice is not always legitimate. For example, if one wants a certain result, one can often rig the methodology to produce that result. One can, if one is testing the estrogenic action of a chemical, use an animal model that is estrogen insensitive to ensure negative results (Wilholt 2009). If one picks a methodology to ensure a certain result, however, one is undermining the reason we value science—that is to allow evidence to speak to us about the way the world is, not the way we might wish it were. The value of a particular desired outcome should not cause the scientist to structure the research so that the particular outcome is assured. Such a direct role for values undermines the reason we value science, and is thus an illegitimate use of values in science, violating scientific integrity. It is, in a sense, another way to fudge the data.

The complex nature of values in methodological choices is thus unavoidable. A direct role for values that keeps scientists from performing ethically unacceptable research is both acceptable and laudable. Even if this is felt as an outside intervention on science, it is not a pernicious politicization, but instead an acceptable political influence on science. On the other hand, a direct role for values in methodological choice that generates a desired predetermined result does represent a violation of scientific integrity, for it clearly undermines the reason we value science. Every scientific study worth its name should have the possibility of producing surprising or challenging results, not merely the outcome that the scientist (or the funder) desires. It is in light of these considerations that the role of values, and the values themselves, in methodological choices should be assessed.

Schematically then, the following terrain can be laid out. Depending on where one is in the scientific process, different roles for values are acceptable or not. When deciding which research to pursue, a direct role of values is fine (although we might want to contest the values involved or the particular choices made) (e.g. Reiss & Kitcher 2009). Once the scientist has moved on to the particular methodologies to be employed in the study, care must be taken that a direct role for values that undermines the value of science is not employed, even if some direct role for values (restricting the scientist to the conduct of ethical research practices only) is acceptable. When the research has begun, and data must be collected, characterized, and interpreted, an indirect role for values is the only acceptable role. A direct role for values here would undermine the reason we value science. Finally, once the research is complete, the data interpreted, and the findings made clear, how the scientist chooses to disseminate or utilize the research can be subject to a direct role for values again. Whether to conduct further research, apply the research in certain ways, or even withhold certain details—because making them public could be seriously harmful for a species, in the case of endangered but hunted species research, or for humanity as a whole, as was debated in the recent H5N1 case—all fall within the purview of a direct role for values. However, as with the methodological choices above, a direct role should not be used to undercut the value of science, by, e.g., withholding unwelcome results. A respect for the value of science, and the nature of the other values, are the crucial issues.

One might object at this point that I have made careful distinctions among the places and roles for values, but not among the kinds of values. The value-free ideal has held, in contrast to the view articulated here, that some values, namely epistemic and cognitive values, are fine throughout the scientific process, particularly at the heart of doing science, and that all other values should be excluded from scientific reasoning (e.g., McMullin 1983; Lacey 1999). I disagree for several reasons.

First, the traditional characterization of epistemic/cognitive values is unrefined. Some of the so-called "values" are more minimum criteria for good scientific work, and as such, can serve as a direct reasons for accepting or rejecting scientific theories (Laudan 2004). The "values" of empirical adequacy and internal consistency are better understood as minimal floors, below which a theory or explanation should not fall, and a failure to meet those demands is a good reason to reject a theory. Other cognitive values, such as simplicity, scope, explanatory power, and predictive power, have suffered from a conflation of two important senses, in that what instantiates them is crucially different, and the value of the cognitive value shifts accordingly (Douglas 2013b). The two different senses are: 1) the value applies to the theory in relationship to the evidence which supports it, and 2) the value applies to the theory on its own. If we are considering the value instantiated in the first sense, where we are considering a theory that is simple with respect to the complex evidence it explains, or that predicts a wide array of evidence, or that explains a broad scope of evidence, etc., then that sense of the value does not fall under my discussion above. Indeed, this sense of value helps us to assess how much uncertainty we think is present in making a claim based on evidence, and thus does come conceptually prior to the indirect role for values described above.

But if we are considering the value instantiated in the second sense, that we have a theory that just appears simple and elegant (irrespective of the evidence that might support it) or that seems to have broad scope in that it might cover a wide swath of phenomena (but whether it actually explains the evidence in that broad swath is as yet undetermined), the account I give above does emphatically apply. This sense of cognitive value, which is the more usual one articulated in the literature (e.g., Kuhn 1977; McMullin 1983) and which is the usual place for philosophers of science to note how they all "pull against each other" has no *epistemic* merit—it tells us nothing about how uncertain we should be or how reliable our inference likely is. It is more of a cognitively pragmatic consideration, that theories or explanations which instantiate these values are easier to work with and thus more likely fruitful. Because of the lack of epistemic bearing, such values should be constrained to the indirect role only at the heart of reasoning. We should use them to assess the acceptability of uncertainty in the following way: if we think the evidence moderate for a claim, and the claim instantiates one of these values in this second sense, we should then utilize this aspect of the claim (that it is simple and thus easy to work with, that it has broad potential scope and thus many potential areas for application and test, etc.) to develop tests quickly and thus either improve the evidential basis or show the flaws of the view. The values in this sense serve as a hedge against uncertainty, and thus might be a reason to find the uncertainty acceptable in the short term, but only if scientists actively utilize the valued aspect of the theory to reduce uncertainty through further development. Social and ethical values can trade against cognitive values in this indirect role. It would be acceptable for some scientists to find the social consequences of error too high and to reject a theory until the evidential basis is strengthened, while other scientists accept the theory because of its cognitive attributes and use them for further testing. In short, a more careful examination of the traditional epistemic/cognitive values reveals important texture with implications for their role in scientific reasoning. The value-free ideal glosses over this texture, and thus allows cognitive values (particularly in the second sense) to play an improper direct role in scientific reasoning.

This is not the only reason I object to the value-free ideal. I also object to it because it ignores scientists' basic general responsibility to carefully consider inductive risk and the consequences of error in their work. But delving into this objection takes us too far afield, and it is developed thoroughly elsewhere (Douglas 2009b, chap. 4). With the value-free ideal set aside, the distinction between direct and indirect roles can articulate the proper functioning of values throughout the scientific process.

This view of values *in* science, developed in light of the value *of* science, can now provide us with a clear definition of scientific integrity. First, as described here, scientific integrity is a quality of individual scientists, their reasoning, and particular pieces of scientific work. Thus, a person, a paper, a report can all be said to have scientific integrity. The crucial requirement for scientific integrity is *the maintenance of the proper roles for values in science*. Most centrally, an indirect role only for values in science is demanded for the internal reasoning of science. When deciding how to characterize evidence, how to analyze data, and how to interpret results, values should never play a direct role, but an indirect role only. This keeps values from being reasons in themselves for choices when interpreting data and results. In addition, values should not direct methodological choices to pre-determined outcomes, nor should they direct dissemination choices to cherry-pick results. This restriction on the role of values, to the indirect role only at these crucial locations in the scientific process, is necessary to protect the value of science itself, given the reason we do science is to gain reliable empirical knowledge. We do science to discover things about the world, not to win arguments. Protecting scientific integrity as so defined thus protects the value of science.

With this definition of scientific integrity, it is clear that many of the classic concerns over scientific integrity in RCR fall under this definition. For example, data fabrication and falsification is the manufacturing of evidence because of a value playing an improper direct role. The scientist wants certain results, and rather than gather actual evidence, makes it up. The value of presenting particular results overrides the value of science, and serves in the direct role for the recording of data. Other ways to violate integrity include cherry-picking evidence when drawing conclusions, ignoring known criticisms, and ensuring that the methods used will produce the desired results. All of these violations involve a value (the value of getting a certain result, usually because of other strong value commitments) playing an improper direct role in the scientific process. Violations of integrity are clear reasons to dismiss the work of a scientist; a scientist that lacks integrity should have no epistemic authority whatsoever. But, conversely, the presence of integrity does not require that we accept the work as reliable. One can still disagree with science or scientists that have integrity; integrity is necessary but not sufficient for reliability.

Although many core concerns are captured by this definition of scientific integrity, other aspects of RCR are not. Plagiarism, for example, is less a violation of scientific integrity, as defined here, than a violation of the norms of assigning credit within the scientific community. As such, it is a violation of a scientist's responsibility to the epistemic community of science and a very serious matter. But it is not a violation of scientific integrity on the narrow view given here, as it does not harm the epistemic content of science. And, as noted above, the proper ethical treatment of human and animal subjects arises from the requirements of the broader society in which science functions, rather than a requirement of scientific integrity. Neither of these serious violations of ethical conduct harm directly the epistemic content of science. Maintaining scientific integrity is but one of the responsibilities of scientists, and is insufficient on its own for RCR. But having this precise and narrow view of scientific integrity can help us see more clearly what should count as politicization.

4 Politicization of science as a violation of scientific integrity

What does this view of scientific integrity mean for our understanding of the politicization of science? Clearly, political forces could cause a scientist, either voluntarily or through coercion, to violate the proper roles for values in science and thus violate scientific integrity. Examples of this include scientists pressured to (or for their own political purposes deciding to) fabricate evidence, cherry-pick evidence, distort results, or stick to a claim even when known criticisms which fatally undermine the claim remain unaddressed. The main intellectual fault in all these cases is failing to be responsive to genuine empirical concerns, because doing so would make one's political point weaker or undermine a cherished ideological perspective. It is to utilize a direct role for values and have that determine one's results. It is to use the *prima facie* reliability and authority of science, which rests on its robust critical practices and evidential bases, and to throw away a concern for the source of science's reliability in favor of the mere veneer of authority. It is to turn science into a sham. No wonder scientists get so upset when violations of scientific integrity occur.

One might worry that it is too difficult to detect this sort of politicization, as it rests on assessing the role of values in reasoning. Can we assess how someone else's reasoning works? Many cases of data falsification have been found looking at published work (Goodstein 2010). Other violations of scientific integrity can be found as well in published or public work. To find such violations, we should examine patterns of arguments. For example, a failure to respond to criticisms raised repeatedly and pointedly is a clear indication of a problem. If a scientist, or a political leader using science, insists on making a point based on evidence even when clear criticisms undermining their use of that evidence have been raised, and they fail to respond to those criticisms, one is warranted in suspecting that the cherry-picked evidence is but a smokescreen for a deeply held value commitment serving an improper direct role, and that ultimately, the evidence is irrelevant.

Violations can also be detected in overt or covert interference with the activities of scientists. The rewriting of science advisory or summary documents so that unwelcome findings are buried and desired findings are generated is another clear, detectable way in which political forces can interfere with scientific integrity. Here, another actor's values run roughshod, in an improper direct role, over evidential considerations. Political actors may not like the results produced by scientists, but their response should not be to declare them by fiat to be otherwise. Instead, politicians can legitimately question whether the evidence is sufficient to support certain policies, whether other policy options might be preferable, or whether value commitments should demand contrary courses of action. One need not accept every piece of scientific work or every report as definitive. But to attempt to alter such findings so that they do support one's preferred political interests is to politicize science by violating scientific integrity.

Defending science from such attempts at politicization requires the kinds of institutional reforms now proceeding as a result of the Holdren Memo (Thomas 2012). Political officials need to know that interfering with scientific reports is unacceptable politicization of science, and that it creates the same kind of damage to scientific integrity as scientists fabricating evidence. Institutional sanctions should be equally severe in both cases. In addition, scientists need to know that they can freely discuss their work and the actual content of it with both other scientists and the general public. Such discussions create the conditions for assessing expertise, its integrity, and its evidential basis.

In general, it would help if the value of science to society, to provide robust empirical knowledge, even if uncertain and changeable, were broadly accepted and understood. It is crucial to keep in mind: 1) that science generally provides the most reliable knowledge available, but also 2) that any given claim may prove mistaken and 3) that values are needed throughout science in the indirect role, to assess whether the evidence is sufficient. Understanding this puts a burden on public officials: if you want to ignore a piece of science, you should say why—what do you think is wrong with it or why it is not relevant to policy. If you want to use a piece of science, you should also say why—why is the study strong enough, what value considerations shape that assessment and the subsequent policy choices. If it were more broadly understood what the value of science is, and what the nature of science is, perhaps it would be harder for science to be successfully politicized, as the demands for public discourse would shift accordingly.

5 Politicization beyond scientific integrity

With this narrow and clear definition of scientific integrity, we can identify politicization which violates scientific integrity. Is this the only way in which science can be politicized? If we maintain scientific integrity perfectly, are all attempts at politicization, of worrisome political influences on science, thwarted? I think the unfortunate answer is no. While many of the most blatant and disturbing efforts at the politicization of science have been targeted at scientific integrity, there are other ways to politicize science that do not strike at scientific integrity.

Consider the fact that a direct role for values is acceptable in the direction of research efforts and the selection and funding of research projects. Because of this fact, political forces could decide that rather than funding bogus research that is gerrymandered to produce desired results (a clear violation of scientific integrity), it would be better for political reasons to simply not fund any research on certain topics, thus discouraging research from being done. Such distortion need not occur through interference with funding agencies. Through the rubric of intellectual property rights, some research can be effectively quashed even if scientists have the needed funds. Biddle 2014 has raised concerns over GMO research in this regard. Restricting research through licensing agreements does not violate the narrow sense of scientific integrity defined in this paper, but it clearly does seem a politicization of science. Political forces can distort which science can and cannot be done.

In order to protect against this kind of politicization, through legitimate roles for values in science, a broader perspective on values in science and the proper functioning of the scientific community is needed. One issue is not just the roles values play but whether the values themselves are acceptable or defensible. In addition, one needs to assess whether a sufficiently diverse range of scientists (to ensure adequate criticisms of each other's work are being raised) are working on a range of projects that do not just serve a narrow set of interests. If power and money draw the efforts of scientists into a narrow range of projects (as seems to have happened in biomedical research, see (Reiss & Kitcher 2009)), society will not be well served. Even if the science being done is performed with perfect integrity, the results may be distorted and politicized simply because they are the only results available. This is a much harder problem to track and assess, and has not been the main area of concern with the politicization of science. But I suspect it will become a key area of debate in the coming decades.

6 Conclusion

By focusing on why we value science, I have provided a clear and coherent definition of scientific integrity. That definition of integrity is to maintain the proper roles for values in scientific reasoning. Values play an important role throughout science, but must be constrained to particular roles at key points in the process. Violations of integrity allow values to displace the importance of evidence in science, thus undermining the value of science. While this definition of integrity no longer encompasses all of the responsible conduct of research, it is sufficiently precise that we can see how science can be politicized through violations of scientific integrity.

With this clarity, we can see both how to detect politicization that violates integrity and how to discourage such politicization. But violating scientific integrity is not the only way to politicize science. One can politicize science at a broader level, by distorting which science is done, so that politically unwelcome projects are never begun. Both how to detect such politicization and what should count as such politicization, given the legitimate interests of society in shaping research efforts, must await further discussion and debate.

Bibliography

Biddle, J. B. (2014). Can patents prohibit research? On the social epistemology of patenting and licensing in science. *Studies in History and Philosophy of Science Part A*, *45*(0), 14–23, doi:10.1016/j.shpsa.2013.12.001.

De Winter, J. & Kosolosky, L. (2013). The epistemic integrity of scientific research. *Science and Engineering Ethics*, *19*(3), 757–774, doi:10.1007/s11948-012-9394-3.

Douglas, H. (2009a). Reintroducing prediction to explanation. *Philosophy of Science*, *76*, 444–463.

Douglas, H. (2009b). *Science, Policy, and the Value-Free Ideal*. Pittsburgh: University of Pittsburgh Press.

Douglas, H. (2013a). The moral terrain of science. *Erkenntnis*, 78(5), 1–19, doi: 10.1007/s10670-013-9538-0.

Douglas, H. (2013b). The value of cognitive values. *Philosophy of Science*, 80(5), 796–806.

Goodstein, D. (2010). *On Fact and Fraud*. Princeton: Princeton University Press.

Holdren, J. (2010). Memorandum for the Head of Executive Departments on Scientific Integrity. URL www.whitehouse.gov/sites/default/files/microsites/ostp/scientific-integrity-memo-12172010.pdf.

IoM/NRC (2002). Integrity in scientific research: Creating an environment that promotes responsible conduct. Tech. rep., National Academies Press, Washington.

Kuhn, T. (1977). Objectivity, value judgment, and theory choice. In *The Essential Tension*, Chicago: University of Chicago Press, 320–339.

Lacey, H. (1999). *Is Science Value Free? Values and Scientific Understanding*. New York: Routledge.

Laudan, L. (2004). The epistemic, the cognitive, and the social. In *Science, Values, and Objectivity*, Machamer, P. & Wolters, G., eds., Pittsburgh: University of Pittsburgh Press, 14–23.

MacKay, C. R. (1995). The evolution of the institutional review board: a brief overview of its history. *Clinical Research and Regulatory Affairs*, 12(2), 65–94, doi:10.3109/10601339509079579.

McMullin, E. (1983). Values in science. In *Proceedings of the 1982 Biennial Meeting of the Philosophy of Science Association*, vol. 1, Asquith, P. & Nickles, T., eds., East Lansing: Philosophy of Science Association, 3–28.

Reiss, J. & Kitcher, P. (2009). Biomedical research, neglected diseases, and well-ordered science. *Theoria*, 24(3), 263–282, doi:10.1387/theoria.696.

Singapore Statement (2010). Singapore statement on research integrity. URL www.singaporestatement.org.

Steneck, N. (2006). Fostering integrity in research: Definitions, current knowledge, and future directions. *Science and Engineering Ethics*, 12(1), 53–74, doi: 10.1007/PL00022268.

Thomas, J. (2012). The slow but deliberate march toward scientific integrity. Science Progress, URL http://scienceprogress.org/2012/05/the-slow-but-deliberate-march-toward-scientific-integrity/.

Union of Concerned Scientists (2004). Scientific integrity in policymaking: An investigation into the Bush Administration's misuse of science. URL www.ucsusa.org.

Wilholt, T. (2009). Bias and values in scientific research. *Studies in History and Philosophy of Science Part A*, *40*(1), 92–101, doi:10.1016/j.shpsa.2008.12.005.

Heather E. Douglas
University of Waterloo
Canada
heather.douglas@uwaterloo.ca

On the Co-Unfolding of Scientific Knowledge and Viable Values

HUGH LACEY

1 Introduction

Poincaré once wrote:

> Ethics and science have their own domains, which touch but do not interpenetrate. The one shows us to what goal we should aspire; the other, given the goal, teaches us how to attain it. So they never conflict since they never meet. There can be no more immoral science than there can be scientific morals. (Poincaré 1920, 12)

I will argue that scientific knowledge and value judgments do "interpenetrate"—not in reductionist or logically linear ways, but so as to unfold together dialectically—reflecting, on the one hand, that values may legitimately affect the methodological decisions that shape scientific research (see section 2) and, on the other hand, that holding values rationally has presuppositions that may be open to scientific investigation (section 3). This is not to deny that scientific knowledge and value judgments are distinct, to maintain that there are relations of logical entailment between them, to reject that the criteria for the cognitive appraisal of scientific knowledge do not incorporate any ethical/social values, or to suggest that scientific knowledge suffices by itself to resolve the value conflicts of our day.

The "touch" of science and values is multifaceted and ubiquitous (Lacey 1999, 12–18). Science "touches" values, e.g., when scientific developments occasion the need for ethically salient deliberations on matters (e.g., risks of technoscientific innovations) that hitherto had no place in the world of lived experience. Values "touch" science in various ways, including when research priorities are chosen. This "touch" may be soft (e.g., conducting science in accordance with professional codes of ethics); or it may have far-reaching consequences, as it tends to have today when (e.g.) priority is given to research that may lead to technoscientific innovations that will contribute to economic growth and competitiveness, to furthering values of technological progress (V_{TP}) and of capital and the market ($V_{C\&M}$).[1] The interaction between science and values is two-way in contexts where the benefits and risks of using technoscientific objects are appraised: values influence what is considered to be a potential benefit or harm, and so what should be investigated; and empirically grounded risk analysis informs endorsing a hypothesis (e.g.) that risks are insignificant, i.e., to making the judgment—after considering the consequences (and their ethical salience) of acting informed by the hypothesis should it be false—that the evidence supporting it is sufficiently strong (despite remaining uncertainties) to legitimate acting in ways informed by it (see section 2.3).

The idea that scientific knowledge and holding values only "touch" and do not "interpenetrate", which has deep historical roots (Lacey & Mariconda 2012), retains a strong grip on contemporary sensibilities. It is reinforced by widely held views, elaborated in Lacey (1999; 2002; 2005a), about scientific methodology and about the nature of values.

Here are versions of the views about scientific methodology:

(i) Proper evaluation of scientific knowledge and understanding is based on empirical data and cognitive criteria that neither presuppose nor imply any ethical/social value judgments.

(ii) Scientific knowledge characteristically is expressed in theories that are investigated in research practices that deploy decontextualizing strategies (DSs).

"Strategies are elaborated in Lacey (1999; 2005a). Under DSs, admissible theories are constrained to represent phenomena as lawful, and this typically involves representing them and encapsulating their possibilities in terms of their being generable from their underlying order: underlying structures, processes, and interactions of their components and levels of organization, and laws that govern them. Representing phenomena in this way decontextualizes them. It dissociates them from any relations they may have with social arrangements and human lives, from any link with human agency, value and sensory qualities, and from whatever possibilities they may gain in virtue of their places in particular social, human and ecological contexts.

---

[1] Holding V_{TP} involves granting high ethical/social value to expanding the human capacity to control natural objects, especially as embodied in technological innovations, to innovations that increase the penetration of technology ever more intrusively into ever more domains of human life, and to the definition of problems in ways that may permit scientifically informed technological solutions (Lacey 2002; 2005a).

Hence, the categories permitted in admissible theories contain none of those used in common discourse to represent these matters; and, since no value terms are used, there can be no value judgments among the entailments of admissible theories. Complementing these constraints on admissible theories, empirical data are selected, sought out and reported using descriptive categories that generally are applicable (or they may be collected, analyzed and stored automatically by mechanized surrogates for observation) in virtue of measurement, instrumental and experimental operations.

Where DSs are deployed—as they are in research that opens up possibilities of technoscientific innovation (Lacey 2012)—value judgments play no role in the (cognitive) evaluation of theories [item (i)], and none are entailed by confirmed scientific knowledge [from (ii)]. Value judgments, thus, are beyond the purview of DS-investigation and its results. Therefore, if it is of the nature of scientific methodology to use DSs, value judgments are completely beyond the purview of scientific evaluation—leaving little alternative to the view that values are just subjective preferences or reflections of personal or group interests. This does not exclude values having some uncontroversial roles in scientific inquiry. One concerns setting priorities for research. What particular values should play this role is contested—but, if values are subjective preferences, this cannot be resolved scientifically. That we have a lot of knowledge of certain phenomena (e.g., the genomes of crop plants and how to modify them), but less of others (e.g., sustainable agroecosystems), reflects the dominance of particular values (V_{TP} and $V_{C\&M}$) that shape interests that today are widely held by scientists, their institutions, funders and employers.

2 Values and the methodological decisions that shape scientific research

A deeper methodological issue lies hidden here. It has been the predominant view throughout the modern scientific tradition that it is of the nature of scientific methodology to use DSs, regardless of the kind of phenomena being investigated (whether investigating structures of plant genomes and how to alter them, or the principles of sustainable agroecosystems). DSs have indeed proved to be remarkably fruitful and versatile, and their reach keeps expanding. Under them, a great deal of knowledge has been obtained of many different kinds of phenomena, of their underlying order and of generalizations about them; and this knowledge has served to inform and to explain the efficacy of countless innovations in medicine, agriculture, information, communications, energy, transport, industry, etc., many of which are valued widely across value-outlooks. Nevertheless, on the whole applied science has not served value-outlooks evenhandedly, but has especially favored interests fostered by V_{TP} and $V_{C\&M}$, so much so that the social viability of some value-outlooks has been seriously undermined.

The values (V_{TP} and $V_{C\&M}$) that are especially well served by applications of DS-results are the same ones that shape research priorities today; and they accord low priority to the investigation of phenomena (e.g.) of the following kinds: risks, especially long-term environmental and social risks of technoscientific innovation; the causal networks in which problems facing the poor, and scientific practices themselves, are located; and alternative practices (e.g., in agriculture, agroecology) that are not primarily based on using technoscientific innovations (e.g., transgenics) (Lacey 2005a; 2008, part 2). These kinds of phenomena are inherently linked with context, and so DSs cannot be adequate for investigating them. Thus, they are not only accorded low priority, but—where scientific methodology is taken to require the deployment of DSs—they are effectively excluded from "scientific" inquiry. That need not mean, however, that empirical investigation of them (in research conducted under context-sensitive strategies—CSs) does not produce results that are confirmed using the same cognitive criteria as are used to confirm results obtained under DSs [item (i), section 1]. Such results are obtained, e.g., in agroecology, the study of agroecosystems and their capacity to be productive, sustainable, protective of biodiversity and social health, and strengthening of the agency, culture and values of local populations (Altieri 1995; Wezel et al. 2009). The strategies (instances of CSs) and results of agroecology are valued highly where, e.g., the values incorporated into the Precautionary Principle—social justice, respect for the full range of human rights, environmental sustainability, equity within and between generations, participatory democracy (UNESCO-COMEST 2005)—are held, contesting V_{TP} and $V_{C\&M}$. Holding such values leads not only to different priorities for research, but to the deployment of a kind of methodological pluralism that includes CSs as well as DSs, and to the replacement of (ii) by something like:

(iii) Scientific knowledge derives from systematic empirically-based inquiry, conducted under strategies that are apt for gaining knowledge and understanding of the phenomena being investigated.

Thus, I suggest, commitment to the view that deployment of DSs is of the nature of "scientific" methodologies [item (ii)] rests, not on appeal to "objective" criteria of cognitive appraisal [item (i)], but on holding the values V_{TP} and $V_{C\&M}$ (for further argument, see Lacey 2009), and so it secretes commitment to value-outlooks that accord high value to technoscientific innovation that serves economic growth (Lacey 2008). Thus, although (non-cognitive) value judgments play no role in the cognitive evaluation of theories, (as distinct from the appraisal of their social value) and none are logically entailed by confirmed DS-results, they may play logically legitimate (although ethically and socially contested) roles connected with adopting strategies and with considering a theory as a desirable candidate for cognitive appraisal. The link between adopting strategies and holding social values is mediated by the social value of the object of investigation. Strategies need to be apt in the light of the characteristics of the object of research. DSs may be deployed to investigate many objects—e.g., planetary movements, subatomic particles—simply because, given their characteristics, they are amenable to investigation under DSs. In these cases, adopting DSs is not linked with holding V_{TP}; and research, conducted under DSs, may follow a trajectory that is not closely linked with V_{TP} (as in "basic science"). It is the virtually exclusive adoption of DSs (and considering them to be essential to "scientific" methodologies) that is associated with holding V_{TP}. It is not that adopting DSs is for the sake of furthering V_{TP}, but that there are mutually reinforcing (dialectical) relations between adopting them and holding V_{TP} (Lacey 1999; 2005a). In contrast, the adoption of agroecological strategies in research is dialectically linked with interest in sustainable agroecosystems, an interest nurtured by values like those that are incorporated into the Precautionary Principle; and other CSs are adopted because of concerns about the legitimacy of using certain innovations.

As stated above, knowledge gained under DSs has led to countless applications and serves to explain their efficacy. But it does not suffice to justify the legitimacy of using them. It cannot, for legitimacy concerns matters of ethical import, and DSs lack categories needed for ethical deliberation. Legitimacy deals with issues about benefits, risks (and uncertainties) and alternative practices that, like agroecology, are informed by knowledge gained largely under CSs. Yet the legitimacy of implementing technoscientific innovations is widely taken for granted. Prima facie, subject to rebuttal in the light of the outcomes of standard risk assessments, efficacy tends to be taken to be sufficient for legitimacy (Lacey 2011); and this is often defended by subtly interweaving the claims: "no risks" and "no alternatives".

2.1 Risks

All parties recognize that using technological innovations occasions risks. What is disputed are their character, extent, seriousness, mechanisms, and manageability under well-designed regulations (and how these vary from case to case). Can serious risks be managed?

According to the proponents of technoscientific innovation for the sake of economic growth, risk management should be based on the results of standard risk assessments (SRAs), where SRAs use strategies (instances of

DSs) that are appropriate for conducting research on direct risks to human health and the environment connected with chemical, biochemical and physical mechanisms, that can be quantified and their probabilities estimated. SRAs investigate potential effects of technological innovations that have been labelled "risks" or "harmful"—"risk" itself being a value-laden term is not an acceptable category in research conducted under DSs—using descriptive categories (e.g., toxicity) that may be used within DSs.

Empirical evidence for "there are no serious unmanageable risks" ["no risks"] would be the failure to find empirical evidence against it—and its reasonable endorsement would depend upon sufficient research of the appropriate kind having been conducted. To get more concretely at what is involved here, consider transgenic (TG) crops (for details, see Lacey 2005a, chap. 9). For the proponents of TGs, research conducted in SRAs is the appropriate kind of research. Then, their affirmation of "no risks" is based on the judgments that sufficient, well conducted SRAs have been performed (one by one) for all the potentially serious risks of using the TG varieties released for commercial use in specific environments, and that their results support that there is sufficient evidence to legitimate acting informed by the claim that none of the hazards risked are both significant and unmanageable. The judgment that there is sufficient evidence available to endorse such a claim, however, itself involves value judgments (Douglas 2009), so that people holding different value judgments may make opposing endorsements without being in conflict with the available data (see section 2.3).

In fact, critics do make opposing endorsements. They question both (1) that sufficient, well-conducted SRAs have been made on the TG varieties that have been released commercially and (2) that, even if they had been, that all the appropriate kinds of research have been conducted. I will discuss only the more fundamental (2).[2] It is a value judgment that evaluating the risks that can be investigated in SRAs exhausts the appropriate kinds of evaluation of risks. The critics challenge it, maintaining that appropriate research for appraising "no risks" needs to be more encompassing than that involved in SRAs, and to be conducted using strategies that are apt to deal with contextual factors. That is because the risks that need to be assessed are those that might result in harmful effects on human beings, social arrangements and ecological systems when TGs are actually used in agroecosystems and their products widely consumed, in the socioeconomic contexts of their actual use and over relevant temporal periods—taking into account that risks may arise from a variety of mechanisms owing to the fact that TGs in use are many kinds of things: not only biological entities, but

---

[2]Here I summarize analysis made in Lacey (2011) where, in addition, (1) is elaborated and discussed in detail.

also entities that embody V_{TP} (since they are products of genetic engineering), components of agroecosystems with worldwide dimensions, and commercial objects whose uses are constrained by claims of intellectual property rights, hence entities that embody $V_{C\&M}$.

SRAs are conducted in experimental and test spaces that can deal only with short term impacts on health and the environment that involve only physical/chemical/biological mechanisms. They cannot take into consideration (among other things) that some of the risks involved are likely to be magnified as TGs are more widely used, that the dominance of TG-oriented agriculture in the global market system may itself pose risks of irreversible hazards, and that the mechanisms of risks can be socioeconomic as well as physical/chemical/biological. Hence, among other things, SRAs do not investigate long-term environmental risks with socioeconomic mechanisms, e.g., destruction of biodiversity or contributing to the increase of greenhouse gases in the atmosphere, and risks to social arrangements: undermining alternative forms of farming, of displacing and impoverishing rural workers, and of bringing the world's food supply increasingly under the control of a few market-oriented corporations, potentially intensifying food insecurity throughout the world. SRAs—short-term experimental studies, conducted under DSs, which are insensitive to potentially relevant variables operative in the many and variable contexts of the use of TGs—can provide no evidence that hazards of the kinds mentioned are not being risked. Even if SRAs were exhaustively conducted, necessary and appropriate kinds of research still remain to be conducted. Furthermore, the point is not simply that there are yet-to-be-investigated risks. It is already well established that some of the potential harms that can be investigated under CSs have come to be, e.g., increased birth defects in some areas where herbicides, which have to be used where certain types of TGs are grown, are widely used (Antoniou et al. 2011).

Controversy about risks involves, therefore, not only disagreement about what hypotheses are well supported by available evidence, and about whether the evidence is sufficiently strong to warrant acting informed by "no risks", but also it is implicated in methodological disagreement (which, in turn, is correlated with conflicts of values): is scientific research limited to using DSs, or can it incorporate methodological pluralism that includes some CSs?

2.2 Alternatives

Proponents of TGs tend to maintain that risk assessment that is more extensive than that provided by SRAs is unnecessary and counterproductive (as well as not "scientific", since it is not restricted to the use of DSs). According to them, risks of the kinds indicated by the critics, if there are any, need not be taken into account—they have no bearing on the legitimacy of using TGs, for (they claim) there are no alternative kinds of farming that could be deployed instead of the proposed TG-oriented ways without occasioning unacceptable risks of food shortages, and that could be expected to produce greater benefits connected with productivity, sustainability, and meeting human needs ["no alternatives"]—TGs are necessary to feed the world—so that the risks that the critics want investigated pale into insignificance in face of the risk of not being able to feed everyone.

The critics counter: Agroecological (and other) methods are being developed that enable high productivity of essential crops (while occasioning less serious risks). They promote sustainable agroecosystems, utilize and protect biodiversity, and contribute to the social emancipation of poor communities; and they are particularly well suited to enable rural populations in developing countries to be well fed and nourished—without their further development current patterns of hunger are likely to continue (Lacey 2005a, chap. 10).

Is "no alternatives" well confirmed empirically? It would be only if it were reasonably endorsed following research that responded to questions about (what I call) the space of alternatives (Lacey 2008):[3]

> What agricultural methods—"conventional", TG, organic, agroecological, biodynamic, subsistence, etc—and in what combinations and with what variations, could be sustainable and sufficiently productive, when accompanied by viable methods of distribution, to satisfy the food and nutrition needs of the whole world's population for the foreseeable future?
>
> Are there alternatives with productive capacity comparable to that of TGs? Alternatives that could satisfy food and nutrition needs in contexts where TG methods may have little applicability—not necessarily a single alternative, but a multiplicity of complementary, locally-specific alternatives, which include agroecology and which simultaneously are: (a) highly productive of nutritive foods, environmentally sustainable and protective of biodiversity; (b) more aligned with, and strengthening of, rural communities and the diversity of their aspirations with place and culture; (c) capable of having an integral role in

---

[3] Although engaging in research posed by these questions (which are expanded in Lacey 2011) might not suit the interests of the proponents of TGs, it does not prejudge the dispute between them and their critics.

producing the food needed to feed the increasing world population; and (d) particularly suited to ensure that rural populations in impoverished regions are well fed and nourished?

The scientific questions that guided the development of TGs by agribusiness corporations and their implementation in farming practices, however, were not like these, but rather like this: What traits that would be useful to agribusiness" goals can be engineered into plants using the techniques of genetic recombination? The genuine success of producing TGs with these traits, and their efficacy, does not provide support for "no alternatives".[4]

TGs were implemented without attention being paid to the questions about the space of alternatives, and "no alternatives" is endorsed by their proponents without taking into account, e.g., the well documented successes of agroecology (Altieri 1995), (Lacey 2005a, chap. 10). Support for "no alternatives" would depend on evidence that there are inherent limits to the potential expansion of agroecological (combined with other) methods—and this cannot be done without engaging in research on these methods, research that cannot be confined to DSs. There is growing recognition of the importance of questions about the space of alternatives, and of the view that a variety of farming methods need to be developed if the food and nutrition needs of everyone are to be met, and of the centrality of agroecology in the variety[5]—but the components of the variety, and how they might be distributed, remain to be settled by empirical inquiry. "No alternatives" can only be scientifically appraised by deploying the appropriate resources of methodological pluralism that includes some CSs.

2.3 Endorsing "no risks" and "no alternatives"

Proponents of TGs usually endorse "no risk" and "no alternatives", i.e., they judge that the available empirical data support these hypotheses sufficiently strongly so as ethically to legitimate action and policy/regulation making that is informed by them. These endorsements would be well made if they were made after considering the consequences (and their ethical

---

[4]Later, an additional question was asked: How can the results of TG research be used in impoverished countries to deal with the problems of small-scale farmers (e.g., production in poor agroecosystems) and their communities (e.g., hunger and malnutrition)? The research that followed has been conducted without investigating the historical and socio-economic context of the problems intended to be addressed by using TGs, so that key variables relevant to successfully solving the problems are not investigated (see Lacey 2005a, chap. 8), (Lacey 2011).

[5]See, e.g., (De Schutter 2010), (Desmarais et al. 2010), (Foresight 2011), (IAASTD 2009), (Pretty 2008), (Royal Society 2009), (Wise & Murphy 2012). None of these publications rejects out of hand a role for TGs in the mix of methods that is needed; some anticipate a very significant role for them, and others a more qualified or minor one.

salience) of acting informed by each one of these hypothesis, should it be false—noting that the more ethically serious the consequences, what counts as "sufficiently strongly" needs to be held to more demanding standards (cf. Douglas 2009). To endorse a hypothesis requires making value judgments about the ethical seriousness of such consequences; in the present case they are derived from holding the value-outlooks, V_{TP} and $V_{C\&M}$. Such endorsements are not instances of confirmed scientific knowledge; a hypothesis may be supported sufficiently strongly by the available data to legitimate action informed by it, without it being part of the stock of established scientific knowledge [see item (i), section 1.1]. Since value judgments cannot (now) be eliminated in making endorsements, they do not meet the standards of established scientific knowledge (Lacey 2005b). Claims about the efficacy of using certain TGs do meet these standards; claims of "no risk" and "no alternatives" do not.

Nevertheless, these endorsements are often presented as if they were items of scientific knowledge—where the value commitments in play (V_{TP} and $V_{C\&M}$) are hidden under the cloud of an alleged "consensus" in the mainstream scientific community, and opposing endorsements (which reflect conflicting values such as those incorporated into the Precautionary Principle, and which draw upon additional data that have no place under DSs) are dismissed as "contrary to the mainstream scientific consensus" (Magnus 2008, e.g.,). In fact there is no consensus in mainstream science about "no risk" and "no alternatives" (see references in note 5); and, if there were today, if would reflect consensus around holding V_{TP} and $V_{C\&M}$, which are not matters that fall into the realm where the judgments of scientists can make a genuine claim to be authoritative. Appeal to consensus here has the effect of hiding the role that values must play in making endorsements. Douglas emphasizes that value judgments cannot be eliminated in making endorsements. Given this, opposing endorsements need not be a sign of one side stepping outside the bounds of science. Douglas wants hidden value judgments to be brought out into the open, and for scientists—after engaging in rational deliberation about the values in play—to assume responsibility for making the relevant value judgments. I think, however, that the primary responsibility of scientists is to exploit the resources of methodological pluralism so that scientific research and results become salient in such rational (and democratic) deliberation.

2.4 Scientific investigation of presuppositions of holding values

Value judgments play a role in methodological deliberations and in making endorsements of, e.g., "no risks". In addition, scientific research is salient for deliberation about values. Contrary to the view of values as subjective preferences, holding values may have presuppositions that are open to empirical inquiry. Holding V_{TP} (see note 1), e.g., presupposes claims, such as that technoscientific innovation provides benefits that contribute towards the well-being of human beings generally, that there are technoscientific solutions to most human problems including those occasioned by technoscientific innovations themselves, and that there are no serious alternative proposals available today to the pursuit of economic growth based on technoscientific innovation (Lacey 2002), (Lacey 2005a, chap. 1). Disconfirming these claims would not logically entail that V_{TP} should not be held; but, were they to be disconfirmed, holding V_{TP} would have no rational backing, and it would be odd to continue to hold them by insisting that they are just subjective preferences or expressions of interest.

These claims also underlie the value judgment that undermining alternative practices, which are not based on technoscience, is not a potential harm that needs to be investigated; and they inform the presumption of legitimacy usually accorded to implementing novel technoscientific innovations. They are widely endorsed; but, although they may be investigated empirically, efforts to do so are rare. Perhaps because they are deeply entrenched as part of the "common sense" of our age. Or perhaps because, in order to investigate them empirically, strategies need to be deployed that are apt for taking into account the social/economic/ecological/historical context of innovations. The outcomes of such investigation, conducted using the resources of CSs as well as DSs, might disconfirm the claims, and thus put into question the value judgments that presuppose them. It is true, of course, that fundamental value conflicts cannot be adjudicated by way of scientific research. Nevertheless, this leaves an important role for scientific research concerning them—for, as just illustrated, holding values may have presuppositions that are open to empirical inquiry, and these may be disconfirmed in the course of research. Treating values as subjective preferences (or even as outcomes of rational but non-scientific inquiry) hides this place where science and values may interpenetrate. That holding values has presuppositions, many of which are open to empirical investigation (provided that methodological pluralism is utilised), opens up space for constructive deliberation and discussion.

There is, however, a major impediment to this space being opened up. Concerning technoscientific innovations (like TGs), where V_{TP} and $V_{C\&M}$ are held, the move from efficacy to legitimacy of use is made, mediated by endorsements of "no risks" and "no alternatives" (Lacey 2008; 2011). But (a) "no risks" (supplemented by "any harmful consequences that occur can be taken care of by additional technoscientific innovations") and "no alternatives" are presuppositions of holding V_{TP}; and (b) holding V_{TP} reinforces adopting DSs exclusively; hence, (c) investigations are dismissed that are conducted under CSs. But this is to dismiss the investigations that might lead to challenges to "no risks" and "no alternatives" and, consequently, to holding V_{TP}. Inside of this closed circle, values cannot be other than preferences; and the preferences of the powerful usually win out.

It helps to bring the role of value judgments out into the open (as Douglas proposes)—not just in connection with endorsements of "no risks", but also in connection with the adoption of methodological strategies. The key, however, is to break the grip of DSs. This does not mean abandoning DSs, and replacing them by different privileged strategies (reinforced by values that conflict with V_{TP}), but treating them as one kind of fruitful strategies (that may be indispensable for all kinds of scientific research today) that needs to be complemented with a variety of other strategies (CSs) for the sake of adequately investigating phenomena that do not yield to DSs. An essential step, therefore, is to gain space for research on the kinds of phenomena discussed earlier connected with risks and alternatives; and the strategies deployed in that research bear mutually reinforcing relations with values such as those incorporated into the Precautionary Principle. The fruitfulness of research conducted under CMs needs to be demonstrated—i.e., shown that it can lead to knowledge and understanding that are positively appraised in the light of available data and the cognitive criteria referred to in (i) (section 1). In view of such problems as climate change, pollution, and growing threats to food security for millions of poor people (all causally linked to technoscientific innovation that serves economic growth), conducting such research today is a matter of urgency.

3 The co-unfolding of scientific knowledge and viable values

While adopting methodological pluralism may be motivated today largely by holding values that conflict with V_{TP}, what is at stake is much more than conflicting interests and values. Showing that some alternative strategies can be fruitful, and provide knowledge that could inform (e.g.) agroecological practices, shows that holding items (i) and (ii) (section 1) simultaneously involves tension. Affirming "no alternatives" on the basis of research conducted under DSs is not in accordance with (i); and, in general, where only DSs are utilized what really are endorsements can easily be misidentified as items of scientific knowledge. Only if (ii) is replaced by (iii) (section 2) is (i) likely to be preserved as a fundamental ideal of scientific practices. In addition, adopting methodological pluralism, in sufficiently inclusive a manner so that strategies reinforced by all currently viable value outlooks are given an opportunity to develop, is necessary for the sake of reinstating another traditional ideal of modern science: scientific knowledge as part of the shared patrimony of humankind, scientific knowledge—considered as a totality—available to be used evenhandedly across viable value-outlooks. Hence, it is not just interests and value-outlooks conflicting with V_{TP} and $V_{C\&M}$ that are well served by the adoption of methodological pluralism (interpreted in the light of (iii)), but the conformity of scientific practices with traditional ideals of modern science.

Conducting scientific practices in accordance with these ideals has implications for the reasoning involved in holding values and making value judgments.

Valuing TGs presupposes "no risks" and "no alternatives". Empirical investigation of these claims may provide compelling evidence against them that would require revision of the valuing of TGs. (It might also provide evidence supporting them, and this would strengthen the value judgment.) Similarly, valuing certain alternatives to TGs presupposes that they have significant productive potential, a claim that might be undermined (or strengthened) by the appropriate empirical research.

Regarding the role of values in methodological deliberations, I indicated that holding values that compete with V_{TP} provides motivation for adopting CSs and for obtaining whatever knowledge that might thereby be generated. However, appeal to the values by itself cannot sustain a research program indefinitely. If, after it has been provided with adequate material support, the research fails to be fruitful and, thus, fails to support (e.g.) that agricultural alternatives can be sufficiently productive, then the presuppositions of the value-outlooks that compete with V_{TP} would be undermined—and the values that motivate conducting research under agroecological strategies would be undercut. Similarly, the presupposition of V_{TP} are opened up to empirical investigation, and then the outcome of the research may strengthen or weaken the grounds for holding them.

These are examples of the co-unfolding of scientific knowledge and viable values. Values and science do not just touch, they interpenetrate in deep and logically admissible ways. If the appropriate range of strategies is not deployed, however, and scientific research is limited to strategies of the kind that are involved in the research that generates efficacious innovations (DSs), the values in play in setting research priorities and making endorsements about risks will not be subjected to the critique that can come from scientific investigation; and they will play their role without being opened to reasoned (and democratic) deliberation, and perhaps go unnoticed. The co-unfolding of scientific knowledge and viable values depends on recognising the essential role of methodological pluralism, incorporating CSs as well as DSs. There are dialectical relations between scientific knowledge (gained using the resources of methodological pluralism) and reasonably held values that perhaps can help to cut through some of the impasses confronted in controversies about technoscientific innovations.

The stated aim of the 14th ILMPS Congress was: to "help deepen our understanding of the most promising orientations in science and even help promote future advances in human civilization". Cultivating awareness of the co-unfolding of scientific knowledge and viable values is one way in which philosophy of science can contribute to furthering this aim.

Bibliography

Altieri, M. A. (1995). *Agroecology: The Science of Sustainable Development*. Bolder: Westview.

Antoniou, M., Habib, M. E. E. M., *et al.* (2011). Roundup and birth defects: Is the public being kept in the dark? URL http://pt.scribd.com/doc/57277946/RoundupandBirthDefectsv5> (accessed April 18, 2012).

De Schutter, O. (2010). Promotion and protection of all human rights, civil, political, economic, social and cultural rights, including the right to development. Tech. rep., UNESCO, URL www.srfood.org/images/stories/pdf/officialreports/ 20110308_a-hrc-16-49_agroecology_en.pdf, Report submitted by the Special Rapporteur of UNESCO on the right to food, Olivier De Schutter, 20 December 2010.

Desmarais, A. A., Wiebe, N., & Wittman, H. (Eds.) (2010). *Food Sovereignty: reconnecting food, nature and community*. Oakland: Food First.

Douglas, H. E. (2009). *Science, Policy, and the Value-Free Ideal*. Pittsburgh: University of Pittsburgh Press.

Foresight (2011). The future of food and farming: final project report. URL www.bis.gov.uk/assets/bispartners/foresight/docs/ food-and-farming/11-546-future of-food-and-farming-report.pdf.

IAASTD (2009). *Agriculture at a Crossroads: Synthesis report*. International Assessment of Agricultural Knowledge, Science and Technology for Development, Washington: Island Press.

Lacey, H. (1999). *Is Science Value Free? Values and Scientific Understanding*. London; New York: Routledge.

Lacey, H. (2002). The ways in which the sciences are and are not value free. In *The Scope of Logic, Methodology and Philosophy of Science, 11th ILMPS Congress*, vol. 2, Gärdenfors, P., Kijania-Placek, K., & Woleński, J., eds., Dordrecht; Boston: Kluwer, 513–526.

Lacey, H. (2005a). *Values and Objectivity in Science; Current Controversy about Transgenic Crops.* Lanham: Lexington Books.

Lacey, H. (2005b). On the interplay of the cognitive and the social in scientific practices. *Philosophy of Science, 72,* 977–988.

Lacey, H. (2008). Ciência, respeito à natureza e bem-estar humano. *Scientiae Studia, 6*(3), 297–327, doi:10.1590/S1678-31662008000300002.

Lacey, H. (2009). The interplay of scientific activity, worldviews and value outlooks. *Science & Education, 18*(6–7), 839–860, doi:10.1007/s11191-007-9114-6.

Lacey, H. (2011). Views of scientific methodology as sources of ignorance in the controversies about transgenic crops. Paper presented at the Conference: 'Agnotology: ways of producing, preserving, and dealing with ignorance', center for Interdisciplinary Research, University of Bielefeld, Germany, May 30–June 1, 2011.

Lacey, H. (2012). Reflections on science and technoscience. *Scientiae Studia, 10,* 103–128, doi:10.1590/S1678-31662012000500007.

Lacey, H. & Mariconda, P. (2012). The eagle and the starlings: Galileo's argument for the autonomy of science—how pertinent is it today? *Studies in the History and Philosophy of Science, 43*(1), 122–131, doi:10.1016/j.shpsa.2011.10.012.

Magnus, D. (2008). Risk management versus the precautionary principle: Agnotology as a strategy in the debate over genetically engineered organisms. In *Agnotology: The Making and Unmaking of Ignorance,* Proctor, R. N. & Schiebinger, L., eds., Stanford: Stanford University Press, 250–265.

Poincaré, H. (1920). *The Value of Science.* New York: Dover, 1958.

Pretty, J. (2008). Agricultural sustainability: concepts, principles and evidence. *Philosophical Transactions of the Royal Society B, 363,* 447–465, doi: 10.1098/rstb.2007.2163.

Royal Society (2009). Reaping the benefits: science and the sustainable intensification of global agriculture. Tech. rep., Royal Society, URL http://royalsociety.org/uploadedFiles/Royal_Society_Content/policy/publications/2009/4294967719.pdf.

UNESCO-COMEST (2005). *The Precautionary Principle.* Paris: UNESCO, (World Commission on the Ethics of Scientific Knowledge and Technology).

Wezel, A., Bellon, S., et al. (2009). Agroecology as a science, a movement and a practice. a review. *Agronomy for Sustainable Development, 29*(4), 503–515, doi:10.1051/agro/2009004.

Wise, T. A. & Murphy, S. (2012). Resolving the food crisis: assessing global policy reforms since 2007. Tech. rep., IATP (Institute for Agriculture and Trade Policy) & GDAE (Global Development and Environment Institute at Tufts University), URL http://iatp.org/files2012_01_17_ResolvingFoodCrisis_SM_TW_0.pdf.

Hugh Lacey
Swarthmore College
USA
Universidade de São Paulo
Brazil
hlacey1@swarthmore.edu.

Scepticism and Verificationism

YEMIMA BEN-MENAHEM

ABSTRACT. Verificationism and scepticism are distinct philosophical positions: whereas the global doubt of the sceptic implicates any form of knowledge, verificationists typically trust some forms of knowledge while denying (purported) others. Nonetheless, the verificationist often joins forces with the sceptic in opposing claims to knowledge that both of them deny. For example, the verificationist may join the sceptic in denying absolute time and absolute temporal relations on account of their being unverifiable. In this paper, however, I examine three examples of verificationist arguments that have been used to counter sceptical arguments: Einstein's equivalence principle, Wittgenstein's rule-following paradox and Putnam's model-theoretical argument. I argue that while verificationism is often used to deny truth and meaning (to the unverifiable), it can also be used to confer meaning and defend truth in the face of the threat of scepticism.

The difference between scepticism and verificationism is obvious: The global doubt of the sceptic implicates any form of knowledge and any means of obtaining it, including mathematical proof and direct observation. By contrast, verificationists typically trust certain forms of knowledge while denying (purported) others. In other words, whereas verificationism is tied to a particular epistemology, scepticism denies the very possibility of knowledge and thus preempts epistemology altogether. Nonetheless, the verificationist often joins forces with the sceptic in opposing knowledge-claims that both of them deny. For example, the verificationist may join the sceptic in denying knowledge claims about the infinite, the external world, other minds, absolute space and time, the past, and so on, on account of their being unverifiable. In such cases, a verificationist epistemology, or verificationist theory of meaning, is put in the service of scepticism about certain domains of purported knowledge. Hume and the logical positivists come to mind as philosophers who combined verificationism and scepticism in this way, denying, for instance, knowledge of causal claims in the case of Hume, and the theoretical apparatus of science, in the case of the logical positivists.

Figure 1.

Illustration n° 1 represents, in my view, the more common connection between verificationism and scepticism. But the verificationist is definitely not always a friend of scepticism. American pragmatists, for example, were attracted to verificationism but strongly objected to scepticism. Thus, even though the typical case is that of an alliance between the verificationist and the sceptic, there are also examples of philosophers who dissociate the two positions, endorsing verificationism while declining scepticism. In this paper I will set aside the friends of scepticism and focus on anti-sceptical positions and their relationship with verificationism. In particular, I wish to draw attention to cases in which a verificationist argument has been used (or could be used) to counter a sceptical position. I examine three very different examples: Einstein's equivalence principle, Wittgenstein's rule-following paradox and Putnam's (early) response to his model-theoretical argument. The first of these (Einstein's argument) has not been advanced as an argument against scepticism and yet, it illustrates an ingenious employment of an argument from underdetermination—a standard sceptical argument—to extract novel empirical knowledge. The second (Wittgenstein's argument) is often read as a verificationist solution to a sceptical paradox. On this reading it shows that verificationism can serve as a defense against scepticism. But Wittgenstein's argument can also be read as a *reduction* argument against verificationism. If the latter reading is accepted, verificationism functions here as an unstable position, a mere stepping stone in an anti-sceptical argument. The third example (Putnam's argument) again illustrates the possibility of mobilizing a verificationist argument against scepticism. Before turning to these examples, it is useful to have before us a schematic outline of positions that seek to combine verificationism with an anti-sceptical stance.

1 Verificationism without scepticism

I have noted that the typical verificationist, the ally of the sceptic, is not the subject of this paper. Similarly, paradigmatic rebuttals of scepticism, such as Descartes', namely, rebuttals that are not committed to verificationism need not concern us here. It is the combination of verificationism with an anti-sceptical stance that I seek to draw attention to. What, then, are the strategies that enable a philosopher to endorse verificationism but resist scepticism, say about the external world? Here are a few possibilities.

1. The most direct strategy is to block the move from verificationism to scepticism by showing that one *can* after all find in experience a foundation strong enough to erect on it the entire edifice of knowledge about the external world. In other words, what is verifiable from the empiricist point of view is sufficient to ground knowledge about physical objects. This is, for instance, the route taken by Quine in *Word and Object* and "Epistemology Naturalized". As is the case with other twentieth century philosophers, Quine's starting point is an account of *meaning* rather than a traditional epistemic foundation. Summarizing Quine's position, Davidson ascribes to him the principle

> [...] that whatever there is to meaning must be traced back somehow to experience, the given, or patterns of sensory stimulation, something intermediate between belief and the usual objects our beliefs are about. (Davidson 1983, 144)

And he adds:

> This is a marvelously ingenious way of capturing what is appealing about verificationist theories without having to talk of meanings, sense-data or sensations; for the first time it made plausible the idea that one could, and should, do what I call the theory of meaning without need of what Quine calls meanings. (Davidson 1983, 145)

Having paid Quine these compliments, however, Davidson immediately turns to criticizing his position:

> But Quine's proposal, like other forms of verificationism, makes for skepticism. For clearly a person's sensory stimulations could be just as they are and yet the world outside very different. (Remember the brain in the vat.) (Davidson 1983, 145)

Davidson accuses Quine of the category mistake (!) of conflating causes and reasons.

> No doubt meaning and knowledge depend on experience, and experience ultimately on sensation. But this is the 'depend' of causality, not of evidence or justification. (Davidson 1983, 146)

This insight and critique make Davidson opt for another way of fending off scepticism, namely "a coherence theory of truth and knowledge", a theory which does not depend on verificationism and will therefore not be further discussed here. A similar critique of the empiricist foundation and a similar accusation of the cause-reason blunder is Wilfrid Sellers' celebrated critique of "the myth of the given". In both of these critiques, a strictly empirical foundation of knowledge about 'reality' is denied.

There are a number of other strategies that make room for a combination of verificationism and anti-scepticism.

2. One could rule out radical forms of scepticism, such as scepticism about the external world and other minds, as positions that are impossible to sanely uphold for more than a few seconds. According to this view, the (self proclaimed) sceptic only pretends to suspend judgment about knowledge claims of this kind, but in fact sees physical objects and other persons as perfectly real, just as real as the non-sceptics see them. Less radical forms of scepticism, however, could still be allowed and be justified by verificationism: e.g., it would be possible to deny the reality of the (so-called) theoretical entities of science, on account of their lying beyond the realm of the verifiable.

3. One can reject the epistemic standards of the sceptic as too stringent and urge, with Austin, that enough (evidence) is enough even if it is not everything. This tack, typically favored by pragmatists, makes it possible to endorse verificationism while rejecting scepticism. What makes this combination possible, however, is that the verificationist standards are somewhat relaxed.

4. Pragmatists also contend that doubt, no less than belief, must be justified, and that justification (of both doubt and belief) presupposes a large body of shared standards and beliefs. While circumscribed doubt about specific knowledge claims makes sense against a background of shared belief, the global doubt of the sceptic violates the very standards of rational inquiry and justification, the very standards in the name of which the sceptic has formulated his case in the first place. Once more, verificationism, properly confined to local rather than global issues, can be maintained.

5. A more radical strategy, of which Michael Dummett has been the most prominent representative, construes truth and meaning in terms of provability. Statements that we can neither prove nor refute lack truth values. Statements for which we are unable to specify conditions under which they would be considered verified or falsified are altogether meaningless. On this view, there can be no unknown truth, and thus, the gap between truth and knowledge that worried the sceptic no longer exists. The outer ring in the above illustration is empty.[1] In other words, precisely because one endorses verificationism, no room at all is left for scepticism. Similar arguments, even if not always couched in terms of a verificationist theory of meaning (as in Dummett), have been advanced by other philosophers. Berkeley repeatedly dissociates himself from scepticism. Scepticism, he argues, presupposes a distinction between the real and the perceived. Once we realize that *esse* and *percipi* cannot be separated, scepticism loses its grip.[2] In a similar vein, Peirce, who went to great length to critique scepticism, explicitly denies the unknowable as meaningless.

> The absolutely incognizable has no meaning...Whatever is meant by any term as 'the real' is cognizable in some degree, and so is of the nature of a cognition, in the objective sense of the term. (Peirce 1868, 238)

Note the difference between strategies n° 2–4 and strategy n° 5. The former have the modest goal of ensuring that the verificationist need not end up as a sceptic. Thus, while denying that verificationism entails scepticism, proponents of strategies 2–4 do not claim that verificationism as such yields an argument against scepticism. By contrast, according to the fifth strategy it is verificationism that implies the denial of truth and meaning to the unverified and unverifiable. Here verificationism plays an essential role in the rejection of scepticism.

Moreover, the fifth strategy avoids scepticism by commending *iconoclasm*. It construes the sceptic as lamenting human limitation—our failure to get in contact with reality. But this image of a reality-beyond-reach, we are told by the iconoclast, is no more than an idol! Nothing is lost when this idol is destroyed. The distinction between scepticism and iconoclasm is fundamental not only for understanding philosophers who argue explicitly against scepticism, but also for the proper understanding of arguments that are all too often interpreted as sceptical, but are in fact iconoclastic. Such is Quine's thesis of the indeterminacy of meaning. Taken as a sceptical argument, it suggests a deficiency or malfunction of language—its failure

---

[1] The sceptic can of course still disagree with the verificationist about the validity and justification of knowledge claims in the inner circle.

[2] See, for example, (Berkeley 1710, par. 87–88), and (Popkin 1983) for an extensive discussion of Berkely on scepticism.

to determine meaning. Interpreted iconoclastically, however (as it should be, in my view), it says that certain theories of meaning are no more than mythologies. Language functions properly without 'meaning' in the sense that these mythologies reify. As we will see, analogous considerations apply to Wittgenstein's rule following paradox.

Let me now turn to the second part of the paper and review three examples of verification arguments employed in ways that do not lead to scepticism, or serve to resist it.

2 Verificationism against skepticism

2.1 Einstein's principle of equivalence

The thesis of the underdetermination of scientific theory poses the following epistemic difficulty. If theories are necessarily underdetermined by observation—to wit, underdetermined not only by observations carried out so far, but by every possible observation—then there could be incompatible theories that fit every observation equally well. Such theories, though incompatible, would nonetheless be empirically equivalent. The term 'underdetermination' suggests an analoy with a mathematical problem in which there are fewer constraints (e.g., fewer equations) than needed to determine a unique solution. Underdetermination implies a gap between theory and reality, and therefore plays right into the hands of the sceptic. Indeed, arguments for the underdetermination of theory, from Poincaré's argument for the underdetermination of geometry by experience to the present, are read as sceptical arguments.[3] As such, they are not expected to have empirical import or yield new predictions. While these arguments are recognized as philosophically significant, their significance is negative in the sense that they show what *cannot* be known, asserted or justified. Thus, they are mute from the scientific perspective. Or so it seems.

Consider, however, Einstein's celebrated thought experiment: An observer in a sealed box makes certain experiments. For example, she drops an apple and notes that it accelerates uniformly towards the floor. This effect, Einstein reasoned, can be attributed either to a gravitational attraction of a large mass lying 'underneath' the box and pulling the apple 'down', or to an 'upward' uniform acceleration of the box. In an attempt to distinguish between these possibilities, the observer goes on to drop two objects of different masses at the same time. But this experiment is equally

---

[3]The question of whether scientific theories are in fact necessarily underdetermined by experience is far from trivial. Quine, the best known proponent of the underdetermination of theory, admitted that he had no general proof of underdetermination. See Ben-Menahem (2006, chap. 6) for details. Recently, a proof has been provided by Hilary Putnam (2012).

inconclusive: The objects hit the floor exactly together and their acceleration can again be attributed either to the acceleration of the box, or to gravity.[4] The correspondence between different masses is guaranteed by the equality (up to a unit) of inertial and gravitational mass, a fact that had been appreciated by Newton and tested by him as well as later scientists with increasing precision. It is this equality which underlies and explains Galileo's law—the independence of gravitational acceleration on the mass of the falling body. The equality implies that as long as our observer remains within the box and continues to perform the same kind of experiment, she will not be able to determine which of the explanations is correct—they are empirically equivalent.

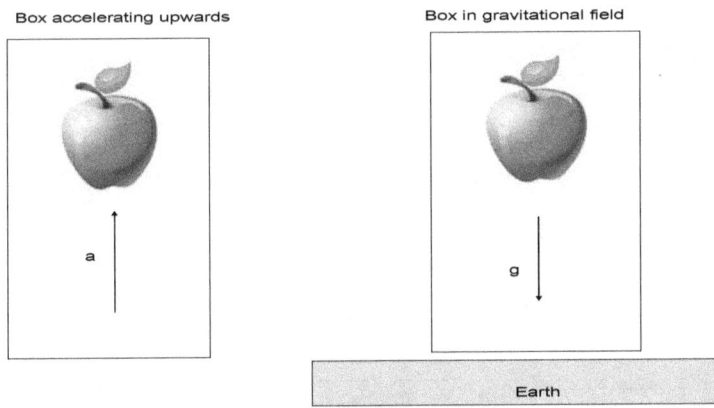

Figure 2.

Einstein's ingenious move was to turn this equivalence into a fundamental principle—the equivalence principle—which holds not only for mechanical phenomena but for all natural phenomena. Locally, all the effects of uniform acceleration are now claimed to be indistinguishable from those of uniform gravitational fields. The explanation of our experiences is thus underdetermined by every observation one could possibly perform within the box.

---

[4]It must be kept in mind that Einstein is referring to a uniform gravitational field; different bodies will behave differently in non-uniform fields, such as the field of a planet, in which case the effects observed in an accelerating frame are distinguishable from the effects of gravity. Further, in a non-uniform field, even a single body of finite dimensions is subject to tidal effects, again distinguishing the two cases.

Elaborating his thought experiment under this generalized equivalence, Einstein imagined the observer to send a beam of light horizontally across the box. Clearly, if the box is accelerating upwards, by the time light reaches the opposite wall, the box would have moved and the trajectory of the beam would have to appear curved. Classically, the beam should not curve in a gravitational field. This experiment could therefore distinguish between the cases. But if the principle of equivalence is true, exactly the same bending of the beam should be observed on either one of the alternative scenarios (see illustration 3).

Here, then, is a new prediction based on the principle of equivalence: light bends in a gravitational field. An argument for equivalence and underdetermination turned out to be a most powerful empirical tool.

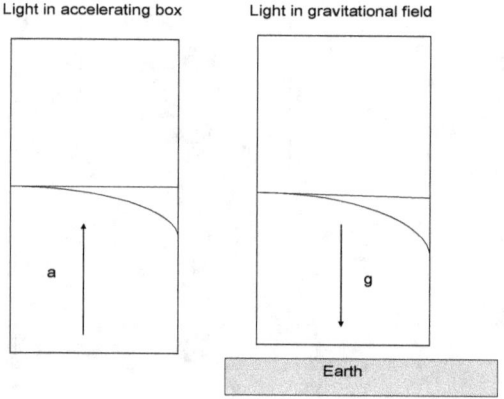

Figure 3.

The application of the principle of equivalent to radiation was the first step on the road to General Relativity. The principle of equivalence further suggested to Einstein a deep connection between gravity and geometry. In both Newtonian mechanics and Special Relativity, inertial frames pick out privileged trajectories, for relative to an inertial frame, free particles and light rays move in straight lines. Since, guided by the principle of equivalence, Einstein sought to unify inertia and gravity, he was led to generalize geometry so as to render inertial and gravitational motion not only physically, but also geometrically, equivalent. The idea was that *gravitational-cum-inertial motion charts the privileged trajectories of a more general geometry*. (In our toy box, the curved path of the beam should represent such a privileged trajectory). The implication is that the geometrical structure of spacetime emerges as a matter of empirical fact! Since the trajectories of particles in a gravitational field are determined by the field, a geometrical structure mapped out by these trajectories is contingent on the structure and strength of the field, and thus on the distribution of the sources that produce the field. Hence the dynamic spacetime of General Relaivity. On the dynamic conception there is no prior geometry, but only the geometry read off the basic physical processes. The 'no prior geometry' vision is the revolutionary core of General Relativity, and taken by many later relativists to be a fundamental constraint on the structure of physical theories in general.

Compare Einstein's elaborate use of his principle of equivalence with Poincaré's argument for the empirical equivalence of different geometries, Poincaré (1902, chap. 3–5). Both of these arguments draw on verificationist sensitivities, for both lead to the conclusion that where there is no experience that can distinguish between two descriptions, there is no fact of the matter as to which of these descriptions is true. But whereas Poincaré utilized his argument as a typical sceptical argument—Einstein turned an argument for equivalence and underdetermination into a lever of empirical innovation. A seemingly sceptical argument has become the cornerstone of a scientific revolution of colossal empirical import.

2.2 Wittgenstein's rule-following paradox

Wittgenstein sums up the problem very concisely in *Philosophical Investigations*:

> This was our paradox: no course of action could be determined by a rule, because every course of action can be made out to accord with the rule. (Wittgenstein 1953, 201)

While it is widely agreed that the rule-following paradox constitutes a sceptical paradox, there is far less agreement on what Wittgenstein's considered a solution to his paradox. On one of the more common interpretations, the solution consists in verificationism: in response to the paradox, Wittgenstein (on this interpretation) replaced his earlier realist understanding of meaning with a verificationist semantics involving assertability conditions rather than truth conditions (see Dummett 1978; Kripke 1982). Whether or not this solution was the one intended by Wittgenstein, it illustrates that verificationism can be understood to preempt scepticism.

Let us nonetheless consider the question of whether this was indeed Wittgenstein's solution. In support of the verificationist interpretation it can be observed that Wittgenstein has indeed expressed sympathy to verificationism in the philosophy of mathematics as well as other philosophical contexts. What speaks against this interpretation, however, is the fact that in the above passage, verificationism is not even mentioned. Rather, Wittgenstein draws the following lesson.

> What this shews is that there is a way of grasping a rule which is *not* an *interpretation*, but which is exhibited in what we call "obeying the rule" and "going against it" in actual cases. (Wittgenstein 1953, 201, italics in the original)

To understand this conclusion, we must reconstruct Wittgenstein's path to the paradox. The starting point, I would like to suggest, was the nature of necessary truth, a problem that occupied Wittgenstein throughout. Wittgenstein certainly rejected the traditional view of necessity as truth in all possible worlds. His objection to this view is explicit in the *Tractatus* and has not changed in later years. For Wittgenstein, so-called necessary truths are not truths, that is, they do not represent facts of any kind. What are the alternatives to the traditional view? At one point Wittgenstein was certainly attracted to a conventionalist account of necessity. The analogy between so-called necessary truths and rules of our own making recurs in his writings. This conventionalist account pleases the verificationist, who notes that while the traditional account faces the formidable problem of how we can know what holds true in all possible worlds, the conventionalist has a simple answer to epistemic worries about necessity: We know the rules because we are the legislators. For a while it seemed to Wittgenstein that the conventionalist account is correct. It was at this point, however, that he discovered the paradox. Rules, according to the paradox, even rules of our own making, do not determine their interpretation. The promise of both conventionalism and verificationism turned out to be is an illusion. Both of these positions are at least as vulnerable to the paradox as the older realist accounts of necessity. Worse, a new form of scepticism, more radical than earlier ones, has emerged. Wittgenstein would certainly not settle for scepticism, a position he deemed senseless from the *Tractatus* to *On Certainty* (Wittgenstein 1977). Instead, he assumed an iconoclastic stance that altogether jettisons theories of meaning, whether they are couched in terms of truth conditions or in terms of assertability conditions. This iconoclasm implies that there is no deeper explanatory level 'underneath' language that justifies and explains its successful functioning. The only clarification we can hope for is internal to language and consists in the identification of inner links and differences between its various uses. Hence, "there is a way of grasping a rule which is not an interpretation". On this reading the conventionalist account of (so-called) necessary truth is not, after all, a solution to the sceptical paradox. In the same vein, a verificationist theory of meaning does not solve the problem either. Rather, verificationism is a transient phase in Wittgenstein's move towards iconoclasm.

2.3 Putnam's model-theoretic argument

Putnam's argument pertains to a mirror image of underdetermination. Rather than different and incompatible theories fitting the same world, we are now confronted with the possibility of different and non-isomorphic worlds (models) satisfying the same theory. Whereas underdetermination raises doubt about truth—there may be no uniquely correct theory, the model-theoretic argument raises a problem about reference: a theory—even an ideal theory— cannot fix a unique reference relation.

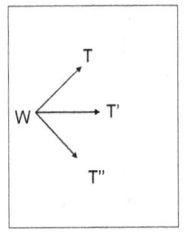 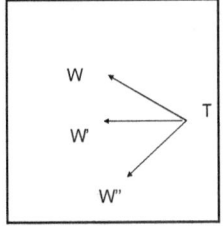

Underdtermination: Different theories fit the same worldy

Putnam's argument: Different worlds satisfy the same theory

Figure 4.

Putnam's writings contain different versions of the model-theoretic argument. The most formal version is based on the Lowenheim Skolem theorem, according to which theories rich enough to include arithmetic are bound to have non-isomorphic models. If one conceives of the relation between our theories and reality as that between a formalism and its models, then a conclusion analogous to that of the theorem follows for the theory-world correspondence: even the best theory, the theory that satisfies every desideratum we can think of, cannot hook on to reality in a unique way. In other words, even if we know exactly which theorems (or sentences) are true, as long as they do not pin down their interpretation, we cannot be sure what they mean.

When first put forward by Putnam, the model-theoretic argument was directed against metaphysical realism. The metaphysical realist, Putnam argued, construes truth as non-epistemic—a theory can meet all our epistemic standards and still be false. It is precisely this non-epistemic construal, Putnam claimed, that makes the metaphysical realist vulnerable to the model-theoretic argument. But Putnam soon realized that the sceptic is equally misguided, for he too upholds a gap between truth and verification. Closing this gap, then, could provide Putnam with a rejoinder to the model-theoretic argument and, at the same time, a defense against scepticism. Indeed, in *Models and Reality* the solution offered to the problem posed by the model-theoretic argument is verificationism—a position that (as we have seen) allows no such gap (Putnam 1983). Verificationism, wherein truth is replaced with what our experiences confirm, or with what our justification procedures warrant, has no place for the metaphysical realist's non-epistemic notions of truth and reality. As Putnam put it in *Reason Truth and History*:

> 'Truth' [...] is some sort of (idealized) rational acceptability—some sort of ideal coherence of our beliefs with each other and with our experiences *as these experiences* are themselves represented in our belief system—and not correspondence with mind-independent or discourse-independent 'state of affairs'. (Putnam 1981, 50, italics in original)

The version of verificationism that Putnam entertained at this time was inspired by Dummett and mathematical intuitionism.[5] For example, Putnam suggested that thinking of models as mathematical constructions rather than abstract mathematical objects would block the sceptical conclusion:

> Models are not lost noumenal waifs looking for someone to name them; they are constructions within our theory itself, and they have names from birth. (Putnam 1983, 25)

In the same spirit, Putnam proceeded to undermine the analogy that created the problem in the first place: our everyday language is not a syntactic structure open to various interpretations. Moreover, Putnam actually broadens the concept of verification—a *practice* such as the use of language can be self-justifying and self-explanatory.

---

[5]There are also significant differences between Putnam and Dummett, e.g., their position on holism, which I cannot discuss here.

To adopt a theory of meaning according to which a language whose whole use is specified still lacks something—namely its 'interpretation'—is to accept a problem which *can* only have crazy solutions. To speak as if *this* were my problem, "I know how to use my language, but now, how shall I single out an interpretation?" is to speak nonsense. Either the use *already* fixes the 'interpretation' or *nothing* can. (Putnam 1983, 24, italics in original)

In Putnam's argument, then, verificationism is the *remedy* to a sceptical problem. I stress this point since *Models and Reality* has often been read as promoting, rather than criticizing, scepticism.[6] Note the resemblance between Putnam's response to his model-theoretic argument and Wittgenstein's response to his rule following paradox. Both of them silence the sceptical doubt by denying the gap between the use of language and its interpretation. But whereas Putnam, at this point, links this solution to verificationism, Wittgenstein has already moved beyond verificationsm to iconoclasm. Interestingly, for Putnam too, the verificationist phase was short lived and he soon 'repented', returning to more realist positions on truth and meaning. These later developments notwithstanding, the lesson of *Models and Reality* is that a practice can confer meaning. This is a Wittgensteinian insight which Putnam could have retained in later years.

The reader who has reached this point may be wondering why Kant has been omitted. The reason is simple: A profound analysis of the crucial role of verificationism in Kant's response to scepticism, can be found in two very recent papers by Carl Posy (2011; Forthcoming). Posy's analysis further supports the conclusion reached here: although often combined with scepticism, verificationism has also been a source of powerful arguments against scepticism.

Bibliography

Ben-Menahem, Y. (2006). *Conventionalism.* Cambridge: Cambridge University Press.

Berkeley, G. (1710). The principles of human kowledge. In *The Works of George Berkeley, Bishop of Cloyne*, vol. II, Luce, A. & Jessop, T., eds., London: Thomas Nelson and Sons, 1949.

Davidson, D. (1983). A coherence theory of truth and knowledge. In *Subjective, Intersubjective, Objective*, Oxford: Clarendon Press, 137–158, 2011.

---

[6]See, for example Van Fraassen (1997) for a reading of the model-theoretical argument as an argument for scepticism rather than a critique thereof.

Dummett, M. (1978). Wittgenstein's philosophy of mathematics. In *Truth and Other Enigmas*, Cambridge, MA: Harvard University Press, 166–185.

Kripke, S. (1982). *Wittgenstein on Rules and Private Language*. Oxford: Blackwell.

Peirce, C. (1868). Some consequences of four incapacities. In *Writings of Charles S. Peirce: A Chronological Edition*, vol. 2, Fisch, M., Kloesel, C., & Moore, E., eds., Bloomington: Indiana University Press, 211–242, 1982.

Poincaré, H. (1902). *Science and Hypothesis*. New York: Dover, 1952.

Popkin, R. (1983). Berkeley and Pyrrhonism. In *The Skeptical Tradition*, Burnyeat, M., ed., Berkeley: University of California Press, 377–398.

Posy, C. (2011). On the finite. In *Logic and Knowledge*, Celluci, C., Grosholz, E., & Ippoliti, E., eds., Newcastle upon Tyne: Cambridge Scholars Publishing, 329–357.

Posy, C. (Forthcoming). Realism, reason and reference. In *Library of Living Philosophers: Hilary Putnam*, Hahn, L., ed., Chicago: Open Court.

Putnam, H. (1981). *Reason, Truth and History*. Cambridge: Cambridge University Press.

Putnam, H. (1983). Models and reality. In *Realism and Reason. Philosophical Papers*, vol. 3, Cambridge: Cambridge University Press, 1–24.

Putnam, H. (2012). A proof of the underdetermination 'doctrine. In *Philosophy in an Age of Science*, Putnam, H., De Caro, M., & Macarthur, D., eds., Cambridge, MA: Harvard University Press, 270–276.

Van Fraassen, B. (1997). Putnam's paradox. In *Mind. Causation and World: Philosophical Perspectives*, vol. 11, Tomberlin, J., ed., Oxford: Blackwell, 17–42.

Wittgenstein, L. (1953). *Philosophical Investigations*. New York: Macmillan.

Wittgenstein, L. (1977). *On Certainty*. Oxford: Basil Blackwell.

Yemima Ben-Menahem
Department of Philosophy
The Hebrew University of Jerusalem
Israel
yemima.ben-menahem@mail.huji.ac.il

Mathematical Abstraction, Conceptual Variation and Identity

JEAN-PIERRE MARQUIS[1]

ABSTRACT. One of the key features of modern mathematics is the adoption of the abstract method. Our goal in this paper is to propose an explication of that method that is rooted in the history of the subject.

1 Introduction

The main purpose of this paper is to sketch a theory of mathematical abstraction as it appeared in 20th century mathematics. Indeed, mathematics in the 20th century is marked by what is called the abstract approach. To wit:

> One of the amazing features of twentieth century mathematics has been its recognition of the power of the abstract approach. This has given rise to a large body of new results and problems and has, in fact, led us to open up whole new areas of mathematics whose existence had not even been suspected. (Herstein 1975, 1)

This quote is taken from the opening page of a standard textbook in abstract algebra. I would not say, however, that one of the amazing features of modern mathematics is the "recognition" of its power. For that suggests that the abstract approach existed before the twentieth century and that mathematicians then came to realize how powerful the approach is. In fact, the abstract approach was *created* in the late 19th and early 20th centuries and it was *developed* in the 20th century. The recognition of its power came along with its development.

---

[1] I gratefully acknowledge the funding received from the SSHRC. I want to thank the organizers for inviting to present this work at the CLMPS 2011. The paper was written in its present form while I was visiting the University Paris-Diderot. I want to thank Jean-Jacques Szczeciniarz and the University Paris-Diderot for their hospitality.

Together with its power, Herstein emphasizes the capacity of the abstract approach to bring to existence whole new areas of mathematics. As is well known, the face of mathematics changed radically during the 20th century, largely because of the abstract approach. The existence of these new areas of mathematics goes hand in hand with the power of the approach. Were these new areas not powerful—no matter how one understands that latter notion—, they would not be taken to refer to genuinely new mathematical objects.

If we were to ask mathematicians what is the abstract approach, they would probably point to obvious examples. Herstein, for one, would simply invite us to study his textbook: after a brief introduction to set theory, mappings and the integers, follow chapters on group theory, ring theory, vector spaces and modules, fields and linear transformations. These are all familiar topics in contemporary mathematics. But why are they part of the abstract approach? My goal in this paper is to elucidate what it is for a mathematical theory to be considered abstract and how this feature contributes to the approach that turned out to be and still is so powerful. Along the way, we should be able to clarify what this power amounts to in conceptual terms.

Given that there is in contemporary philosophy of mathematics a large literature surrounding the nature of abstract entities, let me state explicitly what I am *not* doing in this paper.

First, I am not entering the *ontological* debate. What I have to say bears no relationship with contemporary discussions surrounding abstract entities or the abstract/concrete distinction, e.g., that abstract entities are causally inert or lack any spatio-temporal coordinates, etc. (For a general discussion of the issues involved in this debate, in particular issues related to the nominalism/platonism debate, see for instance (Burgess & Rosen 1997).) The whole discussion I am about to launch unfolds entirely *within* the realm of abstract entities, no matter how these are defined. It is my profound belief that abstraction in mathematics is solely an epistemological issue and that the abstract character of mathematics is *not* an ontological property but rather derives from epistemological features of mathematical knowledge itself. I am not so much concerned with abstract objects than with the *process* of abstraction and the abstract *method*. Some mathematical objects, or rather mathematical concepts, are *abstracted*. They do not inherit a dubious ontological status for that reason. Mathematicians also talk about concrete mathematical entities and, by the latter, they don't mean an abacus or a compass.

> This text attempts a different approach, letting the abstract concepts emerge gradually from less abstract problems about

geometry, polynomials, numbers, etc. This is how the subject evolved historically. This is how all good mathematics evolves—abstraction and generalization is forced upon us as we attempt to understand the "concrete" and the particular. (Solomon 2003, 3)

Thus, I am after a specific process that we find (mostly but not exclusively) in mathematics and that plays a key role in the development of contemporary mathematics.

Second, even though I have pushed aside the ontological issues involved, it is undeniable that abstraction is a multifaceted and polysemous concept. Abstraction has a long history, it is basically as old as Western philosophy itself. On the contemporary scene, it comes in various flavours and textures. It is sometimes analysed as being psychological in nature, at other times, as being purely logical and at other times, as being epistemological. It is also central in the teaching of higher mathematics.[1] My hope is to unearth what seems to me to be key features of *mathematical* abstraction as it actually developed in the 20th century.[2]

Third, I am focussing here on abstraction and the abstract method in the *practice* of mathematics and not in the foundations of mathematics or in its logical analysis. I am *not* merely claiming that the abstract character of modern mathematics emerged as a by-product from the usage of the axiomatic method within *formal* systems and that it is the formal aspect of language which is responsible of the abstract character of mathematics. That approach would equate being abstract with being formal, the latter term referring to formal languages. In other words, I will not identify being abstract with what can be studied apart from any particular interpretation. This is certainly one possible and plausible interpretation of the abstract nature of modern mathematics and, in fact, there is a grain of truth in that picture, as I will try to show. Although it contains a part of the analysis, it fails to include some important parts, in particular the inherent dynamics and recursive aspect of the abstraction process in modern mathematics.

---

[1] For the historical aspects, see (Cleary 1995; Walmsley 2000; Jesseph 1993), for the psychological aspects, see, for instance (Piaget 1977; Barsalou 2003; 2005; Houdé 2009), for the logical components, see (Lorenzen 1965; Fine 2002; Tennant 2004; Antonelli 2010), for some epistemological and logical components, see (Weyl 1949; Pollard 1987; Arbib 1990; Simon 1990; Ferrari 2003) and, finally, for pedagogical reflexions on the subject, see (Piaget 1977; Dubinsky 1991; Frorer *et al.* 1997; Hazzan 1999; Mitchelmore & White 2004).

[2] I recommend Sinaceur's interesting discussion of the various facets of mathematical abstraction (see Sinaceur 2014). Unfortunately, time constraints did not allow me to incorporate elements of her analysis in my present work.

In a nutshell, in the practice of mathematics, the abstract method progressively became a method with many different functions: it was used to *solve* problems, to introduce new concepts and organizing principles, and even to install norms of construction. Furthermore, the method is such that it works in a recursive fashion. I believe that the abstract nature of modern mathematics is better captured by a faithful description of the process of abstraction inherent to the actual historical development of modern mathematics and not merely by the description of axiomatic systems (together with their underlying logic). The latter are a result of the former.

2 Very brief historical remarks

I have to limit myself to sketchy and impressionistic remarks in this paper. For, the history is complex and convoluted. The roots of the abstract method certainly go back to the end of the 18th century with Euler, Lagrange and others, and the beginning of the 19th century with Gauss, Galois, Abel, Dirichlet, Riemann, Dedekind, etc. My aim is not to unearth these roots, but merely bring to the fore certain elements that were inherent to the genesis of the abstract approach. We fortunately have some serious studies that allow us to have a good grasp of the main historical components involved.[3]

It seems reasonably safe to claim that the abstract approach made its *official* and *general* appearance in 1930, in a famous and extremely influential book, namely Van der Waerden *Moderne Algebra*. Here is the opening sentence:

> The "abstract", "formal", or "axiomatic" direction, to which the fresh impetus in algebra is due, has led to a number of new formulations of ideas, insight into new interrelations, and far-reaching results, especially in *group theory, field theory, valuation theory, ideal theory*, and the *theory of hypercomplex numbers*. (van der Waerden 1991, ix)

The first three words of the book are, to my mind at least, striking: "abstract", "formal", or "axiomatic". Notice the "or"; it is not an "and". It is as if van der Waerden considered them to be almost synonyms and perhaps, since they are in quotes, not quite clear. I do believe, however, that he was quite clear about the fact that the axiomatic direction was taking a new orientation, breaking away from its traditional philosophical

---

[3]See, for instance, (Bernkopf 1966; Browder 1975; Dieudonné 1981; Birkhoff & Kreyszig 1984; Wussing 1984; Kleiner 1996; 1999a;b; Corry 1996; Smithies 1997; Corry 2000; Epple 2003; Corry 2007; Dorier 1995; 2000; Gray & Parshall 2007; Moore 1995; 2007).

status. The identification between the abstract, formal and axiomatic was common in the 1930s and it remains so to this day. I will give only one quote, but it would be easy to multiply them:

> It is abstraction—more than anything else—that characterizes the mathematics of the twentieth century. There is both power and elegance in the axiomatic method, attributes that can and should be appreciated by students early in their mathematical careers and even if they happen to be confronting contemporary abstract mathematics in a serious way for the very first time. (Watkins 2007, ix)

When van der Waerden wrote his book, he did not make a clear distinction between the formal, the abstract and the axiomatic. They all came together, fused and converged into one general method which was considered new and powerful. It is easy to show that the formal, and the axiomatic are not necessarily tied to the abstract method. In fact, the abstract method does not *necessarily* rely on the formal and the axiomatic methods. But historically, the formal and the axiomatic methods were combined in a certain manner by mathematicians and it became modern mathematics. Once this combination was found and shown to be elegant and powerful, it transformed the way mathematics was done and progressively led to a second wave of abstraction that came in the 1950s and 1960s and that is still going on today, once again transforming mathematics profoundly.

It is well-known that the first discipline to appear as an abstract theory was group theory. It will be enough for me to underline but one element in this historical process.

> The mathematical literature of the nineteenth century, and especially the work of decisive importance for the evolution of the abstract group concept written at the century's end, make it abundantly clear that that development had *three* equally important historical roots, namely, the theory of algebraic equations, number theory, and geometry. Abstract group theory was the result of a gradual process of abstraction from implicit and explicit group-theoretic methods and concepts involving the interaction of its three historical roots. I stress that my inclusion of number theory and geometry among the sources of causal tendencies for the development of abstract group theory is grounded in the historical record and is not the result of a backward projection of modern group theoretic thought. (Wussing 1984, 16)

This is an interesting and surprising empirical fact: *three* equally important domains had to be available for the abstraction to take place. It seems to be a minimum, at least at this early stage. When only two are available, mathematicians will rather consider an analogy or a generalization, not an abstraction. One telling example of this is the work of Dedekind and Weber on algebraic numbers and algebraic functions (see Corfield 2003, chap. 4). This appears to be an important cognitive component of the story. In fact, even for the whole field of algebra, it seems that three different and equally important theories had to be available for the community to consider that algebra as a whole could go in the direction of the abstract method. However, I should stress that it might not be *necessary*, it might simply reflect a cognitive trait in some individuals or communities.

> Her [i.e., Noether] work had a greater overall impact on algebra than Steinitz's, if only due to the fact that, having appeared about ten years later, it showed that Steinitz's program applied not only for the particular case worked out by him, but for many other significant cases as well. Group theory was thus the first algebraic discipline to be abstractly investigated, and field theory the first discipline that arose from the research of numerical domain into an abstract, structural subject. The study of ideal theory in an abstract ring consolidated the idea that a more general conception lay behind all this: the conception that algebra should be concerned, as a discipline, with the study of algebraic structures. (Corry 1996, 251)

Corry here suggests that after three examples, mathematicians tend to generalize. I would say that, in this particular case, they were ready to abstract. Be that as it may, from the algebraic side, we have the development of group theory, field theory and ring theory, the latter accompanied by ideal theory and module theory. There were developments on the geometric side that were also important. Thus, metric spaces appeared early on the scene in the work of Fréchet, who was soon followed by Hausdorff on topological spaces and Banach on Banach spaces. The history of vector spaces is more convoluted but certainly belongs here. Finally, two theories that have a somewhat different paths but that certainly belong to the picture, if only because they bring in different components to it, the theory of Boolean algebras and lattice theory. It is also worth mentioning at this point that Bourbaki considered that there were *three* mother structures: order structures, topological structures and algebraic structures. Underlying this abstract method, one finds, of course, set theory and, to a certain extent, logic. In the second wave of abstraction, the most important and salient example of the use of the abstract method is certainly the categorical foundations of algebraic geometry provided by Grothendieck and his school in the 1960s. The other striking example along these lines is Quillen's work in homotopical algebra which can be seen as the bedrock of abstract homotopy theory.

Let us come back to the formal and the axiomatic methods and their role in the rise of the abstract method. Historically, both a formal standpoint and the axiomatic method were available. I claim that both were diverted from their original purposes and became key components of the abstract method, to the extent that the latter was more or less identified with them, as we have seen. Let us consider them briefly in turn.

2.1 Symbolic formalism and algebra

Algebra is customarily associated with the manipulation of signs, letters, that are used to represent quantities and are manipulated according to given, explicit rules. Nowadays, we take for granted various symbolic conventions and rules of manipulation associated with various calculus. Needless to say, the introduction of these symbolisms has itself an intricate and philosophically important history.[4] One driving analogy emerged towards the end of the 18th century between rules of manipulations of arithmetic and rules of manipulations of differentials and operations in general. It goes back at least to Lagrange and was developed by Lacroix, Arbogast, Brisson, Franais, Servois on the French side, and using the work by the French as a springboard, by Woodhouse, Babbage, Peacock, Gregory, Boole, DeMorgan on the English side. It became known as symbolic algebra or the calculus of operations. But the key element is that it became a *formal* method.

> Symbolic algebra represented a movement away from algebra as universal arithmetic to a purely formal algebra. It emphasized the importance of structure over meaning, and acknowledged what has been called the *principle of mathematical freedom*. This principle implies that algebra deals with arbitrary, meaningless symbols, mathematicians create the rules regarding the manipulation of those symbols, and the interpretation follows rather than precedes the algebraic manipulation. (Allaire & Bradley 2002, 403)

It has been argued that this view goes back in the philosophical literature at least to Berkeley.[5] One striking expression of this view is found in the British algebraist Peacock:[6]

---

[4] See, for instance, (Serfati 2002; 2005).
[5] See, for instance, (Detlefsen 2005).
[6] The view clearly goes back to Woodhouse as early as 1803. But it seems that his book had almost no impact, apart from the fact that Babbage apparently learned a lot

> Algebra may be considered, in its most general form, as *the science which treats of the combinations of arbitrary signs and symbols by means of defined through arbitrary laws*: for we may *assume* any laws for the combination and incorporation of such symbols, so long as our assumptions are independent, and therefore not inconsistent with each other [...] (Peacock 1830, 71, 78)

It is easy to multiply the quotes of the so-called Cambridge algebraists. I will restrain myself to two:[7]

> [...] symbolical algebra is [...] the science which treats of the combination of operations defined not by their nature, [...] but by the laws of combination to which they are subject [...] [W]e suppose the existence of classes of unknown operations subject to the same laws. (Gregory 1840, 210) quoted by (Allaire & Bradley 2002, 404)

And in Boole:

> They who are acquainted with the present state of the theory of Symbolic Algebra, are aware, that the validity of the processes of analysis does not depend upon the interpretation of the symbols which are employed, but solely upon the laws of their combination. (Boole 1847, 3) quoted by (Allaire & Bradley 2002, 400)

These sound extraordinarily modern to our ears. However, we have to be very careful not to read our conception of algebra, in particular abstract algebra in them, for it is definitely not. To mention but one clear case, Peacock would not include in algebra a non-commutative system, since it differs as such from arithmetic.[8]

For our purposes, it is sufficient to underline one aspect of the theory: it is seen as a *general* method, not as an *abstract* method. As Allaire & Bradley puts it "What can be proved for a class generally, holds for all specific operations in that class" (Allaire & Bradley 2002, 407). But, I

---

from it. It should also be noted that Peacock was one of his students (see Koppelman 1971).

[7]For more, see, for instance, (Koppelman 1971; Allaire & Bradley 2002).

[8]I should add that, in this respect, I disagree with the view proposed by Koppelman, who claims that the work done by the English algebraists fostered "an abstract view and clearly influencing many of the men who were to give, in the 1840's the beginnings of an abstract definition of algebra" (Koppelman 1971, 188). I would say that they developed a *formal* view of algebra and not an *abstract* view. I hope the next sections will allow the reader to see why I would make this nuanced claim.

hasten to add, a (limited but genuine) form of abstraction appeared in the writings of some mathematicians of that period. Here is one of the most striking passage under the pen of the French mathematician Servois:

> Along the way, other links between the differential, the difference, variation and numbers emerged; it was necessary to find its cause, and everything is fortunately explained, when after having stripped, by a severe abstraction, these functions of their specific qualities, one only has to consider the two properties they have in common, being *distributive and commutative between them*. Servois, quoted by (Koppelman 1971, 175) (my translation)

However, this seems to be the exception rather than the norm. The emphasis, at that time, was rather on the analogy underlying operations and numbers.

Many developments in mathematics in the 19th century contributed to the emergence of various shades of formalism: projective geometry, non-euclidean geometry, complex numbers, quaternions, octonions and hyper-complex numbers and, of course, the theory of invariants. One of the legacies of the 18th century was the status of negative numbers! In each case, there were problems attributing sense and reference to the symbols manipulated or, in the case of invariants, particularly in the so-called algorithmic school of Paul Gordan, there were series of manipulations that were used in order to obtain results which could not be justified except as pure rules of computations.[9]

It is therefore not hard to find both in the view or image and in the body of algebra the formal component explicitly mentioned.[10] Thus, in two important works by Weber we find a clear and undeniable endorsement of symbolic formalism. First, in his book on Galois theory, one reads:

> In the following an attempt is made to present Galois theory of algebraic equations in a way which include equally well all cases in which this theory might by used. Thus we present it here as a direct consequence of the group concept illuminated by the field concept, as a *formal structure completely without reference to any numerical interpretation of the elements used*. (Weber, 1893, 521) quoted by (Corry 1996, 36) (my emphasis)

---

[9]See, for instance, (McLarty 2011).
[10]The distinction between the image of a discipline and its body comes from (Corry 1996), who attributes it to (Elkana 1981).

We find a similar claim in the abridged edition of his famous textbook on algebra:

> In analysis one is accustomed to understand a "variable" as a sign which takes successively different values. Algebra uses the word variable as well but in a different sense. Here it is a mere calculating symbol [*Rechnungssymbole*] with which one operates by the rules of calculating with letters [*Buchstabenrechnung*]. (Weber 1912, 47), quoted by (McLarty 2011, 105)

What is the point? As the foregoing quotes show, towards the end of the 19th and at the beginning of the 20th century, it was becoming possible to divorce symbols and their rules from a specific, fixed content, a definite meaning. Algebra was, in some of its areas, considered to be formal. Notice that I did not say abstract, for I claim that this is different. By 1910, field theory and group theory were already considered to be abstract and for good reasons. But the abstract method was not quite in place yet.

2.2 The axiomatic method and the abstract method

Mathematicians talk about the axiomatic method and the abstract method as if they were interchangeable. Of course, this is simply false. The axiomatic method as nothing to do *per se* with the abstract method. Suffice it to mention Euclidean geometry, the paradigmatic example of an axiomatic theory. Euclidean geometry is certainly not considered to be an example of the abstract method.

The main point to make here is that, historically, the axiomatic method was the only known mode of presentation that could perform the function required by the abstract method: to provide a clear statement of a set of properties chosen in the process of abstraction.[11] It is well-known that this was not the main nor the only function of the axiomatic method in the late 19th and early 20th centuries. Nowadays, other modes of presentation can be and are used, e.g., the graphical language of sketches in categorical logic.

3 Mathematical Abstraction and the abstract method: putting the pieces together

What I am going to describe constitutes, it seems to me, the main route to the abstract method. It should be kept in mind that all four components have to be present for the process to be a full process of abstraction. The order is not crucial.

First, there is a domain of distinct types of entities, at least three distinct types of entities, which becomes a domain of variation *within which there*

---

[11] For more on the axiomatic method in 20th century mathematics, see Schlimm (2013).

are nonetheless invariant features. It is crucial that the three domains be considered to be sufficiently different, that the domain of variation be a domain of *significant* variation. This interplay of variation and invariance opens the door to the possibility of abstracting.

Second, for the invariant features to be abstracted, one has to take a formal stance with respect to the individual objects of the various domains. In other words, one has to forget what one is talking about, the meaning of the signs involved and treat them purely formally. Thus, the abstract and the formal are sometimes confused.

Third, the invariant features are abstracted and, at this stage, one needs a method to present these features and be able to investigate them in an autonomous fashion. It is at this moment that the axiomatic method appears to be exactly what one needs. The axioms capture the invariant features and one then uses logic to investigate what can be known from them. Notice that in some cases, the axioms might stipulate properties that are not obviously in the distinct domains as such. It is once the property is enunciated in a language that one can, in some cases, convince oneself that, indeed, it is a property of the entities given.

Fourth, a new criterion of identity for the abstracted entities has to be discovered and fixed. In turn, fixing a criterion of identity is possible if and only if linguistic resources are available for these properties and criteria of identity to be expressed. The new criterion of identity is almost always discovered after the abstract entities have been introduced, for a developed theory has to be available in order to make sure that the criterion of identity has the right properties. The invariant component is from then on circumscribed clearly and independently of the original entities. These are seen to be instances of these new types and are studied as such, that is, there is a shift of attention from the old criterion of identity and its associated properties to the new criterion and its associated properties. Almost always, it is then possible to discover and construct new, unforeseen instances of these new abstract entities. Thus, the domain of variation can expand and is never fixed once and for all. In more philosophical terms, once the new types have been fixed, known examples become tokens of the type and new, unforeseen tokens can be constructed or discovered. However, very quickly the shift of attention draws mathematicians towards intrinsic problems, or one might say pure problems, of the new field, for instance, problems of classification or decompositions into well-organized patterns will become central.

Let me underline immediately that arbitrary sets and functions between them played a key part in the development of the abstract method, particularly in the fourth step. Once the focus shifts towards the abstract entities themselves, one needs to talk of unspecified elements that are determined by the properties stated by the axioms. Sets and functions introduced earlier were perfect candidates for that role.

3.1 Domain of significant variation

I now need to clarify what I mean by a domain of variation and a domain of *significant* variation. The best way to introduce these ideas is by giving an example.

One of the very first cases of an extraordinarily successful abstraction in the history of modern mathematics is certainly that of metric spaces, introduced by Fréchet around 1906 in the context of functional analysis.[12] What I want to emphasize in this case is the range of the domain of variation and the fact that it is a domain of significant variation. I think it is fair to say that at that point, mathematicians did not think in terms of abstract sets in the sense of a collection of faceless points. In the context of geometry, mathematicians were thinking of manifolds, either as subspaces of spaces of real or complex points. In the case of Fréchet, he was dealing with these usual manifolds, namely $\mathbb{R}, \mathbb{R}^2, \ldots, \mathbb{R}^n, \ldots, \mathbb{C}, \mathbb{C}^2, \ldots, \mathbb{C}^n$ together with functions between them on the one hand, and infinite-dimensional functional spaces together with operators between them on the other hand. In his thesis, Fréchet gives four examples of functional spaces that satisfy his axioms. See (Fréchet 1906) or Taylor (1982). Here they are.

1. Let J be a closed interval of the real line $\mathbb{R}$ and consider the space $\mathbb{R}^J$ of continuous functions $f : J \to \mathbb{R}$. A metric on $\mathbb{R}^J$ is defined by

$$d(f,g) = max(|f(x) - g(x)|) \qquad \forall x \in J.$$

2. Consider the space $E_\infty = \mathbb{R}^\mathbb{N}$ of infinite sequences $x = (x_1, x_2, \ldots)$ of real numbers. A metric on E_∞ is given by

$$d(x,y) = \sum_{n=1}^{\infty} \frac{1}{n!} \frac{|x_n - y_n|}{1 + |x_n - y_n|}.$$

3. A space of parametrized curves in $\mathbb{R}^3$ with the standard Euclidean metric between points. Using the latter, Fréchet defines a metric between the curves.

4. Finally, let A be a complex plane region whose boundary consists of one or more contours. Let $\{A_n\}$ be a sequence of bounded regions such that $A_n \subset int(A_{n+1})$ and $A_n \subset int(A)$ and such that

---

[12]It is known that Fréchet knew about the case of groups and that it provided at least guidelines and a model of what could be achieved by moving up the ladder of abstraction.

any given bounded region in the interior of A is in the interior of some A_n for n sufficiently large. Consider the space $\{f : int(A) \to \mathbb{C} | f$ is holomorphic$\}$ and let

$$M_n(f,g) = max(|f(z) - g(z)|) \qquad \text{when } z \text{ is in the closure of } A_n.$$

The metric between two such functions is then defined by

$$d(f,g) = \sum_{n=1}^{\infty} \frac{1}{n!} \frac{M_n(f,g)}{1 + M_n(f,g)}.$$

Although I haven't described the third example in detail, what is striking is how different they are from one another and, perhaps even more so, from the spaces of points $\mathbb{R}^n$ and $\mathbb{C}^n$. The main point is this: if we did not know about the metric involved in each case, we might not think that these entities have something in common. Indeed, we are accustomed to attribute certain properties to real functions: continuity, differentiability, roots, maximum, minimum, etc., we represent the graph of a function as a one-dimensional path in the codomain, thus as something that necessarily has a length, we think of a real function as a systematic relation of dependence between two or more properties, as a quantity that varies according to a certain pattern or whose variation depends on a another variation. A function is essentially thought of as being dynamic. The four examples given by Fréchet are of this kind. A (real) point is, well, a point. It has none of the properties of a function. Thus, the properties of the elements of $\mathbb{R}$ and even $\mathbb{R}^n$ are incommensurable with the properties of the elements of a functional space. I want to insist on the fact that given the properties of functions and given that we think of functions with their properties, it is hard to conceive of a *space* of functions, that is treating the latter as being points. It is as if we were trying to think of the properties of functions and forget about them at the same time. Of course, as soon as we have succeeded in thinking of them as spaces, we stumble upon what is certainly seen as being the main difference between these spaces and the usual spaces of points: the examples given above are *infinite* dimensional. Thus, we also have two different types: finite dimensional spaces on one side and infinite dimensional spaces on the other.

We immediately see how the introduction of functional spaces increases substantially the domain of variation. There is also a considerable amount of variation between the four examples themselves. It is hard to see what infinite sequences of real numbers might have in common with parametrized curves in three dimensional Euclidean space, for instance. They seem to belong to different categories of thought. It is only when they are thought as being genuine spaces that we allow ourselves to attribute them similar properties. Here is how Fréchet himself came to characterize the general situation:

> At first sight such an undertaking might be considered as absurd. How can we speak of a geometry in a space whose "points" are of an undefined nature, when we do not know if the elements are numbers, curves, surfaces, functions, series, sets, etc.? (Fréchet 1951, 152)

By introducing an abstract level of analysis, one can specify the domain of application of geometrical ideas. It is important to note that this range might turn out to be much larger than anticipated. By considering all these cases as being genuine spaces, one has at the same time a language and a universe of interpretation for these terms in which it makes sense to consider these seemingly different geometric entities as being nonetheless entities of the same type. Notice that it is impossible at this stage to think of the abstraction process in terms of an equivalence relation. One has to have the properties that will be abstracted in order to define the criterion of identity between the abstract entities. In other words, the criterion of identity can not be given *a priori* but is derived from the theory. In fact, many of the relevant properties of the spaces will only emerge while the theory is constructed and developed.

The very same analysis can be given for group theory, field theory and ring theory. I cannot present the details in such a short paper. I will merely give pointers towards the relevant features in each case. Describing the domain of variation and seeing how significant that variation is turns out to be rather easy. It is important to keep in mind that, at this stage, what I want to underline is *not* what these domains have in common, that is that they are groups or fields or rings, but, on the contrary, how much they differ at a very basic mathematical level.

As we have already seen in the foregoing quote about the genesis of group theory, there had to be three different theories, the theory of algebraic equations, number theory and geometry, for the abstract point of view to emerge as such. These three domains, from the point of view of the practice, when one consider the nature of the entities and their properties in each case, are, in some sense, orthogonal. Algebra, number theory and geometry: these were, in the 19th century, about different entities having different properties and studied with different methods altogether. One would not think of transferring properties of algebraic equations to numbers – how many roots does it have? –, or properties of numbers to geometric figures – is this triangle prime? Thus, once again, one has to systematically *ignore* most of what one has learned about these entities, how one ought to think about these entities and their properties.

The case of fields is just as clear. Weber gave an axiomatic presentation of the concept of field in 1893. However, in that paper, his goal was not to develop field theory, rather he found it convenient to use the concept in his presentation of Galois theory. But for the record, it is worth mentioning that Weber includes in his examples algebraic numbers (number theory), algebraic functions (algebraic geometry), Galois's finite fields (algebra) and Kronecker's "congruence fields" $K[x]/(p(x))$, where K is a field and $p(x)$ is irreducible over K (algebra). Notice the variation already, but that is not quite enough, for in these cases, one can rather think in terms of analogies between the various domains. It is nowadays acknowledged that the *abstract* theory of fields appeared on the scene with the publication of Steinitz's groundbreaking paper on the algebraic theory of fields in 1910. As Steinitz himself explicitly acknowledges, it was Hensel's p-adic numbers that sparked his investigation.

> I was led into this general research especially by Hensel's *Theory of Algebraic Numbers*, whose starting point is the field of p-adic numbers, a field *which counts neither as a field of functions nor as a field of numbers in the usual sense of the word*. Steinitz, quoted by (Kleiner 1999b, 861) (my emphasis)

One should show why the field of p-adic numbers introduces a *significant* variation. I will unfortunately have to rely on Steinitz's words in the context of the present paper.

The history of abstract ring theory is convoluted and would deserve a whole section in itself. We can set aside Fraenkel's work on rings, since although it constitute an important step towards the theory, it fails to do so for interesting reasons that we simply cannot cover here. (But see Corry 2000, for a nice analysis.) In a sense, one of the problems of ring theory was precisely that the domain of variation was too wide and varied for the construction of the *theory*. Two separate historical strands leading to abstract ring theory have to be distinguished: commutative rings and non-commutative rings. Commutative ring theory originates from algebraic number theory, invariant theory and algebraic geometry and it is this strand that led to Noether's ground breaking work. Non-commutative ring theory comes from the theory of hypercomplex number systems, nowadays called finite dimensional algebras, and there are numerous different cases of these. It would be necessary to focus our attention on Noether's work, but we have to leave this to another study. (See (Kleiner 1996), (Swetz *et al.* 1995), (Corry 2000) and (McLarty 2011) for instance.)

These examples illustrate clearly what it is to start with a domain of significant variation. In all cases, we have mathematical systems that have different, even incompatible properties, e.g., being finite/infinite dimensional, being discrete/continuous, etc. I should point out that in these particular examples, the systems considered are build from below so to speak, that is from specific elements, their properties and operations on these elements or relations between them. Once they are looked at from the abstract point of view, these elements and their specific individual properties become totally superfluous. Thus, what varies fundamentally, at this stage, is that each element has, so to speak, a myriad of properties, a whole individuality. In the process of abstraction, these specific individual properties are almost all ignored in favor of properties that relate these individuals together, properties of parts and how they are related to one another and to the whole. Finding the latter property is not a trivial matter and very often new properties, relational properties, have to be found and emerge during the abstraction process itself. Furthermore, these systems certainly cannot, at first, be considered as being even *possibly* identical, not even as being instances of a unique type with *its* criterion of identity, different from all the specific criteria used for the individual systems. It is impossible to tell that some of these systems might turn out to be *identical* when considered as instances of a new type.

One last remark about a domain of variation is necessary. Axiom systems automatically yield a range of variation, at the syntactical level. But it is seldom fruitful. It took a very long time and a considerable amount of ingenuity before mathematicians considered it possible to obtain a significant domain of variation from the axioms of Euclidean geometry. The strategy here is simple: simply ignore some of the axioms and see whether you get something interesting. But this strategy seldom yields genuinely interesting results. One might simply get a more general framework that does not perform any real work. However, as in the case of Euclidean geometry, what might be taken to be a sterile enterprise can reveal vast and unforeseen possibilities. Hilbert's axiomatization of Euclidean geometry is a remarkable example of a successful, systematic, organization of a domain that, at the same time, characterizes adequately specific domains *and* deals with a domain of variation properly. There are other cases in algebra, e.g., monoids and groups or rings and commutative rings, but also in other fields, e.g., generalized cohomology theories like K-theory, where deleting an axiom still captures a rich domain of variation. The fact is, this method, if it is a method at all, rarely yields interesting fruits: subtracting an axiom at random does not necessarily provide a new, interesting theory. In all the cases we have just mentioned, the domain of variation was already known when the axioms were set up and therefore one knew, in some sense, which axioms could be removed fruitfully. As far as I know, removing an axiom in the definition of a topological space does not yield any interesting *geometric* system.[13] The same is true for the notion of category (in contrast with the notion of group) and probably many others as well.

4   The point of the abstract method

When Stephan Banach introduced the spaces that now bear his name, he justified the use of the abstract method thusly:

> The aim of the present work is to establish certain theorems valid in different functional domains, which I specify in what follows. Nevertheless, in order not to have to prove them for each particular domain, I have chosen to take a different route [...]; I consider sets of elements about which I postulate certain properties; I deduce from them certain theorems, and I then prove for each particular functional domain that the postulates adopted are true for it. Banach, quoted by (Moore 1995, 280)

This is the strategy adopted by most mathematicians afterwards. The abstract method leads to two different methodological levels: first, one proves certain results for the abstract entities themselves, for the types so to speak, and then one shows that domains of interest are tokens of these types and therefore automatically satisfy the properties stated in the theorems proved.

However, this characterization fails to reveal the real import of the method and why it is mathematically and philosophically so important. For, as such, Banach's claim merely says that the abstract approach is a form of generalization and a more economical method. This is indeed the case, but it does not go at the heart of the method, its real strength or power. It should be pointed out that by taking the abstract method, it is sometimes possible to treat a domain of exotic or unusual entities as if they were known. For instance, once p-adic numbers are seen as being a field and that it is possible to prove results about fields from pure field-theoretic properties, one can dispense with trying to manipulate p-adic numbers, with some unusual operations or properties. This is clearly one benefit of the method. But, again, it is not the main force.

The abstract method is taken to yield a *conceptual* analysis of mathematics: one talks one the group-concept, the ring-concept, the vector-space

---

[13] Of course, it might yield an interesting algebraic structure, e.g., an inf-lattice. It is true that Hausdorff included the separability condition that now bears his name in his first axiomatization and that removing it still yield a coherent and interesting geometric notion, although some might still want to debate this last point.

concept. Mathematics is then organized around these concepts which *unify* in a deep way various domains of mathematics that were, and for good reasons, believed to be unrelated.

> In the wake of these developments has come not only a new mathematics but a fresh outlook, and along with this, simple new proofs of difficult classical results. The isolation of a problem into its basic essentials has often revealed for us the proper setting, in the whole scheme of things, of results considered to have been special and apart and has shown us interrelations between areas previously thought to have been unconnected. (Herstein 1975, 1–2)

This is one of the main epistemological claims I want to make here: the introduction of a level of abstraction is seen as a way of clarifying and distilling what, in some cases, has become a complex domain or, in other cases, exhibits similarities, parallels indicating the possibility of an underlying common framework. The previous disjunction is clearly not exclusive. The new abstract level not only simplifies the situation but it also yields a better control and understanding of the concepts involved. As Herstein puts it: it reveals the proper setting for the solution of various problems. The axiomatic method is a part of that process. Axiomatization should be seen, in this light, as a form of design. Axioms capture either a common structure or common properties leading to a better control and understanding of the features at work. The axiomatic method is thus used as a sieve, a filter in these processes. It brings to the fore the Archimedean points upon which solutions to given problems work. What was previously immersed in a mountain of irrelevant details is unearthed and shown to constitute the mechanisms making concepts work together. This is precisely why we feel justified in speaking of abstraction. As I have said, the process leads to new mathematics, conceptually systematic and organized according to clear principles. I claim that this way of using the axiomatic method has evolved in contemporary mathematics to become a standard method.

As any contemporary mathematician knows too well, to work abstractly is to work with mathematical entities in a certain manner. This was already clear to Weber:

> We can [...] combine all isomorphic groups into a single class of groups that is itself a group whose elements are the generic concepts obtained by making one general concept out of the corresponding elements of the individual isomorphic groups. The individual isomorphic groups are then to be regarded as different

representatives of the generic concept, and it makes no difference
which representative is used to study the properties of the group.
Weber in 1893, quoted in (Wussing 1984, 248)

It should be said, however, that in some cases, one wants to keep track
of specific *isomorphisms* between groups and they are just as significant as
the groups themselves. Weber is nonetheless expressing an extraordinarily
modern point of view in 1893.

It is tempting to reduce abstraction to a particular case of generalization.
Generalization is usually assumed to be a clear and simple process: it is
purely logical and consists in inferring a universally quantified proposition
$\forall x P(x)$ from a list of particulars having a property $P(a), P(b), \ldots, P(n)$.
Let me immediately emphasize the fact that this simply does *not* cover
all cases of generalizations that occur in mathematics. A simple example is
provided by the concept of integral and its various generalizations in the last
half of the 19th century (see Villeneuve 2008, for details). For one thing, an
integral is an operation and it is not propositions about the integral that
were generalized but the operation itself. This is but one example. At the
conceptual level, it is the relationships between abstraction and generalization that have to be clarified. It certainly seems possible to generalize
without abstracting. Think of various theorems that are generalized although without leading to more abstract results. For instance, the passage
from the definition of continuity of a function $f : \mathbb{R} \to \mathbb{R}$ at a point to
the notion of continuity over an interval $[a, b] \subset \mathbb{R}$ is a simple generalization that is certainly not an abstraction. The same could be said for the
generalization of theorems of real analysis to theorems of complex analysis.
Abstraction seems to always involve a form of generalization.

Bibliography

Allaire, P. R. & Bradley, R. E. (2002). Symbolical algebra as a foundation for calculus: D. F. Gregory's contribution. *Historia Mathematica*, *29*(4), 395–426, doi:10.1006/hmat.2002.2358.

Antonelli, G. A. (2010). Notions of invariance for abstraction principles. *Philosophia Mathematica. Series III*, *18*(3), 276–292, doi: 10.1093/philmat/nkq010.

Arbib, M. A. (1990). A Piagetian perspective on mathematical construction. *Synthese*, *84*(1), 43–58, doi:10.1007/BF00485006.

Barsalou, L. W. (2003). Abstraction in perceptual symbol systems. *Philosophical Transactions: Biological Sciences*, *358*(1435), 1177–1187, doi: 10.1098/rstb.2003.1319.

Barsalou, L. W. (2005). Situating abstract concepts. In *Grounding Cognition: The role of perception and action in memory, language, and thought*, Pecher, D. & Zwaan, R., eds., New York: Cambridge University Press, 129–163.

Bernkopf, M. (1966). The development of function spaces with particular reference to their origins in integral equation theory. *Archive for History of Exact Sciences*, *3*, 1–96 (1966), doi:10.1007/BF00412288.

Birkhoff, G. & Kreyszig, E. (1984). The establishment of functional analysis. *Historia Mathematica*, *11*(3), 258–321, doi:10.1016/0315-0860(84)90036-3.

Boole, G. (1847). *The Mathematical Analysis of Logic*. Cambridge: Macmillan, Barclay, & Macmillan.

Browder, F. E. (1975). The relation of functional analysis to concrete analysis in 20th century mathematics. In *Proceedings of the American Academy Workshop on the Evolution of Modern Mathematics (Boston, Mass., 1974)*, vol. 2, 577–590.

Burgess, J. & Rosen, G. (1997). *A Subject With No Object: Strategies for Nominalistic Interpretation of Mathematics*. Oxford: Oxford University Press.

Cleary, J. J. (1995). *Aristotle and Mathematics: Aporetic Method in Cosmology and Metaphysics*. Philosophia Antiqua, Leiden; New York: E.J. Brill.

Corfield, D. (2003). *Towards a Philosophy of Real Mathematics*. Cambridge: Cambridge University Press.

Corry, L. (1996). *Modern Algebra and the Rise of Mathematical Structures*, Science Networks. Historical Studies, vol. 17. Basel: Birkhäuser Verlag.

Corry, L. (2000). The origins of the definition of abstract rings. *Gazette des Mathématiciens*, *83*, 29–47.

Corry, L. (2007). From *algebra* (1895) to *moderne algebra* (1930): changing conceptions of a discipline—a guided tour using the Jahrbuch über die Fortschritte der Mathematik. In *Episodes in the History of Modern Algebra (1800–1950)*, History of mathematics, vol. 32, Providence, RI: American Mathematical Society, 221–243.

Detlefsen, M. (2005). Formalism. In *The Oxford Handbook of Philosophy of Mathematics and Logic*, Shapiro, S., ed., chap. 8, Oxford: Oxford University Press, 236–317.

Dieudonné, J. (1981). *History of Functional Analysis*, North-Holland Mathematics Studies, vol. 49. Amsterdam: North-Holland Publishing Co.

Dorier, J.-L. (1995). A general outline of the genesis of vector space theory. *Historia Mathematica*, *22*(3), 227–261, doi:10.1006/hmat.1995.1024.

Dorier, J.-L. (2000). Epistemological analysis of the genesis of the theory of vector spaces. In *On the Teaching of Linear Algebra, Mathematics Education Library*, vol. 23, Dordrecht: Kluwer, 3–81, doi:10.1007/0-306-47224-4_1.

Dubinsky, E. (1991). Reflective abstraction in advanced mathematical thinking. In *Advanced Mathematical Thinking, Mathematics Education library*, vol. 11, Tall, D., ed., Kluwer, 95–126, doi:10.1007/0-306-47203-1_7.

Elkana, Y. (1981). A programmatic attempt at an anthropology of knowledge. In *Sciences and Cultures*, vol. 5, Mendelsohn, E. & Elkana, Y., eds., Dordrecht: Reidel, 1–76, doi:10.1007/978-94-009-8429-5_1.

Epple, M. (2003). The end of the science of quantity: foundations of analysis, 1860–1910. In *A History of Analysis, Hist. Math.*, vol. 24, Providence, RI: American Mathematical Society, 291–323.

Ferrari, P. L. (2003). Abstraction in mathematics. *Philosophical Transactions: Biological Sciences*, *358*(1435), 1225–1230, doi:10.1098/rstb.2003.1316.

Fine, K. (2002). *The Limits of Abstraction*. Oxford: Oxford University Press.

Fréchet, M. (1906). Sur quelques points du calcul fonctionnel. *Rendiconti del Circolo Matematico di Palermo*, *22*(1), 1–74.

Fréchet, M. (1951). Abstract sets, abstract spaces and general analysis. *Mathematics Magazine*, *24*, 147–155.

Frorer, P., Hazzan, O., & Manes, M. (1997). Revealing the faces of abstraction. *The International Journal of Computers for Mathematical Learning*, *2*(3), 217–228, doi:10.1023/A:1009756617451.

Gray, J. J. & Parshall, K. H. (Eds.) (2007). *Episodes in the History of Modern Algebra (1800–1950), History of Mathematics*, vol. 32. Providence, RI: American Mathematical Society, Papers from the workshop held in Berkeley, CA, April 2003.

Gregory, D. F. (1840). On the real nature of symbolical algebra. *Transactions of Royal Society of Edinburgh*, *14*, 208–216.

Hazzan, O. (1999). Reducing abstraction level when learning abstract algebra concepts. *Educational Studies in Mathematics*, *40*, 71–90, doi: 10.1023/A:1003780613628.

Herstein, I. N. (1975). *Topics in Algebra*. Lexington, MA: Xerox College Publishing, 2nd edn.

Houdé, O. (2009). Abstract after all? Abstraction through inhibition in children and adults. *Behavioral and Brain Sciences*, *32*, 339–340, doi: 10.1017/S0140525X0999080X.

Jesseph, D. M. (1993). *Berkeley's Philosophy of Mathematics*. Science and Its Conceptual Foundations, Chicago: University of Chicago Press.

Kleiner, I. (1996). The genesis of the abstract ring concept. *The American Mathematical Monthly*, *103*(5), 417–424.

Kleiner, I. (1999a). Field theory: from equations to axiomatization. I. *The American Mathematical Monthly*, *106*(7), 677–684.

Kleiner, I. (1999b). Field theory: from equations to axiomatization. II. *The American Mathematical Monthly*, *106*(9), 859–863.

Koppelman, E. (1971). The calculus of operations and the rise of abstract algebra. *Archive for History of Exact Sciences*, *8*(3), 155–242, doi:10.1007/BF00327101.

Lorenzen, P. (1965). *Formal Logic*. Dordrecht: D. Reidel.

McLarty, C. (2011). Emmy Noether's first great mathematics and the culmination of first-phase logicism, formalism, and intuitionism. *Archive for History of Exact Sciences*, *65*(1), 99–117, doi:10.1007/s00407-010-0073-y.

Mitchelmore, M. C. & White, P. (2004). Abstraction in mathematics and mathematics learning. In *PME-28*, vol. 3, 329–336.

Moore, G. H. (1995). The axiomatization of linear algebra: 1875–1940. *Historia Mathematica*, *22*(3), 262–303, doi:10.1006/hmat.1995.1025.

Moore, G. H. (2007). The evolution of the concept of homeomorphism. *Historia Mathematica*, *34*(3), 333–343, doi:10.1016/j.hm.2006.07.006.

Peacock, G. (1830). *A Treatise on Algebra*. Cambridge: J. & J.J. Deighton.

Piaget, J. (1977). *Recherches sur l'abstraction réfléchissante*, vol. 34–35. Paris: Presses Universitaires de France, 1st edn.

Pollard, S. (1987). What is abstraction? *Noûs*, *21*, 233–240.

Schlimm, D. (2013). Axioms in mathematical practice. *Philosophia Mathematica*, *21*(1), 37–92, doi:10.1093/philmat/nks036.

Serfati, M. (2002). Analogies et "prolongements" (écriture symbolique et constitution d'objets mathématiques, de Leibniz à L. Schwartz). In *De la méthode*, Colloq. Sémin., Besançon: Presses Universitaires Franc-Comtoises, 271–318.

Serfati, M. (2005). *La Révolution symbolique*. Transphilosophiques. [Transphilosophies], Paris: Éditions PÉTRA, la constitution de l'écriture symbolique mathématique. [The construction of mathematical symbols], With a preface by Jacques Bouveresse.

Simon, P. (1990). What is abstraction and is it good for? In *Physicalism in Mathematics*, Irvine, A., ed., Dordrecht: Kluwer, 17–40, doi:10.1007/978-94-009-1902-0_2.

Sinaceur, H. (2014). Facets and levels of mathematical abstraction. *Philosophia Scientiae*, *18*(1), 81–112, doi:10.4000/philosophiascientiae.914.

Smithies, F. (1997). The shaping of functional analysis. *The Bulletin of the London Mathematical Society*, *29*(2), 129–138, doi:10.1112/S0024609396002305.

Solomon, R. (2003). *Abstract Algebra*. Belmont, CA: Thompson Brooks/Cole.

Swetz, F., Fauvel, J., et al. (Eds.) (1995). *Learn from the Masters!*, Classroom Resource Materials Series, Washington, DC: Mathematical Association of America.

Taylor, A. E. (1982). A study of Maurice Fréchet. I. His early work on point set theory and the theory of functionals. *Archive for History of Exact Sciences*, *27*(3), 233–295, doi:10.1007/BF00327860.

Tennant, N. (2004). A general theory of abstraction operators. *The Philosophical Quarterly*, *54*(214), 105–133, doi:10.1111/j.0031-8094.2004.00344.x.

van der Waerden, B. L. (1991). *Algebra. Vol. I*. New York: Springer-Verlag, 7th edn., based in part on lectures by E. Artin and E. Noether, Translated from the seventh German edition by Fred Blum and John R. Schulenberger.

Villeneuve, J.-P. (2008). *Types de généralisations et épistémologie des mathématiques: de l'intégrale de Cauchy à l'intégrale de Lebesgue*. Ph.D. thesis, Université de Montréal.

Walmsley, J. (2000). The development of Lockean abstraction. *British Journal for the History of Philosophy*, *8*(3), 395–418, doi:10.1080/096087800442110.

Watkins, J. J. (2007). *Topics in Commutative Ring Theory*. Princeton, NJ: Princeton University Press.

Weber, H. (1912). *Lehrbuch der Algebra. Kleine Ausgabe in einem Bande*. Braunschweig: Vieweg u. Sohn.

Weyl, H. (1949). *Philosophy of Mathematics and Natural Science*. Princeton: Princeton University Press.

Wussing, H. (1984). *The Genesis of the Abstract Group Concept*. Cambridge, MA: MIT Press, a contribution to the history of the origin of abstract group theory, Translated from the German by Abe Shenitzer and Hardy Grant.

Jean-Pierre Marquis
Département de philosophie
Université de Montréal &
CIRST
Montréal
Canada
Jean-Pierre.Marquis@umontreal.ca

Communicating and Trusting Proofs: The Case for Foundational Proof Certificates

DALE MILLER

It is well recognized that proofs serve two different goals. On one hand, they can serve the *didactic* purpose of explaining why a theorem holds: that is, a proof has a *message* that is meant to describe the "why" behind a theorem. On the other hand, proofs can serve as certificates of validity. In this case, once a certificate is checked for its syntactic correctness, one can then trust that the theorem is, in fact, true. (For additional discussions of these two aspects of proof, see, for example, Asperti 2012; MacKenzie 2001).

In this paper, we argue that structural proof theory and computer automation have matured to such a level that they can be used to provide a flexible and universal approach to *proof-as-certificate*. In contrast, the notion of *proof-as-message* is still evolving and deals with structures, such as diagrams and natural language texts (Nelson 1993), that are not yet well formalized.

Since the notion of proof-as-certificate is at times strongly debated in the literature, we discuss in Section 1 several aspects of proof in order to identify those situations in which certification by proof can prove valuable. After that discussion, we use the rest of this paper to outline more specifics of how proof theory can be used to provide for a foundational approach to the design of a universal notion of proof certificate.

1 Characterizing proofs and their roles

To understand the roles and the nature of proof, we need to take a step back and review why proofs exist and how they are used. A key aspect of proofs seems to be that they are documents that are communicated within a group of individuals (possibly separated in both space and time) in order to inspire trust.

1.1 Societies of humans and machines

Communication takes place within various "societies" comprised of individuals dedicated to common ends: such individuals can be human or mechanical. Admitting machines into such societies seems sensible in the many modern situations where computers are making decisions and are reacting to other individuals to further the goals of a society. We list here various kinds of societies of agents and some possible goals for them: while such societies may have several goals, we select here those goals for which a notion of proof plays an important role.

1. A *sole mathematician* writes an argument that convinces herself and she then moves to address new problems. In this small society, a proof is a communication between the mathematician at one moment (the time she developed the proof) and some future time (when she works on the next problem). A goal of such a sole mathematician is to continue to develop a line of mathematical research.

2. A collection of *mathematician colleagues* searches for beautiful and deep mathematical concepts. The energies of such a group are put into finding good definitions and connections among ideas.

3. An *author of a mathematics text and his readers* is a society that is typically distributed by both geography and by time: the readers are located in a future after the text is written. The goal of this society is to have a successful *one-way communication*: that is, the author must be able to communicate with readers without getting feedback on how successful was the communications.

4. A group consisting of *programmers*, who are writing code for a popular operating system, and *users*, who are attempting to use that operating system on their computers, has a goal of producing quality software that the users find convenient and secure.

5. A *group of programmers, users, mobile computers, and servers* can form a society that exchanges money for various services (e.g., email, news, backups, and cloud computing).

Notice that in example 4, machines are not meant as individuals of the group: instead, they are tools used by the individuals. On the other hand, it seems appropriate to classify smart phones, electronic banking systems, and software servers all as individuals in example 5 since the choices and decisions that they take affect the goals of the society.

1.2 Proofs as documents communicated within societies

By logical formulas we mean the familiar notion of syntactic objects composed of logical connectives, quantifiers, predicates, and terms: these have, of course, proved useful for encoding mathematical statements and assertions in computational logic. Proofs can be seen as one kind of document that is communicated within a society of agents (human or computer) with the purpose of instilling trust in an assertion (written as a logical expression). We return to the example societies in the previous section and illustrate roles for proofs in them.

1. The only communication possible within a society consisting of a *sole mathematician* involves that mathematician telling a future instance of herself to trust that a certain formula is a theorem. If at some point in the future, that mathematician trusts her proof, she might take certain actions, such as developing consequences of that theorem.

2. Consider a group of *mathematician colleagues* such as the one featured in Lakatos's *Proofs and Refutations* 1976. This society interacts within a lively and narrow spacial dimension with the agents sitting together discussing. The individuals also interact across time, of course, as new examples, counter-examples, definitions, and proofs appear. The goal of such a society of mathematicians might be to "develop deeper insights and understanding of geometry". The group exchanges messages and makes presentations. Proofs in this setting are generally informal since the energies of the group are put into exploring and discovering definitions and connections among ideas.

3. A society involving the *author of a mathematics text* and his *readers* generally involves a one-way communication: the readers will have the text only after the book is written and the readers may be physically and temporally remote from the author. A good example of such an author and text is, of course, Euclid and his *Elements*, which has been an important text for the communication of deep results about geometry to readers for two millennia.

4. A *group of programmers and users* of an operating system might need to circulate among its members many kinds of documents: bug reports from users should alert programmers to things that need to be fixed; programmers release new versions of software components; programmers exchange programs, scripts, and interfaces; etc. Some of these documents, such as interfaces, probably contain typing information, which can often be seen as formulas for which the program is a proof:

type checking is then a simple kind of proof checking for simple assertions about the program. In addition, certain parts of an operating system can be so critical to the proper functioning of the operating system that a formal proof of some correctness conditions might be required: for example, it might be desirable for certain guarantees about device drivers (low level code used to control devices attached to a computer) to be formally verified by, say, a model checker (Ball et al. 2004).

5. A *group of programmers, users, mobile computers, and servers* can be seen as a society involving machines as individuals since the decisions and actions they make can help the groups achieve its goals. For example, a mobile phone might be expected to maintain certain security policies and this might mean that certain mobile code might not be downloaded to the phone. As a result, certain services might not be available to the user of that phone and some income for those services might be lost. If the infrastructure behind the movement of code allows for proofs to be attached to mobile code, the phone may allow the execution of mobile code if the phone could check that the attached proof proves certain security assertions of the code. The development of such an infrastructure has been studied under the title "proof carrying code" (Necula 1997).

As these examples illustrate, societies circulate a wide variety of documents in order to help meet their goals. Of these many documents, proofs can be roughly identified as those that inspire trust in one agent of the conclusions drawn by another agent. One might acquire trust in a program in a number of ways that do not use proofs: for example, one's trust in a program might be inspired by the fact that over its lifetime, no one has found errors in it: while such evidence is an important source of trust, it is not a document nor a proof.

1.3 Formality of proofs

Proofs can be divided into those that are informal and those that are formal.

We generally expect that *informal proofs* are readable by humans and are didactic. We also expect that they do not contain all details and that they may have errors. Informal proofs are circulated within societies of humans where they can be evaluated in a number of ways: Is the proof proving something interesting? Are the assumptions the right ones? Are the proof methods appropriate? Is this situation an example or a counterexample? If an informal proof is evaluated highly enough, more might be done with it: it might be written for a broader audience and it might be formalized.

Typically, an informal proof will be made "more formal" when the group of people with which it is intended to communicate becomes larger and more diverse (involving greater separation in time and space).

A *formal proof* is a document with a precise syntax that is machine checkable: in principle, an algorithm should make it possible to "perform" the proof described in the document. We shall not assume that formal proofs are human readable or that they contain "explanations" of why a formula is actually true. Trusted computer tools are used to check proofs so that other human or machine agents come to trust the truth of a formula.

1.4 Revisiting criticisms of proofs-as-certificates

Given the discussion about proofs above, it seems useful to now revisit some of the criticisms often leveled at proofs-as-certificates.

Consider, for example, two different societies discussed by Lakatos in *Proofs and Refutations* 1976. One such society is Euclid and the readers of his *Elements*. Here, Lakatos criticizes this text for "its awkward and mysterious ordering" of definitions and theorems. Euclid's text is notable for the society that it has served: given the vast number of readers of the *Elements* that have been distributed over both space and time, it seems that some of that text's success comes, in part, from its formal (sometimes unintuitive) structure which increased its universality. Another society famously considered by Lakatos is that of a small society of mathematicians with limited distribution in time and space. In such a setting, communications can be informal and the society of mathematicians is more involved in an exploration of truth and good mathematical design. Even though these two societies involve only humans, their different distribution in time and space leads to rather different requirements on proofs-as-documents.

Consider now a society of agents involved with building and using an operating system. Clearly, the quality of the operating system is important: it should perform various duties correctly as well as maintain certain security standards. Such a society is highly dynamic: new features are added and others are removed; bugs are discovered and patches are issued; and the operating system must allow for extension to its function by allowing new device drivers to be added or new executable code to be loaded and run. In such a setting, it seems futile to expect that there is a unique formal specification of the operating system to which members of the society are attempting to find a formal proof. None-the-less, informal proof and formal proof could still have some role to play among some agents of this society. Some programmers may want informal proofs that their programs satisfy certain requirements while other programmers might want to have completely formal proofs involving possibly weak properties of some other programs.

In light of this description of a society working to develop an operating system, consider some of the criticisms of formal methods raised by De Millo, Lipton, and Perlis in (Millo et al. 1979). They argued, for example, that formal verification in computer science does not play the same role as proofs do in mathematics: this certainly does not seem problematic because of the differences among the many agents in this society. For example, informal proof may play an important role among some agents while formal proof may play an equally important role among other agents. Those components of an operating system that are static parts of many generations of such a system (such as, for example, sorting algorithms, file system functions, and security protocols) may need to be trusted at a level that formal verification could provide. Those components that are dynamic, experimental, and constantly changing would not be sensible targets for formal verification. De Millo, Lipton, and Perlis state that

> Outsiders see mathematics as a cold, formal, logical, mechanical, monolithic process of sheer intellection; we argue that insofar as it is successful, mathematics is a social, informal, intuitive, organic, human process, a community project.

Given the richness of societies that are part of building large software systems, it seems clear that both views of proofs are important and both serve important roles.

If we allow for machine-to-machine communications of proofs, then formal proof can play a central role. The *proof carrying code* project of Lee and Necula (Necula 1997; Necula & Lee 1998) illustrates just such a situation. In that setting, a society of agents contains at least two machine agents, one that provides executable code and the other that is charged with permitting the accumulation of new code as long as that code maintains certain security assurances. Ensuring that security assurances are maintained requires some knowledge about the executable code. Examples of such assurances are that the code does not access inappropriate memory cells or that a typing discipline is maintained: e.g., that a "string" object is not transformed into, say, an "electronic wallet" object. The approach described by Lee and Necula requires that the executable code is paired with a formal proof that that code satisfies the necessary assurances: such a proof can be checked prior to accepting to execute the code.

To underline again the different roles of proof in different societies consider the following statement from Lakatos (1976):

'Certainty' is far from being a sign of success, it is only a symptom of lack of imagination, of conceptual poverty. It produces smug satisfaction and prevents the growth of knowledge.

While this criticism of formal proof sounds appropriate for those charged with the discovery of mathematical concepts, it is not a valid criticism (nor was it intended to be) of those building safety critical software where formal proof can play an important role in establishing certainty (MacKenzie 2001).

1.5 Formal proofs and machine agents

While much of the value of proofs comes from sharing and checking them, the current state of affairs in computational logic systems makes exchanging proofs the exception instead of the rule. Many theorem proving systems use proof scripts to denote proofs and such scripts are generally not meaningful in other theorem provers: they may also fail to denote proofs for different versions of the same prover. There is also a wide variety of "evidence of proofs" that appear in computational logic systems: these can range from proof scripts to resolution refutations and tableau proofs to winning strategies in model checkers. When one theorem prover does accept proofs from another prover, the bridge built between those two provers is generally ad hoc: see, for example, (Fontaine *et al.* 2006) where proofs from an SMT prover are translated into proof scripts understandable to the Isabelle prover.

In the remainder of this paper, we turn our attention to a *foundational* approach to designing proof certificates to be universal and amenable to communicating and checking.

2 Formulas and logical interpretation

Before describing proof certificates in more specifics, we fix the language of formulas and inference rules that will hopefully allow a wide range of logics and proofs to be encoded naturally. In fact, Church's *Simple Theory of Types (STT)* (Church 1940) provides a syntactic framework for unifying propositional, first-order, and higher-order logics. Such formulas allow quantification at all higher-order types which in turns allows for rich forms of abstractions to be encoded. This framework also comes with an elegant and powerful mechanism for binding, quantification, and substitution by its incorporation of the simply typed λ-calculus into its equational theory. A remarkable feature of STT is that by making simple syntactic restrictions to the types of constants, one can restrict STT to propositional logic or to (multisorted) first-order logic. It is also immediate to add to formulas modal, fixed point, and choice operators. This choice of a framework for specifying formulas is not only one of the oldest such frameworks but also a common choice in several modern theorem proving systems.

Our approach to proofs of formulas departs from the simplistic setting of Church's original proposal where Axioms 1-6 described the logical core of higher-order logic and the remaining axioms enable mathematical theories by introducing extensionality, infinity, and choice. Instead, we mix Church's approach to formulas, bindings, and λ-calculus with the sequent calculus proofs provided by Gentzen for classical and intuitionistic logics 1969 and by Girard for linear logic 1987.

For the rest of this paper, we shall assume that we will be using a single language of logical formulas (namely, STT) and a single framework for describing proofs (Gentzen style sequents). We shall not, however, assume that the reader is intimately familiar with either of these two formalisms.

It is worth noting that we are not proposing to use the LF framework (Harper *et al.* 1993) for specifying proof systems and proofs. While LF can easily accommodate the formulas of Church's STT, the design of LF as a dependently typed λ-calculus fixes a particular proof structure, namely, natural deduction for intuitionistic logic. We shall use the flexibility of the sequent calculus to allow many different forms of proof to be *performed* without the necessity of encoding them as any particular kind of term structure.

3 Two desiderata for proof certificates

We shall now use the term "proof certificate" to mean a document that should elaborate into a formal proof via the efforts of a proof checker. We list now the first two of four desiderata for proof certificates.

> **D1:** *A simple checker can, in principle, check if a proof certificate denotes a proof.*

Proof checkers should be simple and well structured so that they can be inspected and possibly proved formally correct. The correctness of a checker should be much easier to establish than the correctness of a theorem prover: in a sense, a proof checker removes the need to have trust in theorem provers. The separation of proof generation from proof checking is a well understood principle: for example, Pollack (1998) argues for the value of independent checking of proofs and the Coq proof system has a trusted kernel that checks proposed proof objects before accepting them (The Coq Development Team 2002). Proof checking is likely to be at times computationally expensive, so different proof checkers may perform differently depending on the resources (say, memory and processors) to which they have access.

> **D2:** *The format for proof certificates must support a wide range of proof systems.*

In other words, a given computational logic system should be able to take the internal representation of the "proof evidence" that it has built and output essentially that structure as the proof certificate. This one proof certificate format should be be able to encode natural deduction proofs, tableau proofs, and resolution refutations, to name a few. Thus, if a system builds a proof using a resolution refutation, it should be possible to output a certificate that contains an object that is roughly isomorphic to that refutation.

A theorem prover is said to satisfy the "de Bruijn criterion" if that prover produces a proof object that can be checked by a simple checker (Barendregt & Wiedijk 2005). Desiderata **D1** and **D2** together imply a "global" version of the de Bruijn criterion: if every theorem prover can output a proper proof certificate, then any prover can trust any other prover simply by using a trusted checker. The tension between "simplicity" of the checker (**D1**) and the "flexibility" of the certificates (**D2**) is clearly a challenge to address. Section 4.1 briefly describes an approach to addressing this tension by identifying "macro" and "micro" inference rules and the rules that allow micro rules to be assembled into macro rules.

Before presenting two additional desiderata, we examine two implications of desiderata **D1** and **D2**.

3.1 Marketplaces for proofs

Formal proofs of software and hardware are developing some economic value. For example, some professional and contractual standards (for example, DefStan 00-55 of the UK *Defence Standards* (Ministry of Defence 1997)) mandate formal proofs for software that is highly critical to system safety (see (Bowen 1993) for an overview of such standards). The cost of going to market with a computer system containing an error can, in some cases, prove so expensive that additional assurances arising from formal verification can be worth the costs. For example, an error in the floating point division algorithm used in an Intel processor proved to be extremely costly for Intel: more recently, formal verification has been used within Intel to improve the correctness of its floating point arithmetic (Harrison 1999).

Where there is economic value there are opportunities for markets. If proof certificates satisfy desiderata **D1** and **D2**, it should be possible to develop a marketplace for proofs in the following sense. Assume that the ACME company needs a formal proof of its next generation safety critical system (such as might be found in avionics, electric cars, and medical equipment). ACME can submit to the marketplace a formula that needs to be proved: this can be done by publishing a proof certificate in which the entire proof is elided. The market then works as follows: anyone who can fill the hole in that certificate in such a way that ACME's trusted proof checker can validate it will get paid. This marketplace can be open to anyone: any theorem prover or combination of theorem provers can be used. The provers themselves do not need to be known to be correct. The people submitting completed proof certificates must also try to ensure that the ACME proof checker, with its restrictions on computational power, can perform the checking: otherwise they would not be paid for their proof certificate.

If someone working in the marketplace finds a counterexample to a proposed theorem, then that person should also get paid for that discovery. Similarly, partial progress on proving a theorem might well have some economic values. A comprehensive approach to proof certificates should formally allow counterexamples and partial proofs: we will not pursue these issues here.

3.2 Libraries of proofs

Once proof certificates are produced they can be archived within libraries. In fact, libraries might be trusted agents that are responsible for checking certificates. Since such checking is likely to be computationally expensive in many cases, libraries might be designed to focus significant computational resources (e.g., large machines and optimizing compilers) on proof checking. Once a proof certificate is checked and admitted to a library, others might be willing to trust the library and to use its theorems without rechecking certificates. To the extent that formal proofs have economic value, libraries will have economic incentives to make certain that the software that it uses to validate certificates is trustable. If someone else (a competing library, for example) finds that a non-theorem is accepted into a library, trust in that library could collapse along with its economic reason for existing. Libraries can also provide other services such as searching among theorems and structuring collections of theorems.

4 Two more desiderata for proof certificates

We shall now present two additional desiderata.

> **D3:** *A proof certificate is intended to denote a proof in the sense of structural proof theory.*

By "structural proof theory" we mean the literature surrounding the analysis of proofs in which the restriction to analytic proofs (e.g., cut-free sequent proofs or normal natural deductions) still preserves completeness. For references to the literature on structural proof theory, see (Gentzen 1969; Prawitz 1965; Troelstra & Schwichtenberg 1996; Negri & von Plato 2001). Checking a certificate should mean that a computation on the certificate should yield (at least in principle) a formal proof in the sense covered by that literature.

This desideratum insists that certificates can be related to a well studied notion of proof and, as such, it should be possible to apply many well known and deep formal results from proof theory (cut-elimination, normalization, constructive content, etc) to certificates. For example, proof certificates might support the extraction of witnesses and, hence, programs: given a (constructive) proof of $\forall x.A(x) \supset \exists y.B(x,y)$ and a proof of $A(c)$, these two proofs together (via their certificate format) might be expected to yield a witness d such that $B(c,d)$ holds. Similarly, one might hope to do *proof mining* (Kohlenbach & Oliva 2003) with or *program extraction* (Bates & Constable 1985) from proof certificates stored in a library. By using such sophisticated techniques for manipulating proofs, it should be possible to build browsers of certificates that would allow humans to interact with proof certificates in order to get a sense of their "message" (see Section 1).

Our final desideratum (**D4** below) addresses the fact that formal proofs can be large and that certificates must, somehow, allow proofs to be redacted. Large proofs will tax computational resources to store, communicate, and check them. Thus, any definition of proof certificates must provide some mechanism for making them compact even if the proof they denote is huge. One approach to making proofs smaller could be "cut-introduction": that is, examine an existing proof for repeated subproofs and then introduce lemmas that account for the commonality in those subproofs. In this way, lemmas could be proved once and the various similar subproofs could be replaced by "cutting-in" instances of that lemma. There are clearly situations where cut-introduction can make a big difference in proof size. Proof certificates must, obviously, permit the use of lemmas (clearly permitted by desideratum **D3**). But this one technique alone seems unlikely to be effective in general since proofs without cuts (without lemmas) can be so large that they cannot be discovered in the first place. Our fourth desideratum suggests another way to compress a proof.

D4: *A proof certificate can simply leave out details of the intended proof.*

Things that can be left out might include entire subproofs, terms for instantiating quantifiers, which disjunct of a disjunction to select, etc. Thus, proof checking may need to incorporate proof-search in order to check a proof certificate that left out some details. As a result, proof checkers will not just check that all requirements of inference rules match correctly. Instead, they will need to be logic programming-like engines that involve unification and (bounded) backtracking search. An early experiment with using logic programming engines to reconstruct missing proof information was reported by Necula & Lee (1998).

This desideratum forces the design of proof certificates in rather particular directions. While the other desiderata seem general and even obviously desirable, this fourth desideratum is the most distinctive in our proposal here.

4.1 Flexible description of proof systems

Taken together, desideratum **D2** and **D3** require that we can provide a rich set of *inference rules* similar to the *analytic* rules (introduction, elimination, and structural) used in proof theory. One way to achieve such richness is to identify a comprehensive set of "atoms" of inference as well as the rules of "chemistry" that allow us to build the "molecules" of inference. We briefly describe how such an approach might work; see (Miller 2011) for more specifics.

The atoms of inference The sequent calculus provides an appealing set of primitive inference rules: these include the introduction of one logical connective and the deletion and copying (weakening and contraction) of formulas. Gentzen used this setting to distinguish classical and intuitionistic logic simply as different restrictions on structural rules 1969. Linear logic (Girard 1987) provides a finer analysis of the roles of introduction rules and the structural rules: this analysis provides additional atoms of inference by, for example, separating connectives into their multiplicative and additive forms. The decomposition of the intuitionistic implication $B \supset C$ into $!B \multimap C$ is another example of this finer analysis of logical connectives. In order to capture inductive and co-inductive reasoning (including model-checking-like inference), the atoms of inference should also include fixed points and equality (Baelde 2008; 2012; McDowell & Miller 2000). Since the trusted proof checker needs to only implement the atomic inference rules, the checker can be simple in its design, thus satisfying **D1**.

The molecules of inference Without any additional discipline, the structure of the atomic inferences within sequent calculus proofs is chaotic: the application of one inference can have little relationship with the application of any other inference rule. A well studied discipline for organizing atoms of inference into the *molecules of inference* is provided by the technical notion of a *focused proof system* (Andreoli 1992; Liang & Miller 2009; 2011). These proof systems attribute "polarity" to atomic inference rules. Atoms of the same polarity can stick together to form molecules: atoms of different polarities form boundaries between molecules. The resulting collection of molecules of inference form a proper proof system since they satisfy such properties as cut-elimination. In this sense, the resulting molecules of inference satisfy desideratum **D3**. There is also flexibility in how polarities are attributed so it is possible to "engineer" the set of molecules to cover a wide range of proof evidence, thus satisfying desideratum **D2**. Finally, when details of a proof are elided in a proof certificate (desideratum **D4**), the proof checker will need to conduct a search and that search should be understood as being conducted at the molecular and not atomic level: when filling in details to a proof, one should not be searching for new molecules via new combinations of atoms.

Since adequately representing one proof system within another proof system is central to our design of proof certificates, we expand on this topic next.

4.2 Three levels of adequacy

When comparing two inference systems, we follow Nigam & Miller (2010) by identifying three "levels of adequacy". The weakest level of adequacy is *relative completeness*: a formula has a proof in one system if and only if it has a proof in another system. Here, only *provability* is considered. A stronger level of adequacy is that of *full completeness of proofs*: the proofs of a given formula are in one-to-one correspondence with proofs in another system (such a correspondence must also be compositionally described). If one uses the term "derivation" for possibly incomplete proofs (proofs that may have open premises), an even stronger level of adequacy is *full completeness of derivations*: here, the derivations (such as inference rules themselves) in one system are in one-to-one correspondence with those in the other system. When claiming equivalences between proof systems, one should describe the level of adequacy of the associated correspondence: in general, we shall strive to always have the engineered macro inference rules (molecules of inference) encode target proof systems at the third and most demanding level of adequacy. These degrees of adequacy appear to correspond roughly to Girard's proposal (2006, chap. 7) for three levels of adequacy based on semantical notions: the levels of *truth*, *functions*, and *actions*.

The third level of adequacy (which, of course, implies the other two levels) is particularly significant here since it provides a sensible means for addressing desideratum **D4**. If a proof certificate elides an entire subproof then the proof checker will need to reconstruct that subproof. The designer of the proof certificate presumably has elided that subproof because he feels that it is an easy proof for the proof checker to discover. This impression is only useful, however, if the search conducted by the proof checker (which strings together the atoms of inference) can be related directly to the search for the elided proof. This match must hold for successful applications of inference rules as well as for failing applications of inference rules. The notion of full completeness of derivations allows making this match.

5 Mixing computation and deduction

Proofs and computations have, of course, a great deal in common. The Curry-Howard Isomorphism views certain (constructive) proofs as programs. Here, we are interested in another connection between proofs and computation: that is, during checking of (or performing) a proof, certain computations must be made. For example, a condition on a step in a proof might require that a certain number evenly divides another number: such a condition can be established by a straightforward computation.

Proof checkers can be divided into those that rely solely on determinate (functional) computations and those that permit the more general notion of non-deterministic (relational) computation. Proof checkers of proofs in typed λ-calculi generally rely on extensive uses of β-reduction. Via the *deduction modulo* approach to specifying proof systems, theories can, at times, be turned into functional computations that sit within inference rules (Dowek *et al.* 2003). The Dedukti proof checker (Boespflug 2011) implements deduction modulo by compiling such computations into a functional programming language.

On the other hand, there are proof checkers that are built using non-deterministic search principles and that employ logic programming engines. For example, some of the early proof checkers (Appel 2001; Appel & Felty 1999; Necula & Lee 1998) involved in the *proof carrying code* effort used logic programming based on (subsets of) higher-order logic (Miller & Nadathur 2012). These systems experimented with backtracking search (sometimes, even within the unification process). In one paper, the non-determinism inherent in a Prolog-based proof checker was resolved by supplying the checker with an oracle that was responsible for having all the answers to the question "I have several choices to consider, which should I take?" (Necula & Rahul 2001).

While placing significant amounts of computation (either functional or relational) into inferences seems necessary for capturing a wide range of proof certificates, this integration comes with some costs. First, one must accept that a compiler and a runtime system for a programming language are part of the trusted core of a proof checker. While compilers and interpreters for both functional and logic programming implementations are well understood, their presence in a proof checker will certainly complicate one's willingness to trust them. Second, proof checkers running on different hardware could have rather different resources available to them: thus, the computation required to check a proof might be available to one checker and not to another. This problem could be addressed by having a network of trusted libraries of proofs: such libraries could publish theorems only after their own proof checkers have checked a given proof certificate. Libraries could, of course, have computational resources available that might not be available on, say, desktop or mobile computers.

The use of proof search (non-deterministic, logic programming) to do proof checking may introduce some special issues of its own. Most logic programming engines (the efficient ones) usually come with a depth-first search strategy for building proofs. This style of search is notoriously poor when dealing with problems related to deduction. Since a proof checker is only rechecking or reconstructing an object that is already known to exist, the proof certificate could come with useful bounds on how much search needs to be done in order to reconstruct a particular, elided subproof. For example, a depth-bound for a depth-first-search process should be a natural value to estimate when eliding an existing proof object. Another issue with using a non-deterministic proof checker is that there can be a mismatch between finding *the* proof or finding *a* proof: theorems generally have many proofs. Since a proof certificate might elide information, there is no guarantee that the proof the checker reconstructs is the original proof. This discrepancy does not appear to be serious since the proof checker will, at least, find a proof.

6 Conclusion

A proof is often expected to explain why a given theorem is true: such explanations are generally informal and flow from human to human. On the other hand, a proof can also serve as certification: such certificates are generally formal objects and flow from machine (the prover) to machine (the checker). We have advanced four desiderata for proof certificates and have outlined how results from structural proof theory and the automation of logic can be used to build certificates satisfying those desiderata. The resulting approach to proof certificates is based on *foundational* rather than *technological* considerations. There are several important consequences of having foundational proof certificates. First, one must not trust theorem provers but only proof checkers: since checkers are based on simple and universal proof principles, they should be much easier to trust. Second, open markets for proofs can exist where those who need a proof can unambiguously request a proof of a theorem (via an empty certificate) and can unambiguously check that a proposed proof is, in fact, correct (using a trusted checker). Third, since proof certificates are not based on changing technological considerations but on a permanent foundation, libraries of proofs are possible. Such libraries offer the possibility to become trusted proof checkers as well as agents for structuring theories.

Note. Since this paper was first written in July 2012, the paper (Chihani et al. 2013) has appeared: this paper provides details about how one can use focused proofs in classical first-order logic to specify foundational proof certificates for a number of proof systems.

Acknowledgments. We thank the anonymous reviewers for their comments on a earlier draft of this paper. This research has been funded in part by the ERC Advanced Grant ProofCert.

Bibliography

Andreoli, J.-M. (1992). Logic programming with focusing proofs in linear logic. *Journal of Logic and Computation*, *2*(3), 297–347, doi:10.1093/logcom/2.3.297.

Appel, A. W. (2001). Foundational proof-carrying code. In *16th Symp. on Logic in Computer Science*, 247–258.

Appel, A. W. & Felty, A. P. (1999). Lightweight lemmas in Lambda Prolog. In *16th International Conference on Logic Programming*, MIT Press, 411–425.

Asperti, A. (2012). Proof, message and certificate. In *Intelligent Computer Mathematics – Proceedings of AISC, DML, and MKM 2012, Lecture Notes in Computer Science*, vol. 7362, Jeuring, J., Campbell, J. A., et al., eds., Springer, 17–31, doi:10.1007/978-3-642-31374-5.

Baelde, D. (2008). *A linear approach to the proof-theory of least and greatest fixed points*. Ph.D. thesis, École Polytechnique, URL www.lix.polytechnique.fr/ dbaelde/thesis/.

Baelde, D. (2012). Least and greatest fixed points in linear logic. *ACM Transactions on Computational Logic*, *13*(1), 2:1–2:44, doi:10.1145/2071368.2071370.

Ball, T., Cook, B., et al. (2004). SLAM and static driver verifier: Technology transfer of formal methods inside Microsoft. In *Proc. Integrated Formal Methods, 4th International Conference, IFM 2004, Canterbury, UK, Lecture Notes in Computer Science*, vol. 2999, Boiten, E. A., Derrick, J., & Smith, G., eds., Springer, 1–20.

Barendregt, H. & Wiedijk, F. (2005). The challenge of computer mathematics. *Transactions A of the Royal Society*, *363*(1835), 2351–2375.

Bates, J. L. & Constable, R. L. (1985). Proofs as programs. *ACM Transactions on Programming Languages and Systems*, *7*(1), 113–136.

Boespflug, M. (2011). *Conception d'un noyau de vérification de preuves pour le $\lambda\Pi$-calcul modulo*. Ph.D. thesis, École Polytechnique.

Bowen, J. P. (1993). Formal methods in safety-critical standards. In *Proc. 1993 Software Engineering Standards Symposium*, IEEE Computer Society Press, 168–177.

Chihani, Z., Miller, D., & Renaud, F. (2013). Foundational proof certificates in first-order logic. In *CADE 24: Conference on Automated Deduction 2013*, Bonacina, M. P., ed., no. 7898 in Lecture Notes in Artificial Intelligence, 162–177.

Church, A. (1940). A formulation of the simple theory of types. *Journal of Symbolic Logic*, 5(2), 56–68, doi:10.2307/2266170.

Dowek, G., Hardin, T., & Kirchner, C. (2003). Theorem proving modulo. *Journal of Automated Reasoning*, 31(1), 33–72, doi:10.1023/A:1027357912519.

Fontaine, P., Marion, J.-Y., et al. (2006). Expressiveness + automation + soundness: Towards combining SMT solvers and interactive proof assistants. In *TACAS: Tools and Algorithms for the Construction and Analysis of Systems, 12th International Conference, Lecture Notes in Computer Science*, vol. 3920, Hermanns, H. & Palsberg, J., eds., Springer, 167–181, doi:10.1007/11691372_11.

Gentzen, G. (1969). Investigations into logical deduction. In *The Collected Papers of Gerhard Gentzen*, Szabo, M. E., ed., Amsterdam: North-Holland, 68–131, translation of articles that appeared in 1934-1935.

Girard, J.-Y. (1987). Linear logic. *Theoretical Computer Science*, 50, 1–102.

Girard, J.-Y. (2006). *Le Point aveugle: Cours de logique: Tome 1, Vers la perfection*, vol. 1. Paris: Hermann.

Harper, R., Honsell, F., & Plotkin, G. (1993). A framework for defining logics. *Journal of the ACM*, 40(1), 143–184, doi:10.1145/138027.138060.

Harrison, J. (1999). A machine-checked theory of floating point arithmetic. In *12th International Conference on Theorem Proving in Higher Order Logics, Lecture Notes in Computer Science*, vol. 1690, Bertot, Y., Dowek, G., et al., eds., Springer, 113–130.

Kohlenbach, U. & Oliva, P. (2003). Proof mining in L_1-approximation. *Annals of Pure and Applied Logic*, 121(1), 1–38, doi:10.1016/S0168-0072(02)00081-7.

Lakatos, I. (1976). *Proofs and Refutations*. Cambridge: Cambridge University Press.

Liang, C. & Miller, D. (2009). Focusing and polarization in linear, intuitionistic, and classical logics. *Theoretical Computer Science*, 410(46), 4747–4768, doi:10.1016/j.tcs.2009.07.041.

Liang, C. & Miller, D. (2011). A focused approach to combining logics. *Annals of Pure and Applied Logic*, *162*(9), 679–697, doi:10.1016/j.apal.2011.01.012.

MacKenzie, D. (2001). *Mechanizing Proof.* Cambridge, MA: MIT Press.

McDowell, R. & Miller, D. (2000). Cut-elimination for a logic with definitions and induction. *Theoretical Computer Science*, *232*(1–2), 91–119, doi:10.1016/S0304-3975(99)00171-1.

Miller, D. (2011). A proposal for broad spectrum proof certificates. In *CPP: First International Conference on Certified Programs and Proofs, Lecture Notes in Computer Science*, vol. 7086, Jouannaud, J.-P. & Shao, Z., eds., Springer, 54–69, URL www.lix.polytechnique.fr/Labo/Dale.Miller/papers/cpp11.pdf.

Miller, D. & Nadathur, G. (2012). *Programming with Higher-Order Logic.* New York: Cambridge University Press, doi:10.1017/CBO9781139021326.

Millo, R. A. D., Lipton, R. J., & Perlis, A. J. (1979). Social processes and proofs of theorems and programs. *Communications of the Association of Computing Machinery*, *22*(5), 271–280.

Ministry of Defence, U. K. (1997). UK Defence Standardization. URL www.dstan.mod.uk/, defStan 00-55.

Necula, G. C. (1997). Proof-carrying code. In *Conference Record of the 24th Symposium on Principles of Programming Languages 97*, Paris: ACM Press, 106–119.

Necula, G. C. & Lee, P. (1998). Efficient representation and validation of proofs. In *13th Symp. on Logic in Computer Science*, Los Alamitos, CA: IEEE Computer Society Press, 93–104.

Necula, G. C. & Rahul, S. P. (2001). Oracle-based checking of untrusted software. In *POPL*, 142–154.

Negri, S. & von Plato, J. (2001). *Structural Proof Theory.* Cambridge; New York: Cambridge University Press.

Nelson, R. B. (1993). *Proofs Without Words: Exercises in Visual Thinking.* Washington: Mathematical Association of America.

Nigam, V. & Miller, D. (2010). A framework for proof systems. *J. of Automated Reasoning*, *45*(2), 157–188, doi:10.1007/s10817-010-9182-1.

Pollack, R. (1998). How to believe a machine-checked proof. In *Twenty Five Years of Constructive Type Theory*, Sambin, G. & Smith, J., eds., New York: Oxford University Press.

Prawitz, D. (1965). *Natural Deduction.* Uppsala: Almqvist & Wiksell.

The Coq Development Team (2002). *The Coq Proof Assistant Reference Manual Version 7.2*. Tech. Rep. 255, INRIA, more recent versions may be obtained from the site http://coq.inria.fr/.

Troelstra, A. S. & Schwichtenberg, H. (1996). *Basic Proof Theory*. Cambridge: Cambridge University Press.

Dale Miller
INRIA-Saclay & LIX/École Polytechnique,
Palaiseau
France
dale.miller@inria.fr

The Rise of Post-Genomics and Epigenetics: Continuities and discontinuities in the history of biological thought

MICHEL MORANGE

1 Introduction

Twentieth-century historians and philosophers of science have emphasized the importance of discontinuities in the development of the sciences. Karl Popper outlined the role of falsification of pre-existing theories and models. Thomas Kuhn extended the scope of these transformations to the replacement of a full paradigm by a new incommensurable one—with all the difficulties generated by the absence of a precise definition of the word "paradigm". In contrast, social studies of science have focused on controversies, and the way they orient scientific research in one or other direction.

Many historians of biology have already pinpointed how difficult it is to apply these models of scientific development to the biological sciences. The absence, or at least the scarcity, of theories in biology, and the questioned existence in this discipline of paradigms and paradigm shifts, make the use of Popper and Kuhn's descriptions problematic. Even more seriously, because it also challenges the approach that has been favoured in the social studies of science, the existence of continuities seems particularly obvious in the biological sciences.

The mechanical conception of life is already present in the writings of Aristotle and Galen, far before Galileo Galilei and Descartes made it explicit. It has still a major place in the descriptions of molecular and cell biologists: macromolecules are compared to engines and nanomachines (Block 1997; Alberts 1998).

But the enzymatic (chemical) conception of life also has a long history. Aristotle already compared what happens during the development of an organism to the action of ferments. This comparison was widely exploited by alchemists during the Renaissance. The enzyme theory of life, according to which all the properties of organisms can be explained by the presence and action of enzymes—catalyzers of chemical reactions—was dominant in the first part of the 20th century (Olby 1974). The recent enthusiasm for epigenetics (Jablonka & Lamb 2005) is clearly the legacy of this tradition, pushed to one side for decades by the mechanistic models of molecular biology.

Recurrent use of similar models and questioning may have a shorter lifetime, but can nevertheless be quite significant. This is the case of the thermodynamic vision of biological phenomena, from its inception in the 19th century. Not only was the thermodynamic behaviour of organisms questioned—do they violate the second law of thermodynamics?—but also the models of thermodynamics permeated other branches of biology, from evolutionary biology to embryology. The "fitness landscapes" of Sewall Wright are similar, not in their meaning but in the way they are represented, to the energy landscapes of thermodynamicists. The epigenetic landscape drawn by Conrad Waddington in the 1940s, derived from the representations of Wright and designed to describe the cellular differentiation pathways, brought with it this strong connection with thermodynamics.

Continuity in biological thought can also be the continuity of debates and oppositions. The mechanical and chemical (enzymatic) explanations of biological phenomena have recurrently been opposed one to the other, in different forms and avatars, throughout the history of biology. Similar types of oppositions have existed, between reductionism and holism (Jacob 1973), and between a direct and indirect effect of the environment on the evolutionary transformations of organisms (Lamarckian vs. Darwinian explanations), epigenetics somehow being the last avatar of the Lamarckian tradition. Gerald Holton describes these recurrent oppositions as an alternation of *themata* (Holton 1978).

During the second part of the 20th century, biology seems to have abandoned the *"longue dure"* of its interpretative frameworks in favour of a frantic rhythm of transformation. Whereas the period extending between the 1930s and 1950s corresponded to the triumph of the Modern Synthesis and molecular biology, oppositions to Modern Synthesis have progressively developed since the 1970s, and the recent rise of Evo-Devo and epigenetics is claimed to have challenged its foundations. Since the 1990s, the question of the death of molecular biology has regularly been raised (Morange 2008). New disciplines such as systems biology and synthetic biology have emerged that emphasize their differences from the previous approaches to biological phenomena.

My purpose in this contribution is to examine the nature of these recent transformations. Do they represent radical novelties and discontinuities, or are they the re-emergence of long-term oppositions, the avatars of previous models that disappeared?

I will successively consider four new research fields: synthetic biology, systems biology, Evo-Devo and epigenetics. The goal is to position these new disciplines in the historical landscape of biological thought.

2 Synthetic biology

Although scientific projects to synthesize artificial forms of life appeared at the beginning of 20th century (Keller 2002; Pereto & Catala 2007), synthetic biology as a new active field of research emerged *circa* 2000, and gained visibility through a small series of spectacular, iconic experiments (and scientists!) (Morange 2009). A good example was the construction by Michael Elowitz and Stanislas Leibler of a "repressilator", a functional module which, introduced into bacteria, generated in them a circadian rhythm (Elowitz & Leibler 2000).

What characterizes synthetic biology, and distinguishes it from genetic engineering, is its rational engineering spirit. Fully functional modules, and not isolated genes, are introduced into organisms. The construction of these modules is preceded by a long process of modelling. Efforts are made to create pieces (biobricks) that can be reused for different projects.

What is the meaning of the emergence of synthetic biology (Morange 2013)? Synthetic biologists frequently argue that they want to do things differently, and better, than evolution. Instead of tinkering as evolution does, they engineer new simple and perfectly functional devices. Whereas in the mechanical tradition, organisms have always been considered as perfect machines, synthetic biologists emphasize the defects of the devices present in organisms, and in this way show the limits of the mechanical conception of life.

These were the proclaimed ambitions of some synthetic biologists. But does the work performed in synthetic biology correspond to this heroic image (O'Malley 2009)? Synthetic biologists use directed evolution, *i.e.*, a combination of random variations and selection, to increase the efficiency of the systems that they have built. More generally, they exploit as much as possible the lessons that can be learnt from the observation of natural systems, organisms.

Therefore, synthetic biology can also be seen as the achievement of the mechanistic programme of biology. Synthetic biology is the encounter between the connection of knowledge and action that was placed at the root of modern science by Francis Bacon, and the successful recent description of organisms and cells as assemblies of macromolecular mechanical devices. Synthetic biology is the completion of the project of early molecular biologists to explain functions by the description of the macromolecules that are involved in their realization. Its belated rise is the direct consequence of the time that has been necessary to accumulate enough information on the complex biological systems at the molecular level. Not only is synthetic biology the last step in the molecular description of organisms, but its eventual successes will also be the proof that the molecular level is the favoured level of organization for organisms.

3 Systems biology

Systems biology has close relations with synthetic biology: the latter depends upon the descriptions provided by the former. And both share the same ambiguity in their relations with evolutionary biology.

A major apparent difference is that systems biology has deeper roots in the history of biological thought. Kant is regularly mentioned as one of its founders, and in the 20th century von Bertalanffy explicitly proposed a "general system theory" (1968). Therefore, the recent emergence of systems biology is frequently seen as a return to a more holistic vision in biology, and the abandonment of the most reductionist programmes of molecular biology.

But is this the case? It would probably be more appropriate to speak of "molecular systems biology" to designate the numerous studies proliferating today. These studies are totally dependent upon the precise molecular descriptions that have been obtained in previous years. Obviously, the extant form of systems biology is more the legacy of molecular biology than a new alternative way to do biology!

4 Epigenetics

Epigenetics is one of the richest fields of current biological research, but also one of the most complex. An historical account of its emergence would require a full study.

Most of the epigenetic studies done today concern chemical modifications of DNA (methylation) and of the proteins surrounding it, *i.e.,* chromatin (for the proteins, this consists of a complex ensemble of different chemical modifications that have diverse effects).

These modifications alter the level of gene expression. They are reversible, but nevertheless stable enough to be transmitted during cell division, and in some rare cases through generations. They can respond to changes in the environment.

These epigenetic marks were first described in the 1960s and 1970s, but their importance in development as well as in some diseases such as cancer only became obvious at the beginning of the 2000s.

But epigenetics has deeper roots in the history of biology. The name clearly refers to epigenesis, and the adjective epigenetic is common to epigenesis and epigenetics. Epigenesis is a global (and vague) theory of development stating that the organism is progressively built during embryogenesis. Its importance cannot be fully understood independently of the antagonistic model of development, preformationism, which flourished for one century (1650-1750), and according to which the organism is preformed in the male sperm or female egg, and only grows during embryogenesis.

The word "epigenetics" was coined by Conrad Waddington in 1940 to designate a new science of embryology in which the major role of genes would be fully acknowledged (Waddington 1940). Waddington pictured the action of genes as the creation of a complex "epigenetic landscape" in which cells progressively acquired their differentiated properties through gene action.

"Epigenetic inheritance" was also the expression progressively used to designate, from the end of the 1940s, hereditary phenomena that were independent of the genes. In this case, the use of the adjective "epigenetic" can be understood, in relation to the criticisms directed at genetics, to be a return to the preformationist conception of development.

Finally, at the end of the 1950s, when the coding role of the genes was progressively elucidated, the adjective "epigenetic" was used by David Nanney to designate changes affecting genes in their level of expression, and not in their coding properties (Nanney 1958).

The word epigenetics, used today to designate a precise ensemble of mechanisms, has not totally lost the different meanings that have successively been attached to it and to its adjective. It explains many ambiguities and confusions generated by its present-day incautious use. For instance, in most cases and in contrast to what is frequently written, an epigenetic modification is not inherited.

How can we explain the success of epigenetics? The first (good) scientific reason is that epigenetic modifications are involved in many important processes such as development, genomic imprinting (the different expression of the same genes when they are transmitted by the father or by the mother), and many diseases. These modifications are essential to the understanding and mastery of the reprogramming of the nucleus that occurs during cloning. Epigenetic modifications, and regulation of gene expression by microRNAs, another fashionable area of research, mechanistically are narrowly linked.

Epigenetics is also located at the crossroads of various biological disciplines: genetics, developmental biology, but also evolutionary biology and ecology.

But epigenetics is, most of all, the re-emergence of many oppositions to genetics that remained dispersed, but now have found a banner under which to gather. These oppositions originated at the end of the 19th century when August Weismann and many others proposed a material and corpuscular model of heredity. The current success of epigenetics is related to the different meanings that have accreted to the word during its historical development, and which it still more or less retains. The adjective "epigenetic" can be used to say that a biological process is independent of the genes, such as the fine tuning of the synaptic connections, inherited independently of the genes, or open to the environment—the lactose genetic regulatory system is now considered by many as an epigenetic system. Therefore, those opposed to genetic determinism, to the particulate conception of the gene, and to the "isolation" of the genetic material from the environment find in models and rare epigenetic data arguments in favour of their heterodox views.

Epigenetics today is simultaneously a set of new mechanisms of gene regulation and the return to different conceptions opposed to those of genetics. In its models and practices, epigenetics is closer to biochemistry than to genetics and molecular biology. Somehow, this can be seen as the belated revenge of biochemists against molecular biologists. The only significant issue is to know whether this fragile coalition between new mechanisms and old conceptions will hold. Personally, I would be ready to bet that mechanisms will remain, but that the complex mixture of resentments towards genetics will progressively dissolve without producing any dramatic change in the present explanatory framework of biologists.

5 Evo-Devo

Evo-Devo can be considered as the belated encounter between embryology, rebaptized developmental biology in the 1970s, and evolutionary biology. The Modern Synthesis, the wedding between the Darwinian model of evolution and population genetics, did not include embryology, for reasons which are beyond the scope of this contribution. The trigger to the development of Evo-Devo was the discovery that genes involved in the first steps of embryogenesis have been structurally and functionally conserved during evolution (Morange 2011). Characterization of these developmental genes, of the way they are organized in complex networks, of the DNA sequences that control their expression, and of their targets has become the main activity of developmental biologists.

To perceive the novelty of Evo-Devo, it is necessary to recall the conceptions that preceded the discovery of developmental genes. Though some genes and mutations affecting development were described in the early days of genetics, most genes of an organism were considered to participate, in one way or another, in its development. Evolution consisted of the invention of new processes—does the development of a *Drosophila* have anything in common with that of a mammal?—and for the same reasons the genes and genetic mechanisms involved were expected to be different.

Clearly, Evo-Devo imposed a new view of development and evolution, focused on the existence of a small set of genes and genetic mechanisms, highly conserved and recombined during evolution.

Evo-Devo has much in common, at the epistemic level, with epigenetics. Its recent success is due to the unexpected discovery of new general mechanisms of development. But it also offered an opportunity for models that had been marginalized during previous decades to make an astounding return to the limelight.

One excellent illustration is to be found in the work of Eric Davidson. His pioneering molecular studies on the sea urchin in the 1960s led him forty years later to propose the first precise description of the gene regulatory networks controlling the development of this organism.

But the interpretation of these data by Davidson is totally opposed to the main statements of evolutionary theory. For him, the main driver of evolution is mutation, not natural selection. The nature of mutation determines whether the resulting organism will be a new variety or a new phylum (Erwin & Davidson 2009). Evolution is saltationist: one must distinguish, as Richard Goldschmidt did, micro- and macromutations, which are of a different nature (Goldschmidt 1940). This is a return to the mutationist view of evolution advocated by Hugo de Vries at the end of the 19th century. Two major components of the Modern Synthesis are set aside: the continuous nature of variations and the creative role of natural selection.

Not all supporters of Evo-Devo share Davidson's drastic opposition to the Modern Synthesis. But most would probably agree that in the last two decades molecular tools have revealed the richness and diversity of variations involved in evolution: at the genetic level, from point mutations to genome duplication, but also at the phenotypic level with the discovery of molecular mechanisms conserved for the building of organisms. These new findings have to be incorporated, in one way or another, into a renewed Modern Synthesis. It is impossible to imagine that the abstract models of population genetics will not be dramatically altered by the accumulation of molecular descriptions and will remain, in their present form, the last word on evolution.

6 Conclusions

This rapid tour of currently emerging disciplines in biology shows that there is no obvious novelty in them, but a complex weaving of newly described mechanisms, and re-emerging old models.

Is such a picture characteristic of biology? Maybe the use of common language in biology facilitates this *"longue dure"*, this persistence of models and interpretations. But my feeling is that similar situations probably exist in other disciplines.

What happens cannot be described as a simple alternation of competing models. Systems biology is not the simple return to a holistic form of biology, or epigenetics the end of genetic determinism. Everything looks as if there was a kind of reservoir, a "purgatory", where past scientific models and ideas not retained in the current body of science await a new life. When new observations are made and new mechanisms described, these past models are extracted and collated with the new data to see whether they might be of some help in their interpretation.

Acknowledgments

I am indebted to Dr. David Marsh for critical reading of the manuscript.

Bibliography

Alberts, B. (1998). The cell as a collection of protein machines: Preparing the next generation of molecular biologists. *Cell*, *92*, 291–294, doi:10.1016/S0092-8674(00)80922-8.

Block, S. M. (1997). Real engines of creation. *Nature*, *386*, 217–219, doi:10.1038/386217a0.

Elowitz, M. B. & Leibler, S. (2000). A synthetic oscillatory network of transcriptional regulators. *Nature*, *403*, 335–338, doi:10.1038/35002125.

Erwin, D. & Davidson, E. H. (2009). The evolution of hierarchical gene regulatory networks. *Nature Reviews/Genetics*, *10*, 141–148, doi:10.1038/nrg2499.

Goldschmidt, R. (1940). *The Material Basis of Evolution*. New Haven: Yale University Press, 1982.

Holton, G. (1978). *The Scientific Imagination: Case studies*. Cambridge: Cambridge University Press.

Jablonka, E. & Lamb, M. J. (2005). *Evolution in Four Dimensions: Genetics, epigenetics, behavioral and symbolic variation in the history of life*. Cambridge: The MIT Press.

Jacob, F. (1973). *The Logic of Life: A history of heredity*. Princeton: Princeton University Press.

Keller, E. F. (2002). *Making Sense of Life: Explaining biological development with models, metaphors, and machines.* Cambridge, MA: Harvard University Press.

Morange, M. (2008). The death of molecular biology? *History and Philosophy of the Life Sciences, 30,* 31–42.

Morange, M. (2009). A new revolution? The place of systems biology and synthetic biology in the history of biology. *EMBO Reports, 10,* 50–53.

Morange, M. (2011). Evolutionary developmental biology: Its roots and characteristics. *Developmental Biology, 357*(1), 13–16, doi:10.1016/j.ydbio.2011.03.013.

Morange, M. (2013). Comparison between the work of synthetic biologists and the action of evolution: Engineering versus tinkering. *Biological Theory, 8*(4), 318–323, doi:10.1007/s13752-013-0134-y.

Nanney, D. (1958). Epigenetic control systems. In *Proceedings of the National Academy of Sciences,* vol. 44, 712–717.

Olby, R. (1974). *The Path to the Double Helix.* London: Macmillan.

O'Malley, M. A. (2009). Making knowledge in synthetic biology: Design meets kludge. *Biological Theory, 4,* 378–389.

Pereto, J. & Catala, J. (2007). The Renaissance of synthetic biology. *Biological Theory, 2*(2), 128–130.

von Bertalanffy, L. (1968). *General System Theory.* New York: George Braziller.

Waddington, C. H. (1940). L'épigénotype. *Endeavour, 1,* 18–20.

Michel Morange
Centre Cavaillès
République des savoirs: lettres, sciences, philosophie USR 3608
École normale supérieure, Paris
France
michel.morange@ens.fr

Evidence-Based Medicine and Mechanistic Reasoning in the Case of Cystic Fibrosis

Miriam Solomon

> Feminist objectivity makes room for surprises and ironies at the heart of all knowledge production; we are not in charge of the world. (Haraway 1991)

Cystic fibrosis (CF) is an autosomal recessive genetic disorder that occurs primarily but not exclusively in people of European descent. It is one of the most common life-shortening genetic diseases (1 in 25 people of European descent are CF carriers; 30,000 people with CF live in the USA). It was first identified in 1938 by the American physician Dorothy Andersen, on the basis of a cluster of clinical symptoms affecting the lungs, pancreas, liver and other organs. Lung inflammation and frequent infections are the most serious problems, apparently caused by the buildup of thick mucus. Life expectancy has greatly increased, from a few months in 1950 to 37 years in 2008. The increase in life expectancy has been due to the gradual accrual of treatments addressing symptoms of CF, starting with antibiotics, chest percussion and supplementary digestive enzymes and continuing more recently with bronchodilators, saline mist, Pulmozyme and other treatments. The gene responsible for CF was identified on the long arm of chromosome 7 in 1989 by a team led by Francis Collins. It is called the CFTR gene—cystic fibrosis transmembrane conductance regulator—which codes for a protein regulating chloride transport across cell membranes. The most common mutation is deltaF508 (2/3 of cases) but there are over 1,500 known mutations.

Treatments for CF are complex and time consuming, taking several hours per day, as well as regular professional monitoring and adjustment. By the 1970s it became evident that CF patients do best when taken care of by multidisciplinary teams (physicians, nurses, social workers, physical therapists, etc.) at specialized CF Care Centers, and this has become standard in the developed world. In the USA, the Cystic Fibrosis Foundation plays an important role in co-ordinating both research and care; similar organizations operate in other countries and Cystic Fibrosis Worldwide operates at an international level. The well-known surgeon/writer Atul Gawande praises these institutions, writing that "Cystic Fibrosis care works the way we want all of medicine to work" (Gawande 2004). He is impressed with the systematic and evidence-based delivery of healthcare. He also reports that in the best treatment centers—those with aggressive adherence to evidence-based guidelines—life expectancy is 47, ten years longer than average.

The staples of CF treatment are not particularly high tech, nor specific to CF treatment. They were developed for other uses. Chest percussion several times daily to loosen mucus, oral and inhaled prophylactic antibiotics, bronchodilators, hypertonic saline mist, supplementary digestive enzymes, ibuprofen, Pulmozyme and aerobic exercise are recommended even for those who are unsymptomatic. For CF, it is better to prevent symptoms than to treat them when they occur. In advanced disease, additional intravenous antibiotics, inhaled oxygen, ventilators, and even lung and liver transplantation are the standard of care.

How were these therapies discovered? They were not created specifically for CF, but for other conditions with similar symptoms. They address the symptoms—the distal end(s) of the causal chain(s) starting with the defective CFTR gene. They do not address the proximate end of the causal chain: the defective gene and its products. Each proposed intervention was tested, and clinical trials have improved in quality over time (as they have in the rest of medicine). Some proposed interventions turned out not to be effective, for example, sulfa drugs for infection, high humidity (mist) tents for loosening mucus and corticosteroids for reducing inflammation. There was no way to figure out in advance that antibiotics would do better than sulfa drugs, hypertonic saline mist better than water mist, and ibuprofen better than corticosteroids. Our current knowledge about CF treatment is the product of more than fifty years of clinical trials. Knowledge of CF therapies is evidence-based, in the precise sense intended by "evidence-based medicine". Of course, not all trials were perfectly randomized double-masked controlled trials, especially in the early years, but EBM (properly understood) includes the observational and other kinds of trial that provided knowledge about effective interventions. Much recent work on symptomatic therapies for CF has the highest quality of evidence. The results of clinical trials for symptomatic therapy have been combined in systematic reviews such as a comprehensive pulmonary therapy review produced by the Cystic Fibrosis Foundation (Flume et al. 2007) and many recent Cochrane reviews of particular therapies. It is the implementation of this knowledge in system-based care that Atul Gawande commends.

At the same time as these improvements in clinical care for CF were discovered, knowledge of the genetic mechanisms producing the disease advanced in more basic research. After the identification of CF gene on the long arm of chromosome 7 in 1989, there was widespread optimism that gene therapy replacing the defective CFTR gene with a normal CFTR gene was around the corner, and that CF would disappear before the end of the 1990s (Lindee & Mueller 2011). The goal was to correct the "root cause" of CF by correcting a basic mechanism. There was success in both mouse models and *in vitro* models of gene therapy for CF in the early 1990s. CF was one of the first genetic diseases for which gene therapies were designed, and early human trials used an inhaled adenovirus to try to transfer healthy CFTR genes to lung cells. None of these interventions made it to stage 3 clinical trials: there were immediate problems with both effectiveness and safety. The grand hopes of an immediate cure for CF faded. Researchers moved on to other genetic conditions, such as OTC deficiency. The OTC trial at the University of Pennsylvania ended in 1999 with the death of an experimental subject, Jesse Gelsinger, and since then gene therapy research has been proceeding with more modesty and greater caution. We do not know how far away a solution to the technical difficulties of gene therapy is.

Identification of the genetic mechanisms of CF has been applied more successfully in the development of carrier, prenatal and newborn genetic testing. However, the result of this work has been to create *more* uncertainty about prognosis, treatment, and even definition of disease, since it turns out that clinical disease depends on both the type of mutation of the CFTR gene (there are more than 1500 known types of mutation) and on other, mostly unknown, factors such as modifier genes. It turns out that there are individuals with two defective genes who do not suffer clinical disease at all, as well as individuals with one defective gene who have clinical manifestations of CF. It is not possible to predict the severity of the disease from the type of mutation (there is some correlation, but clearly other factors play a role), and there is considerable phenotypic variability (Lindee & Mueller 2011).

Lindley Darden recently said, with some exasperation, that the genotype-phenotype relations in CF are "more complex than anyone studying a disease produced by a single gene defect had reason to expect" (Darden 2010). In fact, evidence of complexity was already there in the late 1980s, by which time it was known that CF exhibits considerable genetic and phenotypic variation, with poor correlations of genotype to phenotype. Even simple Mendelian inheritance does not mean simple molecular mechanisms.

The tendency to oversimplify genetics has been with us since the birth of Mendelism, and perhaps reached its peak during the 1990s with the excitement surrounding the human genome project and its anticipated commercial applications.

What is a mechanism? A standard account that is helpful to mention here is the Machamer *et al.* (2000) account:

> Mechanisms are entities and activities organized such that they are productive of regular changes from start or set up conditions to finish or termination conditions.

At a more general level, mechanistic theories are a common kind of scientific theory. Mechanistic explanations have a *narrative* form in that they are often linear, sequential, causal, and extended in time. They are typically deterministic narratives that, in the case of CF, have fuelled hope by highlighting opportunities for interrupting the regular chain of events with causal interventions. This is the case for both the distal end of the causal chain, already discussed, and for the proximal end—the defective gene and the biochemistry of the CFTR protein. Although our knowledge of some basic mechanisms in CF has grown, we do not understand these mechanisms fully—for example we do not know whether the excess mucus is produced by an inflammatory response or by chloride transport problems. Full mechanistic understanding of CF would of course include understanding the functional abnormalities in each of the more than 1500 mutations that can cause the disease, as well as the role of modifier genes.

Despite the last 25 years of increasing understanding of the "root" mechanisms of CF, the gains in life expectancy have come from addressing the distal end of the chain of causes, and have not depended on knowledge of molecular mechanisms. Evidence-based knowledge about CF care is knowledge about non-specific and mostly low tech interventions that increase lifespan from infancy well into middle age. So our most precise knowledge is about our crudest interventions. This epistemic irony is perhaps only temporary. Although so far gene therapy for CF has failed, there are other proposed interventions targeting basic mechanisms that show promise.

The first proposed intervention is the use of drugs designed to correct the mutated CFTR protein so that it can restore chloride transport to normal. With over 1500 ways of making mistakes in the CFTR protein, it is likely to be a lengthy project to complete. However, there is a promising start with the development of "chaperone" drugs such as Kalydeco (formerly VX770) and Ataluren, which are designed to repair the proteins produced by the G551D and deltaF508 mutations (about 4% of CF patients have the G551D mutation; 66% have the deltaF508 mutation). Kalydeco was FDA approved in January 2012, and PTC therapeutics recently reported positive stage 3 clinical trials for Ataluren. Other "chaperone" drugs, such as VX809 and VX661, are currently in clinical trials.

Another idea for a CF therapy targeting basic mechanisms is to create or enhance an alternative chloride transport channel that does not depend on the CFTR protein. The drug Denufosol was designed to do just this, and did well in early stage 1 and 2 clinical trials. If the drug had worked, it would probably have been appropriate for all CF patients, no matter which underlying mutation of the CFTR gene they have. Unfortunately, Inspire Pharmaceuticals announced in January 2011 that Denufosol did not show efficacy in stage 3 clinical trials.

Even with apparently good mechanistic understanding, it is not possible to know in advance which proposed interventions will work. In this sense, there is no difference between intervening at the proximal or the distal end of the causal chain. We saw already that many reasonable suggestions about interventions at the distal end did not work. A general problem with mechanistic accounts is that they are typically incomplete, although they often give an *illusion* of a complete narrative. Incompleteness is the consequence of there being mechanisms underlying mechanisms, mechanisms inserted into mechanisms, background mechanisms that can fill out the mechanistic story, and mechanisms that can hijack regular mechanisms. There are possible black boxes everywhere in mechanistic stories, despite an easy impression of narrative or causal completeness. Since we do not have a Theory of Everything, it is not possible to know in advance whether or not a particular mechanistic intervention will have the intended result. Mechanistic reasoning produces suggestions, which then have to be tested.

It is well known that EBM puts "pathophysiological rationale" at the bottom of the evidence hierarchy, alongside epistemically suspect strategies such as "intuition" and "expert consensus" (Evidence-Based Medicine Working Group 1992). Mechanistic reasoning is one kind of pathophysiological rationale, one that is particularly valued in this age of molecular genomics. So EBM expresses skepticism about predictions from our understanding of mechanisms. This has led to the impression—voiced by Jeremy Howick (2011) and others—that hardnosed practitioners of EBM might just as soon do away with the speculative ideas of mechanistic theories. If practitioners of EBM think this, they are mistaken (I cannot find any clear examples of such practitioners of EBM; Howick himself produces a spirited defense of "mechanistic evidence"). Perhaps their error comes from placing mechanistic reasoning in an evidence hierarchy, suggesting that it is a kind of evidence. In my assessment, although there can be evidence for and against mechanistic theories, mechanistic reasoning is not itself a kind of evidence. Mechanistic reasoning is a kind of speculation about causal interactions. It is a popular method of theorizing, but it is neither a part of, *nor a competitor with,* EBM.

Another way of putting this is to say that *there is no such thing as "mechanistic evidence",* only "mechanistic reasoning" and "evidence for the existence of mechanisms". Jeremy Howick tries to insert knowledge about mechanisms into the evidence hierarchy, terming it "mechanistic evidence" (Howick 2011) and trying to find the grade of evidence that it best fits. I think this strategy confuses evidence and theory. Reasoning about mechanisms is theoretical reasoning.

I propose a way of thinking about EBM and mechanistic reasoning in the case of cystic fibrosis (and expect that it applies in other cases as well) that will be familiar to traditionally trained philosophers of science. This is to use something like the traditional framework of "context of discovery" versus "context of justification". Mechanistic reasoning is a strategy for context of discovery, because it produces suggestions for causal interventions that may be helpful in treatment of a disease. Evidence-based medicine refers to the stage of clinical trials, especially stage 3 clinical trials that provide evidence for or against (and thereby justify or disconfirm) suggestions that come from mechanistic reasoning. In the case of cystic fibrosis, mechanistic reasoning and EBM work at different stages of inquiry. They are both important methodologies (ways of reasoning). But sometimes they point in different directions, which is probably why some view them as alternative and even competing methodologies. They point in different directions when clinical trials that are expected by mechanistic reasoning to work fail, and when interventions are discovered "empirically" that defy, or await, mechanistic explanation.

In the early years of the evidence-based medicine movement there was so much excitement about the potential for systematic standards of evidence to improve the practice of medicine that other methods such as mechanistic reasoning looked inferior. Hence the statement from the Evidence-Based Medicine Working Group in 1992:

> Evidence-based medicine de-emphasizes intuition—and pathophysiological rationale [...] and stresses the examination of evidence from clinical research [i.e., randomized controlled trials].
> (Evidence-Based Medicine Working Group 1992)

The early practitioners of EBM thought that they had a new kind of objectivity, and that older methods such as clinical intuition and pathophysiological rationale, which also claimed objectivity, were much more fallible. However, this comparison of the *reliability* of EBM and other kinds of reasoning obscures the fact that EBM is not a complete method of science. We don't have a choice to use EBM everywhere and thereby get more reliable science (bracketing for the moment the problems that John Ioannides and others have found with the reliability of RCTs—see for example (Ioannidis 2005). EBM has its place in what philosophers used to call "the context of justification", i.e., testing a proposed therapy. EBM cannot get going unless a therapy is proposed. How do we come up with new therapies? We have several methods, but most of them involve pathophysiological reasoning and preliminary experimentation and tinkering. We go through brainstorming and modeling, and then in vitro trials, animal in vivo trials, stage 1 and stage 2 clinical trials to determine safety and therapeutic levels, before getting to the stage 3 clinical trials that provide the evidence for EBM conclusions. Often (as happened, for example, with gene therapy for CF) we do not get to stage 3. I think of pathophysiological reasoning and early experimentation as part of what philosophers used to call "context of discovery". Thinking about mechanisms aids discovery in science. (I differ from the traditional philosophers' accounts in not leaving discovery entirely up to dreams and other "irrational" influences, which I think too weakly constrain creative thought.)

So EBM should not discount mechanistic reasoning, unless it wants to bite the hand that feeds it. In the more recent EBM literature, there is some recognition of this, using different terminology. For example, at least one prominent EBM researcher, Donald Berwick, claims that EBM has "overshot the mark" by denegrating "the kind of discovery that drives most improvement in health care", by which he means early trials and tinkering. Berwick recommends "broadening" EBM with these more wide ranging practices, which he calls "pragmatic science" (Berwick 2005). He advocates methods such as taking advantage of local knowledge, using open-ended measures of assessment and "using small samples and short experimental cycles to learn quickly". EBM has not formalized all the tools that we need to do science, only some of them. Berwick recommends that we not only encourage, but *publish* results in "pragmatic science".

Perhaps Berwick would have used the term "translational medicine" if it had been available to him. The term appears first in the literature in 2003 and does not become common until about 2006. Translational medicine primarily focuses on what is called "T1" research, which is going from research to basic clinical applications (Woolf 2008). (T2 is the dissemination of clinical research to clinical practice, sometimes referred to as "quality improvement".)

Translational medicine—unlike EBM–does not make claims about the objectivity of its results. This is not surprising, since translational medicine occupies the same epistemic position as what is traditionally termed "context of discovery". As the logical empiricists noticed, we do not have a logic of discovery. Objectivity is not claimed for translational medicine, I think, in order to avoid censoring the creative impulse.

Translational medicine initiatives are aimed at encouraging creativity by housing researchers from different disciplines together and providing resources (money, time) to do the necessary brainstorming, tinkering and pilot studies. I find it remarkable that good old fashioned physical proximity of diverse researchers (a tried and true method of encouraging creativity) is the core of the implementation of translational medicine. Mechanistic reasoning is a cognitive resource that is frequently used by basic researchers to theorize about a domain and make hypotheses about possible interventions.

Why a *new* term, "translational medicine", for the high risk, innovative aspects of research that have always been necessary? Ten years of insistence on the methodological rigor of EBM may have obscured the fact that EBM is not a complete scientific method yet it denigrated other methods. A new technical term gives a progressive and scientific-sounding label to what is needed, official encouragement for the messy work of creativity.

Bibliography

Berwick, D. M. (2005). Broadening the view of evidence-based medicine. *Qual. Saf. Health Care, 14*(5), 315–316, doi:10.1136/qshc.2005.015669.

Darden, L. (2010). Mechanisms, Mutations and Rational Drug Therapy in the Case of Cystic Fibrosis, unpublished manuscript.

Evidence-Based Medicine Working Group (1992). Evidence-based medicine. A new approach to teaching the practice of medicine. *JAMA : The Journal of the American Medical Association, 268*(17), 2420–2425, doi: 10.1001/jama.1992.03490170092032.

Flume, P. A., O'Sullivan, B. P., et al. (2007). Cystic fibrosis pulmonary guidelines: Chronic medications for maintenance of lung health. *American Journal of Respiratory and Critical Care Medicine, 176*(10), 957–969, doi: 10.1164/rccm.200705-664OC.

Gawande, A. (2004). The bell curve. *The New Yorker, December 6*, 82.

Haraway, D. J. (1991). *Simians, Cyborgs, and Women : The reinvention of nature.* New York: Routledge.

Howick, J. (2011). *The Philosophy of Evidence-Based Medicine.* Chichester: Wiley-Backwell.

Ioannidis, J. P. (2005). Contradicted and initially stronger effects in highly cited clinical research. *JAMA : The Journal of the American Medical Association, 294*(2), 218–228, doi:10.1001/jama.294.2.218.

Lindee, S. & Mueller, R. (2011). Is cystic fibrosis genetic medicine's canary? *Perspectives in Biology and Medicine, 54*(3), 316–331.

Machamer, P., Darden, L., & Craver, C. (2000). Thinking about mechanisms. *Philosophy of Science, 67*(1), 1–25.

Woolf, S. H. (2008). The meaning of translational research and why it matters. *JAMA : The Journal of the American Medical Association, 299*(2), 211–213, doi:10.1001/jama.2007.26.

Miriam Solomon
Philosophy Department
Temple University 022-32
Philadelphia, PA 19122
USA
`msolomon@temple.edu`

Risk Management and Model Uncertainty in Climate Change

ROGER M. COOKE

ABSTRACT. This paper explores the issues of risk management and model uncertainty in climate change. Examples are drawn from the Integrated Assessment Model DICE, but apply more generally. Techniques for dealing with model uncertainty include structured expert judgment with independent expert panels, stress testing, canonical variations and probabilistic inversion.

1 Introduction

The standard economic approach to analyzing the climate change problem has been to search for efficient abatement policies. Many Integrated Assessment Models (IAMs) achieve this objective by maximizing the present value of consumption, equating the marginal benefits of abatement in terms of reduced climate damages with the marginal costs of reducing emissions. Every trader, banker, and investor knows that maximizing expected gain entails a trade-off with risk. In finance, there are many of ways of addressing the risk, such as hedging, insuring, maintaining capital reserves, short selling, throwing a financial Hail Mary pass, or doing nothing and rolling the dice. The standard economic approach does not consider the risks associated with a climate policy that maximizes expected gain. According to the theory of rational decision (Savage 1972), preferences can always be represented as expected utility, hence from this viewpoint, any aversion to risk could be folded into the rational agent's utility function. This theory, recall, applies to rational *individuals*; groups of rational individuals do not comply with the axioms of rational decision theory.

Weitzman (2009) has recently called attention to the risks of climate change,[1] arguing that current approaches court probabilities on the order

---

[1] Weitzman (2009) estimates a probability on the order of 0.05 of 10 C or more warming in 200 years, as a result of doubling pre-industrial greenhouse gases, and a probability of 0.01 of 20 C or more warming. High warming could trigger positive feedback effects,

of 0.05 ∼ 0.01 of consequences that would render life as we know it on the planet impossible. One may quibble with Weitzman's numbers, but they are not outside the realm of mainstream climate science, and no scientist argues that consumption as we know it would persist in a 10 C warmer world. What is the plan to manage this tail risk, risk from extreme events in the tail of a distribution?[2]

The fact is that 'professional risk taking organizations' do manage risk, and not by bending the utility function of a representative consumer. Rather, they employ techniques like probabilistic design and quantitative risk analysis to quantify risk and optimize expected gain under a risk constraint. Nuclear power plants are currently designed not to exceed a yearly probability of serious core damage of 10^{-4} per reactor per year. Banks and Insurance companies under the Basel II and Solvency II protocols sequester capital reserves to cover a one-in-200 year loss event. The Dutch sea dykes employ probabilistic design to hold the yearly inundation probability under 10^{-4}. Ultimately, managing societal risk is a problem of group decision, not of maximizing expected utility of an individual rational agent.

Risk management shifts the research question from 'how does the optimal abatement level change for different parameter values?' to 'how does our policy choice fare under the distribution of future conditions we may face?' As such, it places the quantification of uncertainty in the foreground. Uncertainty quantification is more than putting distributions on model's parameters. The antecedent question is 'is it the right model?' This is the question of model uncertainty.

It is a truism that all models are false. We speak of model uncertainty if multiple models exist for describing the phenomena which cannot be ruled out. In that case uncertainty might not be well captured by putting distributions on the parameters of *one* model. Moreover, if uncertainty quantification is done by independent experts, we cannot presume that they all

---

leading to even higher temperature increases. For example, the Paleocene-Eocene thermal maximum is speculated to have released one sixth of the carbon content of methane hydrate estimated to underlie world oceans today, and raised mid latitude mean annual terrestrial temperatures by 4-5 C, having profound effects of evolution of the world's fauna (Gingerich 2006).

[2]Ackerman *et al.* characterize the impact of thisshift as follows: "How would this perspective change our approach to climate economics and policy choices? Economics would find itself in a humbler role, no longer charged with determining the optimal policy. Instead,there remains the extremely complex and intellectually challenging tasks—for which the tools of economics are both appropriate and powerful—of determining the least-cost global strategy for achieving those [risk management] targets [...]" (Ackerman *et al.* 2009a;b). Stern (2008) writes: "As economists, our task is to take the science, particularly its analysis of risks, and think about its implications for policy."

subscribe to one and the same model. Techniques for coping with model uncertainty discussed here are:

stress testing

canonical variations

structured expert judgment.

Stress testing means checking that the models behave reasonably when parameters are given extreme, though possible, values. Failing a stress test indicates model uncertainty. Canonical variations are used to explore the model space for alternative models. Gone are the days when quantification of the uncertainties was left to the modelers themselves. At the state of the art, quantification is done by domain experts in a rigorous and transparent manner. That entails that experts quantify their uncertainty over model independent phenomena that are observable in principle if not in fact. This uncertainty must be pulled back onto the parameter space of any given model, and the degree to which the observable uncertainty can be captured by a model's parameter uncertainty forms an important aspect of model evaluation. If there were no uncertainty, there would be no recourse to expert judgment. In as much as structured expert judgment involves quantifying uncertainty, probabilistic inversion may be seen as fitting models to expert judgment data.

These features were deployed in the uncertainty analyses of accident consequence models for nuclear reactors conducted jointly by the European Union and the U.S. Nuclear Regulatory Commission. With 2,036 elicitation variables assessed by 69 experts spread over 9 panels and a budget of $7 million 2010 dollars (including expert remuneration of $15,000 per expert), this suite of studies is representative of the state of the art. It established a benchmark for expert elicitation, performance assessment with calibration variables, expert combination, dependence elicitation and dependent uncertainty propagation—features largely absent from the IAM discussion. The techniques discussed here have been extensively applied and exhaustively reviewed. For a summary see a special (RPD 2000), complete references may be found in (Cooke & Kelly 2010). For good order, there are also Bayesian approaches to model uncertainty, anchored in the Bayesian inference paradigm.

Shifting to a risk management focus will also suggest corresponding changes in policy approaches. Instead of seeing the policy problem as one of pricing the world's greatest externality, the policy question is how much are we willing to spend to buy down the risk of catastrophic impacts and what is the most cost-effective way to do so? After all, we can't short the planet.

This paper is not an application of uncertainty analysis, nor is it an exhaustive catalogue of techniques for coping with model uncertainty. It argues that model uncertainty should be taken seriously in the IAM community, and presents techniques successfully deployed in the past for doing so. That said, it is likely that the climate change problem will pose new challenges which can only be met by marshalling the best efforts from diverse disciplines.

A concluding section summarily compares the approaches to risk in the climate change debate and in the banking and insurance sectors.

2 Stress testing: Pollyanna and Chicken Little

Uncertainty analysis often takes models outside their comfort zone where they are—hopefully—empirically grounded. Kann & Weyant (2000) note that many models are not calibrated for extreme values of their inputs since they are structured to provide close estimates for small perturbations of parameter values. Stress testing is preformed to check that the models remain realistic and capture the relevant possibilities as their parameters are fed values at the "limits of the possible". Pollyanna and Chicken Little both sit at the table. If either of these conditions fails, then uncertainty cannot be represented via distributions over the models parameters.

Stress testing is illustrated with the economic growth dynamics used in many IAMs. IAMs specify economic damages as a function of temperature change, and model use growth dynamics to model their impact on output and utility. For example, damages $\Omega(t)$ at time t induced by temperature change $T(t)$ relative to pre-industrial mean temperature are represented in DICE as factor that reduces economic output:

(1) $\Omega(t) = 1/[1 + 0.0028T(t)^2]$.

The standard Cobb Douglas production function expresses output $Q(t)$ as a function of total factor productivity, $A(t)$ (a parameter evolving over time capturing technological change), capital stock, $K(t)$, and labor (assumed to be the population), $N(t)$. Temperature induced damages $\Omega(t)$ and abatement efforts $\Lambda(t) \in (0,1)$ reduce output:

(2) $Q(t) = \Omega(t)[1 - \Lambda(t)]A(t)K(t)^\gamma N(t)^{1-\gamma}$.

Capital in the next time period is the depreciated capital of the previous time period (at rate δ), plus investment (output minus consumption):

(3) $K(t+1) = (1-\delta)K(t) + Q(t) - C(t)$.

Substituting (2) into (3) and replacing the difference equation with a differential equation, this growth model reduces to a Bernoulli equation solved by Jacob Bernoulli in 1695. If we put $T(t) = \Lambda(t) = 0$, then (3) reduces to a common growth model whose empirical basis is a subject of a debate (see, e.g., Romer 2006, chap. 3) which the IAM community has generally skirted. As standard macro economics texts (Romer 2006; Barro & Sala-i Martin 1999, e.g.,) do not reference this fact, a derivation is given in the appendix.[3]

It is important to notice that, according to (3), the rate of change of capital depends *only* on current values of the quantities in (2). There is no other "stock variable" the accumulation of which could influence capital growth. To give a prosaic example, letting $K(t)$ denote cyclists' altitude on a mountain, does the rate of change of K depend only on the current K and the current added energy of peddling, regardless whether the cyclist is going up or down hill? Authors who suggest that growth might be path dependent draw attention to this assumption (Stern 2008; 2009). We use the term "Bernoulli dynamics" to describe any growth model based on the Bernoulli equation.

An illustrative stress test of Bernoulli dynamics focuses on the growth model (3) without climate damage. Assume there is no temperature change, no abatement and take constant savings rate constant at 20%. Total factor productivity and population are held constant at their initial values in DICE, and we use DICE values for other parameters.[4] Figure 1 shows two capital trajectories. The first trajectory starts with an initial capital of 1\$, that is, $\$1.5 \times 10^{-10}$ for each of the earth's 6.437×10^6 people. The second trajectory starts with an initial capital equal to ten times the DICE2007 initial value. The limiting capital value is independent of the starting values—with a vengeance: the two trajectories are effectively identical after 60 years. The same will hold for any intermediate starting capital. If this were true we would have needed no Marshall plan, and after 5000 years of civilization we shouldn't expect differences in per capita output. Such obviously unrealistic consequences underscore the need for circumscribing the empirical domain of application of the Bernoulli dynamics. Regardless whether the model adequately describes small departures from an equilibrium state, its use for long term projections inevitably entails this sort of behavior.[5] The IAM DICE computes out to 2595, and the IAM FUND (Anthoff & Tol

---

[3]There is a good Wikipedia page:
http://en.wikipedia.org/wiki/Bernoulli_differential_equation.

[4]DICE uses: $\delta = 0.1$, $A = 0.02722$, $N = 6514[10^6]$, $K(0) = 137[10^{12}\$]$, savings rate $= 0.2$, and $\gamma = 0.3$.

[5]Although DICE uses a capital depreciation rate of 10%, D. Romer suggests $3 \sim 4\%$ (Romer 2006, 25); in this case convergence would occur in 150 years, after 60 years,

2008) goes out to 3000. Putting uncertainty distributions on the model's parameters will not change that. Output is determined by (2) and shows

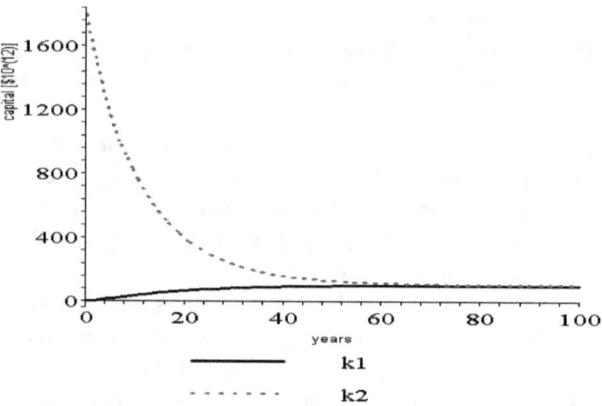

Figure 1. Two capital trajectories with DICE with default values, no temperature rise, no abatement $K1(0) = 1\$$ and $K2(0) = 1800$ *trillion*$

the same behavior. To dispel the suggestion that Figure 1 is an idle theoretical exercise, Figure 2 computes the output trajectories for the paths in Figure 1, this time using the DICE2009XL software.[6] The behavior at zero is influenced by the granularity of the 10 year time steps in DICE, but the rate of convergence is similar.

There has been much discussion about the form of the damage functions in IAMs, and several modelers, (Pizer 1999; Nordhaus 2008, e.g.,), have put distributions on model parameters and presented the Monte Carlo results as uncertainty analysis. In light of the evident lack of realism in the fundamental growth dynamics, even without temperature damage, one must question the wisdom of using such models to advise and inform climate policy, without considering model uncertainty.

3 Canonical model variation

It is often noted that simple models like the above cannot explain large differences across time and geography between different economies, pointing to the fact that economic output depends on many factors not present in

---

capital with $1 starting value would be slightly more than half that of the $1800 trillion trajectory.

[6]Dice2009_EXCEL_051109aa, accessed Monday, July 19, 2010.

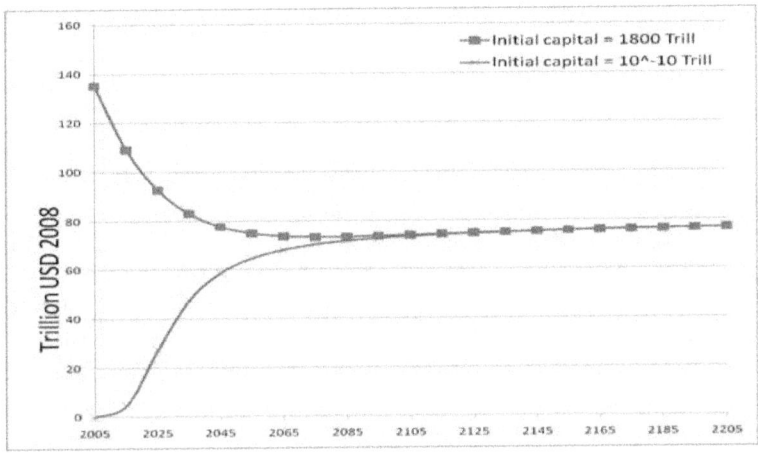

Figure 2. Output gross of abatement cost and climate damage ($trill 2000 USD) Base case, no temperature damage, no abatement, constant population, constant total factor productivity (0.0307951), initial output from production function and DICE defaults for other parameters (DICE 2009 EXCEL version).

such simple models. To "save the phenomena" researchers have proposed enhancing the basic model with inter alia social infrastructure, government spending, human capital, knowledge accretion, predation and protection, rent seeking, extortion and expropriation, see (Romer 2006, chap. 3), (Barro & Sala-i Martin 1999, chap. 12). Interest in geographical covariates has recently been rekindled (Dell *et al.* 2009; Nordhaus *et al.* 2006; Nordhaus 2006). Before adding epicycles to the simple Cobb-Douglas model, it is well to consider whether other growth dynamics with comparable prime facie plausibility can be formulated within this restricted modeling vocabulary.

We illustrate with one variation based on the following simple idea: Gross World Production (GWP [trill USD 2008]) produces pollution in the form of greenhouse gases. Pollution, if unchecked, will eventually destroy production allowing pollution to recede, where after production can resume. This simple observation suggests that Lotka-Volterra dynamics might provide a perspective which an uncertainty analysis ought not rule out.

Greenhouse gases are modeled with the carbon cycle in DICE. It is often said that emissions [GTC] are a fixed fraction of GWP (Kelly & Kolstad 2001). To see where this leads, we take the emission fraction fixed at 0.1, taken from the period 2015–2025 in DICE. Greenhouse gases, converted to *ppmC*, determine the equilibrium temperature rise above pre-industrial levels according to:

(4) $\quad T(GHG(t)) = cs \times ln(GHG(t)/280)/ln(2).$

where cs is the climate sensitivity parameter (the use of equilibrium as opposed to transient temperature is a simplification that could be easily removed). Real GWP has grown at an annual rate of $\beta = 3\%$ over the last 48 years (this includes population growth). Dell et al. (2009) argue that rising temperature decreases the growth rate of GWP. Using country panel data, within-country cross-sectional data and cross country data they derive a temperature effect which accounts for adaptation. On their analysis, yearly growth, after adaptation, is lowered by $\delta = 0.005$ per degree centigrade warming. This gives the following system, where ε is the ratio of emissions [GTC] to output [Trill USD 2008].

(5) $\quad GHG(t+1) = 0.988 \times GHG(t) + 0.0047 \times Biosphere(t) + \varepsilon \times GWP(t)$

(6) $\quad Biospere(t+1) = 0.9948 \times Biosphere(t) + 0.012 \times GHG(t) -$
$\quad\quad\quad\quad\quad\quad\quad\quad 0.0005 \times DeepOceans(t)$

(7) $\quad DeepOceans(t+1) = 0.999 \times DeepOceans(t) - 0.0001 \times Biosphere(t)$

(8) $\quad GWP(t+1) = [1 + \beta - \delta \times (T(GHG(t)))]GWP(t).$

The first three equations reflect the carbon cycle in DICE, while the last equation incorporates warming-induced damages on economic growth. Initial values for the atmospheric GHG stock, terrestrial and shallow ocean biosphere C stock and deep ocean C stock are taken from DICE.[7] If T were linear in GHG, this would be a simple Lotka-Volterra type non-linear dynamical system. To appreciate what this means, write the change in GWP as $\beta GWP(t) - \delta.(T(GHG(t)))GWP(t)$. The increment $\beta GWP(t)$ is reduced by a damage term. For fixed asset level the damage in GWP would be proportional to GHG, and for fixed GHG the decrement is proportional to the asset level. Of course $T(GHG)$ is not linear, but the morphology of a simple non—linear system is still at work. As T gets large, GWP declines at an increasing rate; that is, GWP collapses. This conclusion will not surprise readers of Jared Diamond (2005), but to the best of the author's knowledge there is no macro-economic model for collapse.

---

[7]The transfer coefficients in DICE are converted to yearly rates. Of course, the transfer coefficients in the carbon cycle would not remain fixed over long time scales owing to features like die off of the ocean's phytoplankton, deforestation, and release of methane from the oceans.

Several authors[8] have suggested that climate damages might hit capital stocks. If capital has a positive real rate of return, say β, and capital growth is negatively impacted by a term linear in $T(GHG(t))$, then a similar system would result. Capital replaces GWP in equation (5) and GWP in equation (2) is replaced by a production function giving output as a function of capital, labor and productivity. This approach is not used here, as the empirical literature supporting a climate effect on growth is concerned with GWP and not capital, but its behavior is very similar to that described below.

Figure 3 shows GWP and GHG as functions of time out to 500 years, with all variables at their nominal values. GWP collapses. Greenhouse gases also recede, but not to their initial level; as the carbon stock in the biosphere and the deep ocean has gone up, and these reservoirs serve as a source to the atmosphere long after industrial emissions have declined. Evidently, different ways of modeling the impact of climate change damages give qualitatively different predictions, and steady state values may not be relevant for current policy choices. Neither theoretical nor empirical evidence excludes the Lotka-Volterra type of interaction between damages and production presented here. A credible uncertainty analysis should fold in this and other possibilities, which brings us to the next point of examining a distribution of future conditions for a given policy choice.

3.1 *Les jeux sont faits*

At Monte Carlo casinos, you place your bets before the roulette wheel is spun. Similarly, the policy questions we face under the large uncertainties and tail risks of climate change are not addressed by re-optimizing IAMs at parameter values sampled from a distribution. We are not afforded a peek at the true values of uncertain parameters before choosing a consumption-abatement path. Instead, the risk manager seeking to safeguard our future must assess the distribution of possible outcomes once a policy path is chosen. Policies can then be adapted to keep the risk within tolerable bounds.

---

[8]See (Fankhauser & Tol 2005) and (Ackerman et al. 2009a;b). Nordhaus (1999) discussed the possibility that climate induced damages hit capital stocks: loss of coastal assets due to sea-level rise, destruction of crops, devaluation of cold-weather capital like ski resorts, and accelerated obsolescence of indoor climatization systems. More serious might be shifting rainfall patterns necessitating new water infrastructure, deteriorating social infrastructure due to migration, war, predation, etc., and destruction of natural capital in the form of fertility of soils and fecundity of oceans (Boyce et al. 2010). Rapid temperature changes over annual or decadal scales, so-called "climate flickering," would likely make impacts on capital worse (Hall & Behl 2006). In addition Sterner & Persson (2008) show that climate damages could also alter relative prices; if this is the case it can justify more restrictive emissions controls. Note that the discount rate plays no role in our discussion.

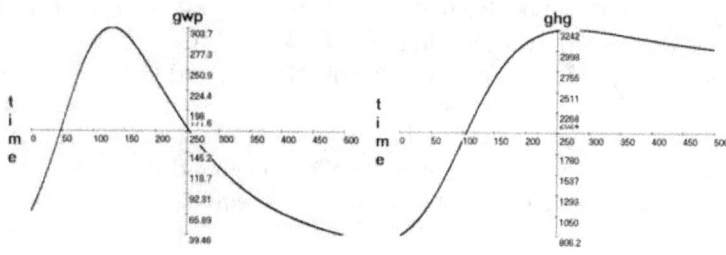

Figure 3. The impact of climate damages on GWP (left) and greenhouse gases [GTC](right).

Much realism can be added to the Lotka Volterra model. It is easy to see that decreasing the constant emission rate ε will postpone but not prevent the GWP collapse; the humps are merely shifted to the right in Figure 3. Similarly, decreasing the damage rate δ will allow us to get richer before GWP collapses; the humps get higher. A different fate within this simple model can be achieved only if, sooner or later, the emission rate effectively goes to zero. (Averting collapse could also be achieved if the damage rate went to zero, but with constant emission rate, this would lead to temperatures at which life is unsustainable.) In DICE's "no policy" base case, the emission fraction $[GTC/GWP]$ goes from 0.13 to 0.011 in 200 years. The required reductions would depend on new technologies whose existence is uncertain. To capture this uncertainty with a simple model, we replace ε with the time dependent emission factor $0.1 \times exp(-t \times \alpha)$, where t is time in years, and α is log uniformly distributed on $[10^{-6}, 10^{-4}]$. Thirty samples from the distribution of this emissions factor are shown in Figure 4.

All paths start at 0.1, and the emission factor after 250 years ranges from 0.097 to 0.005. Of course, how these different emission paths effect GWP will depend on all the other uncertainties. Choosing 'ball park' distributions for these parameters,[9] Figure 5 shows thirty paths for GWP and temperature

---

[9]Climate sensitivity is epistemic (uncertain, but constant through each run, $t = 0\ldots500$) and Beta distributed on $[1, 15]$ with parameters $(4, 24)$, mean 3. The damage parameter δ, log uniform distributed on $[0.004, 0.01]$, and the emission rate parameter α, log uniform on $[10^{-6}, 10^{-4}]$ are also epistemic. Lower values of δ lead to temperatures at which life is unsustainable, pointing to a weakness of linear damage models in this context. Only the intrinsic growth rate of production is aleatoric (independently sampled

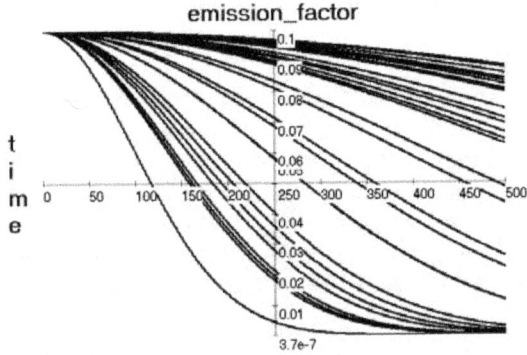

Figure 4. Thirty Emission factor paths [GTC/ $Trill USD2008].

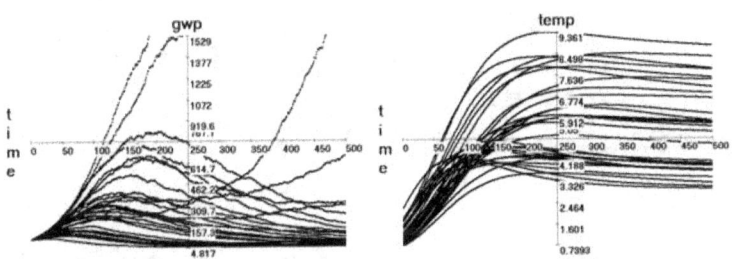

Figure 5. Thirty GWP (left) and Temperature (right) paths.

For the most aggressive reduction paths in Figure 4, GWP enjoys uninterrupted growth, on other paths GWP collapses; the timing and height at collapse depend on all uncertain parameters. Maximum temperatures range from 9.4 C to 3.7 C. The happy growth rates in Figure 5 arise if the emission reduction rate is very aggressive, the climate sensitivity is very low, and the damage rate δ is very low; and none of these factors by itself is sufficient. Note that the costs of the different emission reduction policies are not reflected in the GWP paths.

---

for each time t) and Beta distributed on [0.01, 0.06] with parameters (5, 7) and mean 0.03. In Figures 4 and 5, all these distributions are independent.

The issue of dependence has been largely absent from discussions of uncertainty in the IAM community. These uncertainties are subjective and dependencies may be caused by putative physical coupling (e.g., high climate sensitivity might be caused by feedbacks that also exacerbate damage rates, high damage rates may drive aggressive emissions reductions). Alternatively, dependence may result from pooling experts' assessments (a mixture of independent distributions is not generally independent) or from post-processing (see next section). The graphs in Figure 6 use the same marginal distributions as Figure 5 but now with a dependence structure reflecting a positive dependence between cs and δ and a negative dependence between δ and α the exact manner in which this is done are not of interest here. The dependence has significantly lowered our uncertainty in temper-

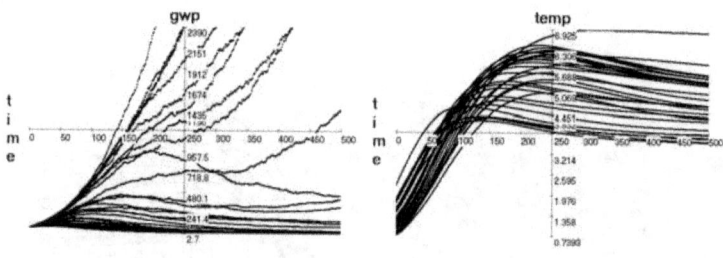

Figure 6. Thirty GWP (left) and Temperature (right) paths with dependence.

ature, maximum temperatures now range from 4.4 C to 6.9 C, instead of 3.7 C to 9.4 C. About one third of our growth paths are happy. Another way to appreciate the impact of dependence is to consider two emission fraction reduction policies shown in Figure 7. The lower policy is the most aggressive policy in Figure 4. It reduces the emission fraction from 0.1 to 0.05 in 117 years. The second milder policy reaches 0.05 in 185 years.

Figure 8 shows that the aggressive policy under the independence assumption is roughly comparable to the milder policy under dependence, with regard to maximum temperature. Of course, other dependence structures could lead to very different pictures, and the independence assumption is not generally conservative. The figures merely serve to illustrate that dependence cannot be ignored. For good order, these sketches do not constitute an analysis of uncertainty; they are arguments for doing a proper analysis of uncertainty.

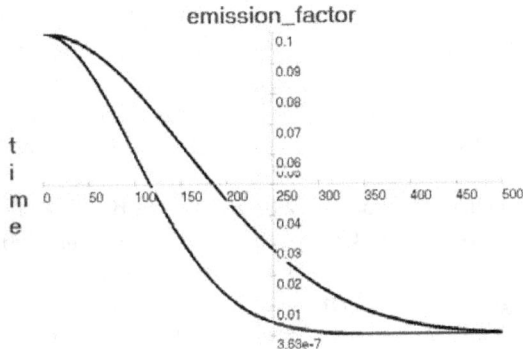

Figure 7. Two emission factor reduction policies.

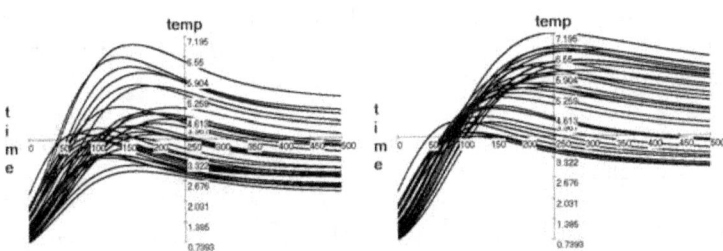

Figure 8. Temperature scenarios for aggressive reduction policy with independence (left) and for milder policy with dependence (right).

4 Structured expert judgment for quantifying uncertainties

Uncertainty analysis with climate models must be informed by the broad community of climate experts - not simply the intuitions or proclivities of modelers - through a process of structured expert judgment.[10] Experience teaches that independent experts will not necessarily buy into the

---

[10] Nordhaus' expert survey on climate change (Nordhaus 1994) may be compared with the protocols followed in the EU-USNRC studies (Cooke & Goossens 2000). Notably absent in Nordhaus' study is validation of expert performance.

models whose parameter uncertainties they are asked to quantify. Hence, experts must be queried about observable phenomena, results of thought-experiments if you will, and their uncertainty over these phenomena must be 'pulled back' onto the parameters of the model in question. This process is analogous to the process by which model parameters would be estimated from data, if there were data. The new wrinkle is that data are replaced by experts' uncertainty distributions on the results of possible, but not actual, measurements. The 'pull back' process is called probabilistic inversion, and has been developed and applied extensively over the last two decades, (see Kraan & Bedford 2005; Kurowicka & Cooke 2006; Cooke & Kelly 2010). In general, an exact probabilistic inverse does not exist, and the degree to which a model enables a good approximation to the original distributions on observables forms an important aspect of model evaluation. The details of the expert judgment process are outside the scope of this paper, but three features deserve mention.

1. Experts are regarded as statistical hypotheses, and their statistical likelihood and informativeness are assessed by their performance on calibration questions from their field whose true values are known post hoc. Experts' ability to give statistically accurate and informative assessments is found to vary considerably.

2. Experts' uncertainty assessments can be combined using performance based weights.

3. Dependence, either assessed directly by experts or induced by the probabilistic inversion operation, is a significant feature of an uncertainty analysis.

An application of uncertainty analysis in the climate change arena will doubtless pose new and unforeseen challenges and will require solutions beyond the current state of the art. However, the problems are sufficiently serious to warrant an expenditure of effort at least comparable to best efforts made in the past.

The importance of a structured expert judgment approach to uncertainty quantification was illustrated in the uncertainty analysis of the consequence of accidents with nuclear reactors, mentioned in the introduction (Harper et al. 1995; Cooke & Kelly 2010). Throughout the 1980s, complex models were built to predict the dispersal of a radioactive source term, the resulting health and economic damage, as well as effects of countermeasures. Initially, the modelers, like those at the Kernforschungszentrum Karlsruhe (KfK), quantified the uncertainties in their models themselves. Table 1 shows the ratio of the 95th to the 5th percentiles for the centerline (peak) concentration and lateral spread of airborne radioactive material 10 km downwind, under neutral atmospheric conditions, and also dry deposition velocity. The KfK values are the result of "in house" uncertainty quantification by the modelers themselves. The EU-USNRC values resulted from a structured expert elicitation using 8 international experts vetted by an independent steering committee. The expert elicitation produced a much greater range of uncertainty than the modelers did themselves.

Ratio: 95%-tile / 5%-tiles of uncertainty distributions		
	KfK	EU-USNRC
Peak centerline concentration per unit released, 10 km downwind, neutral stability	3	174
Crosswind dispersion coefficient, 10 km downwind, neutral stability	1.46	11.7
Dry deposition velocity 1 μm aerosol, wind speed 2 m/s	30.25	300

Table 1. Ratio of 95- and 5 percentiles of uncertainty distributions computed by National Radiological Protection Board (NRPB) and Kernforschungszentrum Karlsruhe (KfK), and by the European Union, US Nuclear Regulatory Commission (EU-USNRC).

5 Implications for risk management

A graph as shown in Figure 3 raises the question 'What risk of collapse are we willing to run? How much are we willing to pay to buy down the probability of collapse?' These questions represent a fundamentally different set of concerns than those of a social planner optimizing expected utilities for a representative consumer. Several policy approaches to this range of potential outcomes have been discussed in the literature, including hedging strategies (Manne & Richels 1995) and adaptive strategies (Lempert *et al.* 1996).

Regulations for Banks and Insurance companies, such as the Basel II and Solvency 2 Protocols in the EU, instruct companies to manage the risk of extreme events using a value-at-risk (VAR) framework. A target insolvency probability is chosen or set through regulation, typically 1-in-200 per year, and the firm must maintain capital reserves to cover losses from a 1-in-200 year loss event (McNeil *et al.* 2005). If Weitzman is right, society is currently much more cavalier with the survivability of the planet than banks and insurance companies with their own solvency.

The closest to a solvency constraint in the current climate change arena may be "dangerous anthropogenic interference" (DAI),[11] and debates over how to define it are ongoing. Definitions of DAI run the gamut from crossing some physical threshold, such as the melting of the Greenland ice sheet (e.g., Hansen 2005), to reducing GHG to some level, to individual perceptions of danger (Dessai *et al.* 2004). Defining DAI is not a solely scientific issue, but involves the public and their governments as well (Oppenheimer 2005). Studies have suggested that business-as-usual emissions could result in probabilities of DAI far exceeding 1-in-200 (Mastrandrea & Schneider 2004; McInerney & Keller 2008).

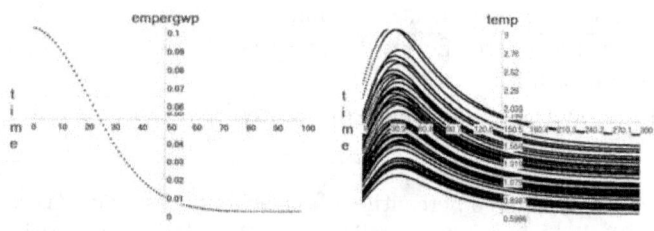

Figure 9. Emission reduction path and 50 temperature paths, all with maximum below 3 C.

For use as a risk constraint in risk management, the probability of the undesired event should be calculated as a function of our policy choices. For illustration, suppose our DAI is defined as mean temperature greater than 3 C. By doing simulations similar to those described above, we can search for an emission reduction policy which keeps the probability of exceeding 3 C warming below a specified threshold, say 0.02. For example, with the model used in Figures 4 and 5, emission policy shown in Figure 9 is found to satisfy this constraint. Of course much is left out of this picture. The cost of this policy is not computed, and there may be more efficient ways of satisfying this risk constraint. We will also learn about the effects

---

[11] Signatories to the United National Framework Convention on Climate Change (UN-FCCC) pledged to stabilize greenhouse gas concentrations "in the atmosphere at a level that would prevent dangerous anthropogenic interference with the climate system. Such a level should be achieved within a time frame sufficient to allow ecosystems to adapt naturally to climate change, to ensure that food production is not threatened and to enable economic development to proceed in a sustainable manner." 193 countries are parties to the UNFCCC, suggesting that DAI is as close to a global consensus on a climate change "solvency constraint" the world is likely to achieve.

of greenhouse gas emissions as we go forward. These models merely serve to illustrate the calculations that would implement a risk management policy.

These reflections challenge us to deploy risk management strategies on a global scale. We suggest this begin with (i) stress testing models, (ii) exploring alternative models, and (iii) quantifying uncertainty in such models via structured expert judgment. We are condemned to choose a climate policy without knowing all the relevant parameters, but we are not condemned to ignore the downside risks of our choices.

Acknowledgement

The author gratefully acknowledges Carolyn Kousky, Ray Kopp, Dallas Burtraw, Thomas Sterner, Robert Kopp, Jan Mares and Dorota Kurowicka for their help and comments.

Appendix

The model (3) reduces to a differential equation solved by Jakob Bernoulli in 1695. Putting

$$C(t) = \mathcal{K}Q(t), (d/dt)K(t) = K'(t)(3):$$

A.1) $K'(t) + \delta K(t) = B(t)K(t)^\gamma; K(t) > 0; B = \mathcal{K}\Lambda(t)A(t)N(t)^{1-\gamma}/(1+\Psi T^p(t))$.

Set $w = K^{1-\gamma}$. Then $w' = (1-\gamma)K^{-\gamma}K'$. Dividing by $K(t)^\gamma$, the equation (A.1) becomes:

$$W'/(1-\gamma) + \delta w = B.$$

Multiply both sides by $(1-\gamma)e^{(1-\gamma)\delta t}$ to get:

$$e^{(1-\gamma)\delta t}w' + (1-\gamma)e^{(1-\gamma)\delta t}\delta w = (d/dt)(e^{(1-\gamma)\delta t}w) = (1-\gamma)e^{(1-\gamma)\delta t}B,$$

If B is constant, the solution is

(9) $\quad e^{(1-\gamma)\delta t}w(t) = (1-\gamma)B \int_{u=o..t} e^{(1-\gamma)\delta u} du + w(0).$

Make the change of variable $x = t - u$, and write this as

(10) $\quad w(t) = (1-\gamma)B \int_{x=o..t} e^{-(1-\gamma)\delta x} dx + e^{-(1-\gamma)\delta t} w(0);$

(11) $\quad$ A.2) $K(t) = [(1-\gamma)B \int_{x=o..t} e^{-(1-\gamma)\delta x} dx + e^{-(1-\gamma)\delta t} K(0)^{(1-\gamma)}]^{(1-\gamma)}.$

If we set $K'(t) = 0$ in A.1, or let $t \to \infty$ in A.2, we find that the steady state value of capital is given by $K* = (B/\delta)^{1/(1-\gamma)}$; this value is approached as $t \to \infty$ regardless of the initial value of capital. If there is a constant temperature rise of T C above pre-industrial levels, then the steady state value becomes $K* = (\Omega B/\delta)^{1/(1-\gamma)}$, where $\Omega(T) = 1/(1 + 0.0028388T^2)$. Steady state output is then $Q*(t) = \Omega K^{*\gamma} AN^{1-\gamma}$. For linear temperature increase up to T in $t = 200$ years, we multiply the integrand in (A.2) by $\Omega(Tt/200)$.

Bibliography

Ackerman, F., DeCanio, S. J., et al. (2009b). Limitations of integrated assessment models of climate change. *Climatic Change*, *95*(3-4), 297–315, doi: 10.1007/s10584-009-9570-x.

Ackerman, F., Stanton, E. A., & Bueno, R. (2009a). Fat Tails, Exponents and Extreme Uncertainty: Simulating Catastrophe in DICE. Tech. Rep. SEI Working Paper WP-US-0901, Stockholm Environment Institute-U.S. Center, Somerville, MA, URL http://sei-us.org/publications/id/40.

Anthoff, D. & Tol, R. S. J. (2008). The climate framework for uncertainty, negotiation and distribution. technical description 3.3, FUND.

Barnett, J. & Adger, W. N. (2003). Climate dangers and atoll countries. *Climatic Change*, *61*(3), 321–337, doi:10.1023/B:CLIM.0000004559.08755.88.

Barrett, S. (2008). The incredible economics of Geo-engineering. *Environmental & Resource Economics*, *39*(1), 45–54.

Barro, R. J. & Sala-i Martin, X. (1999). *Economic Growth*. Cambridge, Ma: MIT Press.

Boyce, D. G., Lewis, M. R., & Worm, B. (2010). Global phytoplankton decline over the past century. *Nature*, *466*(7306), 591–596, doi:10.1038/nature09268.

Cooke, R. M. & Goossens, L. H. J. (2000). Probabilistic Accident Consequence Uncertainty Assessment. Procedures Guide for Structured Expert Judgement EUR-18821, European Commission, Directorate-General for Research.

Cooke, R. M. & Kelly, G. N. (2010). Climate Change Uncertainty Quantification: Lessons Learned from the Joint EU-USNRC Project on Uncertainty Analysis of Probabilistic Accident Consequence Code. Discussion Paper 10-29, Resources for the Future, Washington, D.C.

Dell, M., Jones, B. F., & Olken, B. A. (2009). Temperature and income: Reconciling new cross-sectional and panel estimates. *American Economic Review*, *99*(2), 198–204, doi:10.1257/aer.99.2.198.

Dessai, S., Adger, W. N., et al. (2004). Defining and experiencing dangerous climate change. *Climatic Change*, *64*(1-2), 11–25, doi: 10.1023/B:CLIM.0000024781.48904.45.

Diamond, J. (2005). *Collapse*. London: Penguine Books.

Fankhauser, S. & Tol, R. S. J. (2005). On climate change and economic growth. *Resource and Energy Economics, 27*(1), 1–17.

Gingerich, P. D. (2006). Environment and evolution through the paleocene–eocene thermal maximum. *Trends in Ecology & Evolution, 21*(5), 246–253, doi:10.1016/j.tree.2006.03.006.

Hall, D. C. & Behl, R. J. (2006). Integrating economic analysis and the science of climate instability. *Ecological Economics, 57*(3), 442–465, doi: 10.1016/j.ecolecon.2005.05.001.

Hansen, J. E. (2005). A slippery slope: How much global warming constitutes "dangerous anthropogenic interference"? *Climatic Change, 68*(3), 269–279, doi:10.1007/s10584-005-4135-0.

Harper, F. T., Goossens, L. H. J., et al. (1995). Probabilistic accident consequence uncertainty study: Dispersion and deposition uncertainty assessment. Tech. Rep. NUREG/CR-6244, EUR 15855 EN, SAND94-1453, U.S. Nuclear Regulatory Commission and Commission of European Communities, Washington (USA) and Brussels-Luxembourg.

Kann, A. & Weyant, J. P. (2000). Approaches for performing uncertainty analysis in large-scale energy/economic policy models. *Environmental Modeling & Assessment, 5*(1), 29–46, doi:10.1023/A:1019041023520.

Keller, K., Bolker, B. M., & Bradford, D. F. (2004). Uncertain climate thresholds and optimal economic growth. *Journal of Environmental Economics and Management, 48*(1), 723–741, doi:10.1016/j.jeem.2003.10.003.

Kelly, D. L. & Kolstad, C. D. (2001). Malthus and climate change: Betting on a stable population. *Journal of Environmental Economics and Management, 41*(2), 135–161, doi:10.1006/jeem.2000.1130.

Kraan, B. & Bedford, T. (2005). Probabilistic inversion of expert judgments in the quantification of model uncertainty. *Management Science, 51*(6), 995–1006.

Kurowicka, D. & Cooke, R. (2006). *Uncertainty Analysis with High Dimensional Dependence Modelling*. Wiley.

Lempert, R. J., Schlesinger, M. E., & Bankes, S. C. (1996). When we don't know the costs or the benefits: Adaptive strategies for abating climate change. *Climatic Change, 33*(2), 235–274, doi:10.1007/BF00140248.

Manne, A. S. & Richels, R. (1995). The greenhouse debate: Econonmic efficiency, burden sharing and hedging strategies. *The Energy Journal, 16*(4), 1–38.

Mastrandrea, M. D. & Schneider, S. H. (2004). Probabilistic integrated assessment of "dangerous" climate change. *Science*, *304*(5670), 571–575, doi: 10.1126/science.1094147.

McInerney, D. & Keller, K. (2008). Economically optimal risk reduction strategies in the face of uncertain climate thresholds. *Climatic Change*, *91*(1–2), 29–41, doi:10.1007/s10584-006-9137-z.

McNeil, A. J., Frey, R., & Embrechts, P. (2005). *Quantitative Risk Management: Concepts, Techniques, and Tools*. Princeton: Princeton University Press.

Nordhaus, W. D. (1994). Expert opinion on climatic change. *American Scientist*, *82*(1), 45–52.

Nordhaus, W. D. (1999). The economic impacts of abrupt climatic change. In *Meeting on Abrupt Climate Change: The Role of Oceans, Atmosphere, and the Polar Regions*, National Research Council.

Nordhaus, W. D. (2006). Geography and macroeconomics: New data and new findings. *Proceedings of the National Academy of Sciences of the United States of America*, *103*(10), 3510–3517, doi:10.1073/pnas.0509842103.

Nordhaus, W. D. (2008). *A question of Balance : Weighing the Options on Global Warming Policies*. New Haven: Yale University Press.

Nordhaus, W. D., Azam, Q., et al. (2006). The G-Econ Database on Gridded Output: Methods and Data. Tech. rep., Yale University, New Heaven.

O'Neill, B. C. & Oppenheimer, M. (2002). Dangerous climate impacts and the Kyoto Protocol. *Science*, *296*(5575), 1971–1972, doi:10.1126/science.1071238.

Oppenheimer, M. (2005). Defining dangerous anthropogenic interference: The role of science, the limits of science. *Risk Analysis*, *25*(6), 1399–1407.

Parry, M., Arnell, N., et al. (2001). Millions at risk: Defining critical climate threats and targets. *Global Environmental Change*, *11*(3), 181–183.

Peck, S. C. & Teisberg, T. J. (1995). International CO2 emissions control: An analysis using CETA. *Energy Policy*, *23*(4–5), 297–308.

Pizer, W. A. (1999). The optimal choice of climate change policy in the presence of uncertainty. *Resource and Energy Economics*, *21*(3–4), 255–287.

Romer, D. (2006). *Advanced Macroeconomics*. Boston: McGraw Hill, 3rd edn.

RPD (2000). Radiation Protection and Dosimetry: Expert Judgement and Accident Consequence Uncertainty Analysis. Special Issue.

Savage, L. J. (1972). *Foundations of Statistics*. New York: Dover, 2nd edn., 1st edn. Wiley, 1954.

Sherwood, S. C. & Huber, M. (2010). An adaptability limit to climate change due to heat stress. *Proceedings of the National Academy of Sciences*, *107*(21), 9552–9555, doi:10.1073/pnas.0913352107.

Stern, N. (2008). The economics of climate change. *American Economic Review*, *98*(2), 1–37, doi:10.1257/aer.98.2.1.

Stern, N. (2009). *Imperfections in the Economics of Public Policy, Imperfections in Markets, and Climate Change, Nota di lavoro/Fondazione Eni Enrico Mattei*, vol. 367. Milano: Fondazione Eni Enrico Mattei.

Stern, N. & GB Treasury (2006). *Stern Review : The economics of climate change*. London: HM Treasury.

Sterner, T. & Persson, U. M. (2008). An even sterner review: Introducing relative prices into the discounting debate. *Review of Environmental Economics and Policy*, *2*(1), 61–76, doi:10.1093/reep/rem024.

WDI (2014). World Bank World Development Indicators. URL http://data.worldbank.org/data-catalog/world-development-indicators.

Weitzman, M. (2009). On modeling and interpreting the economics of catastrophic climate change. *Review of Economics and Statistics*, *91*(1), 1–19, doi: 10.1162/rest.91.1.1.

Weitzman, M. (2010). *GHG Targets as Insurance Against Catastrophic Climate Damages*. Cambridge, MA: National Bureau of Economic Research.

Roger M. Cooke
Resources for the Future and Department of Mathematics
Delft University of Technology
The Netherlands
cooke@rff.org

Knowledge and the Creation of Physical Phenomena and Technical Artefacts

PETER KROES

1 Introduction

Most of the phenomena studied in physical laboratories first have to be created, since they do not occur spontaneously.[1] Indeed, as Hacking (1983) has pointed out in his book *Representing and Intervening,* nowadays there is much more to studying the physical world than simply observing it; in contrast to the way Columbus discovered America, the physical phenomena of interest are not 'lying around' to be observed and discovered but have to be brought about by the active intervention of the scientist in experiments. Elsewhere, I have argued that this does not mean that the creation of physical phenomena is similar to the creation of technical artefacts (Kroes 2003). The creation of physical phenomena involves a *weak* form of creation: none of the properties of the physical phenomenon itself is created in the sense that it is the result of a deliberate design decision on the part of the creator. What is designed by the experimenter are the conditions under which (the "theatre") the physical phenomenon occurs. This is not the case for the creation of technical artefacts; they are created in a *strong* sense, that is, many properties of the object created are intentionally designed into it by its creator. This difference between weak and strong creation is reflected in current patent law; only human inventions are eligible for a patent. In spite of the fact that physical phenomena are created, they are not patentable because they themselves are not human inventions. Only technical artefacts are true inventions or true creations of the human mind and as such qualify for a form of protection through intellectual property rights.

In this paper I will further elaborate on this difference between the weak creation of physical phenomena and the strong creation of technical artefacts

---

[1] I thank Maarten Franssen for his comments on an earlier version of this paper.

and I will analyse the role of knowledge about the object of creation in these creation processes. More in particular I will argue that knowledge about what is created plays a different role in the two cases. The creation of physical phenomena is possible without detailed knowledge about what kind of phenomenon is created; further research may be necessary to discover what kind of phenomenon has been created. Building upon Thomasson's theory of artefact kinds, I will argue that this is not the case for the creation of technical artefacts; it is not possible to create an instance of a technical artefact kind without knowing what kind of technical artefact is created (Thomasson 2003; 2007). In order to analyse the role of knowledge (about what is created) in these two kinds of creation processes I will start with analysing and comparing the kinds of knowledge involved in representations of physical objects and technical artefacts (section 2).[2] My conclusion will be that from a representational point of view there are no significant epistemic differences between representations of physical phenomena and of technical artefacts. Then I will turn to an analysis and comparison of the creation of physical phenomena and technical artefacts (section 3). By then, the stage will be set to compare the role of knowledge about the object created in the creation of physical objects and technical artefacts (section 4). The final section summarises the results.

2 Representations of physical objects and technical artefacts

In order to prepare the ground for our analysis of the *role* of knowledge about the object created in creating physical objects and technical artefacts, I will first look at whether there are any significant differences in our knowledge of physical objects and technical artefacts. More in particular I will compare how physical objects and technical artefacts are represented. So, in this section we will look at physical objects and technical artefacts from an epistemic-representational perspective or stance. In order to avoid misunderstanding, it is important to point out from the beginning an ambiguity in the notion of the representation of an object. It may refer to the activity of representing that object or to the result of that activity, namely *a* representation. Here I will focus on the latter meaning of representation, so I will not enter into a discussion of the role of knowledge and whether different kinds of knowledge, know how or skills are involved in making representations of physical and technical objects.

---

[2]Note that I will concentrate on the creation of a particular kind of physical phenomena, namely of physical objects; the reason for this is that it makes the comparison to technical artefacts easier. The following analysis, however, may be extended to the creation of physical processes and their comparison to the creation of technical processes.

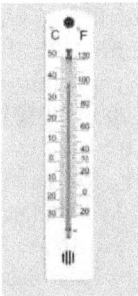

Figure 1. A picture of a material thing

Are there any epistemic differences between a representation of a physical object and a representation of a technical artefact? Consider the picture of Figure 1. It is a picture of a material thing that, depending on whether it is approached from a physical or design stance (Dennett 1987), may be described or represented as a physical object or as a technical artefact. From a physical stance the thing is represented as a *physical* object, i.e., as a thing with only physical features, such as mass, geometrical form, thermal and electrical properties et cetera. As a physical object it does not make sense to say that the object is represented upside down, because as a physical object the thing has no preferred direction (of representation). Similarly, all numbers and marks on the thing (such as the '° C' and '° F' mark) have from a physical point of view no symbolic meaning at all; they are nothing more than contingent configurations of particular physical substances. From the design stance the same thing (the notion 'same' is to be interpreted in an ostensive way) is represented as a *technical artefact* that on top of its physical features also has functional features. As a technical artefact the thing is a physical object (construction) that is based on a human design and has a function, namely to measure temperature. Moreover, various parts of the technical artefact have sub-functions; for instance, the sub-function of the glass bulb containing the red fluid is to conduct heat from the environment to the fluid. The marks are not just arbitrary configurations of particular physical substances; they have symbolic functions. They represent the Fahrenheit and Celsius scales and have been put alongside the glass tube to make it easy to read off the temperature. Whereas from a physical stance the form of the glass tube containing the fluid is a purely contingent matter for which it does not make sense to search for a physical explanation, this is not so from a design stance. As a designed object with a function, certain features of the glass tube may be explained with reference to this function.

Finally, as a technical artefact the thing is represented upside down, because this representation is not in accordance with how this thing as a technical artefact is to be used normally; as a technical artefact the thing comes with a use plan (Houkes & Vermaas 2002; 2010), a notion that does not make sense if the thing is represented from a physical stance.

So from an epistemic point of view the main difference between representing a thing as a physical object and as a technical artefact concerns the representation of functional features. The description of a thing as a technical artefact requires on top of the description of its physical features also a description of its functional features. Since knowledge of these functional features may be as objective as knowledge of its physical features (Searle 1995) this appears not to introduce any significant epistemic differences. This is in line with the widespread and usually implicit assumption that representations of biological organisms, which also involve reference to functional features, do not, because of this reference, introduce any epistemic differences between the physical and biological sciences. If we assume furthermore that this reference to functional features does not introduce epistemic differences when it comes to *making* representations of a thing as a physical object or as a technical artefact, which seems rather plausible, then we end up with the conclusion that from a cognitive-representational point of view there are no epistemic differences between representing, and representations of, a thing as a physical object or as a technical artefact.[3]

3 The creation of physical objects and technical artefacts

In this section I shift my attention from knowledge and representations of physical objects and technical artefacts to creating such objects. I have to pause for a moment on the question what it means to create a physical object or a technical artefact, before I can turn in the next section to an analysis of the role of knowledge about the object of creation in these processes. To start with, note that both an agent representing a physical object and an agent creating a physical object or a technical artefact are creating something. The representing agent creates something abstract, a representation, and the creating agent creates a concrete physical or technical object. Now, the representing agent can truly claim to be the *author* of the abstract representation (Hilpinen 1993), but not of the object repre-

---

[3]This conclusion does not imply that there are no significant differences between the *formal* representation of knowledge of physical and functional features; if indeed functional features of technical artefacts are mind-dependent, as I assume, in contrast to the intrinsic nature of physical features, the formal representation of technical functions may turn out to be much more complicated; for more details, see Kroes (2010).

sented. In general, I will consider an agent to be the author of an object iff the agent is its maker or creator and the object created bears the agent's marks. The expression "bears the agent's marks" means minimally that certain of the features of the created object are the result of (design) decisions by the agent; in other words, the agent could have decided to make certain of its features otherwise. In this sense, Czanne is certainly the author of his paintings. In general, however, the marks do not have to be so specific that it is possible, on the basis of the marks, to trace the individual who made it.

In a representational stance, whether it is a physical or a design stance, the agent is essentially an observer (a 'spectator') with regard to the object represented and has only cognitive interests (we abstract from whatever other interests the agent may have in terms of using the knowledge gained for practical purposes). The cognitive agent is not by definition a passive observer since (s)he may take the thing apart and perform all kinds of experiments in order to arrive at a reliable representation. Nevertheless (s)he is a passive observer in the sense that the object to be represented is given and whatever s(he) does to that object in order to arrive at an adequate representation of it may not leave any trace in the representation of that object (Tiles 1992; Lelas 1993).[4] So the representing agent is not the author of the object represented. With regard to the creation of physical objects the situation between the creating agent and the object created is, *pace* Hacking, similar to the one between the representing agent and the object represented; the creating agent is not the author of the object created. If, for example, an experimental physicist succeeds in creating an electron-positron pair from a gamma ray, then this physicist can hardly be called the author of this electron or positron. If these objects, qua physical objects, would be truly his creations, then this fact would have to show up in an adequate representation of this electron-positron pair. However, all electrons and positrons have the same physical properties and exhibit the same physical behaviour. Neither the particular physical equipment with which this pair of objects was created, nor the experimentalist who performed the experiment leaves any marks on the objects created. From a physical point of view the history of the created objects (by whom and under what conditions they were created) does not matter at all; the creating agent as well as the experimental machinery are all transparent as far as the object created is concerned (Tiles 1992; Lelas 1993). Here we are dealing with a case of *weak* creation; the physicist creates the conditions

---

[4]Of course, this does not mean that the representation does not contain any conventional elements; any representation does, if only due to the chosen mode of representation (linguistic, graphic et cetera).

under which physical objects show themselves spontaneously and their behaviour may be studied, but none of the properties of the created object itself is due to (design) decisions of creating agent. Thus, the experimental physicist cannot claim to be the author of the physical objects created. This may be a valid conclusion when it comes to the creation of elementary particles such as electrons or positrons, but what about an experimental physicist who creates a particular multi-layered sample in order to study its physical behaviour? This sample is clearly the outcome of her (design) decisions: she could have decided to add an extra layer or to change the dimensions of one or more layers. Is this not an object created by her and of which she may rightly claim to be its author? Indeed, she may claim to be the author of this individual object, with its unique causal and intentional history; only this thing has that particular history in which she plays a role. But from a physical point of view this history does not matter and in so far she has merely created a *physical* object, that is, in so far this object is an instance of a particular *physical kind*, she cannot claim to be its author. As an instance of a particular physical kind, none of its features, just as in the case of an electron or a positron, is determined by (design) decisions of its maker. In case somewhere in the universe another instance of the same physical kind would occur naturally, it would show exactly the same physical behaviour. As a physical object (as an instance of a physical kind) it does not bear any marks of its creator and thus has no author. Again, from a physical point of view the causal and intentional history of the sample—why the experimentalist created it this way and not in another—is totally irrelevant. When it comes to the creation of instances of technical artefact kinds the relation between the creating agent and the object created changes radically. To see why, we have to take a closer look at what it means to be and to create an instance of a technical artefact kind. Taking my lead from the definition of an artefact kind in general by Thomasson (2003; 2007), I will define being an instance of a technical artefact kind in the following way (Kroes 2012b, chap. 4): An object x is an instance of the technical artefact kind K iff x is the result of a successful execution of a correct design of a K. Thus, whether an object is a technical artefact of kind K or not depends crucially on its history: if it has the appropriate *intentional* and *physical* history ("is the result of a successful execution of a correct design") it is a technical artefact of kind K. The creating agent must have had the intention to create an instance of a particular technical artefact kind on the basis of a correct design and must have executed that intention successfully. This means that a technical artefact is a mind-dependent object: the intentions of the creating agent are constitutive for being a technical artefact of a particular kind.[5] According to the above conception of technical artefacts, every technical artefact has necessarily an author.[6] Many of the features that define what kind of object is created are the result of explicit or implicit design decisions of its author. With regard to technical artefacts two different kinds of authors have to be distinguished, which I will refer to as the author-inventor and the author-copier. The author-inventor is the agent who creates the first instance of a new technical artefact kind; the author copier is an agent who creates a new instance of an already existing technical artefact kind.[7] As we will see later on, there is a significant difference with regard to the knowledge about the object created of the author-inventor and the author-copier. In contrast to the creation of physical objects, technical artefacts are created in a *strong* sense; they do have an author and are really human creations. Table 1 summarizes some of the similarities and differences between weakly and strongly created objects by comparing an Americium atom (a physical object) and a bicycle (a technical artefact) (for more details, see Kroes 2012a).

Americium Atom	Bicycle
Human-made physical object	Human-made physical object
Not occurring naturally on Earth	Not occurring naturally on Earth
No technical function by itself	Technical function by itself
Not based on intelligent design	Based on intelligent design
Natural object	Technical artefact
Weak creation	Strong creation

Table 1. Comparison of strong and weak creation

Having elucidated the different relationships between the creating agent and the object created in the case of the creation of a physical object and of a technical artefact, I will turn in the next section to an analysis of the role of knowledge about the object created in both cases.

---

[5] As well as its physical features; thus, technical artefacts have a dual nature (Kroes 2012b).

[6] An interesting question, which falls outside the scope of this article, is whether there are any significant differences between the authorship of abstract representations and the authorship of concrete technical artefacts.

[7] As we will see in more detail in the next section, the author-copier must have the intention to create another instance of an already existing technical artefact kind and not just a copy of the physical structure of an instance of that kind; in that case (s)he would be making just a physical object.

4 Knowledge and the creation of physical objects and technical artefacts

In section 2, I argued that there are no significant epistemic differences with regard to knowledge of physical objects and technical artefacts. Now another issue is at stake, namely the *role* of knowledge about the object to be created, be it a physical object or a technical artefact, in the process of creating it. In this respect there is a significant difference, or so I will argue.

Let us first have a look at knowledge and the creation of physical objects. When we think of attempts to create physical objects like the Higgs boson at CERN, then it is clear that the creation of the conditions for physical objects to 'show themselves' may involve the use of massive technology and all the specialized scientific and technological knowledge and forms of know-how that go with the use of that technology. In this respect there appear to be no real differences between the creation of physical objects and the creation of technical artefacts, since also the creation/production of modern-day sophisticated technical artefacts typically requires the use of massive technology. But here my concern is not the knowledge about (how to produce) the conditions under which the (spontaneous) creation of a physical object takes place, but the knowledge about the created object itself. It may often be the case that on theoretical grounds we are able to predict its physical features and its behaviour and that it is then up to the actual experiment to confirm or refute those predictions. Or it may be that the experiment is more explorative without clear ideas what its outcome may be. However that may be, let us concentrate on the situation in which the outcome of the experiment is a real surprise and that a new kind of particle (or phenomenon) appears to be created. In that case, we appear to have succeeded in creating a physical object without much detailed knowledge about it.

However, this raises the question of how much knowledge about the object created is necessary to substantiate the claim that a new kind of physical object has been created (discovered). What knowledge about the object created is minimally necessary to justify such a claim? This is a difficult issue by itself. I will confine myself here to the following remarks. First of all the creation process has to be stable and repeatable in order to exclude that we are dealing with an 'artefact' of the experimental set-up. Secondly, assuming that this condition is fulfilled, what seems minimally necessary is knowledge on the basis of which it is possible to identify a created object as an instance of the new kind. This knowledge may partly consist of knowledge of particular empirical features related to the object itself, but partly, and in extreme cases predominantly, of knowledge about how the object was created. So, also the causal history of how an object came into existence may figure in the identity criteria for the new kind of physical object (for this the stability and repeatability of the creation process is of course crucially important). Thus, it is possible to justify the claim that an instance of a new kind of physical object has been created without detailed knowledge of its specific physical features; it is then up to future research to uncover those features. Thus, in principle it is possible to create a physical object (or phenomenon) without knowing much about what it is that has been created.

The situation with regard to the creation of technical artefacts is markedly different. From our definition of a technical artefact kind it follows immediately that it is not possible to create an instance of a technical artefact kind K without knowledge of the design of a K, that is, without knowledge of the physical and functional features of an instance of the technical kind K. So, one must know what kind of technical artefact is being created in order to be able to create an instance of that kind. To clarify this in more detail I will consider three different cases:

1. an archaeologist who makes a copy of an object that he has come across and that he takes to be a technical artefact but of which he has no idea what its technical function and design is,

2. an author-copier who makes another instance of an already existing kind of technical artefact, and

3. the author-inventor who comes up with the first instance of a new technical artefact kind.

With regard to the archaeologist confronted with an unknown technical artefact the following may be remarked. If for whatever reason he decides to make a copy of that object the copy cannot be another instance of the unknown technical artefact kind. Since the archaeologist has access to the thing as a physical object, what he can do is to make a copy of the thing in so far it is a physical object. But then, irrespective of how close a physical copy it may be of the original artefact, the resulting object lacks the appropriate intentional history to qualify as another instance of the (unknown) technical artefact kind. Note that this archaeologist lacks any criterion for determining which physical features of the original are relevant for making another instance of that technical artefact kind, since he has no clue as to the function of the whole and of its parts. Because of this he might feel forced to make an exact copy close to or at the molecular level (which of course would be nonsensical had he known he was, for instance, making a copy of a steam engine, but not so had he known that he was dealing with a memory stick or some nano-device). Contrast this with the creation of another physical object of a yet unknown kind. As long as enough is known about the created object to be identified as an instance of the relevant physical kind, this is perfectly possible.

Let us turn to the author-copier, who decides to create another instance of an already existing technical artefact kind K. Assuming that she has enough knowledge of the design of the technical artefact kind and that she is able to execute this design successfully, she is creating another instance of that technical artefact kind. "Enough" here means that her knowledge of this kind of technical artefact is *grosso modo* comparable to the knowledge that the author-inventor had when she was creating the first instance of this technical kind. So the object created has the requisite physical and intentional history to be countenanced as a technical artefact of the kind K. *Prima facie* the only difference between this author-copier and the author-inventor is that the latter, by being the inventor of this artefact kind, was the first person ever to make an instance of it. But more is involved; an issue about vagueness pops up. Knowingly or not the author-copier may have a somewhat different design in mind than the author-inventor or may execute the design in a slightly different way. After all, it will be difficult for the inventor-copier to make an exact copy of the original as far as its physical features are concerned. Under what conditions is the copied artefact still a faithful copy of the original and thus another instance of the same technical artefact kind K? Clearly there must be room for minor changes in the design and for minor deviations in its execution. Suppose that the author-copier has the right design in mind and executes this design successfully except for one minor detail because of which the object made will not be able to perform its intended function. Has she then made a technical artefact, albeit a malfunctioning one (for instance, a coffee machine with a malfunctioning on/off switch that may easily be repaired) or has she made nothing more than a complicated physical structure (or a piece of junk)? Here we touch upon the problem of the identity criteria for technical artefact kinds. As Thomasson (2007) remarks, these are rather malleable. In order to allow for some measure of vagueness, our definition will have to be relaxed in the following way: An object x is an instance of the technical artefact kind K iff x is the result of a *largely* successful execution of a *largely* correct design of a K.[8] Of course, in order to put limits on when a technical artefact may be an instance of a particular artefact kind K, limits will have to be imposed on when an execution is largely successful or a design largely correct. I guess that these limits will be highly context sensitive. Note, by the way, that analogous problems about vagueness may occur with regard to physical kinds (e.g., geological formations, minerals, planets).

---

[8] See also Thomasson's definition of an artefact kind 2003.

Finally, we arrive at the author-inventor, the person who makes the first instance of a new technical artefact kind. As remarked by Thomasson (2007), she is in a special position with regard to her knowledge of the object that she creates. The author-copier could in principle be wrong about the kind of technical artefact she is making, because she may have a design in mind that is different from the design the author-inventor had in mind when creating the technical artefact kind. In that case she may actually be the author-inventor of a new kind of technical artefact instead of being an author-copier. The author-inventor, however, cannot be wrong about the kind of technical artefact she is creating, given that she meets the success criteria stated in the definition of creating a technical artefact kind. From an epistemic point of view she is in a special position with regard to the object she has created. She has privileged epistemic access to this object because it is her own creation. With this privileged epistemic position come her baptismal rights. She is entitled to give the new kind of technical artefact a name; this name stands for the necessary and sufficient conditions for being an instance of that kind. These conditions are to be derived from the success criteria stated in the definition. Of course, here again the issue of vagueness comes up.

When we turn from the author-inventor of a technical artefact kind to the creator-discoverer (i.e., the first creator of an instance) of a new kind of physical object a different picture of the epistemic status of the creator with regard to the object created emerges. The creator-inventor may be to a large extent ignorant or greatly wrong about what kind of physical object he has created. Further research may reveal that the object created has very unexpected properties. All of this, of course, does not prevent that the creator-inventor exercises his baptismal rights and gives a name to the newly created kind of physical object. However, in contrast to the naming of a new technical artefact kind by its author-inventor, this naming cannot be interpreted in terms of necessary and sufficient conditions, the reason being the different epistemic relation between the creator-inventor and the object created.

5 Summary

In order to analyse and compare the role of knowledge about the object to be created in creating physical objects and technical artefacts, I have first taken a brief look at the nature of knowledge about physical objects and technical artefacts. I have argued that from an epistemic-representational stance there are no differences. Although knowledge about technical artefacts implies, on top of knowledge of physical features, also knowledge of functional features, both kinds of knowledge can be expressed in propositional form.

Next I have turned to a creational stance, made a distinction between weak and strong creation and have argued that the creation of physical objects involves weak creation in contrast to the strong form of creation involved in the creation of technical artefacts. Taking my lead from Thomasson's conception of artefact kinds in general, I have proposed a definition of technical artefact kinds according to which an object is a technical artefact kind if and only if it has an appropriate physical and intentional history. I have shown that given this definition of technical artefacts it is not possible to create a technical artefact without detailed knowledge of the object to be created, in contrast to the creation of physical objects. It is possible to create a (new) physical object without much knowledge about that object itself. Thus, knowledge about the object to be created plays a different role in the creation of physical objects and of technical artefacts, and this difference in role is intimately related to the two different from of creation involved.

Bibliography

Dennett, D. C. (1987). *The Intentional Stance*. Cambridge, MA: The MIT Press.

Hacking, I. (1983). *Representing and Intervening; Introductory Topics in the Philosophy of Natural Science*. Cambridge: Cambridge University Press.

Hilpinen, R. (1993). Authors and artifacts. *Proceedings of the Aristotelian Society*, *93*, 155–178.

Houkes, W. & Vermaas, P. E. (2002). Design and use as plans: an action-theoretical account. *Design Studies*, *23*(3), 303–320, doi:10.1016/S0142-694X(01)00040-0.

Houkes, W. & Vermaas, P. E. (2010). *Technical Functions: on the Use and Design of Artefacts*. Dordrecht: Springer.

Kroes, P. (2003). Physics, experiments and the concept of nature. In *The Philosophy of Scientific Experimentation*, Radder, H., ed., Pittsburgh: University of Pittsburgh Press, 68–86.

Kroes, P. (2010). Formalization of technical functions: Why is that so difficult? In *Proceedings of TMCE 2010 Symposium*, Delft: Delft University of Technology, 155–166, ancona, Italy.

Kroes, P. (2012a). Nano-artefacts and the distinction between the natural and the artificial. In *Nanotechnologies: Towards a Shift in the Scale of Ethics?*, Kermisch, C. & Pinsart, M.-G., eds., Bruxelles-Fernelmont: E.M.E., 47–61.

Kroes, P. (2012b). *Technical Artefacts: Creations of Mind and Matter – A Philosopy of Engineering Design*. Dordrecht; New York: Springer.

Lelas, S. (1993). Science as technology. *British Journal for the Philosophy of Science*, *44*, 423–442, doi:10.1093/bjps/44.3.423.

Searle, J. (1995). *The Construction of Social Reality*. London: Penguin Books.

Thomasson, A. L. (2003). Realism and human kinds. *Philosophy and Phenomenological Research*, *67*(3), 580–609, doi:10.1111/j.1933-1592.2003.tb00309.x.

Thomasson, A. L. (2007). Artifacts and human concepts. In *Creations of the Mind: Essays on Artifacts and their Representations*, Laurence, S. & Margolis, E., eds., Oxford: Oxford University Press, 52–73.

Tiles, J. E. (1992). Experimental evidence vs. experimental practice? *British Journal for the Philosophy of Science*, *43*, 99–109.

Peter Kroes
Delft University of Technology
The Netherlands
p.a.kroes@tudelft.nl

Mathematics and the New Technologies Part I: Philosophical relevance of a changing culture of mathematics

BENEDIKT LÖWE[1]

1 Mathematics and the new technologies

Mathematicians use their computers every day: they write e-mails, download papers from preprint servers, upload their own research on the same servers, log in to online communities dealing with mathematics to ask questions, they typeset their own papers with the typesetting system LaTeX, etc. But is this use of the computer and the internet relevant for questions of philosophy of mathematics about the nature of mathematics, the relationship between mathematics and the physical world, or the epistemic status of mathematical knowledge?

The traditional answer to this question is: *Not at all.* Traditionally, mathematics is seen as the paradigmatic deductive science endowed with aprioricity and a characteristic lack of spatial or temporal location of its truthmakers. One of the traditional claims is that while the mathematical discipline is a social and historical product, the underlying mathematics itself (and this is all that matters philosophically, for a traditionalist) does not depend on the way it was socially and historically produced.

The new technologies clearly have a formidable and undeniable effect on the research experience of mathematicians (such as the wide availability of papers via the internet, the communication speed and possibility of remote collaboration by the use of e-mail and visual remote connections, computer proof assistants and automated theorem provers, online crowd-sourcing of mathematical ability in order to solve open problems), but according to

---

[1] The author should like to thank the programme committee of CLMPS XIV for inviting him as the chair of the special symposium on *Mathematics and the new technologies*. The author acknowledges the financial support of the European Science Foundation as part of Networking Activity 359 in the EuroCoRes programme LogICCC.

the traditionalist, this effect does not touch the philosophical aspects of mathematics.

However, in recent years, a movement called *Philosophy of Mathematical Practice* has staged a revolt against the traditionalist view. The view that mathematics should be seen as a human cultural product is not new: we find it in books like (Lakatos 1976) and (Davis & Hersh 1971), and more recently in (Hersh 1997) or (Ernest 1998). However, until ten years ago, it was seen as a *maverick* position in the philosophy of mathematics; now it represents a growing part of the philosophy of mathematics community.[1] Philosophers of mathematical practice observe that a number of philosophical statements about mathematics are either empirical statements about mathematicians or at least depend crucially on such statements. As a consequence, any philosophical position that believes in the interplay between the practice of the field studied and its philosophy, cannot ignore the fact that mathematics is a human cultural activity.

Philosophy of mathematical practice is not a homogeneous movement and does not correspond to a uniform philosophical position. For the purposes of this paper, we understand the term "philosophy of mathematical practice" to refer to the meta-philosophical stance that empirical facts of mathematics as practiced can affect philosophical questions and their answers in a philosophically relevant way.

From such a meta-philosophical position, the mentioned "formidable effect" of the new technologies on the research experience of mathematicians might also affect the philosophy of mathematics. On the other hand, even the philosopher of mathematical practice will concede that not every effect on research practice is philosophically relevant. The modern mathematician writes e-mails where Gauss wrote letters; the modern mathematician controls the typography of her papers much more than a mathematician half a century ago, but is also constrained by the rules of the universal typesetting system. Are these changes relevant for philosophy of mathematics? Or, to make the question even more extreme, if a new restaurant is built next to the mathematics department that enables researchers to have dinner and return to their offices to prove more theorems, this restaurant has an effect on their research experience. But is that new restaurant part of the story that the philosophy of mathematics needs or wants to unravel?

Clearly, not all effects of the use of new technologies are philosophically relevant, but in this tripartite paper, we are aiming to show that some of

---

[1]This is best witnessed by a series of proceedings volumes of related conferences (Van Kerkhove & Van Bendegem 2007; Van Kerkhove et al. 2010; Löwe & Müller 2010; François et al. 2011) and the foundation of the *Association for the Philosophy of Mathematical Practice* in 2009. An overview of the motivation behind philosophy of mathematical practice can be found in (Buldt et al. 2008).

them are clearly involved with some of the traditional questions of philosophy of mathematics, in particular the epistemology of mathematics. The three papers correspond to three of the four talks given in the special invited symposium *Mathematics and the New Technologies* at the *Congress for Logic, Methodology and Philosophy of Science* in Nancy on 22 July 2011. Part II (Koepke 2014) corresponds to Peter Koepke's talk entitled *Formal mathematics and mathematical practice*, and part III (Van Bendegem 2014) corresponds to Jean Paul Van Bendegem's talk entitled *Mathematics in the cloud: the web of proofs.*

The following section and Koepke's part II will highlight the effect that automated theorem provers and proof assistants have on the practice of assessing the correctness of mathematical arguments; Van Bendegem's part III will then move to the other side of mathematical epistemology, the *context of discovery* and the use of new technologies in the process of producing new mathematics.

2 A problem in the epistemology of mathematics

Philosophers of mathematics are interested in the status of mathematics as an *epistemic exception* with a type of knowledge being categorically more secure than that of other sciences (Heintz 2000; Prediger 2006). At the other end of the epistemological spectrum, we have the whimsical *knowledge by testimony*, considered *epistemologically vulnerable*.[2] And yet, mathematicians in practice often use knowledge by testimony when they use results from research papers without checking their proofs in detail. How can the epistemic exception of mathematics survive if some of the proofs rely on pointers to the literature? A simple and naïve answer to both questions would be that the deductive nature of mathematics allows referees to check correctness of the proofs of published papers with absolute certainty, and thus the written codification of mathematical knowledge is certain knowledge, relieving us from any qualms about referring to it. However this is very far from the truth; in his opinion piece published in the *Notices of the American Mathematical Society*, (Nathanson 2008) paints a dark picture of the mathematical refereeing process:

> Many (I think most) papers in most refereed journals are not refereed. There is a presumptive referee who looks at the paper, reads the introduction and the statement of the results, glances at the proofs, and, if everything seems okay, recommends publication. Some referees check proofs line-by-line, but many do not.

---

[2] For more details on the epistemological problem of testimony, cf. (Adler 2012).

> When I read a journal article, I often find mistakes. Whether I can fix them is irrelevant. The literature is unreliable.

The mathematical peer review process is lamentably understudied. Geist *et al.* (2010) give a description of the level of scrutiny involved in the peer review process and present two (rather preliminary) empirical studies: while, ideally, referee reports "should address Littlewood's three precepts: (1) Is it new? (2) Is it correct? (3) Is it surprising?" (Krantz 1997, 125), in practice, the level of detail of referee reports varies a lot. Among other things, the results in (Geist *et al.* 2010) show that the level of detail in which mathematical correctness is checked during the peer review process does not at all support the naïve view sketched above. In a survey of mathematical journal editors, only about half of them thought that it is the task of the referee to check the correctness of all proofs. In fact, mathematicians seem to have an almost stochastic view of the correctness checking in the peer review process:

> "Refereed proof" [is not the last word on correctness]: it just means that somebody has seen the paper, and if it is done correctly, he actually went through the proofs, and he believes that it is true, and this is very much biased by the human factor. Let's say [a famous mathematician] comes up with a paper, and I have to referee it, and then Im already preoccupied with the fact that [he] is a very well known mathematician, and so that it probably will be OK. And then there's the time pressure: you have all this stuff that you have to do, and then they ask you to review this 50 page paper, and you are sure that if you are really going to check all the details then you'll reach the conclusion "that's probably OK". You have a tendency to believe that the proofs are correct, and in addition you think "Well, he's publishing it, not I, so it's his responsibility that it is OK". Ideally, this referee has nothing else to do, he knows the subject better than the guy who wrote about it, and he will study it, and say "Yes, this is all correct". So, if the author thinks it's OK, then—let's be pessimistic—it has a probability of 95% to be OK. And so, if the referee checked it and and also thinks it's OK, then this also has a 95% chance of being correct, and so you have a very large probability that it is fine. And that is basically how it works. It's never going to be "full proof". I don't think that this exists.

> [If a] paper was sent to a mathematical journal of high reputation, so, say, *Acta Mathematica*; this tells us something about

the size of the mathematical community involved. That it went to a good journal means that the journal thought of looking for good referees, so it was established more surely than that it would have been sent to the journal of a tiny mathematical society with very few members. This puts the scenario in a framework which makes it very likely that the result is correct.³

Weber & Mejia-Ramos (2011) investigate the techniques that mathematicians use to convince themselves that a proof is correct, and find that they are mostly heuristic techniques as is exemplified in the following quote from one of their test subjects:

> [To understand a proof] means to understand how each step followed from the previous one. I don always do this, even when I referee. I simply don always have time to look over all the details of every proof in every paper that I read. When I read the theorem, I think, is this theorem likely to be true and what does the author need to show to prove it true. And then I find the big idea of the proof and see if it will work. If the big idea works, if the key idea makes sense, probably the rest of the details of the proof are going to work too.⁴

3 The effect of the new technologies on the epistemological question

If the proof checking of human experts is considered so unreliable and just a matter of minimizing the chances that errors are missed, this opens the field for computer-checked proofs. The topic of automated theorem provers in the philosophical literature is mostly discussed as an additional epistemological issue, e.g., in the context of the computer-assisted proof of the four colour theorem (Tymoczko 1979): the elimination of the human expert seems (at least for some philosophers) to reduce our trust in the correctness of the proof. Turning this argument on its head, one could think of replacing the untrustworthy human expert (who has "all this stuff that [he has] to do [and has] a tendency of believe that the proofs are correct") by a more trustworthy machine.

In the setting of *Philosophy of Mathematical Practice*, what would we have to show in order to prove that something has an effect on the philosophy

---

³These statements are from Eva Müller-Hill's interviews with a research mathematician *Interviewpartner 6*: (Müller-Hill 2010, 342–343, 345). The statements are not actual quotations, but text based on the interview transcript and transformed into full sentences. The second paragraph is quoted from (Geist *et al.* 2010, 162–164).

⁴Test subject *M5* quoted from (Weber & Mejia-Ramos 2011).

of mathematics? If there is a believable scenario changing some features of current mathematical practice in which philosophical papers written nowadays would have to be substantially updated in order to meet the standards of philosophical discourse, this could be seen as sufficient to argue that the changes have an effect on the philosophy of mathematics.

Koepke (2014, 7) discusses the potential effect that automated proof checking has on mathematical practice, including the possibility of a new social publishing norm that requires mathematicians to submit a formalized proof with their paper (cf. also (Miller 2014) in this volume). In the following, we shall consider the following *Gedankenexperiment*: let us suppose that at some point in future, mathematicians have universally accepted that every proof has to be submitted with an attachment of a formalized proof in a regimented natural language (such as the language of the *Naproche* system, mentioned in (Koepke 2014, 6.2)) in order to be considered for publication. Correctness is then automatically checked, and the task of the referee focusses on assessing whether the paper is interesting and new. Whether we believe that this is likely to happen or not, is immaterial:[5] all that matters is that it is a possible scenario with a substantially changed culture of mathematical practice.

Let us give two examples to establish that the mentioned scenario has philosophical consequences in the above sense:

The first is the question of unreliable testimony in mathematical epistemology. As discussed in 2, the epistemologist of mathematics has to deal with a major headache: on the one hand, philosophers and mathematicians alike claim that there is an epistemic quality to mathematical knowledge that makes it more reliable than knowledge acquired by the method of induction in other sciences; on the other hand, we see a heuristic practice of checking correctness that defies the firm belief in the objectivity of mathematical knowledge. This discrepancy requires an explanation, as long as the practice of proof checking remains as it is described in (Geist *et al.* 2010; Weber & Mejia-Ramos 2011). In the described possible scenario, the discrepancy would have been resolved, or at least been replaced with a substantially different question. In the possible future in which mathematicians relegate proof checking to machines, there might be other pressing epistemological issues, but the question raised in 2 would have to be rephrased.

Our second example deals with discussions about the philosophical position of formalism: some of the critics of formalism have focused on the fact that formal derivations are far removed from typical arguments given for mathematical correctness (Rav 1999; Buldt *et al.* 2008). According to this line of argument, any version of formalism that focusses on the for-

---

[5] In fact, the present author does not think that this scenario is very likely.

mal derivation as the main object witnessing correctness of a mathematical statement is criticized there are hardly any formal derivations, and due to the dissimilarity between proofs and derivations, it is difficult to see the former as approximations to the latter.[6] The development of bridging tools such as *Naproche* that allow human mathematicians to use a language very similar to natural mathematical language and translate this into a formal derivation will weaken any such philosophical arguments that would then have to be reconsidered. Tanswell (2012) argues that *Naproche*, if fully developed, might reopen some of the discussions about formalism and allow philosophers to redefine formalism as a philosophical position in line with these new developments, and offer novel defenses for such a renewed position.

These two examples show that the effect of the new technologies on mathematical practice is not the equivalent of the new restaurant built next to the mathematics department, but offers genuinely new vistas in the philosophical landscape.

Bibliography

Adler, J. (2012). Epistemological problems of testimony. In *The Stanford Encyclopedia of Philosophy*, Zalta, E. N., ed., Fall 2012 edition.

Buldt, B., Löwe, B., & Müller, T. (2008). Towards a new epistemology of mathematics. *Erkenntnis*, *68*, 309—329.

Davis, P. J. & Hersh, R. (1971). *The Mathematical Experience*. Boston: Birkhäuser.

Ernest, P. (1998). *Social Constructivism as a Philosophy of Mathematics*. Albany: State University of New York Press.

François, K., Löwe, B., et al. (Eds.) (2011). *Foundations of the Formal Sciences VII, Bringing together Philosophy and Sociology of Science, Studies in Logic*, vol. 32. London: College Publications.

Geist, C., Löwe, B., & Van Kerkhove, B. (2010). Peer review and knowledge by testimony in mathematics. In *PhiMSAMP, Philosophy of Mathematics: Sociological Aspects and Mathematical Practice – Texts in Philosophy*, vol. 11, Löwe, B. & Müller, T., eds., London: College Publications, 155–178.

Heintz, B. (2000). *Die Innenwelt der Mathematik. Zur Kultur und Praxis einer beweisenden Disziplin*. Wien; New York: Springer.

---

[6] Also in the formal mathematics community, this has been seen as an obstacle for working towards the possible future of our *Gedankenexperiment*; cf. (Wiedijk 2007): "The other reason that there has not been much progress on the vision from the QED manifesto is that currently formalized mathematics does not resemble real mathematics at all. Formal proofs look like computer program source code."

Hersh, R. (1997). *What is Mathematics, Really?* New York: Oxford University Press.

Koepke, P. (2014). Mathematics and the new technologies, Part II: Computer-assisted formal mathematics and mathematical practice. In *Logic, Methodology and Philosophy of Science. Proceedings of the Fourteenth International Congress (Nancy)*, Schroeder-Heister, P., Hodges, W., et al., eds., Logic, Methodology and Philosophy of Science, London: College Publications, 409–426.

Krantz, S. G. (1997). *A Primer of Mathematical Writing. Being a Disquisition on having your ideas recorded, typeset, published, read, and appreciated.* Providence, RI: American Mathematical Society.

Lakatos, I. (1976). *Proofs and Refutations.* Cambridge; New York: Cambridge University Press.

Löwe, B. & Müller, T. (Eds.) (2010). *PhiMSAMP, Philosophy of Mathematics: Sociological Aspects and Mathematical Practice, Texts in Philosophy*, vol. 11. London: College Publications.

Miller, D. (2014). Communicating and trusting proofs: The case for foundational proof certificates. In *Logic, Methodology and Philosophy of Science. Proceedings of the Fourteenth International Congress (Nancy)*, Schroeder-Heister, P., Hodges, W., et al., eds., Logic, Methodology and Philosophy of Science, London: College Publications, 323–341.

Müller-Hill, E. (2010). *Die Rolle formalisierbarer Beweise für eine philosophische Theorie mathematischen Wissens.* Ph.D. thesis, Rheinische Friedrich-Wilhelms-Universität Bonn.

Nathanson, M. B. (2008). Desperately seeking mathematical truth. *Notices of the American Mathematical Society*, 55(7), 773.

Prediger, S. (2006). Mathematics—cultural product or epistemic exception? In *Foundations of the Formal Sciences IV, The History of the Concept of the Formal Sciences, Studies in Logic*, vol. 3, Löwe, B., Peckhaus, V., & Räsch, T., eds., London: College Publications, 217–232.

Rav, Y. (1999). Why do we prove theorems? *Philosophia Mathematica*, 7(1), 5–41.

Tanswell, F. S. (2012). *Proof and Prejudice: Why Formalising doesn't make you a Formalist.* Master's thesis, Universiteit van Amsterdam, ILLC Publications MoL-2012-07.

Tymoczko, T. (1979). The four-color problem and its philosophical significance. *Journal of Philosophy*, 76(2), 57–83.

Van Bendegem, J. P. (2014). Mathematics and the new technologies, Part III: The cloud and the web of proofs. In *Logic, Methodology and Philosophy of Science. Proceedings of the Fourteenth International Congress (Nancy)*, Schroeder-Heister, P., Hodges, W., *et al.*, eds., Logic, Methodology and Philosophy of Science, London: College Publications, 427–439.

Van Kerkhove, B., De Vuyst, J., & Van Bendegem, J. (Eds.) (2010). *Philosophical Perspectives on Mathematical Practices, Texts in Philosophy*, vol. 12. London: College Publications.

Van Kerkhove, B. & Van Bendegem, J. P. (Eds.) (2007). *Perspectives on Mathematical Practices, Bringing together Philosophy of Mathematics, Sociology of Mathematics, and Mathematics Education*. Logic, Epistemology and the Unity of Science, Dordrecht: Springer.

Weber, K. & Mejia-Ramos, J. P. (2011). Why and how mathematicians read proofs: an exploratory study. *Educational Studies in Mathematics, 76*(3), 329–344.

Wiedijk, F. (2007). The QED manifesto revisited. *Studies in Logic, Grammarand Rhetoric, 10*, 121–133.

Benedikt Löwe
Institute for Logic, Language and Computation
Universiteit van Amsterdam
Amsterdam
The Netherlands

Fachbereich Mathematik
Universität Hamburg
20146 Hamburg
Germany
b.loewe@uva.nl

Mathematics and the New Technologies Part II: Computer-Assisted Formal Mathematics and Mathematical Practice

PETER KOEPKE

1 Introduction

Formal mathematics denotes the programme to carry out all of (pure) mathematics in complete formality: to express notions and statements in a *symbolic language* and to prove statements by derivations in a *symbolic calculus*. Due to the complexities of full formalizations this programme was at first merely an attractive vision, going back to ideas like Gottfried Leibniz's *characteristica universalis* and *calculus ratiocinator*. It was *theoretically* vindicated by Kurt Gödel's completeness theorem (Gödel 1929). In recent years, however, formal mathematics is becoming *practically* feasible, using computer support and automatic theorem proving.

Formal mathematics harmonizes with philosophical standpoints that view mathematics as a deductive science, and in particular with *formalism*. Advances in formal mathematics provide a body of *actual formalizations*, as opposed to the theoretical *formalizability* usually considered in formalism. This may shift the balance between various philosophies of mathematics towards formalism. Advances will also provide proof checking and proving tools for the mathematical practitioner, and they will influence the mathematical practice.

So the argument between conventional philosophies of mathematics and the *Philosophy of Mathematical Practice* may be dependent on concrete answers to questions like: Which proofs can be generated automatically? Can ordinary mathematical proofs, or intelligent but limited modifications thereof, be checked automatically? Can one make the application of formal mathematics just as natural as the use of other mathematical software like computer algebra systems or the LaTeX typesetting software?

So before embarking on philosophical speculations we try to give an impression of the potential of formal mathematics by appraising its current state and likely midterm developments. After a general introduction, we list important formal mathematics systems, in which substantial mathematical results have been proved or proof-checked. These systems use input and output languages reminiscent of programming languages. We suggest to improve the *naturalness* of formal mathematics by using (*controlled*) *natural languages* instead. The exploratory systems SAD and Naproche implement some of these ideas.

We expect that by combining best methods from a variety of systems formal mathematics will become stronger and in particular acceptable and applicable in ordinary mathematical work. This will also have significant philosophical implications.

2 Formal mathematics

Formal mathematics emerged alongside formal logic and modern abstract mathematics. In *The Principles of Mathematics* (Russell 1938, Preface to the First Edition, v) Bertrand Russell enunciates the standpoint of *logicism*:

> [...] that all pure mathematics deals exclusively with concepts definable in terms of a very small number of fundamental logical concepts, and that all its propositions are deducible from a very small number of fundamental logical principles [...].

He then formulates the programme of *formal mathematics*, to be pursued in a subsequent volume (Russell 1938, Preface to the First Edition, p. vi):

> The second volume [...] will contain chains of deductions, from the premisses of symbolic logic through Arithmetic, finite and infinite, to Geometry, [...].

This programme was partially realized by A. N. Whitehead and Russell in *Principia Mathematica* (Whitehead & Russell 1910-1913). Gödel begins his article on the incompleteness theorems by describing the state of formal mathematics at the time (Gödel 1931, 144, translation: 145):

> Die Entwicklung der Mathematik in der Richtung zu größerer Exaktheit hat bekanntlich dazu geführt, daß weite Gebiete von ihr formalisiert wurden, in der Art, daß das Beweisen nach einigen wenigen mechanischen Regeln vollzogen werden kann. Die umfassendsten derzeit aufgestellten formalen Systeme sind das System der *Principia Mathematica (PM)* einerseits, das

Zermelo-Fraenkelsche (von J. v. Neumann weiter ausgebildete) Axiomensystem der Mengenlehre andererseits. Diese beiden Systeme sind so weit, daß alle heute in der Mathematik angewendeten Beweismethoden in ihnen formalisiert, d.h. auf einige wenige Axiome und Schlußregeln zurückgeführt sind.

> *The development of mathematics toward greater precision has led, as is well known, to the formalization of large tracts of it, so that one can prove any theorem using nothing but a few mechanical rules. The most comprehensive formal systems that have been set up hitherto are the system of* Principia mathematica *(PM) on the one hand and the Zermelo-Fraenkel axiom system of set theory (further developed by J. von Neumann) on the other. These two systems are so comprehensive that in them all methods of proof today used in mathematics are formalized, that is, reduced to a few axioms and rules of inference.*

First-order set theory and in particular the Zermelo-Fraenkel system ZFC with the axiom of choice is commonly accepted as the natural foundation of modern structure-orientated mathematics. There is a considerable degree of agreement between ontology and semantics since many basic notions are defined set-theoretically, e.g.:

> A *group* is a *set* together with [...].

By Gödel's completeness theorem (Gödel 1929) there is *complete* agreement between syntax and semantics: every proof can be replaced by a formal derivation (in set theory). These observations underpin the programme of formal mathematics: to *actually* produce formal derivations from informal proofs.

3 On the feasibility of formal mathematics

Principia Mathematica turned out to be a project of unexpected dimensions and difficulties. Only a small part of the intended matter could be covered. Russell wrote in his autobiography (Russell 1998, 155):

> [...] my intellect never recovered from the strain.

Nicolas Bourbaki who worked towards a complete and systematic exposition of mathematics claimed the unfeasibility of complete formalizations (Bourbaki 2004, 10, 11):

> [...] such a project is absolutely unrealizable: the tiniest proof at the beginnings of the Theory of Sets would already require

several hundreds of signs for its complete formalization. [...] formalized mathematics cannot in practice be written down in full [...].

But with the advent of electronic computers, the practical side of long repetitive tasks appeared in a different light. In 1962, John McCarthy wrote (McCarthy 1962):

> Checking mathematical proofs is potentially one of the most interesting and useful applications of automatic computers. Computers can check not only the proofs of new mathematical theorems but also proofs that complex engineering systems and computer programs meet their specifications. Proofs to be checked by computer may be briefer and easier to write than the informal proofs acceptable to mathematicians. This is because the computer can be asked to do much more work to check each step than a human is willing to do, and this permits longer and fewer steps. [...] The combination of proof-checking techniques with proof-finding heuristics will permit mathematicians to try out ideas for proofs that are still quite vague and may speed up mathematical research.

4 Practical systems for formal mathematics

McCarthy's prediction is being realized in formal mathematics. Since the 1950's there have been a number of formal mathematics systems, differing in purpose, techniques, and scope. Automatic theorem provers are intended to find formal deductions for hypotheses given to the system. There are general purpose automated theorem provers for arbitrary (first-order) statements, and specialized provers optimized for specific areas. It was soon realized that automated theorem provers were hardly able to match the abilities of expert mathematicians in finding successful strategies and constructions for proofs of non-trivial statements. This gave rise to systems where human users provide clues for the proof-finding algorithm, either in advance in some dedicated proof language or interactively.

In this section we briefly describe a selection of important formal mathematics systems which are geared towards wide coverage, ordinary mathematical argumentation, and proving prominent theorems. These systems require expert users to master their idiosyncratic languages and commands, and to understand the underlying logical and software mechanisms.

Automath (Automath) was a pioneering large-scale project in formal mathematics, begun in 1967 by Nicolaas de Bruijn. de Bruijn explained in (de Bruijn 1994, 215):

[...] the Automath project tries to bring communication with machines in harmony with the usual communication between people.

L. S. van Benthem Jutting (van Benthem Jutting 1977) demonstrated the applicability of Automath to substantial mathematical theories by transcribing the *Grundlagen der Analysis* of Edmund Landau (Landau 1930) into Automath. Automath contained many important ideas and techniques which were taken over by other projects.

Some parts of formal mathematics have developed in parallel with general computer science. So Automath employed a LISP-like input language, which by today's standards would hardly considered to be "readable".

The problem of "readability" in formal mathematics was addressed by the Mizar system (Mizar), which has been developed by Andrzej Trybulec since about 1975. The Mizar language is related to the ALGOL programming language and intends to capture several features of the common mathematical language. Moreover Mizar allows a more natural proof style by bridging "obvious" proof steps with the aid of an integrated automated prover. The system accepts simple transformations and deductions which are common in ordinary proofs without further justification. Most importantly, Mizar comprises a vast library of checked proof texts which can be used as lemmas for further proving. The library contains material from many fields of mathematics, including the Banach Fixed Point Theorem for compact spaces, Fermat's Little Theorem, the Fundamental Theorem of Algebra, the Fundamental Theorem of Arithmetic, the Gödel Completeness Theorem, the Jordan Curve Theorem, and many more.

Whereas Mizar uses a fixed first-order logic and Zermelo-Fraenkel set theory, the Isabelle project (Isabelle) initiated by Larry Paulson only has a minimal inbuilt logic and can be configured to work with different logics and background theories. One of the largest Isabelle formalizations is that of Gödel's theorem of the relative consistency of the axiom of choice by Paulson (Paulson 2003). Many other substantial theorems have been redone in Isabelle like the elementary proof of the Prime Number Theorem by Jeremy Avigad *et al.* (Avigad *et al.* 2007).

The system Coq (Coq) is built on type theory and intuitionistic logic. The most spectacular Coq formalizations are the proof of the Four Colour Theorem by G. Gonthier (Gonthier 2008), and, very recently, the Feit-Thompson theorem (Gonthier 2012) which is an important part of the classification of finite simple groups.

Higher order logic is the basis of the HOL Light system (HOL light) by John Harrison, in which Harrison has proved theorems like the Fundamental

Theorem of Calculus, Brouwer's Fixpoint Theorem, and the Prime Number Theorem, using an analytical proof.

5 Enhancing the naturalness of formal mathematics

Although formal mathematics theoretically has a universal potential, it has not yet entered mathematical practice. Freek Wiedijk (Wiedijk 2007) states:

> The other reason that there has not been much progress on the vision from the QED manifesto is that currently *formalized mathematics does not resemble real mathematics at all*. Formal proofs look like computer program source code.

An average mathematician does not use any of the existing formal mathematics systems since they do not go along with the usual, or "natural" mathematical experience.

The naturalness of mathematical texts depends on many factors which are related to human abilities and expectations in various areas. Fields of mathematics have developed their own sublanguages of the mathematical language with specific symbols, methods and implicit background assumptions. A text may be directed at an audience with a specific background knowledge and sophistication. These factors will also be appreciated differently by different individuals. So we can only discuss some general aspects of formal systems which affect naturalness.

5.1 Mathematical aspects

Mathematical theories strongly influence their style of presentation. Obviously a theory is more adequate for a natural formalization if it is highly formal anyway. If a theory is based on intuitively well-understood concepts from, e.g., geometry, physics, or social interaction, then the presentation tends to appeal to those intuitions in plain but linguistically involved natural language which may be difficult to analyze. If a theory is built up axiomatically or algebraically the development is usually more formal. In the course of unfolding a theory new intuitions evolve and are employed. So the beginnings of a theory will be more adequate for natural formalizations than advanced parts.

Mathematical texts combine logical arguments with numerical and symbolic computations. Up to now the techniques of formal mathematics have emphasized logical arguments, so one should prefer "logical" theories. Set theory in some appropriate axiomatization is a powerful system for the general formalization of mathematics, and has been used in several formalization projects, e.g., by Mizar.

5.2 Linguistic aspects

The language of mathematics combines natural language with mathematical formulas. Most natural language words and constructs retain their original meanings, but there are some exceptions and extensions. Through definitions, a word like "ring" may get a new, mathematical semantics, which is completely determined by a formal definition. The word, however, retains its standard grammar as a neuter noun with plural form "rings". Usually the choice of defined words is not completely arbitrary, but takes into account natural language intuitions, systematics, and conventions. Also completely new words, patterns of words, and phrases may be introduced.

Concerning the meanings of grammatical constructs, the standard mathematical language tries to be complete and unambiguous. Whereas the coordination with "or" is in natural language often understood as "either-or", the usual mathematical interpretation is the inclusive "or"; an exclusive "or" has to be made explicit by "either-or" and other means. The tendency to avoid ambiguities facilitates the linguistic analysis of the mathematical language.

Mathematical exactness requires an analysis of *every* sentence of a text. The analysis must be intelligable for a human author so that the author can keep control over the process. This necessitates the use of a grammar-based *deep linguistic analysis* instead of, e.g., stochastic methods.

A mathematical text is a *discourse* in the language of mathematics, i.e., a structured sequence of sentences. *Discourse representation theory* (see Kamp & Reyle 1993) provides means to transform a given discourse into a logical representation which retains important structural elements of the text like the scopes of certain constructs or the interdependencies of sentences through pronouns and other anaphora.

One is lead to the definition of *controlled natural languages* (CNL) which are subsets of the natural language of mathematics with a strict formal grammar and formal semantics. A powerful controlled languages with an associated computer implementation is the language *Attempto Controlled English* (ACE) which combines a rich "natural" language with mechanisms of interest for mathematical applications.

5.3 Internal representations

Attempto Controlled English translates input texts into discourse representation structures as an intermediate layer between natural input and its first-order equivalent. There are, however, aspects of proofs which standard discourse representation theory does not model properly, like the order of statements or the scope of assumptions. This motivates the introduction of *proof representation structures* (PRS) which are enriched discourse representation structures able to represent various argumentative and procedural aspects. PRS seem to be crucial data structures to connect natural and formal proofs.

A PRS should contain information on the visibility of relevant assumptions for every statement in the proof. Immediately preceding statements or distinguished main lemmas or theorems are the most probable and "visible" preconditions for a statement so that these should be attempted with higher priority for the proof of the current statement. A good design of visibility criteria can help the automated prover and make proofs more natural in the sense that "obvious" potential premises are selected by the system in a way similar to the tactics of a human prover.

5.4 Logical aspects

In principle all mathematical statements can be translated into first-order statements about sets and the membership-relation. Standard set-theoretic formalizations of mathematical notions like the coding of integers by von Neumann ordinals introduce exponential growth and may not be practically feasable. Therefore intermediate logics should be used which are close to the "natural logic" of mathematical input texts. This requires an efficient (weak) type system so that complex objects or notions can be atomic at some higher level of the type system. This was already described by Bourbaki (Bourbaki 2004, 10):

> [...] it is imperative to condense the formalized text by the introduction of a fairly large number of new words (called *abbreviating symbols*) and additional rules of syntax (called *deductive criteria*). By doing this we obtain languages which are much more manageable than the formalized language in its strict sense. Any mathematician will agree that these condensed languages can be considered as merely shorthand transcriptions of the original formalized language.

5.5 Automated theorem proving

Proofs come with a certain step size or *granularity* depending on the style of proof. Proof checking amounts to the justification of each proof step, either by the argumentative abilities of a human (expert) reader, or by interpolating proof steps by a formal derivation in case of automated proof checking. Ideally automated theorem provers (ATP) like Otter or Vampire should be able to interpolate proof steps of a natural granularity. Experiments with existing formal mathematics systems indicate that this is possible at least in certain contexts.

5.6 Typesetting

Mathematical texts stand out by the elaborate typography for formulas. Systems like TeX and LaTeX enable mathematicians to do mathematical typesetting without expert help. These systems have become *de facto* standards in mathematical publishing and can be considered "natural" formats for communicating mathematics. Natural formal mathematics should accept those formats.

6 Examples of natural formal mathematics systems

6.1 *System for Automated Deduction*: The SAD project

The SAD project (SAD) is based on a controlled natural language for mathematics called ForTheL (Formula Theory Language), which goes back to the 1960's and was further developed by Alexander Lyaletski, Andrei Paskevich, and Konstantin Verchinine (Verchinine *et al.* 2007). SAD is designed to approximate parts of common mathematical language and argumentation. Several frequent and useful phrases and methods of proof have been implemented with appropriate first-order semantics. The language includes a soft type system which is akin to the naive typing often found in mathematical texts. The proof checking process is devided into two layers: a *reasoner* attempts to identify inferences which to humans appear immediate or trivial; if the reasoner fails, the proof search is delegated to some automated theorem prover. Although SAD is only a small prototypical system, it allows for surprisingly natural mathematical texts. The following is an excerpt from a proof that the square root of a prime number is irrational:

> Theorem Main.
>
> For all nonzero natural numbers n,m,p if p * (m * m) = (n * n) then p is compound.
>
> Proof by induction. Let n,m,p be nonzero natural numbers.
> Assume that p * (m * m) = (n * n). Assume that p is prime. Hence p divides n * n and p divides n. Take q = n / p.
> Then m * m = p * (q * q). Indeed p * (m * m) = p * (p * (q * q)). m < n. Indeed n <= m => n * n <= m * m.
> Hence p is compound.
>
> qed.

The frugal ASCII appearance of ForTheL texts can easily be improved by putting a LaTeX layer on top of the language. Here is an original excerpt from an SAD + LaTeX proof of the infinitude of prime numbers which comes rather close to textbook versions:

Theorem 1. *The set of prime numbers is infinite.*

Proof. Let A be a finite set of prime numbers. Take a function p and a number r such that p lists A in r steps. ran$p \subseteq \ ^+$. $\prod_{i=1}^{r} p_i \neq 0$. Take $n = \prod_{i=1}^{r} p_i + 1$. n is nontrivial. Take a prime divisor q of n.

Let us show that q is not an element of A. Assume the contrary. Take i such that $(1 \leq i \leq r$ and $q = p_i)$. p_i divides $\prod_{i=1}^{r} p_i$ (by MultProd). Then q divides 1 (by DivMin). Contradiction. qed.

Hence A is not the set of prime numbers. ∎

6.2 *Natural Proof Checking*:
The Naproche project

Whereas SAD achieves an impressive but limited degree of linguistic naturalness with a carefully crafted small controlled language, the Naproche project (Naproche) aims at an analysis and formal approximation of extensive parts of the full natural language of mathematics. The project set out by analysing mathematical texts using annotations, formal grammars and discourse representations (see Koepke & Schröder 2002; 2003; Cramer & Schöder 2012; Cramer et al. 2011). The fact that formal semantics in linguistics usually leads to representations in first-order logic is advantageous for mathematical texts (see Cramer et al. 2009). In the Naproche software, first-order representations are transformed into queries to automatic theorem provers (ATP) in order to check whether statements in mathematical texts are logical consequences of previously established facts (see Cramer et al. 2010a;b).

The grammars and formats of the linguistic analysis define a controlled language of accepted sentences, the Naproche language. Like Automath, the Naproche project also takes Landau's *Grundlagen* (Landau 1930) as a benchmark text to be reformulated and checked. This has been done for the first two chapters of the book, and we give a sample of a representative theorem and the beginning of its proof, taken from the translation (Landau 1966):

Theorem 4, and at the same time **Definition 1:**

To every pair of numbers x, y, we may assign in exactly one way a natural number, called $x + y$ (+ to be read "plus"), such that

1. $x + 1 = x'$ for every x,
2. $x + y' = (x+y)'$ for every x and every y.

$x + y$ is called the sum of x and y, or the number obtained by the addition of y to x.

Proof: A) First we will show that for each fixed x there is at most one possibility of defining $x+y$ for all y in such a way that

$$x + 1 = x'$$

and

$$x + y' = (x+y)' \quad \text{for every} y.$$

Let a_y and b_y be defined for all y and be such that

$$a_1 = x', \quad b_1 = x',$$
$$a_{y'} = (a_y)', \quad b_{y'} = (b_y)' \quad \text{for every} y.$$

Let $\mathfrak{M}$ be the set of all y for which

$$a_y = b_y.$$

I)
$$a_1 = x' = b_1;$$

Hence 1 belongs to $\mathfrak{M}$.

II) If y belongs to $\mathfrak{M}$ then

$$a_y = b_y,$$

hence by Axiom 2

$$(a_y)' = (b_y)',$$

therefore

$$a_{y'} = (a_y)' = (b_y)' = b_{y'},$$

so that y' belongs to $\mathfrak{M}$.

Hence $\mathfrak{M}$ is the set of all natural numbers; i.e., for every y we have

$$a_y = b_y.$$

The argument, proving the uniqueness of an addition function on the natural numbers, is rather subtle since it uses higher-order arithmetic. This requires some (background) theory of sets and functions, which is not made explicit in the Landau text. The Naproche system includes such a background theory (Cramer 2012) so that the proofs get cleaner and don't have to appeal to "the possibility to define" certain terms. Here is a checked rendering of the Landau argument in the current version of the Naproche system:

> Theorem 4: There is precisely one function $x, y \mapsto x + y$ such that for all x, y, $x + y$ is a natural number and $x + 1 = x'$ and $x + y' = (x + y)'$.
> Proof:
> A) Fix x. Suppose that there are functions $y \mapsto a_y$ and $y \mapsto b_y$ such that $a_1 = x'$ and $b_1 = x'$ and for all y, $a_{y'} = (a_y)'$ and $b_{y'} = (b_y)'$.
> Let $\mathfrak{M}$ be the set of y such that $a_y = b_y$.
> $a_1 = x' = b_1$, so 1 belongs to $\mathfrak{M}$.
> If y belongs to $\mathfrak{M}$, then $a_y = b_y$, i.e., by axiom 2 $(a_y)' = (b_y)'$, i.e., $a_{y'} = (a_y)' = (b_y)' = b_{y'}$, i.e., y' belongs to $\mathfrak{M}$. So $\mathfrak{M}$ contains all natural numbers. Thus for all y, $a_y = b_y$.
> Thus there is at most one function $y \mapsto x+y$ such that $x+1 = x'$ and for all y, $x + y' = (x + y)'$.

Note that this text can be seen as a stricter version of Landau's argument. Due to the natural language features of Naproche and the built-in function theory the reformulated text is as short and readable as the original.

7 Perspectives of formal mathematics

Against the background of the state of formal mathematics as sketched above I propose a sequence of theses, leading from safe ones already substantiated to more speculative ones. In section 4 we saw:

1. Formal mathematics has become an established and active research area.

2. Formal mathematics is already covering a wide range of substantial mathematical results.

There are singular points where current mathematical research uses formal mathematics, e.g., the flyspeck project (flyspeck) of Thomas Hales to construct a formal proof of the Kepler conjecture, or the work of Vladimir Voevodsky in homotopy theory, using the Coq proof assistant. Thus:

3. Formal mathematics is beginning to interact with research mathematics.

4. Formal mathematics could become part of mathematical practice.

In line with Wiedijk's analysis of the current role of formal mathematics we hold that:

5. The acceptance of formal mathematics in mathematical practice will depend on the naturalness of its application.

Section 5 identified areas and proposed methods for the improvement of naturalness. This will involve the combination of best methods from various, already existing systems:

6. The naturalness of strong formal mathematics can be increased considerably.

Therefore:

7. Formal mathematics will become part of mathematical practice.

But it seems too early to make predictions on the degree of coverage and acceptance of formal mathematics tools in the day to day work of future mathematicians. Some practioners of formal mathematics like Jeremy Avigad, Kevin Donnelly, David Gray, and Paul Raff hold (Avigad et al. 2007):

> On a personal note, we are entirely convinced that, although there is a long road ahead, formal verification of mathematics will inevitably become commonplace. Getting to that point will require both theoretical and practical ingenuity, but we do not see any conceptual hurdles.

On the other hand one can expect resistance by mathematicians who feel that they would lose the traditional freedom of mathematical presentation, which can be very sloppy and even formally false in "inessential" or "trivial" places. To allay the reservations of traditional mathematicians, formal mathematics systems have to offer rich and natural interfaces, and there has to be reasonable added value for the user.

7.1 A scenario: Formal mathematics and textbook mathematics

Many attempts in formal mathematics are directed towards a register of mathematical discourse described as *textbook mathematics*. This involves extensive texts, a systematic development of some limited area of mathematics, and a rather detailed renderings of proofs. The prerequisites of such texts should be simple, and everything else is introduced within the text, preferably in a Definition–Theorem–Proof style.

Let us assume that formal mathematics is able, within the next decade, to handle some such texts: experts which understand the mathematics and the formal mathematics system reformulate chapters of textbooks into texts which are very similar in typesetting, language, and logical structure to the original text, but which are also checked for correctness by the system. The feasibility of this scenario will depend on the kind of mathematics to be handled (see 5.1).

What are the consequences of such developments? Obviously one could then have textbooks, which are readable like standard textbooks, but which are completely correct (we don't want to discuss the remote possibilities of computer and software faults at this place). This may be a relief to authors and referees. A referee could concentrate on main ideas instead of checking tedious details. On the other hand the demand for formalization may force some mathematically unnatural or superfluous issues into the presentation. Computer proof checking will provide possibilities to explore logical dependencies within the text which are not explicitly mentioned: the automated checker can produce a log of its proof (a "proof object" of some kind) which can be searched for information generated during checking. So the checkable textbook text is like a surveyable surface, under which one could explore different layers of logical detail.

Most mathematical research articles combine some high level reasoning with extensive low level arguments, often of some "combinatorial" kind. Although the high level reasoning may be far above the abilities of formal mathematics systems, combinatorial arguments sometimes have a textbook style as described above. One might consider writing "textbook arguments" with the help of formal mathematics systems to assist authors, referees, and readers. Often the high level reasoning is familiar to experts and proceeds along established intuitions of the field. By contrast, combinatorial arguments are sometimes difficult to grasp and intuite, so that a validity check may be welcomed by everybody involved. In this way, formal mathematics designed for the textbook level might also enter research mathematics.

The introduction of such techniques will depend on decisions and trends within the wider mathematical community. As an example, the systems TEX and LATEX could manifest themselves since they gave authors support and control of a process that previously could only be managed by a longwinded iterative process of approximations to the desired typeset result. Further benefits were given by the small footprint of the data files, the openness of the formats, the quality of the software, and other factors. Within a few years TeX and LaTeX have become a *de facto* standard which is now made essentially mandatory by publishers.

8 Philosophical perspectives

The development of formal mathematics may be viewed as a strengthening of the formalist position. The above proof of the infinitude of primes is not only a text that communicates number theoretic ideas to a fellow mathematician, and which **could** be fully formalized. In a rich formal system, including automated theorem proving, the text **is** already a formal text. Does this indicate some analogy with Richard Montague's *English as a Formal Language* (Montague 1974)?

In the discussion of informal versus formal proofs their seemingly huge dissimilarity is a decisive aspect. Hannes Leitgeb (Leitgeb 2009) writes:

> why not think of "formally provable(-in-T)" (for some instantiation of "T") as a Carnapian explication of "informally provable"? The answer is simple: because it is not. According to Carnap, whatever explicates an explicandum must be as similar as possible to the latter, but as our comparison from above has shown, formal provability and informal provability are just too dissimilar to satisfy this criterion.

But if in the case of the infinitude of primes T is taken to be the abovementioned system SAD + LaTeX informal and formal proof may coincide so that at least in certain situations "formally provable(-in-T)" might be a Carnapian explication of "informally provable"!

Strengthening formalism will affect the balance between the main positions in the philosophy of mathematics and may have far-reaching consequences. In his MSc thesis (Tanswell 2012) Fenner Tanswell has argued that Naproche could be a tool for overcoming the philosophical objections to formalism and develop a new type of formalism.

On the other hand it may be too early to start this discussion in detail. So let me just mention one issue with respect to the Philosophy of Mathematical Practice: The current way of checking mathematical correctness, rather than being meticulous logical checking, has been described by philosophers of mathematical practice as a complicated process based on a network of trust in intuitions, published papers, authorities, refereeing processes, etc. This system will change once formal certificates are available for parts of the mathematical research and dissemination process. Initially certificates will be seen as a welcome extra justification, until they will become mandatory, at least for certain kinds of arguments. Does this mean that certain observations of the Philosophy of Mathematical Practice concerning the shakyness of the present network of trust will become outdated in the long run?

Bibliography

ACE. Attempto Controlled English. http://attempto.ifi.uzh.ch/.

Automath. www.cs.ru.nl/~freek/aut/.

Avigad, J., Donnelly, K., et al. (2007). A formally verified proof of the prime number theorem. *ACM Transactions on Computational Logic (TOCL)*, *9*(1), 1–23.

Bourbaki, N. (2004). *Theory of Sets*. Berlin: Springer.

Coq. http://coq.inria.fr.

Cramer, M. (2012). Implicit dynamic function introduction and its connections to the foundations of mathematics. In *Philosophy, Mathematics, Linguistics: Aspects of Interaction*, St. Petersburg.

Cramer, M., Fisseni, B., et al. (2010a). The Naproche Project – Controlled natural language proof checking of mathematical texts. In *Controlled Natural Language 2009, Lecture Notes in Computer Science*, vol. 5972, Fuchs, N., ed., Berlin; Heidelberg: Springer, 170–186, doi:10.1007/978-3-642-14418-9_11.

Cramer, M., Koepke, P., & Schröder, B. (2011). Parsing and disambiguation of symbolic mathematics in the naproche system. In *Intelligent Computer Mathematics, Lecture Notes in Computer Science*, vol. 6824, Davenport, J. H., Farmer, W. M., et al., eds., Berlin; Heidelberg: Springer, 180–195, doi:10.1007/978-3-642-22673-1_13.

Cramer, M., Koepke, P., et al. (2009). From proof texts to logic. Discourse representation structures for proof texts in mathematics. In *From Form to Meaning: Processing Texts Automatically*, Chiarcos, C. et al., ed., Tübingen: Narr.

Cramer, M., Koepke, P., et al. (2010b). Premise selection in the Naproche System. In *Automated Reasoning, IJCAR 2010, Lecture Notes in Computer Science*, vol. 6173, Giesl, J. & Hähnle, R., eds., Berlin; Heidelberg: Springer, 434–440, doi: 10.1007/978-3-642-14203-1_37.

Cramer, M. & Schöder, B. (2012). Interpreting plurals in the Naproche CNL. In *Controlled Natural Language, Lecture Notes in Computer Science*, vol. 7175, Rosner, M. & Fuchs, N. E., eds., Berlin Heidelberg: Springer, 43–52, doi: 10.1007/978-3-642-31175-8_3.

de Bruijn, N. G. (1994). Reflections on Automath. In *Selected Papers on Automath, Studies in Logic and the Foundations of Mathematics*, vol. 133, R.P. Nederpelt, J. G. & de Vrijer, R., eds., Elsevier, 201–228, doi: 10.1016/S0049-237X(08)70205-2.

flyspeck. http://code.google.com/p/flyspeck.

Gödel, K. (1929). Über die Vollständigkeit des Logikkalküls. In *Kurt Gödel*, Collected Works – *Vol. I: Publications 1929–1936*, Feferman, S., ed., New York: Oxford University Press, 60–101, 1986.

Gödel, K. (1931). Über formal unentscheidbare Sätze der *Principia mathematica* und verwandter Systeme I. In *Kurt Gödel*, Collected Works – *Vol. I: Publications 1929–1936*, Feferman, S., ed., New York: Oxford University Press, 144–195, 1986.

Gonthier, G. (2008). Formal proof – the four-color theorem. *Notices of the AMS*, 55, 1382–1393.

Gonthier, G. (2012). Public email.

HOL light. www.cl.cam.ac.uk/ jrh13/hol-light/.

Isabelle. www.cl.cam.ac.uk/research/hvg/isabelle.

Kamp, H. & Reyle, U. (1993). *From Discourse to Logic*. Dordrecht: Kluwer.

Koepke, P. & Schröder, B. (2002). Natürlich formal. In *Computational Linguistics – Achievements and Perspectives*, Willée, G. et al., eds., Sankt Augustin: Gardez!-Verlag, 184–189.

Koepke, P. & Schröder, B. (2003). ProofML – eine Annotationssprache für natürliche Beweise. *LDV Forum*, 18, 428–441.

Landau, E. (1930). *Grundlagen der Analysis*. Leipzig: Akademische Verlagsgesellschaft.

Landau, E. (1966). *Foundations of Analysis*. New York: Chelsea Pub., 3rd edn., translated into English by F. Steinhardt.

Leitgeb, H. (2009). On formal and informal provability. In *New Waves in Philosophy of Mathematics*, Bueno, O. & Linnebo, Ø., eds., Basingstoke; New York: Palgrave Macmillan, 263–299.

McCarthy, J. (1962). Computer programs for checking mathematical proofs. In *Recursive Function Theory: Proceedings of the Fifth Symposium in Pure Mathematics*, Decker, J. C. E., ed., American Mathematical Society, 219–227.

Mizar. http://mizar.uwb.edu.pl.

Montague, R. (1974). English as a formal language. In *Formal Philosophy: Selected Papers of Richard Montague*, Thomason, R. H., ed., New Haven: Yale University Press, 247–270.

Naproche. http://naproche.net.

Paulson, L. C. (2003). The relative consistency of the axiom of choice mechanized using Isabelle/ZF. *LMS Journal of Computation and Mathematics*, *6*, 198–248, doi:10.1112/S1461157000000449.

Russell, B. (1938). *The Principles of Mathematics*. W. W. Norton, 2nd edn., 1st ed., Cambridge: University Press, 1903.

Russell, B. (1998). *Autobiography*. Hoboken: Taylor & Francis.

SAD. http://nevidal.org.

Tanswell, F. (2012). *Proof and Prejudice: Why Formalising doesn't make you a Formalist*. Msc thesis, Universiteit van Amsterdam, ILLC Publications MoL-2012-07.

van Benthem Jutting, B. (1977). *Checking Landau's "Grundlagen" in the Automath system*. Ph.D. thesis, Eindhoven University of Technology.

Verchinine, K., Lyaletski, A., & Paskevich, A. (2007). System for automated deduction (SAD): A tool for proof verification. In *Automated Deduction – CADE-21, Lecture Notes in Computer Science*, vol. 4603, Pfenning, F., ed., Berlin Heidelberg: Springer, 398–403, doi:10.1007/978-3-540-73595-3_29.

Whitehead, A. N. & Russell, B. (1910-1913). *Principia Mathematica*. Cambridge: Cambridge University Press.

Wiedijk, F. (2007). The QED Manifesto revisited. *Studies in Logic, Grammar and Rhetoric*, *10*(23), 121–133.

Peter Koepke
Mathematical Institute
University of Bonn
Germany
koepke@math.uni-bonn.de

Mathematics and the New Technologies Part III: The Cloud and the Web of Proofs

Jean Paul Van Bendegem

"New technologies, new mathematics" might seem a defensible slogan, as we try to show in this threefold set of papers and likewise "New mathematics, new philosophy of mathematics" will hardly be doubted by mathematicians and philosophers alike, but the difficult question is whether transitivity is applicable here so that we can conclude that new technologies also produce new philosophical questions and problems. The previous papers (Löwe 2014, this volume) and (Koepke 2014, this volume) support to some extent the idea that transitivity is possible: the peer review process, automated theorem proving, rewriting procedures, formal proof checking and so forth are convincing examples. The same goes, I believe, for experimental mathematics[1] where number crunching can lead to unexpected results, that would not have been available without the sheer computational power required,[2] or where the visualization of geometrical shapes can inform us about particular properties of that shape.[3] In this paper another example will be presented that further supports the derived slogan. It will deal with networks and knowledge distributed over such networks. More precisely, the use of blogs, networks and discussion within an internet community or, as we refer to it today, "in the cloud" will be discussed. Do such structures alter mathematical practice—for that is what I will focus on rather than mathematical results on their own[4]—and thereby introduce new philosophical questions

---

[1] See, e.g., (Baker 2008) and (Borwein & Devlin 2009).

[2] A famous example is Goldbach's conjecture. This has been checked up into the billions but, apart from the fact that the conjecture has been verified for all these numbers, the graph of the function $G(2n)$ = the number of ways $2n$ can be written as the sum of two primes shows a clearly strictly increasing function. The shape of that graph could generate some hypothesis about the behaviour of G.

[3] The best known examples of course being all fractal structures.

[4] I will not go into details here but the philosophy of mathematical practice is a relatively new branch in the philosophy of mathematics that focuses on the whole mathe-

and problems? Before addressing the larger question, it will be helpful to have a brief look at a particular example, namely the Polymath project.

1 Polymath: a short presentation[5]

In January 2009 mathematician Timothy Gowers, a Fields medalist, opened a website (http://polymathprojects.org/), accessible for everyone, mathematicians and non-mathematicians alike, announcing that he was searching for a proof of a particular mathematical statement. The "invitation" was to join him in that search. Anyone could post a message about almost everything on the condition that it was somehow related to the proof search. In fact, a set of ground rules was announced to avoid the whole enterprise becoming all too chaotic. The hope was to find a proof and, if that were to occur, to publish the proof through the usual existing channels, namely mathematical journals, using a pseudonym, itself not an uncommon practice. This description is not essentially different from normal mathematical collaboration, except for the large number of people, including laypeople, involved. It remains to be seen whether this means that this approach is substantially new or just a matter of scale. That being said, let us first have a look at the problem itself.

The problem Gowers launched on the website is known as the Density Hales-Jewett Theorem (DHJ) for $k = 3$ at first, but later generalized to arbitrary k. This problem is part of the field of Ramsey theory, involving the combinatorics of colouring problems. The typical format of such problems is that "Given a so-and-so structure of sufficiently large size, then there will always be substructures that have a particular property". We shall say that such a property is *unavoidable*.

More specifically, DHJ for $k = 3$ states the following. Let the following be given:

a set $K = \{1, 2, 3\}$ (the parameter k is the size of K, $\#K = k$)

a set $N = K^n$, i.e., the set of all words of length n, using K

Next we need four definitions:

A *variable word* is an element i of N where some places are replaced by variables, thus $k_1 k_2 \ldots k_i x k_{i+2} \ldots k_n$ is a (one-place) variable word

A *filled-in word* is a variable word where all variables have been replaced by the same element of K

---

matical process and not merely the endresults. See for a first impression, (Mancosu 2008) and (Van Bendegem 2004).

[5]This paper is related to (Allo *et al.* 2013). The Polymath project is more fully discussed there and is presented in reduced form here.

A *combinatorial line* is a non-empty subset I of N that contains all filled-in words for all elements in K of a given variable word. *Example*: if we take $n = 6$, then:

the subset $\{122132, 122232, 122332\}$ is a combinatorial line, as is the subset $\{112132, 212232, 312332\}$.

Define the density d of a subset M of N by $d = \#M/\#N$

The DHJ for $k = 3$ says this: For every $d > 0$, there exists an n such that every subset M of N with density at least d contains a combinatorial line.

The "unavoidable" property here is the presence of a combinatorial line. So the theorem says that no matter how low the density of a particular subset, if the words made on the basis of the alphabet can be sufficiently long, there will always appear a combinatorial line.

In addition, one very special feature needs to be mentioned, namely that a proof already existed.[6] However, this proof relied on methods and techniques from domains far away from combinatorics, among other things, ergodic theory. So, as often happens in mathematical research, although one has a proof of the theorem, nevertheless this does not prevent mathematicians from searching an alternative[7] and, more importantly, an *elementary* proof, i.e., a proof using the concepts, proof methods and techniques of the domain itself.

What happened after the opening of the website? First, apart from Gowers, mathematician Terence Tao (UCLA), also a Fields medalist, joined the enterprise. After 6 weeks, 39 contributors had contributed 1228 comments (after every 100 comments, summaries were made by Gowers to keep an overview), not only a proof was found, but it became immediately clear how it could be generalized for arbitrary k. The proof has been published under the pseudonym: *D. H. J. Polymath* (which makes one think of course of other fictitious names in the history of mathematics, the most famous one no doubt being Nicolas Bourbaki).[8] Surely the most striking feature of the whole process is that "amateurs", both inside and outside of the mathematical community (so, e.g., high school teachers are here considered to be amateurs) could and did participate. Whether we should be as enthusiastic as Jacob Aron—see (Aron 2011)—in *New Scientist* and claim that this will "democratise the process of mathematical discovery" or as Michael

---

[6] See (Furstenberg & Katznelson 1991).

[7] An extreme example is Pythagoras' theorem for which at present some four hundred proofs exist. The website www.cut-the-knot.org/pythagoras/index.shtml lists nearly hundred basic variations.

[8] See (Polymath 2010).

Nielsen (2012), who states that "The Polymath Project is a small part of a much bigger story, a story about how online tools are transforming the way scientists make discoveries" (Nielsen 2012, 3), is of course another matter. The question to be dealt with here is whether philosophers of mathematics should be as enthusiastic as Aron and Nielsen about this phenomenon.

2 Yes, but is it philosophically relevant?

The answer that will be given to the question in the title above is basically one argument (scheme) that will be developed stepwise. Let us start with some simple premises that no one will doubt (although at this stage no statement has to be made about their philosophical relevance):

> (P1) Resources required for problem-solving available to mathematicians are finite.

In most cases the major resource will be time but not exclusively so. It must also involve, e.g., the (creative) capacities of the mathematicians involved and the externally available computing power (think, e.g., of the already mentioned rich area of experimental mathematics). All of these elements are clearly finite. What I am appealing to here, is nothing but the economical properties and aspects of problem-solving, that economists are perfectly aware of, as they are aware of the finiteness of resources or, to use their preferred term, the scarcity of goods.[9]

> (P2) There exist (many) mathematical problems that are beyond the resources of an individual mathematician or even a fixed group of mathematicians.

This premise can be supported in different ways. The first one is quite simply of an evidential nature. We have faced and are still facing with mathematical problems that either involve the use of computer programs, such as the four-colour theorem or the sphere packing problem, and pose a problem as to their correctness (see (Koepke 2014, 409–426) of this set of papers), or are amazingly long such as the well-known classification theorem for finite simple groups, estimated at fifteen thousand pages (although since then serious attempts are being done to reduce that number, down to some five thousand pages). Of course, one might argue that no matter how we

---

[9]Economical features are to be found everywhere in mathematics if one cares to look for it. Even a simple formula such as

$$\sum_{i=1}^{n} i = \frac{n(n+1)}{2}$$

reduces the computational power required to add the first n numbers.

got there, we do in fact have a classification theorem so it must have been in the range of what mathematicians can achieve after all. True, but it does indicate that we did at least move beyond the individual mathematician's resources and had to move to the community or group level.

The second way—personally my favourite—is an "absolute" argument, relying on a Gödelian argument. Take any mathematical theory M and its language L in an axiomatic formulation. Look at all the statements whose length is n, i.e., the statements that consist of n symbols. If there were a computable function $K(n)$ bounding those proofs for various n, then one would indeed have a positive solution to the *Entscheidungsproblem*: determine the length n of the statement, compute $K(n)$, and then try all proofs of length $K(n)$. But since the *Entscheidungsproblem* is not solvable in that way, the function $K(n)$ cannot be bounded by any computable function. Therefore $K(n)$ must have some enormous growth, when n becomes very large. This can be interpreted as an indication, that already $K(10)$ $K(20), \ldots$ will be enormous. But this is only a heuristic argument, like saying that certain computations take a long time because an algorithm is (in the limit) exponentially complex. Such proofs, if encountered, will pose a challenge to any group of mathematicians. The argument can be easily extended to mathematical communities in the sense that, if any mathematical problem can be settled by a group of mathematicians with a size bounded by some finite number and with finite resources, likewise bounded, then the argument can be repeated. Such a group would become (in a sense) a decision instrument. We repeat however that the argument can only be seen as an additional argument for the first way to support the premise because it could very well be that, for "modest" n, the problem does not really manifest itself and hence its impact would be very small if non-existent.[10]

To the extent that these two premises are indeed acceptable and defensible, the following intermediate conclusion is then rather straightforward:

> (IC1) Given the finite resources available, some mathematical problems will either not get solved or not get solved easily and/or quickly.

This by itself is not sufficient to conclude that the resource boundedness makes (parts of) mathematical practice philosophically relevant. After all, we never get all mathematical problems solved anyway as there are an infinite number of them.[11] The mere fact that at any specific time we have only solved a finite number of mathematical problems cannot be a conclusive argument at all. More is needed and that is the role the next three

---

[10]With thanks to Peter Koepke for having pointed out this possibility.

[11]To which should be added that most problems do not get solved for being not interesting and not worth the waste of the mathematicians' time.

premises are supposed to play. All three of them are, I assume, simple and straightforward and they too find their basis in the study of mathematical practice.

> (P3) In many cases, a solution to a mathematical problem introduces new concepts.

A general argument in support of this premise is that any question or problem relies on some presuppositions some of which have to do with the mathematical structure the question or problem is about. Once one has the natural numbers, the prime number concept follows easily, whereas, e.g., the concept of all natural numbers that in a decimal representation have seven sevens in them seems not interesting at all.[12] Or, to put it in different terms, any mathematician when asked about a particular concept in his or her field of expertise, will be able to answer the question what the relevance of that concept is. Very often the answer will be that it allows you to formulate this or that problem in a convenient, perhaps even explanatory[13] way. This characteristic of concepts can be extended to proofs as well.

> (P4) In many cases, a solution to a mathematical problem introduces new proof methods.

Mathematicians have a range of proof methods at their disposal that are easily recognizable as they often have a specific name: proof by mathematical induction, proof by cases, proof by reductio (ad absurdum), proof by infinite descent, [...] In many cases these proof methods were developed because of a particular problem and later on it turned out that the same proof method could be applied to other problems. In that sense, an uninteresting problem can nevertheless possess a quite interesting proof.

> (P5) The development, relevance and use of concepts and proof methods is one of the core themes in the philosophy of mathematical practice and of mathematics.

This is, of course, the crucial premise to reach the conclusion. Apart from the obvious empirical fact that the above statement is true—it is sufficient to look at the literature in the philosophy of mathematics, both "purely" philosophical and foundational, to see how much attention is given to these

---

[12]Which is not to say that all such questions and problems are irrelevant. Whether or not the number π is a normal number, in the sense that all digits have the same frequency of occurrence, is considered to be an interesting problem but is clearly connected to a particular representation.

[13]See (Mancosu 2008) for a nice discussion about explanation in mathematics, an important and difficult topic.

topics, see (Rav 1999) for an excellent analysis—there is the negative argument: what else would philosophers of mathematics talk about? Both elements, concepts and proof methods, belong to the essence of what it is what mathematicians do and hence should be a topic of reflection for philosophers.

If these three statements seem acceptable, then stringing them together, we arrive at a second intermediate conclusion, namely:

> (IC2) The fact that problems get solved, implies that their solutions have a (potential) impact on the philosophy of mathematics.

Finally, if we put the two intermediate conclusions together, we arrive at the final conclusion, which states that:

> (C) The fact that we have to deal with finite resources for our problem-solving capacities has a direct (potential) impact on and is (potentially) relevant for the philosophy of mathematics.

A direct corollary of this conclusion is that:

> (Cor) The ways in which finite resources are distributed over a problem-solving community (of mathematicians) is directly (potentially) relevant for the philosophy of mathematics.

All this being said, even if the reasoning presented in this section is acceptable, it still remains to be shown that the Polymath case is such a case that might change our views on certain philosophical questions. This raises another question that I will briefly address in the next section, namely, whether there are ways to investigate such a claim. In general: suppose you are confronted with a particular way mathematicians have tried, successfully or not, to solve a particular mathematical problem, should their strategy invite us to have a different look at certain philosophical questions? I think this question can be positively answered and, more specifically, what I have in mind are formal models of shared or distributed knowledge.

3 Formal modelling as an additional argument

The literature on the topic of shared or distributed knowledge is quite extensive and I will not try to present a survey here. I will briefly comment on some approaches that for different reasons are directly relevant, ranging from multi-agent systems for obvious reasons, including argument and dialogue structures to describe the interactions between the members of a community and a formal approach of Lakatos' method of mathematical discovery. Before doing that, let me sketch in a few words the informal idea.[14] A network of mathematicians can be described as a Kripke model. We have a set M of worlds, in this case the mathematicians and a relation R on $M \times M$ that tells us what the communication channels are between them.[15] One thing stands out as quite obvious: given what R is, the community M will be able to solve or not solve certain problems. Think of extreme cases: surely if everybody is in touch with everybody else much more information will flow between them, compared to a structure where one mathematician is addressed by all others who themselves have no contact with one another, corresponding to an inward-pointing star-like structure. Take a simple example: suppose that a mathematical problem P can be decomposed into two problems P_1 and P_2 such that solving both these subproblems solves the original problem. In the first case all mathematicians can have a go at the subproblems, whereas in the second case, if someone manages to solve P_1, someone else P_2, then only the mathematician in the center will know that the original problem has been solved, as the two mathematicians who have solved the subproblems cannot communicate with one another. In short, how the community is organized should make a difference as to their problem-solving capacities, as is stated in (Cor) above.

An illustration of the first approach is the recent presentation of multi-agent systems in Dunin-Keplicz and Verbrugge (2010). The reason for this choice is that they discuss the specific situation where the agents are searching for a proof (Dunin-Keplicz & Verbrugge 2010, 91–97). The language they develop involves such elements as GOAL(agent, action), in the case of theorem proving obviously GOAL*(i, prove(theorem(T)))*, the beliefs each agent has, expressed by a belief-operator BEL(agent, statement), involving in this case whether or not the agent believes he or she can contribute to the finding of the proof. On this basis the team leader can put together his or her team and develop a plan that involves, among other things, ways of dividing or splitting up the given problem. In their approach it basically comes down to the reduction of the search for the full proof to the search for proofs of a set of lemmas, the idea being that, once all lemmas have been proven, thereby the original theorem has been proved. The execution of this

---

[14]I have been playing around with this informal idea for some time as early as 1985, see (Van Bendegem 1985).

[15]There is an interesting link to be explored here, namely, the study of small worlds, see, e.g., (Watts 1999). Here the object of study is to describe networks and develop measures for the length of the chains that connect two members in the network. Small changes in such a network can have a tremendous effect on the efficiency of communication in terms of speed.

plan also involves a means-and-ends analysis, in this case the possibility to check a proof and establish its correctness. In their own words:

> There is a division of the theorem T into lemmas such that for each of them there exists a proof, constructed by the lemma prover and checked by the proof checker. Also, there is a proof of the theorem T from the lemmas, constructed by the theorem prover, which has been positively verified by the proof checker. (Dunin-Keplicz & Verbrugge 2010, 94)

Of special interest in their approach is that plans can always be reconfigured, dependent on the state of affairs. In the case of theorem proving, one of the obvious obstacles is that an agent who committed him- or herself to prove one of the lemmas does not succeed (because of shortage of time or, in my terms, because the economic resources have been exhausted). In that case the commitments and beliefs of the agents involved are checked again to see whether another agent can take over the task. Another obvious obstacle is that the proof checker finds a mistake in the proposed proof. All taken together, this model comes pretty close to real-life scenarios. This formal description could be—up to a number of special issues that I will discuss a bit further—easily applied to the Polymath Project, where we have clearly two team leaders—Timothy Gowers and Terrence Tao—and where the other participants believe they can contribute something to the overall problem. It also raises the interesting question whether the Polymath Project should maintain the social structure it has at present. The teamwork approach, sketched very roughly here, suggests a regular update to see whether a reorganisation at a certain point in time is needed or not.

An additional feature is that their framework also deals with dialogues and argumentations, next to and apart from proofs. The main object is to determine under what circumstances and conditions an agent i who believes A can persuade an agent j to accept A. One possibility is on the basis of trust. But the object of a dialogue can also be to seek information from an agent. What is worth mentioning is that the sources they refer to concerning dialogue and argumentation theory are such authors as Erik Krabbe and Douglas Walton.[16] This is to be sure a quite different approach than the recently developed one in terms of argumentation systems, see (Besnard & Hunter 2008) for an overview, where the focus is on attacks and counterattacks, on the weight of an argument and, especially, on conflicting arguments and how to resolve them. At present it seems less clear how this could be easily applied to answer the main question of our

---

[16] Both authors have an impressive publication record so I will only mention a joint work, namely (Walton & Krabbe 1995).

contribution, namely in what ways different social structures can lead to different mathematical developments because of different problem-solving capacities. I will not explore this road any further here.[17]

A few words should also be said about the "founding father" of the study of mathematical practice, Imre Lakatos, whose *Proofs and Refutations* (1976) marked the beginning of the study of mathematics in its actual historical development. Although his proposed method has been both criticized and extended in several ways, it is worth mentioning that a few authors have tried to formalize the Lakatosian method and to connect it with recent developments in theory change and development ((Pease 2007)[18] and (Başkent 2012)).

Nevertheless in order to come to a comprehensive theory of how problems are distributed and how they get solved in a group or community setting, some additional features will have to be dealt with. To round off this section, I just list three of them:

> It must be clear that more complex structures are needed than the lemma-theorem relationship. Especially the other direction, so to speak, should be dealt with. Think of the case where several theorems have been proved and a generalization is proposed that brings the theorems together in an overarching framework but that requires that several theorems have to be reformulated. This process of reformulation strikes us as an important element to understand how mathematical change comes about.

> What needs to be looked at as well are all possible relations between proofs. Sometimes analogies between different proof methods are important—this, incidentally, were comments often made in the Polymath project where suggestions were made to look at a particular proof method as source of inspiration for the proof searched for—or between the same proof method used in different mathematical contexts.

> Above all, any such model should include concept formation. How and why do certain concepts arise and others don't? Do concepts keep their relevance or do they in some cases "disappear"? Is it possible to define the fruitfulness of a concept? Typical examples are of

---

[17]Although it should be mentioned that Andrew Aberdein has been investigating for some years now the use of argumentation theory in mathematics, see (Pease & Aberdein 2011) and (Aberdein & Dove 2013), but this deserves a separate treatment in another paper.

[18]Of special interest is the fact that Pease has recently also contributed, together with Ursula Martin to the study of the Polymath project. See (Pease & Martin to appear).

course mathematical constants. To give but one specific example: all mathematicians share the feeling of puzzlement that the number π appears in the outcome of the summation of the inverse squares of the natural numbers, namely $\pi^2/6$.

4 Conclusion

What has been presented here is, first and foremost, a philosophical exercise. Starting from a specific real-life case study I have tried to formulate a general philosophical argument to show or at least support the hypothesis that social structures do matter to the development of mathematics and thereby also affect the problem agenda of the philosophers of mathematics. That being said, it should not be excluded that laboratory experiments can be done. Imagine two groups of students that have been evaluated beforehand in such a way that, as far as mathematical capabilities are concerned, they are sufficiently comparable, i.e., the individual characteristics do not differentiate between them. Organize the two groups in a different social structure, e.g., one group with a central authority to whom everybody has to report and who is the only one to have an overview and one group where everybody has access to everybody else. Although one might think that the second group could, maybe should be more successful, this is not necessarily so as they run the danger to get stuck in too many details that everybody is offering to the whole group. This thought in itself makes the experiment interesting and, as it happens, there are sources that can be used, namely the work being done in experimental economics, especially where game theory is concerned. This brings us back to cooperation, collaboration and competition, basic social relations in any social group, including that of the mathematicians.

Bibliography

Aberdein, A. & Dove, I. J. (2013). *The Argument of Mathematics.* Dordrecht; New York: Springer.

Allo, P., Van Bendegem, J. P., & Van Kerkhove, B. (2013). Mathematical arguments and distributed knowledge. In *The Argument of Mathematics, Logic, Epistemology, and the Unity of Science,* vol. 30, Aberdein, A. & Dove, I. J., eds., Dordrecht; New York: Springer, 339–360, doi:10.1007/978-94-007-6534-4_17.

Aron, J. (2011). Math can be better together. *New Scientist, 210*(2811), 10–11.

Başkent, C. (2012). A formal approach to Lakatosian heuristics. *Logique et Analyse, 55*(217), 23–46.

Baker, A. (2008). Experimental mathematics. *Erkenntnis, 68*(3), 331–344, doi: 10.1007/s10670-008-9109-y.

Besnard, P. & Hunter, A. (2008). *Elements of Argumentation.* Cambridge, MA: MIT Press.

Borwein, J. & Devlin, K. (2009). *The Computer as Crucible. An Introduction to Experimental Mathematics.* Wellesley: A. K. Peters.

Dunin-Keplicz, B. & Verbrugge, R. (2010). *Teamwork in Multi-Agent Systems. A Formal Approach.* New York: Wiley.

Furstenberg, H. & Katznelson, Y. (1991). A density version of the Hales-Jewett theorem. *Journal d'Analyse Mathématique,* $57(1)$, 64–119, doi: 10.1007/BF03041066.

Koepke, P. (2014). Mathematics and the new technologies, Part II: Computer-assisted formal mathematics and mathematical practice. In *Logic, Methodology and Philosophy of Science. Proceedings of the Fourteenth International Congress (Nancy),* Schroeder-Heister, P., Hodges, W., et al., eds., Logic, Methodology and Philosophy of Science, London: College Publications, 409–426.

Lakatos, I. (1976). *Proofs and Refutations: The Logic of Mathematical Discovery.* Cambridge: Cambridge University Press.

Löwe, B. (2014). Mathematics and the new technologies, Part I: Philosophical relevance of a changing culture of mathematics. In *Logic, Methodology and Philosophy of Science. Proceedings of the Fourteenth International Congress (Nancy),* Schroeder-Heister, P., Hodges, W., et al., eds., Logic, Methodology and Philosophy of Science, London: College Publications, 399–407.

Mancosu, P. (Ed.) (2008). *The Philosophy of Mathematical Practice.* Oxford; New York: Oxford University Press.

Nielsen, M. (2012). *Reinventing Discovery. The New Era of Networked Science.* Princeton: Princeton University Press.

Pease, A. (2007). *A Computational Model of Lakatos-style Reasoning.* Ph.D. thesis, School of Informatics, University of Edinburgh, URL http://hdl.handle.net/1842/2113.

Pease, A. & Aberdein, A. (2011). Five theories of reasoning: Interconnection and applications to mathematics. *Logic and Logical Philosophy,* $20(1-2)$, 7–57.

Pease, A. & Martin, U. (to appear). Seventy four minutes of mathematics: An analysis of the third Mini-Polymath project. In *Proceedings of the Symposium on Mathematical Practice and Cognition II,* Birmingham, preprint at http://homepages.inf.ed.ac.uk/apease/papers/seventy-four.pdf.

Polymath, D. H. J. (2010). Density Hales-Jewett and Moser Numbers. In *An Irregular Mind (Szemerédi is 70), Bolyai Society Mathematical Studies,* vol. 21, Bárány, I. & Solymosi, J., eds., New York: Springer, 689–753.

Rav, Y. (1999). Why do we prove theorems? *Philosophia Mathematica*, 7(1), 5–41, doi:10.1093/philmat/7.1.5.

Van Bendegem, J. P. (1985). A connection between modal logic and dynamic logic in a problem solving community. In *Logic of Discourse and Logic of Discovery*, Vandamme, F. & Hintikka, J., eds., New York: Plenum Press, 249–262.

Van Bendegem, J. P. (2004). The creative growth of mathematics. In *Logic, Epistemology and the Unity of Science*, *LEUS*, vol. 1, Gabbay, D., Rahman, S., et al., eds., Dordrecht: Kluwer Academic, 229–255.

Walton, D. & Krabbe, E. (1995). *Commitment in Dialogue: Basic Concepts of Interpersonal Reasoning*. Albany: State University of New York Press.

Watts, D. J. (1999). *Small Worlds. The Dynamics of Networks between Order and Randomness*. Princeton: Princeton University Press.

Jean Paul Van Bendegem
Vrije Universiteit Brussel
Center for Logic and Philosophy of Science
Ghent University
Belgium
`jpvbende@vub.ac.be`

Technological Paradigm Conceptions in the 1980s

IMRE HRONSZKY

ABSTRACT. This article makes some introductory remarks about the conceptual significance of Kuhn's *Structure of Scientific Revolutions* (Kuhn 1962) to understanding of the dynamic of technology change. The second part of the article concerns different versions of the concept of technological paradigms.

1 Introductory remarks on interpreting the *Structure*

Numerous trials were carried out to extend paradigm conception to areas other than science. Gerry Gutting (1980) rightly identified different uses of the paradigm with respect to

1. analogical utilization,
2. naming any new theory a paradigm—as a generalization without any specific content,
3. using it as a term for any radically new thing.

Some researchers sought to apply the *Structure* analogically. They claimed that the cognition dynamics in their fields were analogous to those of science in some essential respects. These include topics like indigenous beliefs, the arts, fine art, history, mathematics, etc.[1] However, philosophers of science provided different interpretations of the paradigm concept in science. More than that, it was assessed from contradictory perspectives. This paper deals with one special interpretation line and looks for its realization in analogical applications of the paradigm conception to technological development.

---

[1] It is rather difficult to carry out such trials. One has to take into account the obvious essential differences of all these activities in every defining issue. But these trials also call for identification of basic similarities of self-regulatory mechanisms in cognition modes of most different human activities.

1.1 Margaret Masterman's recognition

It is useful to start this article with some versions that had been developed for the conception of a scientific paradigm as outlined by Kuhn. This helps to compare what happened in its application to the cognitive dynamics in technology. I start with Margaret Masterman's reconstruction of the message of the *Structure*. This reconstruction was based on her systematic remarks made during the notorious debate between Kuhn and the Popperians in 1965 (Masterman 1970). Masterman first put the central emphasis on the claim that Kuhn had changed the basic mission of investigation in the philosophy of science. The correct task was turning first to working science, to reconstruct what it really makes in practice, without any preceeding normative bias of what science really is. The philosophy of science is to embed in the learning process how to improve this practice in this activity. Masterman's main message was that Kuhn described science as a historically valid self-corrective system. Systematic, from normative bias liberated observation of the practice of how science works is the most important basis to understand it by reflexive learning. This last is learning from engaging in the practice and making the alleged learning explicit. Further, it is criticizing it and further urging to experiment with this learning in the practice, as a correcting factor. Masterman emphasized: instead of looking at science as a mechanism of setting and solving falsifiable "problems" coming from application of new metaphysics to science, instead of setting mere "problems", Kuhn offered looking for and select "puzzles" to be solved by identifying a "puzzle" setting and solving mechanism, a paradigm. (As the paradigm approach defines it, puzzles have solutions in the paradigms that allow them to be set up, except some non-predictable counter examples that are most important for the mechanism of paradigm changes.) Problems that can be formulated but not transformed into puzzles should be assessed either as non-scientific problems or as scientific problems awaiting later successful transformation into puzzles because of their complexity. We can rationally expect that most puzzles can be solved, because they are formulated accordingly. Paradigms offer a mechanism responsible for progress, including cumulativity (within the paradigm). But, on the other side of their essence, by way of self-induced emerging and deepening crises, they offer a mechanism to overcome the existing paradigm by a new one. Kuhn demonstrated that science was a profession that mostly goes about its work in a normal way but with a self-corrective mechanism is able to systematically lead to its radical renewal, said Masterman.

Masterman reconstructed the systemic nature of the paradigm conception as some sort of a socio-cognitive entity, without naming it as such, as cognitive activity of a community of specialists where the social, the community has essential cognitive function. She emphasized that there are three interdependent, essential functions of a paradigm.[2] First, it works as community behaviour, as a set of habits. These are intellectual, verbal, behavioural, or technical. Following these habits is part of successful problem-solving activities.[3] Second, a paradigm is a research instrument. This conceptualizes puzzles and their solutions, and also supplies a basis to assess the solutions offered. Third, it provides for "metaphysics" by the interpretation of the results describing what the investigated *is*. As Masterman emphasized, Popper envisioned science as a falsification mechanism to move radically from one theory to another, whereas Kuhn envisioned science as a working puzzle-solving tool system that is interpreted as the valid metaphysics as long as it is successful.

Masterman added some different things to Kuhn's premises. She especially emphasized that normal science emerges, when an "artefact" is found that can work as a puzzle-solving instrument. Such artefacts are those analogies that become successful enough. They are "tricks" rendering possible to systematically follow some direction. In addition to their previously proven success, insights emerge as to how they can be further applicable. First, unavoidably moving chaotically in the new cognitive situation, in the pre-paradigmatic phase there is a task to find a "trick", a concrete analogy that works as a cognitive instrument, promising enough to accept to explore the investigated subject. These tricks can serve as exemplars to direct further research.

This understanding needed the dynamics of analogical thinking to be outlined, working explicitly in the pre-paradigm phase, but, as Masterman emphasized, also being unavoidably present in the normal science phase when valid formal generalizations and hypothetico-deductive techniques produce a main bulk of knowledge. In Masterman's interpretation, the dynamics move by developing and exploring something, that I'd like to call oxymoron, but that is at the start a hypothetically promising formation. Researchers try to bring into harmony the A' picture (the "metaphysics of A", "what A is") of A with the instrumental actions and data from working on B to

---

[2]She was a computer linguist and, unfortunately became extremely notorious among philosophers of science because she identified not less than 21 different meanings of the term paradigm in the *Structure*. These meanings belonged to the three functions she identified and expressed partial aspects of them. Kuhn had a complex relation to her and her interpretation.

[3]This allows the identification of paradigm communities. She emphasized the unavoidable role of analogical reasoning in cognition made by a paradigm community. Thus, she opened the path to understanding paradigm conception from a socio-cognitive perspective. This means that, because of the unavoidable analogical nature of reasoning, decisions of a (special) community are unavoidable constituents of cognition to make conclusions practically binding.

provide for the B' valid picture. Analogies provide a concrete direction, a way of seeing how a specifiable artefact is extendable. It belongs to the nature of analogical reasoning that it is only finite in extensibility and is only extensible by "inexact matching".

In this way, the essential capability of paradigms to produce some anomalies that unavoidably lead to crises begins to be explained: one refers to the unavoidable analogical reasoning nature of the activity.[4] Nature, and the unavoidable emergence of crises, began to be explained from within the nature of paradigms, as systematic analogical thinking. In this understanding it is a socio-cognitive undertaking, it is impossible to separate the cognitive use of the paradigms from the cognitive agents who employ them. (Valid cognitive agents are the paradigm communities of scientists.) Hence, it is impossible to reconstruct the paradigm change as a consequence of any sort of logical necessity, but good reasons emerge to make the change as decision by a (new) community. Instead of following logical necessity, when they feel unable to further try to extend the old analogy, some researchers decide to turn to a new paradigm. Choosing a new paradigm is of a (as I would say, a limited) socio-cognitive nature. (It is a separate topic to ask whether some role and what sort remains for similar decisions in the normal phase for decision making.)

In Masterman's interpretation, Kuhn thought in the *Structure* that the paradigm dynamics were "valid about all real science (basic research, applied, technological, are all alike). Namely they were normally habit governed, puzzle-solving activities" (Masterman 1970, 60).

1.2 Joseph Rouse's practice perspective on the *Structure*

There are basic dichotomies in understanding what science is and opposite interpretations of the paradigm conception could be carried out, according to which member primacy is given in the series of the different category pairs. These choices are essential to fix the interpretation of paradigm conception. Decisions over them are unavoidable and lead to opposite results. I first turn to the dichotomy of knowledge vs cognitive practice. Joseph Rouse put the question pointedly in 1987 (Rouse 1987). He emphasized that two fundamentally different interpretations of Kuhn's work were possible. One of them was scarcely grasped in 1987, let alone achieved. But that one is the real Kuhnian transformation to understand science. This is quite a different concept of science, as a special sort of practice, instead of what he calls the epistemological concept. That last concept approaches

---

[4] Every language construction is based on the use of analogies, including science. In Masterman's opinion, tests are effected without changing the puzzle-solving character in the whole working process of a paradigm in its normal phase.

science from its results, as an activity aimed at a special sort of mental representation. As Rouse summarized, this has decisive consequences:

> The main task for scientific research is to provide for a non-contradictory representation.
>
> The most important results are the paradigmatic theory and explicit knowledge.
>
> The typical place of research is what Rouse calls the "observatory".

Hence, counter-examples are taken to be the most important issues in the process of scientific cognition, and the main activity is overcoming counter-examples by appropriate theory building. It facilitates an inclination to doctrinaire behaviour.

In contradiction to this interpretation, a practical conception of science can be formulated:

> The experimental laboratory is the most important locus of research.
>
> Normal science is research where, based on earlier successful practice, scientists know their way around. They have a practical grasp, exemplary ways and skills of conceptualising and intervening. This grasp is based on practical dealing with issues, on acquiring some capacity of using analogical reasoning, on the use of instruments, etc.
>
> Instead of reaching the level of an overarching theory, the main teleology of action is to extend the practical capacity to solve further problems. (Theories help to achieve this task.) This is why it is most important to have working exemplars that show how to go further *per analogiam*. Different working exemplars will lead in different directions.

Science produces specific sorts of "artefacts" (a complexity of material and mental elements) in this conception and works on them with a specific goal in mind. Science is a shared practice of cognitively exploring and exploiting those artefacts that are produced for that reason. In a shared practice aimed to find exemplars, anomalies aren't counterexamples, but indicators that something may be inappropriate in a practical skill. Anomalies will only be solved so far that the practice, based on the solution, can be continued. Conflicts do not occur between irreconcilable theories, but between "scientific ways of life". A particular evolutionary way is embodied in the practice of science. This includes revolutions that produce scientific knowledge by constructing revolutionary new objects of cognitive investigation (both material and thought processes). Furthermore, special knowledge of them is worked out and their validity assessed in the system of societal interactions.

There is also a different interpretation from the usual one of the nature of a crisis. The most important problem to concentrate on is: "What can the new exemplar/model for a new manipulation area be?" It is most important to reach a practical consensus that would allow some new practice to be continued. A consequence of the change in perspective and turning to the practice interpretation outlined by Rouse leads to the attribution of decisive importance to instrumental and experimental revolutions over revolutions in the theory of the paradigm dynamics. Science is then first of all seen as a practical-theoretical capacity of producing phenomena experimentally (it is "power") and understanding them, including the ability to conduct reasoned experiments. A consensus is only needed in as far as it is felt to be needed in practice. Science and technology are seen as special sorts of interconnected material and cognitive action that, as for any sort of action, make things and learn how far it is necessary to change knowledge by interpreting this process. Joseph Rouse identified a decisive reification perspective in interpreting Kuhn and contrasted it with the idea of science as a cognitive practice of producing phenomena and understanding them and their production.[5]

1.3 Further reifications in interpretations of paradigm dynamics

Masterman emphasized that steady interaction of three constituents: acting as a community, working with special instruments, and providing for "metaphysics", together make paradigm dynamics. This cognitive activity as a special sort of practice aims both at an instrumental activity to systematically gather information after having found a "trick" to orient the search process and at intellectualizing the acquired knowledge as "metaphysics", i.e., as temporarily valid knowledge of reality when it is approached by the acquired new paradigm. Rouse identified in the mainstream philosophy of science one sort of reification, the reconstruction of science first of all as a mental representation instead of leaving the prime role to the practice perspective. By looking for different ingredients of this practice, we can ask how the cognitive and the social-institutional relate to each other in paradigm conception.

---

[5]What is meant by "reification"? Roughly, it is the conceptual technique by which seemingly independent beings are hypostasized from the elements of a dynamic whole in a process. Reification gives a (false) metaphysical meaning to the analytical work.

As Trevor Pinch pointed out (Pinch 1982), both a "conservative" and a "radical" reading was developed to the paradigm conception. As he observed critically: "a distinct use of the term emerged to facilitate the separation of the description of the scientists' social activity from the description of their cognitive activity". Thus, they could be first separately investigated as independent "dimensions" of science were constructed. "Conservative" reading starts with an analytical demarcation of the elements that make the whole and is based on the idea that a description of the social and cognitive activity is simply separable. Cognitive activity is thus supported and promoted by an institutional, organizational "side". Science dynamics necessarily are to be interpreted by the sociology of institutions, but it is senseless even perhaps dangerous to develop any microsociology of cognition itself—according to the "conservative" reading. In contrast to doing this, he called for exploration of "the combined socio-cognitive nature" of scientific activity. In this "combined" view, each part only makes sense within the context of the paradigm as a whole.

What role should be attributed to agents and structures in paradigm dynamics? Another reification reading can be given to the paradigm conception in relation to this question. The usual interpretation has been simplifying and deterministic. According to this interpretation, paradigms are dynamic structures, which determine agents' actions and have rigid periodization of their phases in time. They are taken object-like, independently of time and space; agents simply have to follow them. Arie Rip repeatedly called attention to this possible misreading, including the book chapter written with van den Belt in 1987 to review later. He suggested that a fully-fledged process analysis of science dynamics from an agent's perspective would be needed instead. This would embed the paradigm dynamics in an unceasing agents' interaction, from both inside and outside, by and on the structures, exploring the essential constructive element and its force over the whole lifetime of a paradigm.

Reification in interpreting the autonomy of the science dynamics is a further possibility. One can discuss the autonomy of science in a demarcationistic way. Then the autonomy is either constructed descriptively as a "stylized" perceived fact, as a generalizing descriptive learning from searching history, or it is something to be defended in an isolationistic normative attitude. Normative autonomy is realized by setting an essential boundary to preserve the alleged (supreme) nature of the scientific cognitive enterprise. But autonomy can also be conceptualized in a non-demarcationistic way. Then one can speak about autonomy as being necessarily produced and reproduced flexibly through interactions of a multitude of historically changing mediations (allowing their differing in different branches of the sciences) and through interventions into the autonomous, as some only contextually valid, relative issue.

Kuhn formulated a historical model of self-sustaining autonomy realized by successive communities of specialists, in an enduringly favourable societal environment. In this historical process, from time to time new communities emerge and stabilize within science to solve the, by scientific cognition self-induced crises. They establish and sustain a new cognitive tradition, leaving aside a previous one. In this process of autonomy, repeated historical re-definitions of "what science is" is the exclusive task of scientists, which keeps science autonomous, according to Kuhn. The application of the paradigm conception by conceptualizing heterogeneous, technological, communities that are, as experience indicates, in some measure always heterogeneous and their paradigms (or "cognitive traditions"), immediately challenged the one-to-one use of this sort of autonomy conception and called for an understanding of technological dynamics as being semiautonomous. Adjusting of the paradigm conception to technology in some measure endogenized technological activity into the economic-societal sphere, while preserved its semi-autonomous character in a specific way.

2 Reconstruction of the cognition mode in technology by utilizing the paradigm conception

2.1 Constant's technological traditions and communities

Different sorts of technological paradigm conceptions are assessed in this second part of the article. Some historians, philosophers and sociologists of technology, as well as theoretical economists, applied the paradigm approach to technology albeit with a significant delay with respect to the first publication of the *Structure*. It is well-known that in 1980 Edward Constant summarized a twin conception of a technological community and technological tradition (Constant 1980). He aimed to develop a general middle-level model for technological change and worked out a detailed case study. He produced common boundary objects for historians and philosophers of technology to investigate. Constant claimed that a main task emerged by the 1980s. This was to achieve a systematic two-way communication between historians and philosophers of technology. This could be done by providing for common boundary objects, the ideal typical models of historical dynamics and case studies based on them. These constructions enabled selected characteristics of history to be revealed and understood (dealing with "historians' facts", as Constant referred to the American historian E. H. Carr).

Constant was deeply influenced by Kuhn. Literature mostly calls his construct a technological paradigm. But as we shall see later, just like Dosi, he provided a syncretic model. His general aim for understanding technological cognition was to integrate the model of technological change into Donald Campbell's general model of evolutionary epistemology, as the dynamics of changing technological communities and their cognitive traditions. Furthermore, Constant gave space to Popper's ideas of scientific cognition in the reconstruction of technological cognition.[6] He approached technology as a type of cognitive activity with specific purposes. He took technology as a material and mental practice to be interpreted from the perspective of societal practice. Technology is practised by technological communities. These carry out specific, technological cognitive activities in the form of technological traditions. Furthermore, from time to time they change these traditions in revolutionary ways. The complex information set, which makes up technology, is physically embodied in the hardware and software of which a community of practitioners is master. This knowledge is partly explicit and partly tacit. Different traditions of technological practice emerge as accepted modes of operation from time to time.

> Such traditions encompass aspects of relevant scientific theory, engineering design formulae, accepted procedures and methods, specialized instrumentation, and, often, elements of ideological rationale. (Constant 1980, 10)

> Technological revolution has occurred when a new tradition of practice comprising a new normal technology is initiated [and] when a community of practitioners embraces a new tradition. (Constant 1980, 19)

He identified revolution with the first appearance of a new community, even if this was only very small, committing itself to a new technological tradition in its activity.

Constant emphasized that technological cognitive activity also expressed essential dissimilarities to dynamics in science. Technology investigates its object directly, not "vicariously" (referring to Campbell's term) and the selection processes are more complex. Furthermore, there is a more hierarchical structure in technological communities than in science, "satisficing modes" (H. Simon) in technology are different and also the role of economic criteria.

Constant deliberately avoided referring to his findings as technological paradigms. His reasons were two-fold. He referred to the alleged unacceptably numerous different meanings of the term paradigm and the supposed

---

[6] "The conjecture and testing of scientific theories represents perhaps the highest, most abstruse development of the fundamental blind-variation-selective retention model" (Constant 1980, 7).

difficulties in identification of the exemplar following dynamics in technological cognition. As mentioned earlier, he did not accept puzzle solving in the normal mode of working of a tradition. Instead, he spoke of conjectures and their testing in a Popperian way.[7]

Experts have been trying to assess the extendibility limits of the used technologies and the possibilities opened up by a new technology candidate in history. Provided that there was a background science, by 1980 this could give an important basis to technological prediction more and more often. Constant gave an explanation for this phenomenon. In his reconstruction, he identified "presumptive anomalies". Besides what he called functional failures such as plane crashes notwithstanding the improvement efforts, "presumptive anomalies" may appear. These occur, based on new scientific achievements, either when functional failures with some possible future technological advances can be predicted or when the imagination of a much better, radically different, technology is possible.[8] (The predictability of functional failure and possible new technological constructs that can overcome the old ones are not, of course, necessarily connected.) Constant attributed "logical rigor of theoretically derived presumption" to establishing "presumptive anomalies".[9]

Besides of recognizing the existence of "presumptive anomalies" Constant put emphasis on the originality of his recognition of "technological co-evolution" as a main constraining factor in technological design. He clearly formulated that the concept of technological co-evolution implies more than the "technological disequilibrium" or "reverse salient" of Thomas P. Hughes because also includes the directive force of co-evolutionary constraints and even the constraints set by the hierarchy of retentive and selective processes

---

[7]This is rather a questionable idea if one takes into consideration the way technological artefacts are tested in reality. There is an effort first to demonstrate the realizability of technological artefacts, second to use this knowledge to really create them and third to improve them. Only then is it possible to look for the limits of the realization of the invented constructs and perhaps overcome their capacities. There has not been any observable permanent revolution in the history of any technology and it is difficult to realistically imagine such a concept. Masterman claimed that puzzle solving and testing do not exclude each other in normal phase of paradigm dynamics (Masterman 1980). This should have been something for Kuhn to reconsider also.

[8]The recognition of presumptive anomalies does not result from a deepening crisis as is the case with normal anomalies. It is a preventative act, based on available (new) scientific knowledge to avoid functional failure in time. Recognition of the phenomenon is very important. Its extension to dimensions such as the economical or societal, etc., that require radically new technological capabilities also seems worthwhile. Its characterization by Constant is questionable in some essential features as will be indicated later.

[9]As he characterizes it in his article in the book edited by Rachel Laudan (Laudan 1984).

in higher-level macro systems, because most technologies are hierarchically organized systems.

The most important characteristics of Constant's conception are as follows. Technological communities and traditions of technological practice are two-fold, each components are essential components of the whole. Similarly to the story in philosophy of science of the paradigm conception, Constant actually identified a socio-cognitive activity in technology, meaning that some cognitive habits that identify a technological community exist inseparably from some technological traditions. Furthermore, from time to time technological communities make a decision that some sort of new cognitive practices are to be introduced. Normal and revolutionary phases change in Constant's model. But, in contrast to the homogeneous communities in science, Constant's technological communities are heterogeneous. This change was introduced to take into account an obvious fact in the practice of technological development. Knowledge carriers do not only include research communities of technological experts but also firms, government authorities, etc. and take into account economic and societal requirements for successfully functioning artefacts and technologies, not only narrow technological criteria. This is certainly another basic difference from Kuhn's conception of a scientific paradigm. (A closer analogy to Kuhn's scientific paradigm than the paradigm conception that was really used in explaining technology development would have conceptualized technological functionability *in abstracto*, notwithstanding in any relation to economy or the social issue. This conceptualization has a real but rather limited use in technological practice.) Further, according to Constant, technological communities realise a practice of systematic conjecture and rigorous Popperian testing in the normal phase of their activities.

Constant tried to marry a Popperian cognitive mechanism with a normal technological activity when he understood the latter was a process of setting alleged conjectures and testing them rigorously within an accepted tradition. But Constant tried to find even a more radical place for conjectures. As he claimed, while the community practice provided accepted standards for making conjectures and their rigorous testing, technologists frequently violated the accepted rules. There are frequent break-out trials in the dynamics led by technological traditions. With these we have a further clear difference to the mechanism claimed by Kuhn. Constant tried to identify a specific level of blind-variation and selective retention in the activity of technological communities. He claimed to have found it in making and testing wild conjectures, not only testing the systematically developed conjectures and following the accepted rules. Introduction and understanding of this making of wild conjectures from time to time, repeatedly challenges both those involved in engineering design practice and those reflecting on them. Debates have repeatedly been coming back in the history of technology and engineering sciences from the late 19th century. These debates revolve around the nature of engineering cognition. By attributing some sort of unavoidable creativity to successful engineering design, including setting of wild conjectures many bestow some "art" character on it. When debaters try to repeatedly raise the question "how far is engineering design science or art?" they especially rely on the fact that wild, but successful, conjectures repeatedly appear in engineering design practice.

The general model Constant developed is very strongly bounded to his concrete historical research. This research was devoted to the construction of a new type of aeroplane, the turbojet. As he argued, this construction started with a new scientific recognition in aerodynamics. Significant, perhaps paradigm changes in scientific cognition can systematically serve to prepare paradigm changes in technology, by way of recognizing "presumptive anomalies".[10] With the growing capacity of science working in this direction, this is recognition of the utmost strategic importance—provided it is filtered through a serious assessment, a consideration that I shall return to later. Science-technology nexus in the two-way interaction process emerged and developed in an accelerated way from the mid-18th century onwards. It became more and more institutionalized and widespread in different branches of technology. Part of this process is recognizing the necessity to look for new technological paradigms because of the identification of "presumptive anomalies", or at least for decisions based on an accepted belief that some limits of further development of a traditional technology are impossible to overcome.

Constant succeeded in recognizing a very important sub-class of technological revolutions. But he was less successful in realizing a generally valid middle-level reconstruction of technology dynamics, because he did not include other sub-classes in his considerations, not to mention the questionable unification of Kuhn's model with Popperian features.[11]

---

[10]Constant recognized numerous, important sociological, institutional characteristics of such issues. One important concern is that revolutions based on "presumptive anomalies" may often be originated by outsiders, and this is not by chance. He further wrote in connection with this that old experts will lose their competence to this end.in new technologies. Michael Tushman and Philip Anderson observed a fuller situation some years later, in 1986, when they wrote about competence enhancing vs competence destroying effects, concerning the emergence of different "technological discontinuities" (Anderson & Tushman 1986).

[11]Originally, in the *Structure*, Kuhn enumerated available material instruments among the ingredients of a paradigm and attributed much importance to them. We can say that a (scientific) paradigm has material, conceptual and social ingredients. The conceptual ingredients appear in two functions, theories as working instruments and ontology.

2.2 Rachel Laudan and Constant on the autonomy of technological cognition

While most historians of technology concentrated on detailed descriptive reconstruction of individual inventions, the philosopher of science, Rachel Laudan was one of those who, around 1981, identified the task of reconstructing the dynamics of the cognition mode in technology on a general level. She and those who gathered at a workshop in 1981 belonged to sociology, history or philosophy of technology.[12] Roughly, the model developed was expected to demonstrate two things. Firstly, their research orienting basic belief was that technological development has autonomy in relation to science. It is not and is not becoming "applied science", either. With this they wanted to contrast the quickly spreading "technology is applied science" vision as a dangerously misleading myth. Secondly, the model should show that technology itself necessarily has its own dynamic, similar to that of science.

The results of the 1981 workshop were published by Rachel Laudan as editor in 1984 (Laudan 1984). The workshop addressed the question of the specific nature of knowledge production in technology as the central carrier of the alleged autonomy of technological development. Further, it asked if models of scientific change are relevant to explain technological change. Technological cognition was presumed to be representative for "understanding change and development within technology itself". It was to explore how changes that occur in this dynamic also simultaneously take into account how the rich nexus with science, economy, and society sets up in the autonomous development of "technology itself".

The overall purpose of the workshop and the resulting volume was to direct research towards working out a specific epistemological concept of technology with specific patterns of problem solving activity. The application of the paradigm conception was agreed on to make the comparison.[13]

Constant did not focus on technological revolutions that are initiated and emerge with new technological, craft tricks, not with new scientific recognition. It is useful to look at the revolution in microelectronics around 1960, concerning an invaluable technological revolution of the class that starts with a radically new technological "trick" when microelectronics turned from individual transistors to integrated circuits, especially in their planar realization, as a reminder. With this a new type of artefact and product technology was created that provided for the very basis of the further development in microelectronics and for a radically new research object. This revolution could unfold for a while without support from application-oriented scientific research (Moore 1965). However, later it developed into a complex system of interactions between technology and science.

[12]Participants included Edward Constant, Derek de Solla Price and Gary Gutting.

[13]Due to their different goals, research methodologies and evaluations, a clear differentiation must be made between the characteristics of a paradigm change like dynamics

They found that a turn in philosophical investigations was needed. Assessing history based on learning from real historical practice was seen as essential to be able to find valid long-term patterns.

Concerning the differences between science and technology, Laudan's and her company's starting point was the usual view that science is "innovative eliminating" while technology is "innovative preserving". This means that the ontological validity of the existing paradigm in science will be lost when a new paradigm is accepted. In contrast to this, a new technology provides a new way of functioning or new functions, but the validity of old functioning in its original field remains preserved.

Laudan tried to extend the idea of anomaly by a, not only in term of institutionalization, but what I would like to call cognitively essential contextualization of "technology itself". In her opinion, besides functional and presumptive anomalies, different "social anomalies" can systematically occur among technological cognition anomalies. "Anomalies can come from every corner"—as she said. (This includes even anomalies caused by time pressure or lack of tools.) Edward Constant identified technology in this volume with emphasis on realizing knowledge and practice, required for systems design and fabrication, in technological communities and realizing "social function" in organizations.

> Thus, anomaly for actual technological practice in the functional core of the organization can come from all sorts of different directions. (Laudan 1984, 41)

This recognition was both important and adumbrating. Organizations, firms, a group of firms working together, etc. that develop technologies are always with different parts of their environment in a differently tight interaction. Requirements can really come from every direction and, mostly, sooner or later, the organization has to accommodate them. It is possible to consider the emergence of "social anomalies" and use the term social anomaly in the possible widest sense. Then, over functional failures and presumptive anomalies, an army of "social anomalies" influences technological development. This may include organizational anomalies, anomalies of

---

of technological science and concrete technologies or artefacts such as production technologies and products in microelectronics or modern biology-based pharmaceuticals. The immediate direction of research in technological science is developing science. In contrast to this a technological paradigm immediately aims to pinpoint artefacts or production technologies. Research pace and direction and methodologies are different. This difference is often not taken duly into account in the reflexive literature about technological development. In the 1981 workshop the participants dealt with the dynamics of technology as production.

legal adaptation, etc.[14] But it is questionable first, if they all have cognitive consequences for technology as we understand "technology itself" and second, if trials using the existing technological paradigm to meet "social anomalies" urge paradigm changes, at least in some parts of the cognitive paradigm used for the development of the technology in question.

Laudan and Constant highlighted an essential issue here. But it could be that a "social anomaly" does not lead to a new type of cognitive anomaly. But the reverse can also occur, that the "social anomaly" induced cognitive anomaly will lead to cognitive crisis and the decision will be made to look for a new paradigm. Further, it can also occur, that inappropriateness of the available cognitive form and its recognition emerges on the metalevel, including that of existing cognitive values. To end this inappropriateness, a new cognitive form is needed.[15] It concerns setting up new modes of cognitive efforts, to help construct appropriate new technologies.[16] These efforts will come up against the accepted values, habits, methodologies and functioning of the old technological community. It is possible that the old community will try to resist the new efforts. A societal crisis in technology may then emerge.

New societal and economic expectations may lead to requirements for new cognition modes. These can urge focusing on new research instruments and ontologies in science.[17] These instruments may at least partially have been developed earlier by extending the curiosity in existing sciences. But

---

[14]Constant emphasized that each of the functional modules, legal, marketing, strategic planning, within an organization mediates some special social requirement. One can continue the listing with environment protection, data protection or a security module (the last being against terrorism).

[15]This also applies to some issues when new economic needs are involved in the development of expected knowledge. In practice, these economic needs are formulated with respect to limits of time and space and available tools. For example, as in any branch of industry, microelectronics must regularly provide new generations of technology and artefacts over time, but within very short periods to keep to Moore's law. When a technological (process) paradigm works appropriately, it provides knowledge of whatever sort of origin able to serve the goal of realizing the needed technological artifact (process or thing) in time, etc. Elimination of a shortage in time, space and instruments is attempted by real tinkering, by putting into action a series of locally successful "tricks". Successfully manufactured artefacts contain built in marks of tinkering. Tinkering is built into the fabric of successful artefacts, often once and for all.

[16]An example is the transformation of sewerage technologies to serve sustainable sewerage. This is a telling case of changing endogenization of sewerage technology in economy and society that forced a radical renewal of sewerage technologies. This not only led to recognition of a lack of appropriate knowledge but also a lack of the appropriate cognition form for the new type of technology required.

[17]Kuhn only made a single remark in this direction in the numerous texts he later wrote to explain what the *Structure* "really was about". In this remark, he allowed profound change in the direction of cognition as the effect caused by new societal values to be taken into account for the development of science.

they obtain new nexus to society or economy, including their relevance to the societal, economic etc. requirements. One can point out the appearance of environmental sustainability requirements as an example of changes in the cognitive perspective as well as in existing research and technological paradigms.[18]

The opening paradox remained to be interpreted in summary. What was meant by investigating "technology itself" when the demarcation, as was correctly recognized, is essentially dissolved in some respect? As Laudan herself correctly emphasized, "anomalies can come from every corner", i.e., expressed in her terminology, also from "outside", from science and economy and society. Nevertheless they may have essential effects on the "inside", leading to challenges to change existing paradigms. A relatively autonomous working system, a semi-autonomy, was conceptualized in this way, in terms of goal setting, paradigm dynamism and knowledge pool acquired from the contextualized technological cognition oriented in different levels by feedbacks from both science and economy and society. The semi-autonomous development of problem solving activity in a heterogeneous community was conceptualized. The autonomy of technological cognition was preserved by accepting some sort of paradigm dynamics for the problem solving activity in the explanation. This contextualized autonomy could be contrasted with the applied scientific approach responsible for the pace and direction of technological cognition.

2.3 Giovanni Dosi's technological paradigms

Among all who carried out pioneering work by applying Kuhn's ideas to their own field, be it history, philosophy, sociology or economics of technology, Giovanni Dosi perhaps received most response from his own professional audience (Dosi 1982; 1988). This audience was the rising group of evolutionary economists in the 1980s. The notion of a technological paradigm became a quasi-natural explanatory instrument in that community by the end of the 1980s.[19]

---

[18] One side of the interaction with the "environment" is the emergence of social anomalies of most different sorts in the social texture of an existing technology. The other side is requirements for brand new technologies that will set up in most parts of the new "environment". Some of these are organizational, some psychic, some cognitive requirements, etc, They are directed towards users, regulatory authorities, suppliers, indirectly involved layers.

[19] Christopher Freeman and Carlotta Perez extended efforts in understanding the effects of technology dynamics as dynamics of large systems of technologies on whole economies, by developing the idea of techno-economic paradigms (TEPs) (Freeman & Perez 1988). These paradigms were conceptualized to be able to take into account transformations causing changes in Kondratiev cycles. (It is impossible to include any assessment of the story of searching for TEPs within the strongly constrained frame of this article.)

Recognition of the "residual problem" in economics and the extreme importance of its solution in improving economic explanations shattered the community of economists by the end of the 1950s. The candidate for solution, explanation of the role of "technology" taken broadly, i.e., including organization and management technology, got at the centre of explaining economic change. Evolutionary economics set a new paradigm in economics against the neoclassical approach. It denied the validity of the methodology of investigating first static states and approaching processes from this starting point. It did the same concerning the abstract rationality perspective used to describe decision making, by concentrating on static efficiency criteria. Evolutionary economics intended to understand the long run of economy as evolution, endogenizing technology dynamics as one decisive factor. It was to explain that firms actively adapted in the real world, using "bounded rationality" to unmask and conceptualize real uncertainties and contingencies in the evolutionary process above all, not only to quantify risks. As Nelson and Winter outlined, in the new conceptualization firms modelled as routines' clusters make technological innovations along working trajectories of "search routines" and by the dynamism of changing these "search routines" (Nelson & Winter 1982). Working and change of "search routines" offered similarities to paradigm dynamics.

Explanation trials in economics alternated between push and pull models. In both cases, technology was taken as an external factor of economy. Approaches from the demand perspective presumed a short-cut reactive mechanism directing the development of technology, mostly working by way of institutionalization and financing. In push models it was supposed that, from a cognitive point of view, the dynamics of constructing new technologies will be more and more the progressing exploitation of sciences' application. These provide for an ever-growing set of possible technologies. A selection process based on economic criteria transform some of them into realized technologies.

Dosi was interested in a third model, which was to set to simultaneously overcome the deficiencies of both the push and pull approaches. Besides integrating both push and pull elements into the new model, that model would be based on the unavoidable relative autonomy of technology dynamics, of its essence, the cognitive efforts needed for developing technologies. By trying to give appropriate space for economy and science in directing the dynamics of this cognition mode, he provided for an endogenized understanding of the relatively independent technological development.

Dosi called his model a technological paradigm. But he used in its formulation what he named the "common core" of what Kuhn and Lakatos had said about the dynamics of sciences. He looked at a technological paradigm as a special problem solving activity, instead of identifying technology in the usual way of economists of his time as blueprints, artefacts. Dosi constructed or took over from everyday knowledge of economists a series of "stylized facts" that served the role of explanandum for the technological paradigm conception. Some of these were ability to draw innovative ideas from novel opportunities, stemming from scientific advances or the increasing complexity of research and innovation activities in history. A part of his "stylized facts", understood as the "remarkable historical consistencies" as Nicolas Kaldor who introduced the "stylized fact" perspective and the term in methodology of economic explanations characterized "stylized facts", served for Dosi to see the direction of constructing a model that is contrary to the basic presumptions of neoclassical authors. Dosi excluded from the "stylized facts" that deliberate, optimizing choice in a finite set of alternative technology candidates was a real, historically observable behaviour in technological practice. Technological practice worked differently, as he assured. The allegedly observed "strong internal logic", "cumulativity" in the development of some phase of technology and some other, by him accepted "stylized facts" such as "locality" of first looking for a different approach immediately served the goal of constructing a new explanatory model.

Setting systemic relations of "stylized facts" by the paradigm change model served to explain the unconnected "facts". Beside some explanatory power, their systematic order in the paradigm model provided heuristics and predictability at the general level. An explanatory answer to the set of "stylized facts" serving for the explanandum was the statement that a paradigm form, a structural pressure of patterned dynamics of technological cognition, exists. There is a structural determination, a dynamic systemic nature of technological cognition inside technology and no economic-societal requirement, human activity, institutional and financial support could suspend it. The existence of strong structural connections in the model explains that the cognition process inside technology must have some "internal logic", claimed to have been observed. The technical cognition process serves the goals set by the economy, but to provide the knowledge needed to solve them has to follow its own necessary dynamic pattern that makes it an autonomous process.

In this way, cognition in technology was reconstructed as an ongoing process of changing continuous and discontinuous periods, emergence, stabilization and change of relatively ordered patterns with their "momentum" propelled by a "strong internal logic, despite quite diverse market conditions", as he summarized. The presumption of "strong internal logic" in technological cognition got a "message" function for those who believed in the determining role of factors either on the supply or demand side and was contrasted to their view. Dosi's model considered economic policy and management based on the demand perspective, i.e., that demand side requirements are strictly limited by the laws of the relatively independent technological cognition and hence technological development. On the other hand, it told believers in the supply side factors that even when applied science promises possibilities of new technologies, autonomy of technological cognition, because of its "strong internal logic" is impossible to neglect in the technological realization of the promises.

Dosi gave the following definition:

> A "technological paradigm" contextually defines the needs that are meant to be fulfilled, the scientific principles used for the task and the material technology to be employed. In other words, a technological paradigm can be defined as a "pattern" of solutions of selected techno-economic problems based on highly selected principles derived from natural sciences, jointly with specific rules aimed to acquire new knowledge. (Dosi 1988, 1127)

With this definition, Dosi gave a heterogeneous characterization of a technological paradigm. A technological paradigm essentially includes something from science into its definition, as well as from economy-society. But technologies should be conceptualized as specific problem-solving activities with their own, unavoidable patterns. There is a trajectory in normal technological development, a regular cumulation in one direction. Direction changes are shared effects of economic-societal and cognitive causes but the normal cognition process itself, even when it contains endogenized economic criteria, is autonomous, directed by its own laws.

Dosi used "stylized facts" to develop his explanatory model. This step had important consequences. The explanation was bound to a preceding special simplification act in identifying the facts it explained[20] that gave place to the usual, structurally overdetermined interpretation of paradigm dynamic. In this mass-tailored interpretation of the preformed "facts", the genuine creativity of the technological agent was strongly deprived of its possible, observable creative working by the choice of "facts" to be explained.

There is more, or just less various immediate impact of economy or society on the development of technology and cognition in different branches of technology, at different times. The extension is between conscientious disregard of concrete use among the goals when a new artefact is constructed or, just the reverse, when "lead users" dominate the whole innovation chain. An example of the first case is the construction of the so-called glasbeton.

---

[20] For example cumulation was simply taken some sort of linearity.

This promised a lot of still unknown application possibilities, in time of its first construction. Its normal development phase is the exploration of (concrete) application possibilities, being, of course, dependent on the concrete decisions to look for direction, and with this on content to develop. An example of the second case is the application of the living lab method. The same dependence of the normal development applies here, just requirements of concrete users modify the normal development in specific ways. A technological paradigm is a construct which, between limits, can be differently modified even in its normal phase, concerning its texture (content) and direction. Conscientious choices may be for example strategic instruments in the competition. They may aim to modulate the development direction of an industry's branch, or dictate the pace of development of the next generation artefact.

Technological paradigms are of a hierarchical structure. There are sub-paradigms in a hierarchical order of different levels in a comprehensive paradigm. Some of these may be accelerated by some sort of 'outer' feedback, for example a new scientific recognition, or a recognition in a different branch of industry. They often lead to new, surprising results that may have a strong effect even when the development of the comprehensive paradigm is in its "normal" phase. Other sub-paradigms prove unable to produce the needed solution in time required by the principle of necessary co-evolution. The clear lesson to draw would have been that technological paradigms, just as technological paradigm producing higher level mechanisms, have a contextual autonomy with a multitude of permeable boundaries on different levels, not only on the science, the push, but also on the pull, side. In other words, the normal phase of paradigm development in technology as an autonomous process is repeatedly penetrated by effects from "outside". Further, agents have an essential creative role in constructing any concrete paradigm, even in its normal phase. So, they should be appropriately taken into account in the generalized model. Structural conditions are partly produced not only reproduced by the interaction of agents in the normal phase. It is only justified as a first approximation and repeatedly leads to wrong predictions when autonomy is simply attributed to the normal phase as a cumulation mechanism deterministically following its "internal logic".

Dosi identified two levels of change. One was that of working specific successful "search routines" as technological paradigms with their products that obtain their ex post-market evaluation. The other was the paradigm dynamic of changing search routines, the "mutation generating mechanisms", as he called them, in the selection of which economic and societal values assume an *ex ante* role.

Dosi's modelling produced an appealing construction to challenge, in general, the two extremist approaches, the supply and demand models. Henk van den Belt and Arie Rip initiated a "sociological extension", a contextualization of what they called the Nelson-Winter-Dosi model (van den Belt & Rip 1987). They argued for taking into account sociological consequences of anticipative interiorization of the market needs into the cognitive mechanism and further, of the regulating environment, as built-in anticipative institutions within technology dynamics, namely, interiorization of the "social function" (Constant) of technologies. Systematic development of the so-called test departments that anticipate user requirements and the patenting system are examples of this interiorization, they gave. They outlined in a case study on the synthetic dye industry in the third part of the 19th century how two nexus building institutions interacted with technological research in the normal phase. (This reconstruction of the historical case convincingly showed that the earlier reconstruction of the synthetic dye production case as a successful application of science was misleading.) By extending their perspective, they argued for systematic exploration of "nexus building" between the paradigms and the selective environment.[21]

Concerning the basic differences Dosi claimed to be able to distinguish between scientific and technological paradigm mechanisms. It should be noted that he also insisted on the statement that "competition does not only occur between the "new" technology and the "old" one it tends to substitute, but also among alternative "new" technological approaches" (Dosi 1982, 15). This pluralization of the number of candidates was not only in contrast to the analogy that Kuhn's model provided, but also claimed to be unnecessary to explain the history of technology, as some critiques claimed. Nevertheless, it can be taken as a "stylized fact" that, in many cases, different new technologies may simultaneously offer roughly the same service. Together they demonstrate an important feature that is not only interesting for theoretical analysis but can be most important in management and policy making. Roughly, this concerns the following. It is true that technology is about its functionality and strictly speaking some sort of "incommensurability" exists in their functioning, not only between technologies succeeding each other but also when they co-exist. It is an oversimplified practical approach for some purpose when it is said that the same function can be

---

[21]Nexus building institutions may exert a very strong influence. Concerning the patenting legal regulation in pharmaceuticals, especially recently, the USA has a strong orienting role in pharmaceutical research, because a huge amount of candidates will be refused in the patenting phase of the R&D process. This, earlier ex post realised connection between regulators and developers has recently been changing by *de facto* adding to the patenting process quite regular ex ante discussions along the whole "pipeline" with the producers of drug candidates.

realized by different technologies.[22] In practice this simplification is often not acceptable.

Dosi enumerated scientific principles among the constituents of a technological paradigm. Several criticisms can be advanced. First, Dosi should have referred to the whole armory of science. This includes the wealth of technical instruments used for the synthesis of objects and effects and scientific measuring instruments that are used in technological cognition by transforming them to technological ends.[23] Second, the fact that he only concentrated on scientific principles was perhaps result of a small opportunism with the alleged ideas that technological paradigms gradually become directed science and that science is usually identified first of all with principles and theory, at least among philosophers of science and theoretical scientists.The idea of "two cultures" mutually enriching each other by more and more systematically taking in the use of each other's most different goods was already rather developed at that time (Barnes 1982; Barnes & Edge 1982). This could perhaps have helped provide a less reductionist view of science in developing a conception of a technological paradigm. Fourth, science was seen by Dosi as providing for "notional possibilities", to be selected by narrower engineering and economic-societal criteria.

Dosi succeeded in outlining a model whereby the idea of endogenizing technological cognition into economy and society can be persuasive. He succeeded on the meta-paradigmatic level to contribute to the recognition of the the long-term development of technological cognition as being relatively autonomous but in a double way contextualized undertaking. His reconstruction satisfied the theoretical economists more than could produce a well-developed instrument for historians of technology and economy. This was caused by the over-simplifications his model was based on.

2.4 Additional remark on "presumptive anomalies"

Finally, I wish to make a last remark on "presumptive anomalies". It is based on the story of miniaturization in the production of integrated circuits (ICs). A steady fear accompanied the skyrocketing development of miniaturization in the production of ICs in microelectronics since the early 1970s. The fear was that the unbelievably quick development would very

---

[22] As a simple example I refer to the case that energy can be obtained from many sources. Their production processes are different and the products have at least partially different functions just as they have partially different trade-offs. They are comparable in their functioning but because of the family resemblances, unambiguous comparative calculations, not chosen for some reductive comparative perspective are impossible to make.

[23] Approaching the problems with utilitarian simplification, de Solla Price insisted on the idea that new research in science and technology start repeatedly dancing as a pair to the music of instrumentalities (de Solla Price 1984).

soon lead to an immense bump, a final limit, where the process would hit a "red brick wall". With this bump, that branch of industry would crash unless it turned to a qualitatively new technology in time. This strong fear has been running parallel to the exaggerated enthusiasm over the unbelievable successes microelectronics has made in comparison to any other branch of industry. An exaggerated psychological situation emerged and was regularly reproduced, where as much exact calculation as possible was expected to provide for as much sure knowledge as possible to make the unavoidable management steps calculable and realizable in time to avoid the crash. Calculations were already made in the early 1970s to predict the final limit of usability of the optical photolithography, one of the backbones of IC production, predictions made on the base of wavelength considerations. (They also led to very expansive searches for possible new technologies already from the early seventies.) These calculations marshalled available knowledge from optics and technology into the arena and repeatedly provided a seductive clarity. The history of microelectronics repeatedly led to revised, improved calculations and predictions of the insurmountable limits in the performance and growth of optolithography.[24]

Surprisingly, a repeatedly successful practice of the improvement of the technology itself contrasted with the series of these calculations. Practice in lithography was repeatedly successful with significantly lower values of miniaturization than calculations in principle allowed for at that time. Looking back at history and expressing the typical pride of practical photolithography experts one of the important researchers in photolithography, Chris Mack, ironically referred to the repeatedly calculated "new, new limits" (Mack 2004).

Expressed in terms of technological paradigm conception, the point was the reliability of predicting final limits resulting from presumptive anomalies. Calculated on hypotheses that were set on the basis of physics and the structure of the dominant design, presumptive anomalies helped "visualize" and predict allegedly "objective inherent limits" to the extension of technology. At first sight, based on the knowledge of natural laws, the concept of presumptive anomalies intends to provide technological forecasting with a reliable, rather exact forecasting method. In principle, if we take the exactly predicted limit early into account promises for a strategic comparative advantage in managing technological revolutions. But the real situations are essentially more complicated. It came out in practice that an

---

[24]Radical progress possibilities in production technology after exhaustion of the old technology were predictable with the development of X ray, electron beams or extreme ultraviolet radiation technologies. Trials with X ray and electron beams started in the early 1970s.

over-deterministic interpretation of the role of presumptive anomalies was typical in the forecasts.

The problem essentially is that predicting a "presumptive anomaly" is itself a cognition process depending on hypotheses how future of science and the technology in question are taken into account. Predicting limits based on taking into account a "presumptive anomaly" has to rely on the state of the art of physics and the technology of the issue and it has to extrapolate from this. In 1994, Rebecca Henderson reconstructed the unpredictable development in photolithography that was made through the practical development efforts in both science and technology. Essential characteristics of the extrapolation, some genuine uncertainty and the conditional nature of the extrapolation provided space for successful practical research to falsify in practice the truth of the theoretically calculated limits (Henderson 1995).[25] In the unexpectedly changing context of developments in optics, practical results were regularly reached that refuted the just calculated limits. Development of The so called component and complementing technologies developed and the users (producers of the ICs) requirements diminished, because of better technological capabilities. All this contributed to the success.

While it is undeniable that, in principle science may demonstrate inherent technological limits, and that these are earlier or later insurmountable, reliable use of these predictions for concrete issues may prove complex, depending on a series of accompanying issues. A series of examples shows that predictions of inherent technological limits, that seemingly include only exact data and calculations, repeatedly prove false in practice. All this is in essential connection with the complexity of technologies, hence with the possibility of one-sidedness of "exact" calculations and the unavoidable specula-

---

[25]There is an immense practical danger with the effects of these types of miscalculations. When they are accepted for technology policy or management they can serve self-fulfilling prophecies. Then, they may practically prevent further research in the direction they think they were able to prove the final limit. They may urge "too early" to commit to a very expensive search for a new type of technology. The real case can be even more complicated as the development of the photolithography shows with the 20 year long recent trial to change for EUVL. On the one hand, in lack of new technology in the production process, the practice is forced to prove its real capacity to continue the development with the earlier technology, on other hand the industry is unable to realize a new industrial technology even when it is in principle predictable at least in its science component and even already got some "proof of concept" quality long ago. The reason is that predicted trajectories with some unavoidable component technologies have been impossible to realize at such a complicated new technology as the EUV technology. In this case, not only did the limits of the old technology prove to be falsely assessed, but also the prediction of the trajectory of the new technology proved repeatedly to be false. Unfortunately, the recent ideology of presumptive anomalies still lets us concentrate on the quantitative side of the calculations, on its "logical rigor" (Constant) and not on the assessment of how far unavoidable uncertainty and contingencies may deeply penetrate the calculations.

tive efforts even in a "normal technology" phase of technology development. Explanation of the regularly repeated miscalculations in history of technology reveals that predictions of both future technological possibilities and of "natural, inherent technological limits" involve assumptions that are taken for granted when the predictions are made, but prove false later. Further, many assumptions are tacit, hence unconscious when the extrapolations are made.

Any technological paradigm has a complex structure and moves in interaction with a changing inner and outer environment. This structure includes its own component technologies as well as complementary but unavoidable technologies for the working of the investigated technology. Changing user needs with time are to be added to the issue. The changing environment further includes supply agencies and regulation authorities. Even incremental development of a technological paradigm is a complex issue with the interaction of numerous partial processes among these agents that involve some "genuine", Knightian uncertainty. To say it in the language of paradigms, in normal phase search for "puzzles" occurs but an open set of solutions are candidates, depending on the changes in the inner and outer environment. Assessment of the case of the optical photolithography, with its "unexpectedly long age" (Henderson) shows that the whole issue is embedded in presumptions that easily remain tacit and prove false, just because of the unsolvable essential, "genuine", Knightian uncertainty in real historical practice.

Unquestionably, the normal phase of technological paradigm development has a relatively autonomous character and trajectories can be continued with relative safety. But prediction quality remains problematic when it faces the stretched time requirement that especially some industrial branches have to set, and it loses much practical usability in such extraordinarily quickly developing branches as microelectronics, where, paradoxically, it would be most important. Nevertheless, it is possible to try to be more circumspect. However, both the repeated interactions inside the paradigm and the "outer environment" provide an interactive set of unceasable uncertainty. There is an unavoidable social construction moment in the exact calculations that makes them conditional.

3 Conclusion

This article brought us back to the origin of technological paradigm conceptions, to the 1980s. The idea of a technological paradigm provided a plausible model of autonomous change, of self-regulating mechanisms in *mutatis mutandis* understood "technology itself" that unavoidably endogenized economic (and social) criteria. By different authors the paradigm conception was differently adopted as influenced by its economic and social environment. Conceptualizing the dynamics of technological cognition in terms of heterogeneous communities and their cognitive traditions revealed the specific contextual autonomy, provided the necessary changes immediately required by the technological topic have been made. Reification interpretations of characteristics of paradigm dynamics often accompanied and oversimplified the process of modelling.

Bibliography

Anderson, M. L. & Tushman, P. (1986). Technological discontinuities and organizational environments. *Administrative Science Quaterly, 31*, 439–465.

Barnes, B. (1982). The science-technology relationship: A model and a query. *Social Studies of Science, 12*, 166–172, doi:10.1177/030631282012001013.

Barnes, B. & Edge, D. (Eds.) (1982). *Science in Context. Readings in the Sociology of Science*. Milton Keynes: Open University Press.

Constant, E. W. (1980). *The Origins of the Turbojet Revolution*. Baltimore: Johns Hopkins University Press.

Dosi, G. (1982). Technological paradigms and technological trajectories: A suggested interpretation of the determinants and directions of technical change. *Research Policy, 11*(3), 147–162, doi:10.1016/0048-7333(82)90016-6.

Dosi, G. (1988). The nature of the innovative process. In *Technical Change and Economic Theory*, Dosi, G., Freeman, C., et al., eds., London: Pinter, 221–238.

Freeman, C. & Perez, C. (1988). Structural crises in adjustment, business cycles and investment behaviour. In *Technical Change and Economic Theory*, Dosi, G., Freeman, C., et al., eds., London: Pinter, 38–66.

Gutting, G. (Ed.) (1980). *Paradigms and Revolutions, Appraisals and Applications of Thomas Kuhn's Philosophy of Science*. Notre Dame: University of Notre Dame Press.

Henderson, R. (1995). Of life cycles real and imaginary: The unexpectedly long old age of optical lithography. *Research Policy, 24*(4), 631–643, doi:10.1016/S0048-7333(94)00790-X.

Kuhn, T. S. (1962). *The Structure of Scientific Revolutions*. Chicago: Chicago University Press.

Laudan, R. (Ed.) (1984). *The Nature of Technological Knowledge. Are Models of Scientific Change Relevant?* Dordrecht: Reidel.

Mack, C. A. (2004). The new, new limits of optical lithography. *Proceedings of Emerging Lithographic Technologies VIII, SPIE, 5374*, 1–8, doi: 10.1117/12.546201.

Masterman, M. (1970). The nature of a paradigm. In *Criticism and the Growth of Knowledge*, vol. 4, Lakatos, I. & Musgrave, A., eds., Cambridge: Cambridge University Press, 59–90.

Masterman, M. (1980). Braithwaite and Kuhn: Analogy-clusters within and without hypothetico-deductive systems in science. In *Science, Belief and Behavior*, Mellor, D. H., ed., Cambridge: Cambridge University Press, 61–86.

Moore, G. E. (1965). Cramming more components onto integrated circuits. *Electronics, 38*(8).

Nelson, R. R. & Winter, S. G. (1982). *An Evolutionary Theory of Economic Change.* Cambridge, MA: Belknap Press.

Pinch, T. (1982). Kuhn – the conservative and radical interpretations. *4S Newsletter, 1*, 10–25.

Rouse, J. (1987). *Knowledge and Power: Toward a political philosophy of science.* Ithaca: Cornell University Press.

van den Belt, H. & Rip, A. (1987). The nelson-winter-dosi model and synthetic dye chemistry. In *The Social Construction of Technological Systems*, Bijker, W. E., Hughes, T. P., & Pinch, T., eds., Cambridge, MA: MIT Press, 135–158.

Imre Hronszky
Budapest University of Technology and Economics
Hungary
hronszky@eik.bme.hu

Visionary Communication on Techno-Sciences and Emerging Challenges to Societal Debate: The Case of Synthetic Biology

Armin Grunwald

1 Introduction and overview

In the past decade, visionary communication on future technologies and their impacts on society increased to a considerable extent. In particular, this has been and still is the case in the fields of techno-sciences such as nanotechnology (Fiedeler et al. 2010), human enhancement and the converging technologies (Grunwald 2007) and synthetic biology. Visionary scientists and science managers have put forward far-ranging visions which have been disseminated by mass media and discussed in science and the humanities. These observations allow us to speak of an emergence of *techno-visionary sciences* in the past decade.

This development was accompanied by an also increasing research interest in visionary communication at the interfaces between technology, science, and society. TA (Technology Assessment) exercises (Grunwald 2009a), STS studies (science, technology and society), analyses in applied ethics and ELSI studies (ethical, legal and social implications) were performed and involved many researchers and scholars from social sciences and the humanities worldwide.

Synthetic Biology entered the visionary NEST field (new and emerging science and technology) rather late. Its second World Conference in 2006 brought about first interest among CSOs (civil society organisations) (ETC 2007). ELSI, STS and TA research followed directly with first European and international projects. Today, we can speak of high interest in those research communities and also in policy advice (e.g., Presidential Commission 2010) but only low interest in the media and the public.

Against this background the point of departure of this paper is the experience of an increasing wave of techno-visionary communication on techno-sciences in general. Its intention is to use notions and insights from previous debates for analysing the current and emerging debate on Synthetic Biology. This will be done by introducing some basic notions (section 2), by asking for the specific visions in the debate on Synthetic Biology (section 3) with a focus on changing relations between life and technology (section 4) and by then analysing emerging challenges for the further societal debate (section 5) which lead to some methodological requirements (section 6). [1]

2 The increasing importance of visions in technology debate

The emergence of new visionary and futuristic communication (Coenen 2010; Selin 2008) has provoked renewed interest in the communicative roles played by imagined visions of the future in societal debate. Obviously, there is no distinct borderline between the visions communicated in these fields—called futuristic visions (Grunwald 2007)—and other imagined futures such as *Leitbilder* or guiding visions which have already been analysed earlier with respect to their usage in policy advice and technology development (Grin & Grunwald 2000). The following characteristics may roughly circumscribe some specific properties of futuristic visions (Grunwald 2007; 2009b):

> futuristic visions refer to a more distant future, usually some decades ahead, and exhibit revolutionary aspects in terms of technology and in terms of culture, human behaviour, individual and social issues as well

> scientific and technological advance is regarded as by far the most important driving force in modern society (technology push perspective, technology determinism)

> the authors and promoters of futuristic visions are mostly scientists, science writers and science managers such as Eric Drexler, Craig Venter and Ray Kurzweil

> milestones and technology roadmaps are presented to bridge the gap between today's state and the visionary future state (e.g., Roco & Bainbridge 2003) in order to show that the feasibility of the visionary promises and to distinguish them from Science Fiction

---

[1] This paper builds on previous work of the author and extends it (see Grunwald 2007; 2009b; 2012).

high degrees of uncertainty are involved which lead to severe controversies with regard to societal issues (e.g., Dupuy 2007).

While futuristic visions often appear somewhat fictitious in content, it is a fact that such visions can and will have real impact on scientific and public discussions (Grunwald 2007). We must distinguish between the degree of facticity of the *content* of the visions and the fact that they are used in genuine communication processes *with their own dynamics*. Even a vision without any facticity at all can influence debates, opinion-forming, acceptance and even decision-making. Visions of new science and technology can have a major impact on the way in which political and public debates about future technologies are currently conducted, and will probably also have a great impact on the results of such debates—thereby considerably influencing the pathways to the future in two ways at least:

> Futuristic visions are able to change the perception of present and possible future developments and increase the contingency of the *conditio humana* (Grunwald 2007). The societal and public debate about the chances and risks of new technologies will revolve around these visions to a considerable extent, as was the case in the field of nanotechnology (cf. Brune *et al.* 2006) and as is currently the case in Synthetic Biology (see below). Visions motivate and fuel public debate. Negative visions and dystopias could mobilise resistance to specific technologies while positive ones could create acceptance and fascination.

> Visions have a particularly great influence on the scientific agenda (Nordmann 2004) which, as a consequence, partly determines which knowledge will be available and applicable in the future. Directly or indirectly, they influence the views of researchers, and thus ultimately also have a bearing on political support and research funding. Visions therefore influence decisions about the support and prioritisation of scientific progress and are an important part of the governance of knowledge (Stehr 2004; Selin 2008):

The factual power of futuristic visions in public debate is a strong argument in favour of carefully and critically analysing and assessing them in the fields of techno-visionary sciences (section 6). The rationale of such research and reflective work is to increasing reflexivity and transparency in these debates. The participants in ongoing societal debates should know more about these visions, their content and background, their knowledge base and possible values and interests involved.

This conclusion is supported by calls for a more democratic governance of science and technology (Siune *et al.* 2009) against the concern that futuristic visions contain a mixture of facts and values, allowing them to be used for ideological and interest-based purposes. An open, democratic discussion of techno-visionary sciences is a prerequisite for a constructive and legitimate approach to shaping the future research agenda, regulations and research funding (see below, section 5).

3 Visions of synthetic biology

Synthetic biology has recently turned into a vibrant field of scientific inquiry (Grunwald 2012, chap. 7). The combination of engineering with biology promises to make it possible to fulfill many of the goals expected of nanotechnology in an even easier fashion: while nanotechnology involves the development of materials and machines at the nanoscale, synthetic biology builds on the insight that nature already employs components and methods for constructing machines and materials at very small scales. Synthetic biologists hope, both by employing off-the-shelf parts and methods already used in biology and by developing new tools and methods, to develop a set of tools to hasten the advent of the various promises of nanotechnology (SYNTH-ETHICS 2011). Various suggestions have been made for defining synthetic biology, some of them describing Synthetic Biology as:

> the design and construction of biological parts, devices, and systems, and the redesign of existing, natural biological systems for useful purposes (LBNL 2006).

> the design and synthesis of artificial genes and complete biological systems, and the modification of existing organisms, aimed at acquiring useful functions (COGEM 2006).

> the engineering of biological components and systems that do not exist in nature and the re-engineering of existing biological elements; it is determined by the intentional design of artificial biological systems, rather than by the understanding of natural biology (Synbiology 2005).

A characteristic feature of each of these definitions is the turn to artificial forms of life—whether they are newly constructed or produced via the redesign of existing life—each of which is associated with an expectation of a specific utility. Synthetic biology can be viewed as the continuation of molecular biology and genetics with the means of nanotechnology.

Visions are present in the debate on Synthetic Biology at different levels (SYNTH-ETHICS 2011). They include (1) 'official' visions provided and disseminated by scientists and science promoters, and (2) visions disseminated by mass media including negative visions up to dystopian views.

(1) Many visions of Synthetic Biology tell well-known stories about the paradise-like nature of scientific and technological advance. Synthetic Biology is expected to provide many benefits and to solve many of the urgent problems of humanity. These expectations concern primarily the fields of energy, health, new materials and a more sustainable development. The basic idea behind these expectations is that solutions which have developed in nature could directly be made useful to human exploitation by Synthetic Biology:

> Nature has made highly precise and functional nanostructures for billions of years: DNA, proteins, membranes, filaments and cellular components. These biological nanostructures typically consist of simple molecular building blocks of limited chemical diversity arranged into a vast number of complex three-dimensional architectures and dynamic interaction patterns. Nature has evolved the ultimate design principles for nanoscale assembly by supplying and transforming building blocks such as atoms and molecules into functional nanostructures and utilizing templating and self-assembly principles, thereby providing systems that can self-replicate, self-repair, self-generate and self-destroy. (Wagner 2005, 39)

In analysing those solutions of natural systems and adopting them to human needs the traditional border between biotic and abiotic systems could be transgressed. It is one of the visions of Synthetic Biology to become technically able to design and construct life according to human purposes and ends. Then it is only a small step to the creation of artificial life which seems to be the major vision behind. In spite of the fact that biologists often do not like to be regarded as becoming creators of living systems (Boldt & Müller 2008) this aspect is included in all of the definitions of Synthetic Biology quoted above. The creation of artificial life which is an old dream of humankind seems to be at the core of the visions of Synthetic Biology—as was similar in the field of nanotechnology some years ago. In that field researchers did not like to be perceived as if they would go for shaping the world atom by atom as the National Nanotechnology Initiative indicated (NNI 1999)—but this slogan was very powerful in visionary communication.

The knowledge provided by Synthetic Biology can be used to produce new functions in living systems by modifying biomolecules or the design of cells, or designing artificial cells. The traditional self-understanding of biology in the framework of natural sciences aiming at *understanding* natural processes is reinterpreted by synthetic biology (Ball 2005) as a *new invention* of nature and as the creation of artificial life on the basis of our knowledge about 'natural' life. This transforms biology into an engineering science of a new type (de Vriend & Walhout 2006).

The novelty has to do with designing and controlling self-organizing processes. The utilization of phenomena of self-organization, including the possibility of replicating artificial living entities that have been created or modified by technology, is a central aspect of Synthetic Biology and leads to self-organization becoming increasingly significant:

> The paradigm of *complex, self-organizing systems* envisioned by von Neumann is stepping ahead at an accelerated pace, both in science and in technology [...] We are taking more and more control of living materials and their capacity for self-organization and we use them to perform mechanical functions. (Grinbaum & Dupuy 2004, 292)[2]

This way of thinking, in particular the idea of controlling more and more parts of nature and of our environment, continues basic convictions of European Enlightenment in the Baconian tradition. Human advance includes, in that perspective, to achieve more and more independence from any restrictions given by nature or by the natural evolution and to enable humankind to shape its environment and living conditions according to human values, preferences and interests to maximum extent. Some voices extend this movement in a radical sense and consider humans as 'co-creators' and masters of the 'second evolution'.

> Another example is given by the ribosome present in each cell, which is actually a nano-assembling machine which reads the DNA and translates the code into protein. It works wonderfully in nature. The difficulty is to mimic the idea and to use it in practicable technology. This type of Nanobionic requires a second type of evolution. This evolution II is the whole idea of Nano. (Heckl 2004)

This is, obviously, an extremely far-ranging vision, grounded, on the one hand, on traditional Western modernism calling for complete control over nature. On the other, however, it shows some indication of a completely different vision—the vision of reconciling human development and nature by means of bionics. It means going for a more 'natural' technology (nanobionics) (see section 4.1 for a critical review).

---

[2]This positive formulation leads the authors nevertheless to express serious concerns about risks involved.

(2) In public debate things are more diffuse. Up to now there isn't any big societal debate on Synthetic Biology. In spite of the work of CSOs (e.g., ETC 2007), some high-level policy advice (Presidential Commission 2010) and public interventions of promoters such as Craig Venter public debate is not very intense yet (SYNTH-ETHICS 2011).

Recently, the public debate on nanotechnology, in particular on its more futuristic aspects, was analysed in the DEEPEN project funded by the European Commission. One of the results was that cultural narratives form the background of many of the more visionary public debates and concerns. These narratives are (DEEPEN 2009):

Be careful what you wish for

Opening Pandora's box

Messing with nature

Kept in the dark

The rich get richer and the poor get poorer.

This result shows that public perception of new and emerging technology not only depends on the specific parameters and impacts of that technology but also on underlying cultural attitudes and traditions. It also makes clear that the public debate on nanotechnology is part of an ongoing societal debate on new science and technology. There are archetypical stories deeply rooted in contemporary culture. The question is whether these are at place also in the debate on Synthetic Biology. To answer this question is a task for further research and reflection. It seems likely, that the narrative of 'opening Pandora's box' might be related with the well-known risk issues of 'bio-safety' and 'bio-security'.

In 2005 a high-level expert group on behalf of the European Commission called it likely that work to create new life forms will give rise to fears, especially that of synthetic biologists "playing God" (EGE 2005), see also (Dabrock 2009). The question should be scrutinized seriously, especially since playing God is one of the favorite buzzwords in media coverage of synthetic biology. A report by the influential German news magazine *Der Spiegel* (following SYNTH-ETHICS 2011) titled "Konkurrenz fr Gott" (Competing with God), which is a reference to a statement by the ETC Group ("For the first time, God has competition", 2011). The introduction states that the aims of a group of biologists are to reinvent life, thereby raising fears concerning human hubris. The goal of understanding and fundamentally recreating life would, according to the article, provokes fears of mankind taking over God's role and that a being such as Frankenstein's monster could be created in the lab. This narrative is a dystopian version of the Baconian vision of full control (see above). Might be that these both absolutely diverging visions are only different sides of the same coin.

A further vision which plays a role in public debate is also related with the Baconian ideal. It is challenging the idea of full control over life and evolution by raising the question whether this might imply an increasing technicalization of nature and a loss of dignity of life. If life could be produced in a lab, why should we ascribe dignity to it (see section 4)?

In order to summarize we have to state that Synthetic Biology did not yet arrive at the level of a typical 'hot' societal debate. Instead, there is a field of positive visions, formulated more or less by the promoters, and a field of public debate with a more diffuse picture also including strongly negative visions and dystopian fears. Currently it is not foreseeable whether Synthetic Biology will become a subject of major public interest or will remain in circles of specialised CSOs, science managers, funding agencies, and researchers from biology as well as from TA, STS and ELSI.

4 Changing relations between nature and technology

As is the case in any technology debate the question arises for the 'really' novel issues and aspects of Synthetic Biology in its social and ethical perception. The visions extracted above give a preliminary answer: it is the dissolving borderline between life and technology which leads to really new questions while, for example, issues of bio-safety and bio-security might become factually more important but are well-known in their nature from the debate on genetic engineering.

4.1 Technicalization of Nature or Naturalization of Technology?

The dissolution of the borderline between technology and life can be interpreted in two directions: life could be technicalized, and technology could become more natural. Sometimes the term of nano-bionics is used in order to apply a particular perspective on synthetic biology. Bionics attempts, as is frequently expressed metaphorically, to employ scientific means to learn from nature in order to solve technical problems (von Gleich *et al.* 2007). The promise of bionics is that the bionic approach will make it possible to achieve a technology that is more natural or better adapted than is possible with traditional technology. Examples of desired properties that could be achieved include adaptation into natural cycles, low levels of risk, fault tolerance, and environmental compatibility.

In grounding such expectations, advocates refer to the problem-solving properties of natural living systems, such as optimization according to multiple criteria under variable boundary conditions in the course of evolution, and the use of available or closed material cycles (von Gleich et al. 2007, 30 ff.). According to these expectations, the targeted exploitation of physical principles, of the possibilities for chemical synthesis, and of the functional properties of biological nanostructures is supposed to enable synthetic biology to achieve new technical features in hitherto unachieved complexity, with nature ultimately serving as the model. The principles of bionics find application at the nanoscale (Hampp & Noll 2003).

A closer look at the research process in Synthetic Biology suffices, however, to prevent one from having hasty expectations and exaggerated hopes. The cognitive process attempts to gather knowledge about the structures and functions of natural systems, but from *technical intervention,* not from contemplation or via distanced observation of nature. Living systems are not of interest *as such,* for example in their respective ecological context, but are analyzed in the *relationship of their technical functioning.* Living systems are thus interpreted as *technical systems* by synthetic biology. This can easily be seen in the extension of classical machine language to the sphere of the living. The living is increasingly being described in technomorphic terms:

> Although it can be argued that synthetic biology is nothing more than a logical extension of the reductionist approach that dominated biology during the second half of the twentieth century, the use of engineering language, and the practical approach of creating standardized cells and components like in an electrical circuitry suggests a paradigm shift. Biology is no longer considered "nature at work", but becomes an engineering discipline. (de Vriend & Walhout 2006, 26)

Examples of such uses of language are referring to hemoglobin as a vehicle, to adenosine triphosphate synthase as a generator, to nucleosomes as digital data storage units, to polymerase as a copier, and to membranes as electrical fences. From this perspective, synthetic biology is linked epistemologically to a technical view of the world and to technical intervention. It carries these technical ideas into the natural world, modulates nature in a technomorphic manner, and gains specific knowledge from this perspective. Nature is seen as technology, both in its individual components and also as a whole.

> This is where a natural scientific reductionistic view of the world is linked to a mechanistic technical one, according to which nature is consequently also just an engineer [].Since we can allegedly make its construction principles into our own, we can only see machines wherever we look—in human cells just as in the products of nanotechnology. (Nordmann 2007, 221)

Instead of eliciting a more natural technology *per se* as promised by a bionic interpretation of Synthetic Biology the result of this research signifies a far-reaching technicalization of what is natural. Learning from nature for technical problem solving must of necessity already take a technical view of nature. *Prior to* considering synthetic biology from the perspective of technology ethics or societal debate and assessment, it appears sensible to ask if and how such changes in the use of language and such reinterpretations modify the relationship between technology and life or modify our view of this relationship. A 'hermeneutics' of synthetic biology is needed which should improve our understanding about possibly deep-ranging changes in our view on life.

4.2 Changing Relations between Technology and Life

Concepts such as that of life belong to the core concepts of anthropology, biology, and philosophy and, like as a rule most core concepts, are semantically controversial. Life, in the form in which living beings exist and develop in nature, which is only partially accessible to human influence (e.g., via breeding), is open to numerous interpretations. One that gets special attention is the relationship between technology and life. If synthetic biologists were able to create something that can be called "artificial life", how should we consider these entities? This is not only an interesting metaphysical question about the status of artificial living beings—sometimes called 'bio-facts' (Karafyllis 2006; 2008)—but there is also the ethical question of how we are to treat them: Should we treat them as machines or as living beings? What moral status do artificially created living beings have?

What however is evident from these rather episodic considerations is that synthetic biology provokes precisely this type of question. It does not provoke them simply as academic questions, but rather as questions whose perspective at least has a real background:

> Additionally, synthetic biology forces us to redefine "life". Is life in fact a cascade of biochemical events, regulated by the heritable code that is in (and around) the DNA and enabled by a biological machinery? Is the cell a bag of biological components that can be redesigned in a rational sense? Or is life a holistic entity that has metaphysical dimensions, rendering it more than a piece of rational machinery? (de Vriend & Walhout 2006, 11)

Precisely the increasing degree of opportunities for human intervention in living beings provides the motive for conceptual debates, such as recently in the context of bio-facts (see Karafyllis 2008). Karafyllis saw an inherently necessary link between "life" and "growth" resulting from an autonomous drive and concluded that there can never be artificial life: "We can reject the thesis that man can produce 'life' as an artifact by referring to the fact that aggregation is not growth and gestalt is not habitus" (Karafyllis 2006, 555). If this dictum were taken seriously, it would be easy to clear the table of ethical challenges regarding how to deal with artificial life because the latter would be an absolute impossibility.

In several respects, however, critical questions must be raised as to whether the issue of the possibility or impossibility of artificial life can be determined in a *purely conceptual* manner. The first question is—in view of the challenges posed by synthetic biology—whether the rejection of the possibility of artificial life and artificial living beings is really as clear as the quotation suggests. The assertion that something is generally impossible appears to be very bold. We cannot for ever exclude the chance that linking of individual functions by synthetic biology might result in an overall situation that appears to be alive or very similar to it.

Another argument that challenges Karafyllis' conclusions consists in doubting that a clear distinction can be drawn between being able to influence living systems by means of technology and to newly invent and construct them by means of technology. More likely is rather a continuum of technical interventions in living systems ranging from minimal interventions by breeding via genetic engineering to constructions of Synthetic Biology that might be either largely or entirely new.

Furthermore, Karafyllis' thesis of the impossibility of artificial life would be *irrelevant* in a practical sense. It would be clear then that the term "life" could not be attributed to artificially created "beings". Yet if advances in synthetic biology would lead to the creation of such beings, we would face the question of how to deal with them independently of whether they are classified as living beings or not. A definition of life that is not sensitive to empirical changes in the state of knowledge and skill in biology would also be of little assistance in the hermeneutic analysis of these changes.

While many biologists incline toward a reductionist-materialistic view (obvious for example in the context of converging technologies; (Roco & Bainbridge 2003), some take the holistic position, as shown in statements such as, "While machinery is a mere collection of parts, some sort of 'sense of the whole' inheres in the organism" (Woese 2004). The debate about the relationship between technology and life thus represents another stage in the debate between reductionism and holism which is one of the grand debates in modern societies.

5 Challenges to public debate

Societal debates on issues of technological progress are usually generated by undertaking future investigations, scenarios and reflections (Grunwald

2009a). In particular, the debates on techno-visionary sciences operate with far ranging visions and expectations on the one hand, and with dystopian fear and deep concern on the other (Grunwald 2007). Those 'futures' frequently are highly contested (Brown et al. 2000) and mirror the conflicts in modern, pluralistic societies. It is extremely difficult to make the added value of considering and assessing future developments clear and explicit. Several problems challenge the utility of futures (such as scenarios, predictions, and roadmaps) for providing orientation in debates about science and technology (following Grunwald 2007; 2012):

The arbitrariness problem: A fundamental problem with far-reaching future visions or scenarios is the inevitably high degree to which material other than evidence-based knowledge is involved. In many cases, entire conceptions of the future are simply replaced by assumptions due to a lack of knowledge. Controversies and uncertainties are dominating and lead to conflict. If there were no methods of assessing and scrutinising diverging futures agreed upon, the arbitrariness of futures would destroy any hope of gaining orientation by reflecting on future developments. This was the primary concern resulting from the examination of the debate on 'speculative nanoethics' (Nordmann 2007; Grunwald 2010). The arbitrariness problem constitutes a severe challenge and raises doubts about whether such an endeavour could succeed at all.

The ambivalence of techno-visionary futures: In the field of techno-visionary sciences, the high degree of uncertainty and low level of reliable knowledge mean that this type of communication entails specific risks of communication. If the anticipated future developments of techno-visionary sciences diverge dramatically between paradise and apocalypse, ethical assessments of these sciences will diverge in a similar way: "Tremendous transformative potential comes with tremendous anxieties" (Nordmann 2004, 4). This will then have dramatic consequences for public debate and public perception of techno-visionary sciences. Using metaphors to describe what is radically and revolutionarily new in terms of scientific-technical visions can backfire; an attempt to fascinate and motivate people by suggesting positive utopias can lead directly to rejection and contradiction. The visionary pathos in many technical utopias is extremely vulnerable to the simple question of whether everything couldn't just be completely different—and it is as good as certain that this question will also be asked in an open society. Revolutionary changes promised by new technologies give rise not only to fascination and motivation but also to concern, fear and objection (Grunwald 2009b).

Lack of transparency: Given their considerable impact on the way new technologies are perceived in society and in politics and given that they are an important part of their governance visions should be subject to democratic debate and deliberation. The significant lack of transparency and unclear methodical status of futuristic visions are, however, obstacles to transparent democratic debate. Visions of the future are created in accordance with available knowledge, but also with reference to assessments of relevance, value judgements and interests, and are often commissioned by political and economic decision-makers (Grunwald 2011). Visionary futures are frequently created by scientists and science managers who at the same time are stakeholders with their own interests. One possible scenario is that visionary futures suggested by science could dominate social debates by determining their frames of reference; this would leave the social debate with only aspects of minor importance. In this case, those visionary scientific and technological futures could endanger public opinion-forming and democratic decision-making, thus perhaps constituting a new form of "covert" expertocracy. Against the background of normative theories of deliberative democracy, there is therefore a considerable need to improve transparency in this field.

These concerns and challenges have been formulated using experiences from the public debate on nanotechnology and on human enhancement (Roco & Bainbridge 2003; Grunwald 2007). They also apply to visions and expectations concerning Synthetic Biology. The following quote taken from a visionary paper of synthetic biology supports shows the darkness and intransparency of statements of the future:

> Fifty years from now, synthetic biology will be as pervasive and transformative as is electronics today. And as with that technology, the applications and impacts are impossible to predict in the field's nascent stages. Nevertheless, the decisions we make now will have enormous impact on the shape of this future. (Ilulissat Statement 2007, 2)

This statement expresses (a) that the authors expect Synthetic Biology leading to deep-ranging and revolutionary changes, (b) that our today' decision will have high impact on future development but (c) we do not know at all how those future impacts will look like. If this would be true there wouldn't be any chance to shape the course of Synthetic Biology and to assign responsibilities. Even to speak about responsibility and social expectations and influence wouldn't be without any purpose because there wouldn't be any valid subject to talk about. Any ethics of responsibility would be obsolete: complete non-knowledge about future developments and their relations to today's decision-making would make reflections on the desirability or acceptability of those future developments impossible or would lead to arbitrariness in drawing conclusions for today's attributions of responsibilities.

6 Vision assessment and the governance of knowledge

The factual importance of futuristic visions in the ongoing and increasing debate on Synthetic Biology is the main argument for postulating an early vision assessment in order to allow for more rationality, reflexivity, and transparency in these debates (Grunwald 2009b). An open, democratic discussion on visions of the future is decisive for a constructive solution of the orientation problems described and of allowing legitimised decisions on future research, regulation, and research funding. The requirement for transparency with respect to the projections into the future and to the arguments, premises, and imaginations standing behind them is indispensable. This is, however, as the analysis above has shown, by no means given: The actual relevance of futuristic visions stands against their methodically and epistemologically non-clarified status which shows itself in the uncovered ambivalences. This situation requires that such visions be made the subject of analysis:

> If the future depends on the way it is anticipated and this anticipation is made public, every determination of the future must take into account the causal consequences of the language that is being used to describe the future and how this language is being received by the general public, how it contributes to shaping public opinion, and how it influences the decision-makers. (Grinbaum & Dupuy 2004, 11)

Therefore, new tools for the structuring, interpretation, criticism, rationalisation, and evaluation of futuristic visions are necessary to analyse and assess futuristic visions, especially, their reliability, their degree of reality and expectability, their normative aspects and their impact on the public and political debate. A "vision assessment" was proposed to provide some new capabilities (Grunwald 2009b). It can be analytically divided in several steps which are not sharply separated and not linearly ordered but which serve different sub-objectives and involve different methods:

1. With respect to analysis *(vision analysis)*, it would be a question of disclosing the *cognitive and normative* contents of the visions. The cognitive content includes a transparent reconstruction of what is assumed to come to reality in the futures and what the implications of these realisations would be. Further, the visions' *normative* contents have to be reconstructed analytically: the visions of a future society,

or of the development of human beings, as well as possible diagnoses of current problems, to the solution of which the visionary innovations are supposed to contribute. The contribution of such reflective analyses could consist in this respect in the "clarification" of the pertinent communication: the partners in communication should know explicitly what they are talking about as a prerequisite for more rational communication. It is a matter of society's "self-enlightenment" and of supporting the appropriate learning processes.

2. Vision assessment would, further, include evaluative elements *(vision evaluation)*. These are questions of how the cognitive aspects are to be categorised, how they can be judged according to the degree of realisation or realisability, according to plausibility and evidence, and which status the normative aspects have, e.g., relative to established systems of values, or to ethical standards. The purpose is the transparent disclosure of the relationship between knowledge and values, between knowledge and the lack of it, and the evaluation of these relationships and their implications. On the one hand, one can draw upon the established evaluation methods of technology assessment, which often include a participative component. On the other hand, in the field of Synthetic Biology there are some far-reaching questions in normative respect which stand to discussion and which require ethical and philosophical reflection (see section 4).

3. Finally, it is a matter of deciding and acting *(vision management)*. The question is how the public, the media, politics, and science can be advised with regard to a "rational" use of visions. First, the question of alternatives, either already existing or to be developed, to the visions already in circulation stands here in the centre of interest, in accordance with the basic position of technology assessment of always thinking in terms of alternatives and different options. In this manner, visions based on technology can be compared with one another or with non-technological visions. Finally, it is a question of strengthening reflectivity. Communication on the cognitive and normative backgrounds of visions is also part of "responsible" communication making use of visions, in order to make a transparent discussion possible.

In particular, it would be the assignment of vision assessment to confront the various and, in part, completely divergent normative aspects of the visions of the future directly with one another. This can, on the one hand, be done by analysis and desk research; on the other, however, the representatives of the various positions should discuss their differing judgments in workshops directly with and against one another, in order to lay open their respective premises and assumptions.

Bibliography

Ball, P. (2005). Synthetic biology for nanotechnology. *Nanotechnology*, *16*, R1–R8, doi:10.1088/0957-4484/16/1/R01.

Boldt, J. & Müller, O. (2008). Newtons of the leaves of grass. *Nature Biotechnology*, *26*, 387–389, doi:10.1038/nbt0408-387.

Brown, N., Rappert, B., & Webster, A. (Eds.) (2000). *Contested Futures. A sociology of prospective techno-science*. Aldershot: Burlington.

Brune, H., Ernst, H., et al. (2006). *Nanotechnology—Assessment and Perspectives*, Ethics of Science and Technology Assessment, vol. 27. Berlin; New York: Springer.

Coenen, C. (2010). Deliberating visions: The case of human enhancement in the discourse on nanotechnology and convergence. In *Governing Future Technologies. Nanotechnology and the rise of an assessment regime*, Kaiser, M., Kurath, M., et al., eds., Dordrecht: Springer, 73–87.

COGEM (2006). Synthetische Biologie. Een onderzoeksveld met voortschrijdende gevolgen, Netherlands Commission on Genetic Modification , www.cogem.net/pdfdb/advies/CGM060228-03.pdf.

Dabrock, P. (2009). Playing God? Synthetic biology as a theological and ethical challenge. *Systems and Synthetic Biology*, *3*(1–4), 47–54, doi:10.1007/s11693-009-9028-5.

de Vriend, H. C. & Walhout, A. M. (2006). *Constructing Life. Early social reflections on the emerging field of synthetic biology*. Den Haag: Rathenau Instituut.

DEEPEN (2009). Reconfiguring Responsibility. Deepening Debate on Nanotechnology. Tech. rep., Department of Geography, Durham University, www.geography.dur.ac.uk/projects/deepen.

Dupuy, J.-P. (2007). Complexity and uncertainty: A prudential approach to nanotechnology. In *Nanoethics. The ethical and social implications of nanotechnology*, Allhoff, F., Lin, P., et al., eds., New York: Wiley, 119–132.

EGE (2005). Ethical aspects of ICT implants in the human body. Opinion 20, European Group on Ethics in Science and New Technologies, Brussels.

ETC (2007). ExtremeGenetic Engineering. An Introduction to Synthetic Biology. Tech. rep., The Et-cetera Group, www.etcgroup.org.

Fiedeler, U., Coenen, C., et al. (Eds.) (2010). *Understanding Nanotechnology: Philosophy, Policy and Publics*. Heidelberg: Akademische Verlagsgesellschaft.

Grin, J. & Grunwald, A. (2000). *Vision Assessment : Shaping technology in 21st century society towards a repertoire for technology assessment.* Berlin: Springer.

Grinbaum, A. & Dupuy, J.-P. (2004). Living with uncertainty – Toward the ongoing normative assessment of nanotechnology. *Techné: Research in Philosophy and Technology*, *8*(2), 4–25, Reprint in: Schummer, I. & Baird, D. (Eds.) (2006): *Nanotechnology Challenges: Implications for Philosophy, Ethics and Society.* Singapore: World Scientific, 287–314.

Grunwald, A. (2007). Converging technologies: Visions, increased contingencies of the conditio humana, and search for orientation. *Futures*, *39*(4), 380–392.

Grunwald, A. (2009a). Technology assessment: Concepts and methods. In *Philosophy of Technology and Engineering Sciences, Handbook of the philosophy of science*, vol. 9, Meijers, A. W. M., ed., Amsterdam: Elsevier/North-Holland, 1103–1146.

Grunwald, A. (2009b). Vision assessment supporting the governance of knowledge— The case of futuristic nano-technology. In *The Social Integration of Science. Institutional and Epistemological Aspects of the Transfor-mation of Knowledge in Modern Society*, Bechmann, G., Gorokhov, V., & Stehr, N., eds., Berlin: Edition Sigma, 147–170.

Grunwald, A. (2010). From speculative nanoethics to explorative philosophy of nanotechnology. *NanoEthics*, *4*(2), 91–101, doi:10.1007/s11569-010-0088-5.

Grunwald, A. (2011). Energy futures: Diversity and the need for assessment. *Futures*, *43*(8), 820–830, doi:10.1016/j.futures.2011.05.024.

Grunwald, A. (2012). *Responsible Nanobiotechnology. Philosophy and Ethics.* Singapore: Pan Stanford Pub.

Hampp, N. & Noll, F. (2003). Nanobionics II – from molecules to applications. *Physik Journal*, *2*(2), 56–57.

Heckl, W. M. (2004). Molecular self-assembly and nanomanipulation—Two key technologies. *Advanced Engineering Materials*, *6*(10), 843–847, doi: 10.1002/adem.200404493.

Ilulissat Statement (2007). Synthesizing the Future. A vision for the convergence of synthetic biology and nanotechnology. Views that emerged from the Kavli Futures Symposium 'The merging of bio and nano: towards cyborg cells. Tech. rep., Kavli Futures Symposium, Ilulissat, Greenland.

Karafyllis, N. C. (2006). Biofakte. Grundlagen, Probleme und Perspektiven. *Erwägen Wissen Ethik*, *17*(4), 547–558, doi:10.1007/s11569-007-0007-6.

Karafyllis, N. C. (2008). Ethical and epistemological problems of hybridizing living beings: Biofacts and body shopping. In *Ethical Considerations on Today's Science and Technology. A German-Chinese Approach*, Li, W. & Poser, H., eds., Münster: LIT, 185–198.

LBNL (2006). Lawrence Berkeley National Laboratory. www.lbl.gov/pbd/synthbio/default.htm.

NNI (1999). National Nanotechnology Initiative. Washington.

Nordmann, A. (Ed.) (2004). *Converging Technologies—Shaping the Future of European Societies, High Level Expert Group, EUR / European Commission*, vol. 21357, European Union. European Commission. High Level Expert Group "Foresighting the New Technology Wave", Luxembourg: EUR-OP.

Nordmann, A. (2007). If and then: A critique of speculative nanoethics. *NanoEthics*, *1*, 31–46, doi:10.1007/s11569-007-0007-6.

Pereira, A. G., von Schomberg, R., & Funtowicz, S. (2007). Foresight knowledge assessment. *International Journal of Foresight and Innovation Policy*, *4*, 65–79, doi:10.1504/IJFIP.2007.011421.

Presidential Commission (2010). Recommendations on Synthetic Biology. Tech. rep., The Presidential Commission for the Study of Bioethical Issues (the Bioethics Commission) , Washington.

Roco, M. C. & Bainbridge, W. S. (Eds.) (2003). *Converging Technologies for Improving Human Performance*. Dordrecht; Boston: Kluwer Academic Publishers.

Selin, C. (2008). Expectations and the emergence of nanotechnology. *Science Technology and Human Values*, *32*(2), 196–220, doi:10.1177/0162243906296918.

Siune, K., Markus, E., et al. (2009). Challenging Futures of Science in Society. Tech. rep., MASIS Expert Group – European Commission, Brussels.

Stehr, N. (2004). *The Governance of Knowledge*. New Brunswick: Transaction Publishers.

Synbiology (2005). SYNBIOLOGY: An Analysis of Synthetic Biology Research in Europe and North America. Output D3: Literature and Statistical Review Contract 15357 (NEST), European Commission FP6, SYNBIOLOGY_Literature_And_Statistical_Review.pdf in www2.spi.pt/synbiology/documents/.

SYNTH-ETHICS (2011). Homepage of the EU-funded project "Ethical and regulatory issues raised by synthetic biology". Website, URL http://synthethics.eu/.

von Gleich, A., Pade, C., *et al.* (2007). *Bionik: aktuelle Trends und zukünftige Potenziale*. Berlin: IÖW.

Wagner, P. (2005). Nanobiotechnology. In *Nanoscale Technology in Biological Systems*, Greco, R., Prinz, F. B., & Lane, R., eds., Boca Raton: CRC Press, 39–55.

Woese, C. R. (2004). A new biology for a new century. *Microbiology and Molecular Biology Reviews*, *68*(2), 173–186, doi:10.1128/MMBR.68.2.173-186.2004.

Armin Grunwald
Institute for Technology Assessment and Systems Analysis (ITAS)
Karlsruhe Institute of Technology (KIT)
Germany
armin.grunwald@kit.edu

The Convergence of Law and Ethics: Promises and Shortcomings in the Governance of Emerging Technologies

ELENA PARIOTTI

1 What makes a technology an emerging technology?

Do emerging technologies confront our symbolic order with specific challenges and if so, what does this mean for policy and regulation (Swierstra et al. 2009, 215)? In order to answer this question it is crucial to identify the cluster of features that can be identified as specific to define a technology as "emergent". Those technologies, which are usually called "*emerging technologies*", have been rising several and wide debates not only because they are *new* technologies but due to some further features. These features affect the way they should be regulated and can be detected as follows: (i) the lack of agreed and/or unambiguous definitions. This is true both for nanotechnologies and for synthetic biology; (ii) structural interdisciplinarity and convergence, which is partly one of the causes of the difficulties arising with their definition. As it has been highlighted, "[t]he major change expected from Synthetic Biology is the integration of converging technologies, namely nanotechnology, chemistry, engineering, electronics and biology", so that "analytical and synthetic approaches (such as Systems- and Synthetic Biology) and *in vitro* and *in vivo* (bottom-up and top-down) approaches will tend to converge (Bernauer & Müller 2007, 3); (iii) the blurriness between basic and applied research involved by the development of technologies; (iv) the indeterminacy of applications, in terms of feasibility; (v) the role played, with regard to the assessment of their consequences for the environment, human health and social impact, by the notion of uncertainty rather than the notion of risk. This means that, while any technologies could bring about risks, the diffusion of emergent technologies tends to introduce consequences that are uncertain with regard to the quality and degree of risks for human health and for the environment; (vi) a further relevant feature of some new technologies deals with the indeterminacy of their feasible applications.

2 The challenges to responsibility and the requirements for the regulation of emerging technologies

Concerns on nanotechnologies and synthetic biology deal with biosafety, biosecurity, environmental protection, intellectual property, human rights, and global justice. In the face of the question whether the existing relevant normative framework can be regarded as comprehensive, unambiguous, locally consistent, acceptable and effective, hard regulation shows some obvious limits. Indeed, domestic legislation cannot foresee all the developments of a new technology and risks suffering from either predictable inefficacy or a rapid obsolescence. Soft regulation can offer some clear advantages such as flexibility and ability to involve many subjects and stakeholders in assessing risks.[1]

These characteristics would make managing the scientific uncertainty and rapid technological development possible. It is for the sake of these features that soft regulation has become an important instrument of governance of emerging technologies at the European Union level. Soft regulation rests on the so-called "soft law". By "soft law" it is referred to several normative tools ranging from declarations to recommendations, codes of conduct and self-regulatory instruments, which give a role to different kind of (public and private) actors, are not supported by formal legal sanctions, and nevertheless can have legal effects.

Indeed, soft law can be seen as a suitable instrument for realising the integration between legal modes of the rule of law and modes of globalisation, which is necessary to face the complexity and the uncertainty that characterise emerging technologies. Thanks to its composite character, soft law may help the socialization and the concretisation of values and principles—first of all the precautionary principle—in the legal order and their implementation through mechanisms that can be successful, though informal and voluntary.

The reasons for resorting to soft law in regulation are twofold.

The former order of reasons relates to the new configuration taken by the notion of responsibility as to the regulation of technologies. Firstly, the notion of responsibility tends to be relevant with regard not only to the past conduct but also to the future conduct, it takes a forward-looking nature. Furthermore, the nature of many technological risks responsibility goes beyond individual responsibility, and increases the need to clearly acknowledge

---

[1] For a general inquiry into the features and the advantages of soft regulation, (see Ayres & Braithwaite 1992; Baldwin & Black 2008).

this co-responsibility, in order to make it work within the decision-making process.

The latter order of reasons for resorting to soft law stems from the central role played by the precautionary principle and the insufficiency of the risk assessment approach. The precautionary principle plays a central role in the EU regulatory approach to emerging technologies; it lies at the core of the strategy that faces uncertainty about risks (UNESCO-COMEST 2005, 8) and has an architectonical function to pursue sustainable development (UNESCO-COMEST 2005, 8). Nevertheless, the principle seems to lack conceptual clarity as well as univocal and immediate applicability (UNESCO-COMEST 2005, 23–25). Indeed, over the last years, a strong debate has underlined both the centrality and the shortcomings of this principle. It has been criticised for: (i) Being more a state of the mind that leaves high discretion for decision-makers (Sunstein 2005); (Hull 2007), rather than being a principle that is able to orient regulatory choices (Majone 2002, 93, 106); (ii) its tendency to foster even irrational social concerns, which stick to the "status quo" without proper consideration for potential benefits of the technological development (Majone 2002), (Sunstein 2005, 103).

It seems, in other words, that this principle suffers from a higher degree of vagueness compared to what is typical of principles and that can be overcome more on an implementation level rather than on a definitional or theoretical one. Namely, the several ways of implementing the precautionary principle should be able to make its constructive (and not merely inhibitory) meaning clear in each domain, i.e., its concrete ability to foster a constant improvement of the regulation. As underlined in the European Group on Ethics and Technology (EGE) *Opinion* n. 20, the precautionary principle needs to be understood as a dynamic tool, able to follow the evolution of a given field and to constantly verify that the conditions for the acceptability of a certain innovation are met (EGE 2005, 18).

Therefore, "the precautionary principle is of a 'procedural' rather than 'substantive' nature" (EGE 2005); it is structurally linked to the "new governance" model, according to which governance should be distributed (i.e., open to participation), constructive and dynamic (i.e., always ongoing); reflexive (i.e., able to emend itself).

To the extent that these aspects are emphasized, it is possible to find in the precautionary principle a tool suitable to foster a constructive view of the dialogue among science, society, politics, and law. So understood, however, it is a demanding principle.

The relevance of the role played by the precautionary principle can be explained in the light of the complex meaning that has been taken by the notion of responsibility in the field of emerging technologies. Due to the uncertainty in foreseeing applications and consequences, emerging technologies stress the need for a notion of responsibility that can shape paths and frameworks of co-responsibility among the relevant actors and has a forward-looking nature. The notion of responsibility is relevant, when coming to emergent technologies, in all the dimensions that have been devised (Vincent 2011), from a theoretical point of view, i.e., (i) as an aspect of the morality of the actors involved in Research & Development and as a role responsibility even in absence of legal rules or of formally binding legal rules; (ii) as responsibility towards the causes or towards the consequences of certain processes; (iii) as (moral) responsibility stemming from the power to affect decisions and processes; (iv) as legal liability. In each of these meanings, when emerging technologies are at issue, what seems to characterize the devising of responsibility is the difficulty for any of the actors involved, alone, to have all the needed information to assess the risks that could be associated to certain processes, substances, products (Vincent 2011).

But there is something more. The actors involved in technological development are multiple and different (private and public), all of them can contribute in gathering information that can be relevant to assess the risks and the advantages associated to a given product, process or substance, and all have to be constrained in their behaviour, in order to make choices inspired by the precautionary principle, which in its turn tends to shift from a principle valid at governmental or institutional level into a principle—or at least an approach—able to affect private actors' conduct too. So, responsibility turns out to be linked here not only to causes, not only to outcomes, and should not be translated into liability terms only. It is also linked to the proactive role that the relevant actors are able to play in specific contexts, to the capacity of each actor to act in observance of the precautionary principle, to the knowledge each actor can achieve on the consequences of technological innovation.

What should be stressed is that this notion of *responsibility* increasingly tends to be linked to the horizontal effects of the precautionary approach—i.e., its capacity to orient the private conduct and not only the relationship between public authority and citizenships or the policy making—and to blur the boundaries between the legal and moral spheres and tends, so becoming an essentially hybrid notion.

Thus, the EU model of the governance of science and technology, which is based on openness, participation, accountability, effectiveness and coherence crucially depends on its success in achieving two aims, i.e., the coherent and effective implementation of the precautionary principle and the construction of concrete mechanisms to involve the stakeholders in risk management, by expressing this complex notion of *responsibility*.

Self-regulatory tools seem to have a high potential with regard to both these dimensions. Nevertheless, it should be underlined that public involvement and participation qualify themselves as "effective" as long as they are regarded as an element of substantial inclusion in governance processes of the concerned people rather than as a technique of "ex-post" legitimation of norms content or policies. The involvement of civil society should not take place only or mainly in order to inform citizens or groups, as though the terms of the issue were already established before participation started, but it should instead aim to create genuine co-decisions paths.

This is exactly what makes the resort to soft law in the regulation of emerging technologies complex and strategic at the same time.

3 The overlap of ethics and law within soft law: challenges and promises

Starting from what has been said above, soft law seems to be a suitable tool

a) for implementing the precautionary principle;

b) for helping construe that "framework for action" in which "several actors are called upon to contribute" (UNESCO-COMEST 2005, 38);

c) to warrant transparency and knowledge sharing between firms and the public (UNESCO-COMEST 2005, 41);

d) for putting tasks and roles on all the actors involved in technological development and to organize collective responsibility (Von Schomberg 2009);

e) for establishing common standards, independently from domestic law and trying to overcome the side effects of heterogeneity of domestic regulations.

The European model of governance of science and technology is based on extended openness, participation, accountability (CEC 2001), and the choice to rest at a relevant degree on soft regulation in the field of emerging technologies seems to be a proper way to pursue these aims. On the contrary, a failure to achieve these outcomes would amount to rendering this model a merely rhetorical one (Wynne & Felt 2007, 41). To a certain degree, the main advantage of this model lies in its tendency to merge ethical values and regulatory means. In this sense, soft regulation and Ethical Advisory Boards (EABs) involvement work as means for public reflection and democratic involvement.

It should be stressed that the overlap of legal and ethical principles has to be regarded more as an interplay and so understood it seems desirable.

The reference to the ethical sphere provides the tools at the same time with legitimacy and flexibility: public unease with emergent technologies is often specifically associated with *ethical* concerns, which should also be addressed by regulatory initiatives even though not necessarily or always by hard law, provided that the crystallization of ethical views should be avoided but their development should be left to the public debate.

To assess the "normative mix", i.e., this kind of "internormativity" that has characterized the regulation of emerging technologies up to now, we have to consider that this is a phenomenon that interplays with other processes, such as the "ethicisation of technoscience" and the "burocratisation of ethics".

By "ethicisation of technoscience" it is meant the "abuse" of ethics in approaching technoscience, that is the phenomenon by which the governance of technoscience issues is "increasingly framed through a language of ethics and morality".[2] This phenomenon is partly caused by the overcoming of the idea of self-sufficiency of science in founding rational political decisions on technoscience and should be contrasted by supporting active involvement of ethicists in technological development (van de Poel 2008), (Schummer 2008, 61).

The "bureaucratisation of ethics" tends to be a kind of side-effect of the resort to ethics committee or, more widely, to Ethical Advisory Bodies, in order to orient or support political decisions or regulatory choices. It has been criticized for (a) giving up public debate; (b) crystallizing the framing of relevant ethical issues; (c) reducing the dialogue on the ethical issues by limiting it to the positions of the intellectual establishment (Tallacchini 2009, 282), (Siune *et al.* 2009, 38).

Even if soft law expresses the flexible and ongoing nature of the legal rules, the incorporation of ethical options into tools having also legal relevance may increase the risk of political imposition of collective values. Moreover, an increased recourse to soft law instruments might shift into a form of eluding the legislative process when presented with troubling issues (Pariotti 2009).

Thus, for governance of Science & Technology being able to gain the trust of the public, the input from professional experts alone is not enough and a stronger role of the public is needed (Landeweerd *et al.* 2012). The risk that public participation becomes a mere rhetorical instruments has been clearly underlined (Wynne & Felt 2007), (Siune *et al.* 2009), (Tallacchini 2009). The one-way communication approach between scientists, on the one hand, and public and policy makers, on the other hand, has to be abandoned and

---

[2](Siune *et al.* 2009, 32). See also (Schummer 2008) and the idea of *popularisation of emerging technologies through ethics.*

an alternative approach is being worked out in the literature that looks at science, technology and society as coevolving.

This approach does not offer in itself, however, strict procedural criteria to construe a genuine public involvement and the inquiry should look more in depth into the several recent inputs coming from specific theories of governance of Science & Technology, to work out the idea of public involvement to be concretized by governance (Landeweerd *et al.* 2012). Starting from the idea of the co-evolution of science, technology and society and from the need for bi-directional communication between S&T and society, the most suitable procedural suggestions seem to stem from the model of anticipatory governance, in which the stress is put on the notion of *responsible research and innovation* and on the public engagement (von Schomberg 2012).

This is a governance model that can at least try to pursue a responsible innovation, understood as a transparent and interactive process, in which social actors and innovators look at social desirability of innovative process (or product or effects).

This model, in which the role played by ethics is very important and cannot be reduced to constraining technological advances or social behavior, has the following desirable characteristics: (i) the tendency to incorporate the reference to ethical principles in the design process of technology, by fostering "better integration of ethics at the core of scientific development itself" (Jotterand 2008, 23), so that (ii) it can increase socio-ethical reflexivity (von Schomberg 2012); (iii) the fact that it allows continuous feedback from society; (iv) the ability to promote "knowledge assessment procedures" (von Schomberg 2012), to assess the quality of information within the policy process.

In the light of this aim, I want to underline an open issue within the approach to governance of the emerging technologies supported at the EU level. The effort towards a proper and genuine development of the role of the public in anticipatory governance of the emerging technologies crucially crosses the relationship between risk assessment and risk management. Should they be thought of as separated or interconnected stages? A "clear-cut" separation between these spheres would express "the separation between the realms of sciences and politics, or between 'risk facts' and 'values'" (Wynne & Felt 2007, 32). Regardless of the troubling nature of this point, the distinction between them could be to a certain degree reasonable for underlining and preserving the distinction (not necessarily the separation) between the stage of the construction of the scientific grounds for the decision and the stage of political decision. But in this view the problem remains whether the establishment of the acceptable threshold of risk should belong to the first or to the second moment. Of course, the issue is much more an issue of degree and deals with how much the threshold of risk depends on information given by the scientists or techno-scientists communities and how much it relies on social standpoints.

As it may be controversial that risk assessment is entirely fact-based and not value-based, the uncertainty that characterizes emerging technologies makes it difficult and debatable to keep the social perception of risks out of the definition of the acceptable threshold of risk. Here the quantity and quality of risks is indeterminate, the feasibility of the applications is not always clear, therefore, the foreseeing of the consequences of a given technology tends to be intrinsically socially-driven.

But, provided that we can conclude the European turn in risk governance is based on the assumption that risk assessment is not only a matter of "science-based judgments" but is "intrinsically shaped by social values" (Wynne & Felt 2007, 34), then this theoretical turn has to be transferred into regulation and policies. To achieve this outcome, governance frameworks should foster "new policy debates around the 'precautionary principle', 'public-and stakeholder-engagement' and 'participatory deliberation'" (Wynne & Felt 2007, 33), by acknowledging the relevance of ethical views in technology assessment and by making this relevance properly work within the regulatory process.

This effort takes a specific legitimating role within the European Union, where the governance of emerging technologies is supposed to be shaped at the light of the so-called European values. These values are recognized in the European Charter of Fundamental Rights but still their meaning needs to be determined, not only at formal levels—i.e., at judicial and legislative levels—but also in informal fora, through interpretive practices.

Bibliography

Ayres, I. & Braithwaite, J. (1992). *Responsive Regulation: Transcending the Deregulation Debate.* New York: Oxford University Press.

Baldwin, R. & Black, J. (2008). Really responsive regulation. *The Modern Law Review, 71*(1), 59–94.

Bernauer, H. & Müller, K. (2007). Towards a European strategy for synthetic biology. Tech. rep., Working paper on results available from the other EU-funded projects, TESSY Deliverable D1.1.

EGE (2005). Ethical aspects of ICT implants in the human body. Opinion N. 20, European Group on Ethics in Sciences and New Technologies.

EGE (2007). Opinion on the ethical aspects of nanomedicine. Opinion N. 21, European Group on Ethics in Sciences and New Technologies.

EGE (2009). Ethics of synthetic biology. Opinion N. 25, European Group on Ethics in Sciences and New Technologies.

EGE (2010). General report on the activities of the European Group on Ethics in Science and New Technologies to the European Commission 2005–2010. Luxembourg: Publications Office of the European Union, European Group on Ethics in Sciences and New Technologies.

Hull, G. (2007). Normative aspects of a "substantive" precautionary principle. Social Science and Research Network, URL http://papers.ssrn.com/sol3/papers.cfm?abstract_id=1013357.

Jotterand, F. (2008). Beyond therapy and enhancement: The alteration of human nature. *NanoEthics*, *2*(1), 15–23, doi:10.1007/s11569-008-0025-z.

Landeweerd, L., Townend, D., et al. (2012). Report on the best possible models of governance of science and technology. Tech. rep., EPOCH D 2.6.

Majone, G. (2002). The precautionary principle and its policy implications. *JCMS: Journal of Common Market Studies*, *40*(1), 89–109, doi:10.1111/1468-5965.00345.

Pariotti, E. (2009). Regulating nanotechnology: Towards the interplay of hard and soft law. *Notizie di Politeia*, *94*, 29–40.

Schummer, J. (2008). The popularisation of emerging technologies through ethics: From nanotechnologies to synthetic biology. *Spontaneous Generations: A Journal for the History and Philosophy of Science*, *2*(1), 56–62, URL http://spontaneousgenerations.library.utoronto.ca/index.php/-SpontaneousGenerations/article/view/3529.

Siune, K., Markus, E., et al. (2009). Challenging futures of science in society. Tech. rep., Monitoring Activities of Science in Society in Europe Expert Group (MASIS), European Commission, Brussels.

Sunstein, C. (2005). *Laws of Fear. Beyond the Precautionary Principle*. Cambridge, MA: Cambridge University Press.

Swierstra, T., Boenink, M., et al. (2009). Converging technologies, shifting boundaries. *NanoEthics*, *3*(3), 213–216, doi:10.1007/s11569-009-0075-x.

Tallacchini, M. (2009). Governing by values. EU ethics: Soft tool, hard effects. *Minerva*, *47*(3), 281–306, doi:10.1007/s11024-009-9127-1.

UNESCO-COMEST (2005). *The Precautionary Principle*. Paris: UNESCO, (World Commission on the Ethics of Scientific Knowledge and Technology).

van de Poel, I. (2008). How should we do nanoethics? A network approach for discerning ethical issues in nanotechnology. *NanoEthics*, *2*(1), 25–38, doi: 10.1007/s11569-008-0026-y.

Vincent, N. A. (2011). A structured taxonomy of responsibility concepts. In *Moral Responsibility, Library of Ethics and Applied Philosophy*, vol. 27, Vincent, N. A., van de Poel, I., & van den Hoven, J., eds., Dordrecht: Springer, 15–35, doi: 10.1007/978-94-007-1878-4_2.

Von Schomberg, R. (2009). Organising collective responsibility: On precaution, code of conduct and understanding public debate. keynote lecture at the first annual meeting of the Society for the Study of Nanoscience and Emerging Technologies.

von Schomberg, R. (2012). Prospects for technology assessment in a framework of responsible research and innovation. In *Technikfolgen abschtzen lehren*, Dusseldorp, M. & Beecroft, R., eds., VS Verlag fr Sozialwissenschaften, 39–61, doi:10.1007/978-3-531-93468-6_2.

Wynne, B. & Felt, U. e. a. (2007). *Expert Group on Science and Governance. Taking European Knowledge Society Seriously*. Brussels: European Commission DG Research Science, Economy and Society.

Elena Pariotti
Department of Political Science, Law, and International Studies
University of Padova
Italy
`elena.pariotti@unipd.it`

Grand Narratives, Local Minds and Natural Disasters: Community Response to Tsunami in India

APPUKUTTAN NAIR DAMODARAN

ABSTRACT. This paper focuses on the moment of 're-think' and 're-appraisal' that grips a State and its local communities when confronted with natural disaster of epic proportions like the Tsunami. An unprecedented calamity that struck South India in December 2004, the Tsunami not only unsettled the 'grand narratives' and 'established truths' on ways and means of handling natural disasters in India, but also created a new sense of 're-understanding' on the part of the local communities which fell victims to the disaster. With reference to the case of the tsunami that hit coastal District of Nagapattinam in the southern State of Tamil Nadu in India, the paper argues that policy solutions of a new type could emerge as a result of 'learning from the victims'. The paper explains how one could utilize the lessons of the Tsunami in Nagapattinam to weave a different policy perspective on community risk management.

1 What was the Tsunami in India like? The story of Nagapattinam District

The Tsunami tragedy that hit South Indian coastal areas in December 2004 brought in its wake, a trail of unprecedented destruction of human lives and livelihood systems. Primarily affected by the Tsunami were fishermen communities and coastal agriculturists, most of whom were drawn from the economically and socially weaker sections of society. The epicenter of the Tsunami of December 2004 was the Nagapattinam District in Tamil Nadu. The experience of this zone forms the basis of the arguments and policy analysis advanced in this paper.

Nagapattinam is a peninsular delta district in the State of Tamil Nadu, India. The District is surrounded by the Bay of Bengal on the East and Palk Strait to the South. The District has a coastal line of 165 kms. The inland fresh water area of Nagapattinam spreads for about 1,000 hectares (10 km^2). Marine fishing is practiced in the coastal villages of the district.

The tsunami waves of December 2004 wiped out buildings, infrastructure and importantly the means of livelihood of fishermen and farmers in the District. Fishing vessels were smashed, dismantled and tossed up. The vicious water pushed its way towards the inner harbor where the ships were berthed. It tossed the boats a Kilometer away from their berths. The devastation is stated to have rendered over 39000 people homeless.

The Tsunami destroyed 21 villages in entirety. People who managed to escape the fangs of Tsunami were put up by the State machinery in relief camps.

In 2010-11, 120 people from 8 villages were selected for a detailed assessment by the author. The list included 60 fishermen drawn from 4 villages, namely Arcottuthurai, Pushpavanam, Vanagiri and Karaikalmedu. These 4 villages were badly hit by the disaster. Clearly the risk perception systems of local communities about the Tsunami were based on actual experience with the deprivation caused by the havoc of December 2004. The survey yielded many insights on coping strategies adopted by local communities that survived the calamity. It also yielded information on the success and limitations of the relief and rehabilitation measures initiated by the State machinery through the District Administration in the aftermath of the calamity.

2 Findings from Nagapattinam

Economic activities impacted by the tsunami in Nagapattinam, took time to recover to their regular activity levels of the pre-disaster period. The recovery period varied from one location to another, depending on a range of factors that included level of preparedness, supply of raw materials, availability of distribution/transport networks, requirements of the market place and variations in prices for various products and inputs in the wake of the disaster. Certain segments of the service sectors witnessed increases in demand after the disaster (e.g., power/water suppliers and construction/building firms) whilst others witnessed fall in income.

The interviewed victims felt that their maximum suffering was on account of the delay in asset reconstruction. Since local communities had a better understanding of what their losses were, they came up with solutions to minimize damage to their assets. Indeed a few of their suggestions on how to design rehabilitation packages found acceptance with the local authorities.

For the victims of the Tsunami, the best way of speeding up asset reconstruction was to streamline flow of finances besides providing them with technical assistance to repair their capital assets, notably fishing vessels.

Most of the fishermen (90% of our sample) took loans from private lenders and NGOs to cope with their deprivation. A few of them (10% of our sample) obtained loans from the fisheries department to cope with their crisis. Nearly 74% of the fishermen who took loans in 2005, still held their loan at the time of our survey in 2011. Nearly 44% of the fishermen surveyed, had availed loans averaging Rs. 5000 (100 Euros) while for 53% of the sample the loan amounts ranged from Rs. 1000 (20 Euros) to Rs.5000 (100 Euros).

3 Policy incoherence

In the immediate wake of the Tsunami disaster, a frantic state machinery sent out a series of instructions to the local administration as well as to the local communities hinting at the possibilities of more Tsunami waves hitting the area and requesting local communities to evacuate from the shorelines. A day or two later when the Tsunami waves did not hit as was predicted by a State machinery, the faith of the local communities as well as local State machinery about the robustness of risk assessments conducted by the State was lost (Damodaran 2010). Likewise, though rehabilitation packages were introduced over a period of time (one to three weeks after the calamity) to help the victims, these measures did not meet the satisfaction of the victims. Local communities hit by the calamity corresponded horizontally amongst themselves to work out their own risk assessment and also worked together to devise coping strategies that were in tune with their capabilities.

The initial policy diagnosis on the tsunami by the State, focused on the seismic roots of the problem. Solutions to the problem shifted in their emphasis with elapse of time after the occurrence of the event. In the wake of the tsunami disaster and its glaring aftermath, response strategies advanced by different stakeholders ranged from extreme measures of protection (like construction of sea walls along the entire shore line) to more realistic and ecological measures of protection that involved tree planting.

The reason was that the Tsunami was an unprecedented disaster that threw the existing grand narratives on disaster management into disarray. Grand Narratives lost their credibility in the wake of the uncertainty of the kind that the Tsunami was. Rather in the face of uncertainty of events the victims of Tsunami in Nagapattinam were better able to assess solutions to their problems based on their own systems of experiential knowledge generation. The findings and observations regarding the scale and nature of calamity that befell individual households were not factored in by the official State machinery.

Further, sections of the local community that worked out their own coping strategies spread the word around to their fellow victims within and beyond their villages through horizontal communication systems that relied on organizing local meetings and forums to discuss their problems and solutions. This assumed the shape of horizontal risk communication systems at the local level.

A new set of technological/economic self-coping solutions emerged which were bottom up, enduring and better integrated to local realities. These included (1) Rapid Evacuation Strategies (2) Rapid reconstruction of assets that kept cash inflows of victims least disturbed (3) Economic innovations to increase their financial savings in the shape of better, calamity resistant capital assets (4) Focus on locally devised technological solutions for Asset reconstruction.

These locally engendered strategies were different from the standard recipes offered by 'grand narratives' which prescribed the following measures in the wake of natural disasters. (1) Early evacuation strategies that in practice, increased the 'opportunity loss' for the victims (2) Replacement of assets that were matched the economics of lending by local co-operative financial institutions (3) Provision of ex gratia payments and repayable consumption and production related loans to affected communities to tide over their economic difficulties in the rehabilitation phase (4) Provision of subsidized life and asset insurance schemes to guarantee against losses from similar calamities in future (5) Replacement of existing assets with 'superior assets' that are based on exogenously designed technologies that could withstand Tsunamis in future.

The differences in approaches on the issue of disaster management between the State and the local communities had to do with differences in perception about the calamity. While the interviewed local communities saw the Tsunami as a grand punctuating factor in their day to day life and as an unfortunate event that had to happen to disturb the order of things, for the State, Tsunami was an unusual seismically induced natural disaster that needed swift and standardized, rapid solutions by way of State's intervention to help out local communities. Local community perceptions on the issue were not considered by the State machinery to be an efficient solution as they were unscientific and defied standardized approaches.[1] In other words, the Tsunami was an unusual event in the community life for local communities while for the 'State' it was an unprecedented natural disaster. This approach explains the local community accent on 'restoring the equilibrium' through an altered re-inforced community idiom. The latter stood in sharp contrast to the 're-ordered 'life systems emphasized in State

---

[1] For instance the District level functionary mentioned to the author how traditional wooden canoes relied upon by the numerically preponderant local fisherman were inefficient and fragile. In his view these wooden boats needed to be replaced with robust, motor driven steel boats that could withstand future Tsunamis.

strategies. It was natural for local communities to be disillusioned by State prescriptions for rehabilitation and the State's ideas about alternative life systems.

4 Philosophical issues

4.1 Perceiving the Tsunami

The State

The Tsunami sea waves that struck India, Indonesia and Sri Lanka in December 2004 were viewed by the State machinery as an unforeseen calamity that could not be readily explained in terms of standard paradigms of disaster management and relief. The effort was nevertheless to generate similarities with natural disasters like cyclones in order to re-inforce established precepts on disaster management contained in disaster relief codes. However, despite this effort, there was considerable policy incoherence in the aftermath of the Tsunami, as reflected in prevarication on the strategy for handling the calamity, given its tremendous magnitude. In the aftermath of the disaster, the entire discourse on tsunami in India changed from describing the calamity as a pure random, uncontrollable nature of the event, to analyzing the possible causes. This exercise resulted in the advancement of the thesis on the part of policy makers that the worst effects of the disaster could have been avoided if faulty land use changes on shorelands were prevented, adequate protection was provided to the poor people staying on the shores and the unimaginative coastal zone regulations were recast along with inept disaster management plans that hitherto existed to handle lesser calamities like cyclones. In other words, though seen as an absolutely unforeseen natural calamity, the tragedy was analysed by policy makers in terms of inadequacy of rehabilitation measures and the inadequacy of poverty alleviation measures among coastal communities. Even when it came to relief measures, the discourse changed from building 'concrete sea walls' to planting mangroves plants along coastal areas as 'green walls' to prevent damages to communities from possible recurrence of Tsunami in future. Schemes were introduced to plant mangroves along affected coast lines, which were supported with public funds released by the State machinery through the Forest Departments. However when it came to rehabilitation of affected households, conventional precepts were followed. This assumed the shape of providing more advanced capital assets by replacing existing ones (like wooden canoes of small fishermen) with steel boats which, though robust, were more costly and hence not an economically sustainable asset for small fishermen operating in the narrow sea. The only point of congruence between the victims and the State was with reference to newly constructed dwellings that replaced the thatched huts of the past. Local communities were happy with the new dwellings that the State constructed for them in the post Tsunami phase. The State also introduced insurance systems to back relief management and in the process built, what Dean (1999) refers to, as social citizenship with its associated control systems. The fundamental premise of the State was that natural disaster related risks and their management are exogenous in nature, where the experiences and needs of the local communities had little scientific relevance. This thinking created gaps in expectations on the part of local communities, as they found the grand narrative overbearing and not relevant to their basic needs and economic condition.

Thus the State perceived the Tsunami to be an unusual seismic induced natural disaster that befell a hapless local community, which could not be expected to do anything on its own except fall back upon the State for swift and standardized, rapid solutions by way of interventions.[2] Despite prevarication and poor prognosis about the likelihood of future Tsunamis, the State machinery was of the view that, risks posed by unprecedented natural disasters like the Tsunami were exogenous in nature, and called for 'exogenous solutions'.

Local community

Local communities saw the Tsunami as an inevitable event that had to 'happen' to disturb the order of things. The priority was to return to the 'order of things' through re-affirmation of the community idiom. This perception goes well with what Bourdieu (1999, 173) stated to be the trait of an archaic economy which is 'familiar with the opposition between ordinary and extraordinary occasions'. At a more fundamental level, the opposition between the 'ordinary' and the 'extraordinary event' is based on the philosophical premise underlying the Vishishtadvaita School's central principle that considers creation as the 'evolution of cosmic purpose' and as involving the onset of pralaya (or 'deluge') from time to time. Pralaya is treated under this school of India philosophy as the involution and reversal of the whole process of evolution of the cosmic purpose. Both events are considered by this school as resembling the two alternate states of waking and sleeping states of the Jiva'.[3]

---

[2] This approach resonates with the Hegelian notion of the State being an embodiment of 'love, human caring and unity'. See MacGregor (1996), for a discussion on this idea of 'State as Love'.

[3] Vishistadvaita is associated with the philosopher Ramanuja. This school of philosophy offers a distinctive approach to the understanding of Vedanta. Vishistadvaita 'mediates between monism on the one hand and the theism of Dvaita (Dualism) on the other. It accepts the immanence of God in all beings and the innate spirituality and salavability of all Jiva-s, thus shedding the twin evils of exclusiveness and hatred' (Srinivasachari 1978). The main point here is that the middle road nature of this school of

The more important point is that such violent swings towards destruction create a natural need for people affected by the crisis to expand to a more extended group', beyond immediate family ties. This involves moving beyond affinal and non-affinal ties to embrace even larger communities who have the commonality of a common lived experience (Bourdieu 1999, 179). In other words, every calamity of the kind described here, *re-creates* the common logo that bind local communities together. It creates conditions to re-establish what Wilkinson & Quarter (1996, 119–122) describe as the three pillars of 'community consciousness' (that overcome free riding instincts), 'empowering activities' (including strategies of self-reliance) and 'supportive structures' (community development organizations). In this process of re-creation, the cognitive boundaries of local communities are extended, to embrace all the local communities which were affected by the Tsunami than one's own village or nearby villages. Thus the villages affected by Tsunami underwent a 'Bourdieon' moment mainly due to the Tsunami creating new boundaries of community action and mobilization across different villages. This altered boundary spread beyond the legal- administrative boundaries set by the State. Thus the Tsunami affected villages of Nagapattinam, though widely spread around attempted to reach one another through a horizontal communication process to understand their mutual experiences to fashion out local solutions to cope with such calamities in future.[4]

4.2 Assessing the state-local community gap

Local community perception about the Tsunami and its management was in marked contrast to that of the State. While local communities felt that their experience with the disaster was not taken into account by the State, the State machinery proceeded on the assumption that it was possible to devise standard exogenous solutions to the problem as was the case with

---

thought, has a presence in the cognitive space of ordinary communities such as the one encountered in Nagapapttinam, which makes them believe in a more variegated approach to viewing evolution as alternation of creation and destruction.

[4] As an interviewed leader of an affected village told the author, 'there was a keen desire on the part of all village communities affected by the Tsunami to unlearn the past system of looking towards State succour for handling the problem and look upon voluntary community work to craft solutions'. Strangely here one notices an uncanny resemblance of this 'local awakening' with that of the nation building lore surrounding 'America'. Thus one cannot help noticing a Tocquevillan strain to the process of recreation of community feelings in Nagapattinam. As was the case with America, the local communities of Nagapattinam also 'newly emerged from the wilderness, its virgin soil unburdened by castles and coats of arms (Lapham 2012)'. At least 4 local village communities in Nagapattinam that fell victims of the Tsunami succeeded by 2006 in re-building their tumbled fortunes once again in a Steinbuckian sense, through a process of sharing of experiences and resources and mutual help aided by civil society movements and charitable institutions (Steinbeck 2012).

other natural disasters. Majority of the local communities of Nagapattinam interviewed by the author, desired a return to the traditional order for fear of being forced into sophisticated technological solutions and insurance schemes introduced by the State with its attendant control systems.[5]

Local communities felt that Government schemes of rehabilitation were beyond their reach, cognitively speaking and was certainly beyond the scope of community interventions for fine-tuning and customization. This accounted for their preference for the traditional order, which was nothing but 'a corner solution' (as exemplified in an almost exclusive preference for the traditional way of life) in terms of the paradigm of New Classical Economics.[6]

Government officials at the District Collector's office who were interviewed by the author, on the other hand were of the view that in future, tsunamis needed to be confronted with robust technological solutions. They felt that the demand on the part of fishermen communities was not based on scientific understanding of the issue.

However our point here is that to dismiss local community viewpoints as unscientific and also rule out this perception as being of no consequence to community risk management, would be fatal since people who experience calamities could have their own unique ways of looking at the probability of catastrophic events like the Tsunami and on ways of effectively coping with them (Damodaran & Roe 1998). This perhaps explains why horizontal risk communication systems became an important activity amongst local communities of Nagapattinam in the post-Tsunami phase. Further it was the view amongst sections of local communities interviewed by the author, that risk instruments of the State were beyond their reach and understanding. Many of them felt that in the event of a disaster hitting them in future. They had to fall back on their own means, rather than that of the State.[7]

---

[5] Their world view in some sense mirrores Frithjof Schuon's statement that 'Modern technology is the result of a perspective which, having banished from Nature the gods and the genies and having also by this very fact rendered it profane has ended by allowing it to be profaned in the most brutal sense of the word', (Schuon 1990).

[6] See Yang (2001). New Classical Economics goes beyond the paradigm of neo-classical Economics in its focus on corner solutions rather than interior solution. New Classical Economics also focuses on transaction costs and places relatively more emphasis on economies of specialization and network effects of division of labour to describe production conditions as compared to the concept of 'economies of scale' typically employed by neo classical economics to describe viability of economic activities).

[7] As Hanley et al. (1997, 369–371), state, 'The normal trend is to suppose that risk is exogenous and beyond private action of individuals. But not all risk reduction is achieved or derived from collective action based on public policy. Some of it is endogenous and individuals seek to substitute private action to reduce risk for collectively supplied programmes, private water filtration for example. It also follows as the marginal effectiveness of successive collective provisions declines, the relative effectiveness and therefore

The second cause for the gap between Government's grand narrative on the Tsunami and the perceptions of local communities about the calamity was that the Government machinery could not step in effectively to compensate for the possible failure of the market and supply public goods in the aftermath of the disaster. The Transaction costs of Government schemes were also aggravated by the failure of vertical communication from the Government to local communities. This accentuated the loss of faith in risk assessments done by the State machinery regarding future probabilities of such disasters striking them. This also raised doubts in the minds of local communities about the utility of the schemes of disaster relief introduced by the State. Thus the costs of not providing perfect information also added to the Transaction Costs of State action.[8]

By contrast, horizontal communication amongst local communities was more successful.[9]

5 Bridging the gap through altered policy paradigm

The gap in perception between local communities (who fell victims to the Tsunami in Nagapattnam) and the State machinery on the roots and the fall out of the Tsunami, more fundamentally represents what Goeminne (2011) refers to as the 'science/life world distance' or the difference between thinking and experiencing an object (see McIntyre & Smith 1989; Ihde 2011). As Goeminne states, there is a clear distinction between the 'expert side of Science and Technology in the making', the lay side of our 'everyday dealings with the world' and the role knowledge and artefacts play in the latter. Thus Goeminne advocates participatory approaches to bridge the micro worlds of Science and Technology and the macro-world of our everyday practices. This is not a mere legitimation exercise. Indeed participatory approaches are productive.

In some ways Goeminne's stress on participatory approaches seeks to arrive at the golden mean between the Hegelian notion of 'objective the-

---

the value of private provision increases when self protection and collective protection are perfect substitutes. The faith of local victims of the Tsunami in Nagapattinam cannot be explained better than through the thesis of Hanley and Shogren.

[8] Despite the stellar influence of the Welfare Economics, there is a substantial body of opinion against State intervention in mainstream economic theory. As Medema (2007, 442) states the theoretical apparatus that had demonstrated the failure of the market and established the case for government intervention was wrong headed. This apart, the public choice analysis school (Virginia and Chicago Schools) succeeded in demonstrating Government failure thereby questioning the Sidwickian and Pigouvian optimism regarding government intervention as unfounded.

[9] As Jaspers once appropriately remarked,'Abstracted from communication, truth hardens into an unreality' (Marshall 1958).

ory of knowledge' and the Kierkegaard maxim that 'subjectivity is truth' (Kierkegaard 2004).

As Feenberg states elsewhere, echoing Martin Heidegger, the problem with 'technology', is that it is not about getting an efficient solution, as much as it is about 'choosing a different way of life' (Damodaran 2007).

The local communities of Nagapattnam which were hit by Tsunami saw in the technology solutions offered by the State, a transformative potential that could disrupt their community driven livelihood systems. This notion could have been dispelled, if the Grand narrative listened to the voice of concern of the victims of disaster. A technological transformation that is less disruptive to local lives can emerge as a result. This could be achieved through participatory approaches that involve the State and the local communities. In practical terms, participatory approaches that bridge the 'micro worlds of Science and Technology and the macro-world of our everyday practices' (à la Goeminne), would assume the shape of programs for local mapping of risk perceptions that facilitates horizontal communication of coping strategies. Based on such 'mapping' of local risk perceptions, it is possible for the State machinery to understand the viewpoints of communities regarding the nature of their losses and on the typology of solutions that they desire to bring about to ameliorate their situation. Boats that suit local community working patterns, while being at the same time robust enough to withstand future Tsunamis could be the participative solution that everybody desires. Similarly improved systems of protection of shorelands based on planting of species that suit the needs of local communities would work better for both the State and local communities. Likewise, social insurance systems that recognize the private character of risks faced by Tsunami victims, would work decisively better than conventional insurance schemes that suit the economics of insurance companies. Finally, by documenting local community experiences with the Tsunami calamity, policy memories can be kept alive for a longer time to come. This will be of immense value to the State as it handles the next major calamity in future.

Bibliography

Bourdieu, P. (1999). *Outline of a Theory of Practice*. Cambridge: Cambridge University Press.

Damodaran, A. N. (2007). Heidegger and Marcuse: The Catastrophe and Redemption of History – Andrew Feenberg, Review Article. *Tailoring Biotechnologies*, *1*(2), 34–35.

Damodaran, A. N. (2010). *Encircling the Seamless. India, Climate Change and Global Commons*. New Delhi: Oxford University Press.

Damodaran, A. N. & Roe, E. (1998). Theorising to explain and triangulating to explain away? The art and non-art of multi-method policy research. *Economic and Political Weekly, 33*(1–2), 55–62.

Dean, M. (1999). *Governmentality: Power and Rule in Modern Society.* London: Sage Publishers.

Goeminne, G. (2011). Postphenomenology and the politics of sustainable technology. *Foundations of Science, 16*, 173–194, doi:10.1007/s10699-010-9196-5.

Hanley, N., Shogren, J. F., & White, B. (1997). *Environmental Economics – In Theory and Practice.* Delhi: Macmillan.

Ihde, D. (2011). Husserl's Galileo needed a telescope. *Philosophy and Technology, 24*(1), 69–82.

Kierkegaard, S. (2004). *Training in Christianity.* Vintage Spiritual Classics, New York: Vintage Books, Translated by Walter Lowrie, D.D.

Lapham, L. H. (2012). Hostages to fortune. *Lapham's Quaterly, V*(1), 13–20.

MacGregor, D. (1996). *Hegel, Marx, and the English State.* Toronto: University of Toronto Press.

Marshall, M. W. (1958). *Existentialism: Suspension Bridge of Indian Thought.* Santiniketan: Santiniketan Press, Reprinted from Viswabharathi Quarterly, 28(1–2).

McIntyre, R. & Smith, D. W. (1989). Theory of intentionality. In *Husserl's Phenomenology: A Textbook,* Mohanty, J. & McKenna, W. R., eds., Washington, DC: University Press of America, 147–179.

Medema, S. G. (2007). The economic role of government in the history of economic thought. In *A Companion to the History of Economic Thought,* Samuels, W. J., Biddle, J. E., & Davis, J. B., eds., Malden: Blackwell Publishing, 428–444.

Schuon, F. (1990). *The Feathered Sun: Plains Indians in Art and Philosophy.* Bloomington: World Wisdom Books.

Srinivasachari, P. N. (1978). *The Philosophy of Visistadvaita.* Madras: The Adyar Library and Research Centre.

Steinbeck, J. (2012). 'It takes a village', extract from *The Grapes of Wrath. Lapham's Quaterly, V*(1), 160–161.

Wilkinson, P. & Quarter, J. (1996). *Building a Community Controlled Economy: The Evangeline Co-operative Experience.* Toronto: University of Toronto Press.

Yang, X. (2001). *Economics: New Classical versus Neoclassical Frameworks.* Malden: Blackwell Publishers.

Appukuttan Nair Damodaran
Indian Institute of Management Bangalore
India
damodaran@IIMB.ERNET.IN

Engineering Design between Science and the Art

GERHARD BANSE

ABSTRACT. Research on the nature of engineering design led to important contributions in the last years. These results help to better understand (describe, explain, prescribe, predict) and plan the process and result of the imaginary devising and outlining innovation, and the conditions for the possibility of "designing". There is an ongoing, broad discussion that reveals different points of view, problem solving strategies and directions of answer. Current discussions extend the research made in the history of engineering sciences. One of the main questions has been: "Is the design process more rational and formalized or rather an 'intuitive' process?", or more generally speaking: "Is engineering design science or art?" This has led to the current questioning: "Is it possible to generate a general theory of engineering design?" There are different answers and the difference is related to the field of engineering (mechanical engineering, civil engineering, architecture, industrial design, and so on) on the one hand, and to the type of understanding of science, art, heuristics, the ways problems are being solved on the other hand. (Different interpretations have been given on the role of values and evaluation, types of thinking in design, relationships between logic and heuristics, influence of language, visual thinking and non-verbal knowledge and its representation—e.g., sketches, drawings, models, and what is teachable and learnable in design). This presentation emphasizes that the design process is a "mix" of different components, that it is both science and art. Some historical reminders are also made on the development of the "design theory", first of all in Germany.

1 Preliminary Remarks

Theoretical reflections have accompanied the development and institutionalization of engineering sciences (technology included) from the start, albeit each with different focus on content, with different intensity of the discussion and with different consequences. The so-called "dispute on methods" in the second half of the 19th century is an example. Others are reflections in the discourses trying to clarify the concept of invention in the early 20th century. The autonomous agent of technology and engineering sciences was always central to these disputes, the actors (engineers, engineering scientists) and their teaching and research institutions (polytechnics, technical colleges, universities of technology). Again, the specificity of engineering sciences has been the focus of investigations since the mid-sixties in the last century. Especially the philosophies of technology and science, but also art, sociology, economics and the history of technology have been taking part in this activity. Topics have steadily expanded. They range from the philosophy of technology, as it is the focus in this article, including the relationship between natural and engineering sciences and some forms of scientific knowledge in these fields, to the establishment of a general technology (general engineering science). The purpose of this quickly expanding activity is not solely or even primarily to contribute to a better understanding of engineering sciences, based on the knowledge of structures, processes, relationships, and dependencies, etc., but also to provide for a theoretical basis of the design practice in engineering sciences.

Engineering sciences are sciences of a special sort of action. They deal with technological artefacts, processes, methods of action, and procedures with the objective of supporting human action in many areas by means of technology or by providing knowledge and know-how for this. Ultimately, the knowledge created, systematized, improved, and taught by engineering sciences is a type of knowledge for designing working technological systems and their efficient use. The specifically scientific perspective on technological action serves the systematic investigation on the conditions and extensions of the possibilities of successful action involving use of technology, in order to attain, with good prospects, ever more targets with appropriate technological means.

Engineering sciences pursue three target groups on this abstract level (cf. König 2006):

1. *Design targets*: Engineering sciences try to support and improve technological practice with regard to usability, economic viability, efficiency, safety, functionality, etc., of technological procedures, products or systems. New technological knowledge and invention is provided as a starting point for innovations. Further, a qualitatively advanced education of engineers is ensured for their actions in technological practice. Central criterion of success in engineering sciences is their usefulness for society by supporting technological practice. With regard to their practical aims, engineering sciences demonstrate a high

degree of social relevance. This is a crucial argument for receiving governmental financial support.

2. *Cognitive targets*: Technological knowledge, i.e., knowledge of technological contexts and procedures, of material processing and exploitable physical or chemical processes etc. should be produced, made reliable, systematized and improved. The central criterion for success is here the truth of the relevant knowledge in order to achieve recognition within and outside sciences. This group of criteria is decisive for the reputation of particular persons, groups or institutions in engineering sciences. Scientific excellence is the central value here.

3. *Social targets*: The scientific community of technological disciplines (e.g., in mechanical or civil engineering) is to be consolidated through scientific journals, societies, congresses and scientific institutions. This consolidation results in stabilization and expansion of the knowledge system, and its further specialization and differentiation.

Engineering sciences are a specific "combination" of *recognition* [*Erkennen*] and *formation/design* [*Gestaltung*] (cf. Banse et al. 2006; Banse & Grunwald 2009).

The specific nature of engineering sciences can be summarized in the following three characteristics.

1. Engineering sciences are target-oriented sciences. For them, the central concept is application of scientific knowledge and practical experience to satisfy the "technological needs" of society.

2. Engineering sciences are constructive in character. At the forefront are the intellectual anticipation and evaluation of the structure and the functions of new technological systems and their realization.

3. Engineering sciences are integrating sciences. Since technological systems represent a specific coherence of natural and social components, knowledge is included from the natural sciences, other technological sciences and social sciences in the conception, design, evaluation and optimization of technological systems. Knowledge from all these fields is searched for, considered, assimilated, and integrated into a specific unit in solving concrete technological tasks.

Engineering sciences have developed a plethora of methods:

Methods of research (in analogy to the classical, natural sciences they are a mix of theoretical-deductive and empirical-inductive procedures);

methods of design;

methods of implementation. These include various approaches to create prototypes, demonstrate technical feasibility and finally create the physical hardware that can provide the required capabilities.

The methods of design are the most interesting from the point of view of the philosophy, methodology and "logic" of science. That means that they differ from the classical sciences, because there is a contrast between the *synthesis* of a new system and the *analysis* of an existing one.

2 Understanding Engineering Design

In engineering sciences, design refers to solving problems by means of technological products, systems or procedures. A dedicated implementation of knowledge and know-how from engineering sciences is necessary to achieve this. Technological construction and drafting work form the procedure in which knowledge and design come together. Drafting is thus the methodological location at which the specific relationship of theory and practice in engineering sciences can most clearly be seen (cf. Cross 1989; Gregory 1966; Hubka *et al.* 1988; Hubka & Eder 1992).

Most general descriptions of how technological solutions come about distinguish between the design process, which results in a draft of the solution and the process of carrying out design in practice itself (cf. Hubka & Eder 1992).

Engineering design, in German "Entwurf" or "Konstruktion", is the anticipation of new technological systems by identifying a structure which can realize the required function (cf. Banse & Friedrich 2000b), (esp. Banse 2000):

> required function ⇒ (looking for an) adequate structure (elements & relations)

Function means here "transformation" of the status 1 into the status 2, by using special means. There are two problems.

Problem 1: Mostly some or perhaps many adequate structures are possible that can realize a given, a required function. This brings with it the necessity of a series of evaluations, choices and decisions. An *example* is the production of electric energy by coal power plants or nuclear power plants or photo voltaic, etc.

Problem 2: There are different domains in engineering design. The question raises if these are comparable enough to make a valid generalization An *example* is: mechanical engineering (machines, cars), civil engineering

Automotive Engineering[e], Software[s], Health, Transport and Consumer Products[a], Architecture / Urban Planning[aa]
Civil Engineering (structures) [e], Web Sites[mm], Automotive Styling and Consumer Products[a], Drugs / Pharmaceuticals[ss]
Graphic Media[mm], Aerospace Engineering and Senior management[e], Documentary Film Maker[a]
Artistic Fashion[a], Medical Devices[s], Food[ss], Packaging[a], Architecture[aa]
Architecture[aa], Technical Fashion[a], Automotive Engineering and Senior Management[e]
Electronic Products[e], Furniture Designer[a], Software[s], Course Design[mm]
mm: multimedia; e: engineering; a: artistic / product design; aa: architecture; s: software; ss: science

Table 1. Design Domains; Source: after (Eckert & Schadewitz 2011, 249 (Table 1))

(bridges), architecture (buildings), industrial design (furniture), etc. (see Table 1).

The common general interest in engineering design is:

to understand (*describe, explain*, prescribe, predict), and

to plan the process and result.

of the imaginary devising and outlining of novelty and the conditions for the possibility of "design". Concepts of an engineering design endeavor especially consist in trying to capture or describe two things. The first thing is how can the knowledge, required for the construction process ("design process"), "be organized" from existing stocks or how it can be generated in a process resulting from a deficit of knowledge. The second problem is how the process of "transition" from descriptive knowledge to prescriptive [*handlungsleitendem*] knowledge is realized. Does it include prompts, instructions or principial rules?

That is why there are different descriptions of engineering design:

Some of them are more technological and concentrate on the relationship between the purpose and the means;

Some of them are more systems theoretical and concentrate on the relationship between function and structure, on synthesis of an adequate structure for a given function;

Some are more problem theoretical. They concentrate on the problem solving process, heavily including "wicked" problems (cf. Buchanan 1992; Rittel & Webber 1973); and,

Some are more philosophical. They concentrate on the methods of drawing reductive conclusion from the consequences to the reason. This is the field of "practical syllogism" (cf. Wright 1991).

Engineering Design is a process of a concretization activity that starts with an abstract principle and leads to a plan for a real object with specific structure, form, dimensions, etc.

Various concepts for structuring the design process have been developed. Although widely differing in terminology, they have much in common in terms of their meaning. They emphasize different answers to the same question. This is whether the design process can be characterized as a linear process or whether it essentially consists of feedback loops. Orientation towards feedback loops and iterations is, with the opportunities they provide for sequential phases of learning, overwhelming at present. This is even true for graphic representations, in which a linear presentation is frequently preferred for didactic reasons (cf. Banse 2000; Hubka & Eder 1996). Accordingly, the design process consists of a problem analysis, conception building, planning, and elaboration, and these elements are repeated with an increasing degree of detail and approximation to technological action (cf. (Banse 2000, 64), (Banse & Wendt 1986, 47 ff.), (Eder 2000b, 217), (Ropohl 1999, 258 ff.); see Figures 1 and 2).

Problem analysis includes planning of the objective. This is a question of formulating the requirements on the technological task in such a way that their fulfilment should lead to the solution. Primarily it consists in the formulation of a consistent specification. When the target system has been sufficiently clarified a lack of knowledge is often encountered in the transition phase to concept building and planning about its feasibility, either in principle or within the given boundaries (e.g., the cost framework). So-called feasibility or pilot studies are often used to clarify this problem in planning practice. These attempt a rough preliminary evaluation of a project's risk or the structure of a possible solution concept when there is not enough information available to make a precise evaluation.

The main task of the conception phase is the breakdown of the desired overall function of the technical solution into sub-functions, in systems-analytic terms. Solution principles and basic structures for the realization of these principles can then be ascertained. If possible, this should be done on the basis of available scientific and technical knowledge and know-how. Otherwise the conception leads to the identification of gaps in knowledge and know-how and thus in turn to further technological research needs. The options for solving the problems of sub-functions are then integrated into an overall concept for solving the overall problem of function, or several concept variants are drafted (cf. Ropohl 1999).

This conception forms the basis for a draft to be created to scale up a digital model, increasingly, in the draft phase, in which a technological system is developed. On the one hand, this model is evaluated according to various technological criteria. These include feasibility, security, and function fulfilment, for the realization of the necessary preconditions, material, energy and data flow. On the other hand, it is possible to consider criteria from outside the technical field at this juncture already. At the very least, these would include economic and perhaps also further social, legal or ethical aspects. On this basis, an improved draft or model is drawn up. Time structures are elaborated for temporal sequences, and milestones with verifiable requirements are included in each case which constitute the starting point for projections and project management.

The *elaboration phase* covers further concretization and detailing work, together with optimization of the individual components of the draft and its composition. Systematic connections between the necessary resources are worked out, and finally the documentation for carrying out the work is created, involving among other things design instructions for a technological system. The rough cost estimates are refined on this basis and made as resilient as possible.

The results of the design process are elaborated drafts for a technical solution.

3   Engineering Design (Science or Art?)

The results of any design process are the elaborated drafts of a technological solution. This process uses different methods. There are rational-systematic and intuitive-heuristic methods:

> *Rational-systematic methods.* They start from an analysis of the system under consideration in terms of rational problem-solving methods and provide for appropriate top-down solutions. These methods are suitable for planning.

> *Intuitive-heuristic methods.* They attempt to deploy creativity techniques to find new solutions. These methods are scarcely suitable for planning—they rather build on "ideas", visions and intuition.

The co-existence of these two methods is the background to an interesting explanation and discussion of engineering design by using the terms "*science*" and "*art*" (You may find them in the journals "Design", "Design

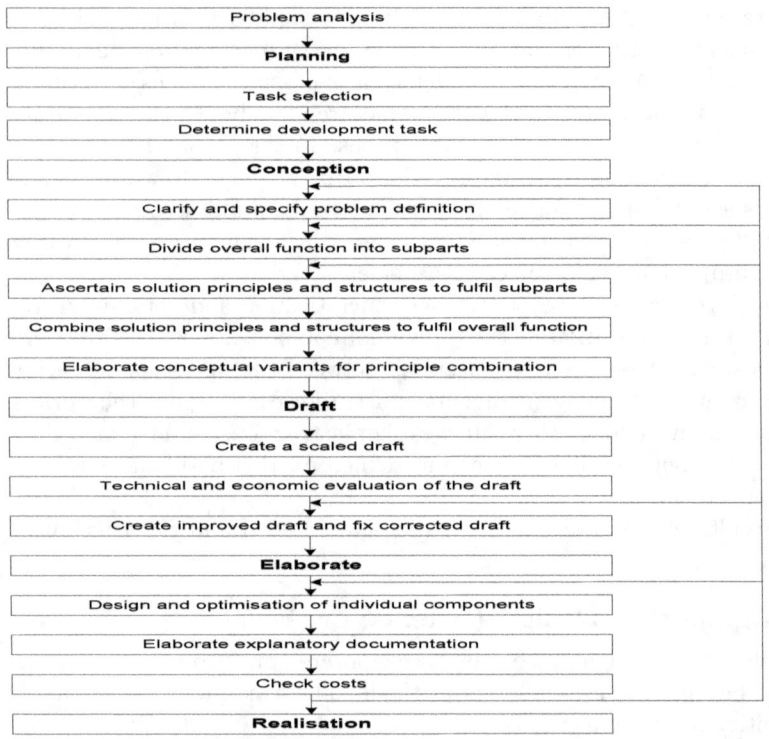

Figure 1. Linear Model of Engineering Design – Source: after (Ingenieure 1977)

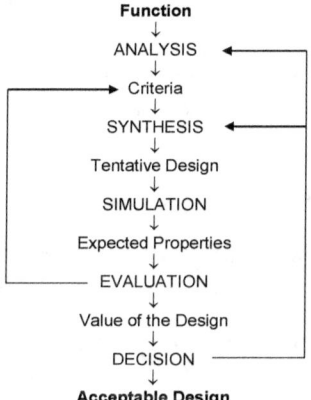

Figure 2. The "Basic Design Cycle"– Source: after (Cross & Roozenburg 1992, 334)

Issues" and "Design Methods", mostly. Concerning the assessment of this discussion (cf. Banse 1999).

To conclude this discussion, it can be said that engineering design is between "art" and "science" and detects the creative and schematic problem-solving processes in engineering action with its two aspects:

> First, the processes and procedures detected are based on imagination. They are synonymous with *art*. They imply intuition, inspiration and imagination, individual experiences and skills, as well as unreflected or traditional routine.

> Second, there is something synonymous with *science*. There is some systematically derived established and reflected knowledge in design, with different degrees of accuracy, especially about the natural "laws" that set the frame for the designing process, as well as its conscious, rule-based, methodologically supported and traceable implementation in artifacts and technologies.

The main question arises as follows: "Is the design process more a rational, formalized process or rather an intuitive' process?" Or to put it in more general terms: "Is design science or art?" (This leads to the question also currently in focus: "Is it possible to generate a general theory of design?")

3.1 Engineering Design (Science or Art?), Historical Remarks

These questions are connected with the rise of the technological sciences and technology as a science-based activity in the middle of the 19th century.

But there were interesting (theoretical, philosophical) reflections on (engineering) design already before that time:

Sokrates (469-399 B.C.): Meutik;

Archimedes von Syrakus (etwa 287-212 B.C.): "heureka" ("I've found it!");

Raimundus Lullus (1232/33-1316): "Ars compendiosa inveniendi veritatem" (1273/75) and "Ars magna Lulli" (um 1300)—idea of an "ars combinandi";

Ren Descartes (1596-1650): Regulae ad directionem ingenii " (nearly 1628, published in 1701);

Gottfried Wilhelm Leibniz (1646-1716): dissertation "De arte combinatoria" (1666)—idea of an "ars inveniendi";

Johann Beckmann (1739-1811): "Entwurf der Algemeinen Technologie" (1806) (Plan of a General Technology)—idea of a "heuristics of invention".

With the rise and development of the technological/engineering sciences a new phase of this discussion began, mostly in Germany and German-speaking countries.

In my opinion, the starting point was *Ferdinand Redtenbacher's* consideration. (He lived from 1809 to 1863. He was the first director of the Polytechnikum Karlsruhe). He looked for a new approach to educating engineers beside the *French method*, based on a strong mathematic-theoretical approach relying on as much geometry as possible, rather combining with it the more empirical *British method*. Basically, he thought that engineers' actions are based both on physical-mechanical knowledge and on specific human abilities, that are only partially teachable. This basic idea was elaborated in the "Resultate fr den Maschinenbau" ("Results for Machine Production", 1848) in which engineering and design involve altogether "principles of mechanics" *and* "sense/mind/feeling" of composition [*Zusammensetzungssinn*], of "configuration" [*Anordnungssinn*] and of "form/model" [*Formensinn*].

A broad discussion started among engineers and engineering scientists, the so-called. "dispute on methods", or *Methodenstreit*. It concentrated on dichotomies, the main components of which were:

"true principles" [*wahre Grundstze*] as opposed to "solid experiences" [*zuverlssige Erfahrungen*], or

"scientific knowledge" vs. "technical ability", or

"calculation" instead of "feeling" or "sense" [*Gefhl*], or

"recognition" and "knowledge" [*Erkennen*] by contrast with "making" and "doing the work" [*Machen*, or "knowing why" and "knowing that" with "knowing how", or

engineering more as a *science* rather than an *art* (cf. Ferguson 1977; 1978; 1992;b).

At one end of these controversies was—as one of the most important contenders—Franz Reuleaux (1829-1905; 1890/91 rector of TH Berlin), the "father of kinematics". As he argued machines could be abstracted into chains of elementary links called kinematic pairs. He wrote in "Theoretische Kinematik" (1875) that solutions for new technological structures can be made or found by deductions ("calculations") which are based on some elementary axioms. He had the dream of an "exact science" of axiomatic structures in the field of engineering.

At the other end was, among others, Alois Riedler (1850-1936; 1899/1900 rector of TH Berlin). His interpretation was that engineering design is more a free "play of imaginations", more an "artistic" process than a "deductive" activity. Its basis is practical experience (gained in universities in "mechanical engineering laboratories"). Decisive difficulties do not lie in the calculations but in the approaches used and their basic assumptions. In "Das Maschinen-Zeichnen" (1896; "Machine Drawing") he introduced modern technical drawing.

The solution to this broad discussion was given by Julius Carl von Bach (1847-1931; 1885-1888 rector of TH Stuttgart). He demonstrated that both components are necessary, and hence that there must be a "good combination" of them in design.

3.2 Engineering Design (Science or Art?), Current Discussion

The discussion on whether engineering design is a deductive science or an intuitive art is still ongoing and arises again periodically as shown in the two examples below.

The first is Claudia Eckert's and Nicole Schadewitz's report on the *Across Design* project, a cooperation between the Engineering Design Centres of University of Cambridge, the MIT and the Open University at Milton

Keynes in 2010 (cf. Eckert & Schadewitz 2011). Their *aim* was to looking for coherence and diversity in the practices of design(ers). The basis for conclusions was learning from four workshops. The *results* were:

1. *"Self-interpretation"*: "scientist" vs. "artist".

2. *Method of design*: method-based, systematic vs. trial-and-error, intuitive.

3. *Kind or type of problem*: overdetermined by too many restrictions, connected with reduction of complexity vs. underdetermined, too many "degrees of freedom", connected with an "interpretation" of the design problem.

4. *Approach*: idealization or simplification vs. holism, related to a "style of design" or a "culture of design".

The effort to achieve a so-called *General Design Theory (GDP)* or *Universal Design Theory (UDT)* is the second example. This theory is to be understood as an axiomatic theory - or a set of such theories - of design, as shown in the three following citations. In 1989 Nam P. Suh wrote: "A general theory for system design is presented based on axiomatic design" (Suh 1998, 189) ; (cf. also Suh 1990). Ralf Lossack and Hans Grabowski wrote in 2000: "The UDT consists of several parts. One very important part of the theory is the kernel of the UDT which is based on an axiomatic approach" (Lossack & Grabowski 2000, 1). In other words, based on some assumptions, or axioms, and some predictions or theorems, one would be able to draw up guidelines for the building of computer-aided design (CAD) systems In the same train of thought, Tetsuo Tomiyama and Hiroyuki Yoshikawa wrote in 1985: "From general design theory, it is deduced that an extensional description method is more suitable for CAD systems than an intensional one, theoretically, but practically both of them should be combined" (Tomiyama & Yoshikawa 1985). Reading this, one would be tempted to say, as the Germans do: "Greetings from Franz Reuleaux!".

4 Conclusions

Many scientific disciplines are interested in the field of engineering design: architecture, art, industrial design, systems theory, engineering, and philosophy of technology and so on. Important contributions in this field have been given by the research in engineering design in the last years. The common interest is to understand, that means to describe, to explain, to prescribe, to predict and to plan the processes and results of the imaginary devising and outlining of novelty or innovation, the conditions of the possibility of "design". That's why there is a broad and interesting discussion that shows different points of views, strategies of problem solving and directions of answers. Rather various themes are discussed. They include the meaning and different kinds of design, design as a process or a result, design in art, in architecture and in engineering, the history of design research, social and scientific needs for research in design, scientific origins and rules of design and contributions from sociology, psychology, education, human biology, and logic, kinds of design knowledge. The main questions are:

"Is the design-theory (is theory in design) more a formalized?" and

"Is it possible to generate a general theory of design?"

There are many philosophical problems in this broad field. Among them are designing as a problem-solving process, the role of values and evaluation in design, types of thinking in design, relationships between logic broadly understood and heuristics, the influence of language, visual thinking and non-verbal knowledge, and their representations in sketches, draws, models and so on, and the teachable/learnable part in design.

Finally, it should be re-emphasized:

1. Engineering design takes place in many different domains: "everything is design(ed)" (Papanek 1971).

2. Engineering design is a process aiming at finding new "means" for a given (technological) function by way of invention.

3. The design process involves different kinds of problems ("wicked"; overdetermined, underdetermined, etc.).

4. Processes of engineering design include different phases, stages, which are combinations of deductive knowledge, ability and experiences. These stages are quite different. Some phases have a more methodical/schematic/systematic character (so called "scientific" stages), and others a more creative/spontaneous/heuristic character (so called "artistic" stages).

5. These discussions have a long history, but still remain very topical. "Old" questions lead to "new" answers.

6. In summary: Engineering design is unavoidably both science *and* art.

7. It depends on the specific "self-interpretation" of the "designer" and the domain of design which of these components are to be considered as more important.

Bibliography

Banse, G. (1999). Research in design theory and methodology. a selective overwiew of recent work in english. In *Stains on the Screen*, Gray, N. & Banse, G., eds., Stamford, Conn.: JAI Press, 149–172, banse, G.: *Design Bibliography*.

Banse, G. (2000). Konstruieren im Spannungsfeld: Kunst, Wissenschaft oder beides? Historisches und Systematisches. In *Konstruieren zwischen Kunst und Wissenschaft: Idee – Entwurf – Gestaltung*, Banse, K., G.; Friedrich, ed., Berlin: Sigma, 19–79.

Banse, G. & Friedrich, K. (Eds.) (2000b). *Konstruieren zwischen Kunst und Wissenschaft: Idee – Entwurf – Gestaltung*. Berlin: Sigma.

Banse, G. & Grunwald, A. (2009). Coherence and diversity in the engineering sciences. In *Philosophy of Technology and Engineering Sciences*, Meijers, A., ed., Amsterdam: Elsevier/North Holland, 155–184.

Banse, G., Grunwald, A., et al. (2006). *Erkennen und Gestalten. Eine Theorie der Technikwissenschaften*. Berlin: Sigma.

Banse, G. & Wendt, H. (1986). *Erkenntnismethoden in den Technikwissenschaften : eine methodologische Analyse und philosophische Diskussion der Erkenntnisprozesse in den Technikwissenschaften*. Berlin: Verlag Technik.

Buchanan, R. (1992). Wicked problems in design thinking. *Design Issues*, *8*(2), 5–21, URL www.jstor.org/stable/1511637.

Cross, N. (1989). *Engineering Design Methods*. Chichester: Wiley.

Cross, N. & Roozenburg, N. (1992). Modelling the design process in engineering and architecture. *Journal of Engineering Design*, *3*(4), 325–337, doi: 10.1080/09544829208914765.

Eckert, C. & Schadewitz, N. (2011). Disziplinen der Produktentwicklung aus der Perspektive des angelsächsischen Raums. In *Wissenschaft im Kontext. Inter- und Transdisziplinarität in Theorie und Praxis*, Banse, G. & Fleischer, L.-G., eds., Berlin: Trafo, 243–254.

Eder, W. E. (2000b). Konstruieren aus der Sicht eines Konstruktionswissenschaftlers. In *Konstruieren zwischen Kunst und Wissenschaft: Idee – Entwurf – Gestaltung*, Banse, G. & Friedrich, K., eds., Berlin: Sigma, 193–218.

Ferguson, E. S. (1977). The mind's eye: Nonverbal thought in technology. *Science*, *197*(306), 827–836.

Ferguson, E. S. (1978). Elegant inventions: The artistic component of technology. *Technology & Culture*, *19*(3), 450–460.

Ferguson, E. S. (1992). Designing the world we live in. *Research in Engineering Design*, *4*(1), 3–11, doi:10.1007/BF02032388.

Ferguson, E. S. (1992b). *Engineering and the Mind's Eye*. Cambridge, MA; London: MIT Press.

Gregory, S. A. (1966). *The Design Method*. London: Butterworths.

Hubka, V., Andraesen, M., & Eder, W. E. (1988). *Practical Studies in Systematic Design*. London: Butterworths.

Hubka, V. & Eder, W. E. (1992). *Einführung in die Konstruktionswissenschaft : Übersicht, Modell, Ableitungen*. Berlin: Springer.

Hubka, V. & Eder, W. E. (1996). *Design Science. Introduction to Needs, Scope and Organization of Engineering Design Knowledge*. Berlin; New York: Springer.

Ingenieure, V. V. D. (1977). *VDI-Guideline 2222 "Konstruktionsmethodik. Konzipieren technischer Produkte"*. Düsseldorf.

König, W. (2006). Geschichte der Technikwissenschaften. In *Erkennen und Gestalten. Eine Theorie der Technikwissenschaften*, Banse, G., Grunwald, A., et al., eds., Berlin: Sigma, 24–37.

Lossack, R. & Grabowski, H. (2000). The axiomatic approach in the universal design theory. In *Proceedings of ICAD2000. First International Conference on Axiomatic Design*, Cambridge, MA: ICAD005, 203–210.

Papanek, V. (1971). *Design for the Real World. Human Ecology and Social Change*. New York: Pantheon Books.

Rittel, H. W. J. & Webber, M. M. (1973). Dilemmas in a general theory of planning. *Policy Sciences*, *4*(2), 155–169.

Ropohl, G. (1999). *Allgemeine Technologie. Eine Systemtheorie der Technik*. München: Hanser, 2nd edn.

Suh, N. P. (1990). *The Principles of Design*. New York/Oxford: Oxford University Press.

Suh, N. P. (1998). Axiomatic design theory for systems. *Research in Engineering Design*, *10*(4), 189–209, doi:10.1007/s001639870001.

Tomiyama, T. & Yoshikawa, H. (1985). Extended general design theory. URL http://md1.csa.com/partners/viewrecord.php?requester=gs&collection=TRD &recid=N8632162AH&q=General+Design+Theory&uid=790893385&setcookie=yes.

Wright, G. H. v. (1991). *Erklären und Verstehen*. Frankfurt-am-Main: Hain, 3rd edn.

Gerhard Banse
Karlsruhe Institute of Technology
Institute for Technology Assessment and Systems Analysis
Karlsruhe
Germany
gerhard.banse@partner.kit.edu

Appendix A: Sections, Plenary Lectures and Special Symposia

A. Logic

A1. Mathematical Logic

Speakers: MARTIN GROHE (Germany – Logic, structure and complexity), JULIA KNIGHT (USA – Comparing classes of structures), JOE MILLER (USA – Beyond the Turing degrees: non-diagonalizability and universal randomness), JUSTIN MOORE (USA – The proper forcing axiom), PAULO OLIVA (UK – Gödel's functional interpretation of classical arithmetic and analysis), PATRICK SPEISEGGER (Canada – Solving equations by quadratures using Model Theory), SIMON THOMAS (USA – Martin's Conjecture and countable Borel equivalence relations)

A2. Philosophical Logic

Symposium on General Proof Theory:

Chair: KOSTA DOŠEN (Serbia – General Proof Theory)
Speakers: WILLIAM LAWVERE (USA – Proof theory and presentation of algebras), PHILIP SCOTT (Canada – Remarks on recent categorical proof theory), DAG PRAWITZ (Sweden – Is there a general concept of proof?)
Individual speaker: DAG WESTERSTÅHL (Sweden – The importance of compositionality)

A3. Logic and Computation

Speakers: STEVE AWODEY (USA – Homotopy type theory), ULRICH BERGER (UK – Coinduction and program extraction in computable analysis), ÉTIENNE GRANDJEAN (France – How tightly close descriptive and computational complexity are: A personal view), CHRISTOF LÖDING (Germany – Uniformization in automata theory)

B. General Philosophy of Science

B1. Methodology and Scientific Reasoning

Symposium on Evolutionary Models in Epistemology and Philosophy of Science:

Chair: ELLIOTT SOBER (USA)
Speakers: HANNES LEITGEB (Germany – Probabilities, conditionals, laws), BRIAN SKYRMS (USA – Evolution of signaling), PETER VANDERSCHRAAF (USA – Changing your spots)
Individual speakers: CARLO CELLUCCI (Italy – Philosophy of mathematics: Making a fresh start), CHRISTOPHER HITCHCOCK (USA – Cause and chance), PAUL HUMPHREYS (USA – The differences between data from simulations and experiments), HANS RADDER (The Netherlands – Does the brain 'initiate' freely willed processes? A critique of Libet-type experiments and their interpretation), WOLFGANG SPOHN (Germany – A priori principles of reason)

B2. Ethical Issues in the Philosophy of Science

Speakers: HEATHER E. DOUGLAS (USA – Scientific integrity in a politicized world), HUGH LACEY (USA – On the co-unfolding of scientific knowledge and viable values)

B3. Historical Aspects in the Philosophy of Science

Speakers: YEMIMA BEN-MENAHEM (Israel – Verificationism and scepticism), MARTIN CARRIER (Germany – On the question dynamics of research: Modes of finding and losing research topics in science)

C. Methodological and Philosophical Issues of Particular Sciences

C1. Logic, Mathematics and Computer Science

Speakers: JEAN-PIERRE MARQUIS (Canada – Mathematical abstraction, variation and identity), DALE MILLER (France – Synthetic connectives and their proof theory)

C2. Cognitive Science (including Linguistics and Psychology)

Speakers: CRISTIANO CASTELFRANCHI (Italy – Grounding social theory on action and cognition), ULRIKE HAHN (UK – Measuring argument strength: A Bayesian approach), FRITZ HAMM (Germany – On anaphora resolution: Some methodological remarks), PHILIPPE SCHLENKER (France – What is dynamic in meaning?)

C3. Biology

Speakers: TIM LEWENS (UK – Pheneticism reconsidered), MICHEL MORANGE (France – The rise of post-genomics and epigenetics: Continuities and discontinuities in the history of biological thought)

C4. Chemistry

Speakers: DAVIS BAIRD (USA – Many ways of knowing), ALFRED NORDMANN (Germany – Knowing and making in an impure science)

C5. Physics

Speakers: CRAIG CALLENDER (USA – The flow of space?), MICHAEL FRIEDMAN (USA – Einstein and the a priori), ROMAN FRIGG (UK – Explaining the approach to equilibrium in terms of epsilon-ergodicity), MIKLOS REDEI (UK – Einstein meets von Neumann: operational separability and operational independence in algebraic quantum field theory)

C6. Medicine

Speakers: ANNE CAMBON-THOMSEN (France – Data driven research and large scale studies in biomedical research: What consequences for data sharing and bioethics in human genetics?), MIRIAM SOLOMON (USA – Evidence-based medicine and mechanistic reasoning in the case of cystic fibrosis)

C7. Environmental Sciences

Speaker: KEVIN ELLIOTT (USA – The nature and significance of selective ignorance in environmental research)

C8. Economics and Social Sciences

Speakers: USKALI MÄKI (Finland – Scientific realism and disciplinary diversity: Revisionist remarks), DONALD MCKENZIE (UK – The credit crisis as a problem in the sociology of knowledge), DON ROSS (South Africa – The evolution and strategic dynamics of individualistic norms)

D. Methodological and Philosophical Issues in Technology

Speakers: ROGER COOKE (USA – Applications of philosophy: The bottom half), PETER KROES (The Netherlands – Knowledge and the design and making of technical artefacts), WENDY PARKER (USA – The target of testing: Models, adequacy and scientific knowledge)

Plenary Speakers

MARCO DE BAAR (The Netherlands – Engineering technical artefacts and scientific instruments), JEREMY GRAY (UK – Henri Poincaré lecture – "The soul of the fact": Poincaré and proof), WILFRID HODGES (UK – DLMPS Executive Committee president's address – DLMPS—Tarski's vision and ours), DAG PRAWITZ (Sweden – Is there a general concept of proof?), HUW PRICE (Australia – Retrocausality—what would it take?), PHILIPPE MONGIN (France – European Philosophy of Science Association affiliated plenary lecture – What the decision theorist could tell the Bayesian philosopher)

Special Symposia

Quantum Information – Conceptual Issues and New Technological Developments

    Chair: DENNIS DIEKS (The Netherlands)
    Speakers: HANS BRIEGEL (Austria – Simulation, computation, and physics What can we learn about the world?), JEFFREY BUB (USA – Einstein and Bohr meet Alice and Bob), ROBERT SPEKKENS (Canada – The invasion of physics by information theory), MAREK ZUKOWSKI (Poland – Bell's Theorem and EPR correlations: The issue, the triumph of the scientific method, misinterpretations, and practical applications)

What Is an Algorithm?

    Chair: HELMUT SCHWICHTENBERG (Germany)
    Speakers: NACHUM DERSHOWITZ (Israel – What is an effective algorithm?), YURI GUREVICH (USA – What's an algorithm?), YIANNIS MOSCHOVAKIS (USA – Panel discussion on "What is an algorithm?")

Mathematics and the New Technologies

    Chair: BENEDIKT LÖWE (The Netherlands)
    Speakers: JEAN PAUL VAN BENDEGEM (Belgium – Mathematics in the cloud: The web of proofs), PETER KOEPKE (Germany – Formal mathematics and mathematical practice), MARTINA MERZ (Switzerland – The Internet: New technology in old bottles?)

International Union of History and Philosophy of Science Joint Commission Symposium

Development of Cognition in Technology and Technosciences

    Chair: IMRE HRONSZKY (Hungary)
    Speakers: GERHARD BANSE (Germany – Design between science and art: Historical remarks), BERNADETTE BENSAUDE-VINCENT (France – Synthetic biology: The construction of a discipline with interdisciplinary contents), APPUKUTTAN NAIR DAMODARAN (India – Grand narratives, local minds and natural disasters: Community responses to tsunami), ARMIN GRUNWALD (Germany – Nano: The end of the selfunderstanding of the classical natural sciences), IMRE HRONSZKY (Hungary – The use of the technological paradigms conception in history of technology and theoretical economics), MARC PALLOT (France – Group cognition within living lab research and innovation: The cycle of experiential knowledge), JOACHIM SCHUMMER (Germany – From synthetic chemistry to synthetic biology: The revival of the verum factum principle), ASTRID SCHWARZ (Germany – Realworld simulation: A conceptual tool for technoscientific field sciences), ARIE RIP (The Netherlands – How to modulate coevolution of technology and society?)

Appendix B: Contributed Papers

A1. Mathematical Logic

MICHAEL ARNDT, LAURA TESCONI – Constructing a proof-tree: An investigation on composition of derivations

RICCARDO BRUNI, PETER SCHUSTER – Approximating Beppo Levi's "principio di approssimazione"

ANDREA CANTINI, LAURA CROSILLA – On constructive set theories with operations and related problems

VALERIA DE PAIVA – Lorenzen games for full intuitionistic linear logic

JAIME GASPAR – Copies of classical logic in intuitionistic logic

VALERY KHAKHANIAN – Properties of universes in realizability models for intuitionistic set theory and its corollaries

TAISHI KURAHASHI – On Kripke frames and arithmetical interpretations for QGL

ALEXANDER KUZICHEV, KAROLINA KUZICHEVA – Two-level version of sequential logic: completeness and consistency aspects

NEWTON MARQUES PERON – A generalization of Dugundji theorem

THOMAS STUDER – Dynamic justification logic

SHEILA R. M. VELOSO, PAULO A. S. VELOSO, PAULA M. VELOSO – On Piaget-like monoids: Monoids for logics

CHRISTIAN WALLMANN – Semantics for Tarskian consequence operations

A2. Philosophical Logic

ALBERT J. J. ANGLBERGER – An axiomatization of Paul Weingartner's 6-valued deontic logic and a result concerning its possible extensions

MICHAEL ARNDT – Localising logical rules in the sequent calculus

JULIO M. ARRIAGA ROMERO – Skeptical doubt, the common doubt and the contextualism of Keith De Rose

GIULIANO BACIGALUPO – Meinong and Husserl on existence

CAN BAŞKENT – Homotopies in classical and paraconsistent modal logics

DIDERIK BATENS – The consistency of Peano arithmetic: Why bother?

PATRICK BLACKBURN, MARIA MANZANO, CARLOS ARECES, ANTONIA HUERTAS – The bird of the hybrid type theory

PAOLA CANTÙ – Peano and Gödel

ROBERTO CIUNI, ANDREAS PIETZ – Which constructive negation for falsificationism?

NICOLAS CLERBOUT – Modal dialogical logic, validity and universal satisfiability

LORENZ DEMEY – Neighborhood semantics for dynamic epistemic logics

VIVIANE DURAND-GUERRIER – An elementary model theoretic perspective in mathematics education

VIRGINIE FIUTEK, SUJATA GHOSH, SONJA SMETS – Higher-order belief change in a branching-time setting

KARINE FRADET – Cooperation in the prisoner's dilemma

BORIS I. FYODOROV – Representation of Bolzano's content inferences with singular terms in the language of predicate logic

TJERK GAUDERIS – Three complications in modelling abduction in science

SVEN OVE HANSSON – Representing a finite mind

TETSUJI ISEDA – A statistical model of vagueness based on supervaluationism

JOHN T. KEARNS – Why blame Aristotle? Rational coherence and the principle of contradiction

NEIL KENNEDY – The ways of modality: On the notion of higher-order modality

BARTELD KOOI – Information change and first-order dynamic logic

ALLARD TAMMINGA, BARTELD KOOI – Lost in translation: The logic of paradox

JUI-LIN LEE – Classical model existence and pure implicational logic

HSIN-MEI LIN – Is the sensitive principle or the safety principle enough?

IANCU LUCICA – La logique des concepts paraconsistents

VIACHESLAV LYASHOV – Explication as specific method of philosophical research

JOSÉ MARTÍNEZ FERNÁNDEZ, RAFAEL BENEYTO TORRES – On some natural four-valued generalizations of weak Kleene logic

TOBY MEADOWS – Truth, dependence and supervaluation

JOKE MEHEUS – Adaptive deontic logics for various types of normative conflicts

JOSÉ M. MÉNDEZ, GEMMA ROBLES, FRANCISCO SALTO – On the variable-sharing property and the axiom mingle

CHIENKUO MI – Truth as a semantic switch

ALBERTO MARIÓ MURA – A partial modal semantics for the Adams logic of indicative conditionals

ALEXANDRU-V. MURESAN – The paradoxical context of logical information: The core of the paradoxical context of information and inference

IONEL NARITA – Modal analysis of strict implication

MICHEL PAQUETTE – Illocutionary logic and social interaction: Speech acts and the conversational record

FABRICE PATAUT – Strong antirealism, logical rules and structural rules

GILLMAN PAYETTE, MASASHI KASAKI – The many dimensions of contextualism in epistemology

CLAYTON PETERSON – On what grounds should we build deontic logic?

THOMAS PIECHA, WAGNER DE CAMPOS SANZ – Constructive semantics and classical logic

VLADIMIR POPOV, VASILYI SHANGIN – In the vicinity of Sette logic

MATEUSZ MAREK RADZKI – Non-Fregean logic and Ludwig Wittgenstein's early insight into application of logic

STEPHEN READ – Proof-theoretic validity

GEMMA ROBLES – The basic constructive logic for weak consistency in the ternary semantics with designated points

ALEKSANDRA SAMONEK – Translation invariance as a criterion of likeness. An analysis of hybrid versimilitude theories

WAGNER DE CAMPOS SANZ, THOMAS PIECHA – The BHK interpretation and extensions of NJ

FABIEN SCHANG – Logic as consequence in opposition

IOAN SCHEAU – The errors of Bertrand Russell

KONSTANTIN SKRIPNIK – Logic as an art and logic as a science: Is it only precedents or tradition?

HARTLEY SLATER – Quine's other way out

WERNER STELZNER – Semantic foundations for the logic of assent

CHRISTIAN STRASSER – An adaptive approach to detachment in conditional logics of normality

CORINA STRÖSSNER – A quantitative logic of normality

KORDULA SWIETORZECKA, JOHANNES CZERMAK – Some calculus of change with S4-necessity

HSING-CHIEN TSAI – Undecidability of some mereotopological structures

WEN-YU TSAI – The swamping problem

MATTHIAS UNTERHUBER – The Ramsey test and Chellas-Segerberg semantics

RAFAL URBANIAK, SEVERI K. HÄMÄRI – Busting a myth about Leśniewski and definitions

FREDERIK VAN DE PUTTE – Adaptive belief contraction

PETER VERDÉE – Towards non-monotonic mathematics: Adaptive logic theories as a pragmatic foundation for mathematics

MATHIEU VIDAL – For a Popperian theory of conditionals

JOSEPH VIDAL-ROSSET – Which core logic?

WEN-FANG WANG – Against classical dialetheism

PAUL WEINGARTNER – Decidable many-valued logic for the application in empirical sciences

BARTOSZ WIECKOWSKI – Lorenzen dialogues and sequent calculus: Equivalence, correspondence, and cut

XUNWEI ZHOU – Material implication v. mutually inverse implication

SYMPOSIUM: Hyperintension, intension, extension – Organizers: Marie Duži, Bjorn Jespersen, Pavel Materna; Speakers: Roussanka Loukanova, Marie Duži, Jens Christian Bjerring, Bjørn Jespersen, Sebastian Sequoiah-Grayson

SYMPOSIUM: The meaning of axioms: From mathematics to logic – Organizers: Alberto Naibo, Mattia Petrolo, Thomas Seiller; Speakers: Denis Bonnay, Gilles Dowek, Alexandre Miquel, Alberto Naibo, Paolo Maffezioli, Sara Negri, Mattia Petrolo, Thomas Seiller, Samuel Tronçon

SYMPOSIUM: New directions in dialogical logics – Organizers: Shahid Rahman, Pierre Cardascia; Speakers: Matthieu Fontaine, Sébastien Magnier, Tiago de Lima, Mathieu Marion, Shahid Rahman, John Woods

SYMPOSIUM: Proof theory, meaning and paradoxes – Organizer: Luca Tranchini; Speakers: Ole Hjortland, Julien Murzi, Luca Tranchini

A3. Logic and Computation

ROY DYCKHOFF – Cut-elimination, substitution and normalisation

MICHAŁ KRYNICKI, JERZY TOMASIK, KONRAD ZDANOWSKI – Logical properties of finite arithmetics

GIUSEPPE PRIMIERO – Modal types and their procedural semantics for contextual computing

SAEED SALEHI – Gödel's incompleteness phenomenon from computational viewpoint

SAM SANDERS – Computing the infinite

TOR SANDQVIST – Computability theory in relation-algebraic form

SYMPOSIUM: Proof systems at the test of computer science: Foundational and applicational encounters – Organizers: Francesca Poggiolesi, Giuseppe Primiero; Speakers: Agata Ciabattoni, Simon Kramer, Frank Pfenning, Lutz Straßburger, Heinrich Wansing

B1. Methodology and Scientific Reasoning

FRED ADAMS – Extended cognition meets epistemology

HOLGER ANDREAS – A structuralist theory of belief revision

SARAY AYALA – How can a purely cognitive philosophy of science deal with social biases? Embodied, situated and distributed cognition to the rescue!

ODED BALABAN – Cartography revisited: A key to understanding scientific knowledge

PHILIPP BALSIGER, MARIANNE RICHTER – What makes an object 'epistemic'? Criteria of relevance for scientific collections and exhibitions

THOMAS BENDA – What objective probability could be
DRAGOS BIGU – A similarity based model of scientific concept formation
MIEKE BOON – Why do we need phenomena? What we can learn from the Engineering Sciences
TOBIAS BREIDENMOSER – Theories of axonal transport in the cell: Empirical evidence against scientific realism
JOSEPH E. BRENNER – A new logic for new technology
MARIA CAAMAÑO – Theory success: Some evaluative clues
RAFFAELLA CAMPANER – Mechanistic and neo-mechanistic accounts of causation: How Salmon already got (much of) it right
EDUARDO CASTRO – Laws of nature and induction
ANGELO CEI – The epistemic structural realist program: Some interference
ANJAN CHAKRAVARTTY – Pluralistic ontologies for scientific realism
SIMONE CHELI – Neither between nor within: Selfhood and otherness in epistemology
RUEY-LIN CHEN – A theoretical analogy: How is Darwin's theory of natural selection analogous to Malthus' theory of population
KAI-YUAN CHENG – Dispositions, conditionals, and ordinary conditions
ANNA CIAUNICA – Higher- and lower-level phenomena: A nonhierarchical approach to fundamental properties
ALBERTO CORDERO – Accumulation of theory parts and meaning variance
SANDRINE DARSEL – What do we learn from case studies?
MAJID DAVOODY BENI – On the ontology of linguistic frameworks: Toward a comprehensive version of empiricism
RICHARD DAWID – A Bayesian model of no alternative arguments
XAVIER DE DONATO RODRÍGUEZ – Idealization, scientific modeling and simulations: A new analysis of idealization as a common framework for the study of models and simulations
FREDERICK EBERHARDT – Scientist vs. nature — priors, strategies and discovery
MATTHIAS EGG – Expanding our grasp: Can causal knowledge save realism from Stanford's new induction?
ANNA ESTANY – The stabilizing role of material structure in scientific practice
LUDWIG FAHRBACH – How to defend scientific realism against the PMI
JOSÉ L. FALGUERA – Identity of scientific concepts and theoretical dependence
CHRISTIAN J. FELDBACHER – A problem for semantic definitions of analyticity
EREZ FIRT – On emergence and causation
GARY FULLER – Narrative explanations
IVAN GAZEAU – Reasoning without language or logic
ALEXANDER GEBHARTER – Determining causal relevancies at event-level
ANNE-SOPHIE GODFROY – How international comparisons transform social reality

SAMIR GORSKY – The logic of surprise: Puzzle, quantum games and information

JOSH HADDOCK – The principal principle, and theories of chance: An account of primitive conditional chance

HELMUT HEIT – Reasons for relativism: Feyerabend on early Greek thought

ROSA MARÍA HERRERA – Metaphors, the solar system and scientific research

PAUL HOYNINGEN-HUENE – The ultimate argument against convergent realism and structural realism: The impasse objection

ANNA IJJAS – Theory vs. interpretation: From a methodological point of view

CYRILLE IMBERT – Collective science: The loss of scientific understanding?

MILENA IVANOVA – Is the relativized a priori incompatible with scientific realism?

SREEKUMAR JAYADEVAN – Theory-talk, meta-theory-talk and metaphysical-talk: Intricacies and pertinence of three levels of discourse in the scientific realism-debate

SAANA JUKOLA – The commercialization of research—A threat to the objectivity of science?

MOLLY KAO – From foundation to function: Rethinking the role of data in science

ASHLEY GRAHAM KENNEDY – Idealization and inference: How false models explain

BERNA KILINC – A generalization of the Condorcet jury theorem

JEFF KOCHAN – Reason, emotion, and the context distinction

MEINARD KUHLMANN – One law, 23 derivations: On the plurality of explanations of Planck's law

THEO A.F. KUIPERS – Refined truth approximation by refined belief base revision

VLADIMIR KUZNETSOV, WOLFGANG BALZER – From philosophy of science to theories of knowledge systems

ERWAN LAMY – How to talk with a skeptic?

JOHANNES LENHARD – A turn in computational modeling: The case of quantum chemistry

FRANÇOIS LEPAGE, CHARLES MORGAN – Two impossibility results about revision of conditional probability

ANNA LEUSCHNER – Pluralism and objectivity: On Longino's and Kitcher's approaches

AIDAN LYON – Why normal distributions are normal

MYHAILO MARCHUK – Externalism, internalism and the conception of the socio-cognitive potentialism

CONOR MAYO-WILSON – Efficient experimentation

DAVID MCELHOES – Difference-making and ontological explanation

ALESSANDRA MELAS – The contemporary notion of chance and Salmon's interactive fork model. An attempt to describe chance by means of some causal criterion

MATTEO MORGANTI – Science-based metaphysics: On some recent anti-metaphysical claims
KUNIHISA MORITA – Pseudo-scientific explanation and scientific explanation
VERUSCA MOSS SIMÕES DOS REIS – The role of philosophy of science in the understanding of "post-academic" science
F. A. MULLER – On an inconsistency in Constructive Empiricism
PEETER MÜÜRSEPP – The aim of science—knowledge or wisdom
ARTO MUTANEN – Interrogative model of inquiry as a logic of experiment
YASUO NAKAYAMA – Scientific progress as increase of expressibility, accuracy and coherence
ILKKA NIINILUOTO – Rethinking belief revision by truthlikeness
ERIK P. NYBERG, KEVIN B. KORB – Conditioning and unfaithfulness
YUKINORI ONISHI – The scientific realism debate from the epistemological viewpoint – Why not consult the theories of knowledge?
INMACULADA PERDOMO REYES – Scientific representation: Uses and interpretation of models
RICHARD PETTIGREW – Accuracy, chance, and the principal principle
WOLFGANG PIETSCH – The limits of probabilism
DEMETRIS PORTIDES – Scientific representation, denotation, and explanatory power
MARIANNE RICHTER, PHILIPP BALSIGER – Visual representation in the light of methodological demands—A critical review of symbol theoretic attempts to operationalize scientific visualization
LÁSZLÓ ROPOLYI – Seven fundamental versions of philosophy of science
ROMAN ROSHKULETS – Metaphysical aspects of postpositivism
EMMA RUTTKAMP – Re-positioning realism
JUHA SAATSI – Scientific realism and inferentially veridical representations
UWE SCHEFFLER, MAX URCHS – Both billiard ball and butterfly?
RAPHAEL SCHOLL – Causal inference, mechanisms and the Semmelweis case
GERHARD SCHURZ – Bayesian confirmation of creationism? On the problem of genuine confirmation
BERTOLD SCHWEITZER – From malfunction to mechanism
DUNJA ŠEŠELJA – Disambiguating the notion of pursuit worthiness
ELLIOTT SOBER, MIKE STEEL – Screening-off (aka the Markov property) and causal incompleteness—a no-go theorem
PETROS STEFANEAS, IOANNIS M. VANDOULAKIS – Proofs as spatio-temporal processes
FABIO STERPETTI – Towards a non-adaptationist approach to mathematics
CASSIANO TERRA RODRIGUES – Deduction, induction and abduction according to Charles S. Peirce: Necessity, probability, discovery

Rafal Urbaniak – Prioritized adaptive logics and the epistemology of thought experiments in physics

Ioannis Votsis – Simplicity as a guide to falsity?

Zenaida Yanes Abreu – Veritistic social epistemology. A reliable proposal?

Jeu-Jenq Yuann – The futility of prescribing what scientists should do: Supplementing van Fraassen's empirical stance with scientific practices

Jesús Zamora-Bonilla, Ana M. Rodríguez – Confirmation, verisimilitude, and acceptance

Kevin J.S. Zollman, Erich Kummerfeld – Conservatism in scientific research: A new problem

Symposium: The interpretation and scope of models of complex systems – Organizer: Christopher Pincock; Speakers: André Ariew, Steven O. Kimbrough, Christopher Pincock, Randall Westgren;

Symposium: Integrity and diversity of traditions and trends in today's philosophy of science – Organizer: Andrei Rodin; Speakers: Hourya Bénis-Sinaceur, Elena Mamchur, Jonathan Regier, Andrei Rodin, Jean-Jacques Szczeciniarz

Symposium: Calibration in scientific practice – Organizer: Léna Soler; Speakers: Léna Soler, Catherine Allamel-Raffin, Catherine Dufour, Jean-Luc Gangloff, Emiliano Trizio, Frdric Wieber, Eran Tal, Jonathan Livengood

B2. Ethical Issues in the Philosophy of Science

Katarzyna Gan-Krzywoszynska, Piotr Lesniewski – On rationality, irrationality and counterrationality in dynamics of knowledge

Kelly Ichitani Koide – Scientific methods and strategies of research: A plurality of paths to the objectives of science

Nicolas Lechopier – Epistemethics: Lessons from an ethnographic study of global health research ethics concerning the articulation between research and practice

Anna Leuschner – Scientific credibility in the public exemplifying climatology: Why it is important, how it is challenged

Vladimir Lobovikov – Science, episteme and mathematical ethics (A law of contraposition of episteme in algebra of formal ethics)

Masahiro Matsuo – Where the opposition to value-free science should be revised

Witold Strawinski – Philosophy of science and ethical issues — from a Warsaw perspective

B3. Historical Aspects in the Philosophy of Science

Oscar Joao Abdounur – Compounding ratios, theories of ratio and geometry in theoretical music in the 16th century

Basak Aray – Neurath on pictures, language and international communication

FRANCESCA BIAGIOLI – Between Kantianism and empiricism: Otto Hölder's philosophy of geometry

MATTEO COLLODEL – Carnap's vision or: How we can learn from the past and enlighten the future of the philosophy of science

DAVIDE CRIPPA – Mathematics and the purity of methods: Some historical considerations

SILVIA DE BIANCHI – Symmetry and the enigma of space and time. Reflections on the origin of gauge theory

JEAN-PIERRE FERRIER – Quality and practice in mathematics from Hilbert to Grothendieck

ALEXANDER FURSOV – Theory underdetermination: The history of science perspective

ADAM GROBLER – Two traditions of conventionalism

MASAKI HARADA – Kant, Fichte and algebraic operations: Philosophy of algebra according to Jules Vuillemin

SHINJI IKEDA – Gödel and Leibniz on concepts and relations

SIMCHA KOJMAN-ROZEN – Changes in the perception of time in Victorian scientific theories: Lyell, Darwin and Maxwell

ARTUR KOTERSKI – The unimportance of Quine's Two Dogmas of Empiricism

DANIEL B. KUBY – A source of Feyerabend's decision-based epistemology: Hugo Dingler's voluntarism

DANA JALOBEANU – Constructing natural historical facts: Baconian methodology in Newton's first paper on light and colors

HENNIE LOTTER – Evolution as metaphor for scientific progress

IVICA MARTINOVIC – With Bokovic against Kant: Ivan Krstitelj Horvath on space and time

ZURAYA MONROY-NASR – Cartesian forces in a soulless physics

RAFFAELE PISANO – Historical epistemology Notes on Archimedes, Torricelli and Sadi Carnot

JOAO PRINCIPE – Kantian aspects of Poincaré's epistemological thoughts on XIXth century physics

DAGMAR PROVIJN – A study of analogical reasoning based on William Harvey's problems and analogies

PASCALE ROURE – History of science and language criticism: A cross-referenced reading of Ernst Mach and Fritz Mauthner

GEORG SCHIEMER – Semantics in type theory

DAVID J. STUMP – Poincaré's two types of conventionalism

HASSAN TAHIRI – Ibn al-Haytham's 'al-Shukuk' or the art of controversy: How the eleventh century Arabic scientist's arguments changed astronomy forever

JANOS TANACS – Some semantic considerations for the conceptual transition from Euclidean to non-Euclidean geometry

IOANNIS M. VANDOULAKIS – On A. A. Markov's attitude towards Brouwer's intuitionism

SCOTT WALTER – Beyond Poincaré and Einstein: A. A. Robb's theory of space and time

GABOR ZEMPLEN – Methodological remarks on knowledge-production and text-production: Newton's optical controversy and methodological shifts

RENATA ZIEMINSKA – Inconsistency of ancient skepticism

SYMPOSIUM: A plurality of currents in today's historical epistemologies – Organizers: Karine Chemla, Koen Vermeir; Speakers: Koen Vermeir, Nadine de Courtenay, Martin Kusch, Emily Grosholz, Thomas Sturm

SYMPOSIUM: Poincaré, Philosopher of science: A historical and philosophical approach – Organizer: Augusto J. Franco de Oliveira; Speakers: Rosário Laureano, María de Paz, Isabel Serra, Olga Pombo, Nuno Jerónimo, Augusto J. Franco de Oliveira

SYMPOSIUM: Confronting French roots and current historical epistemologies – Organizer: David Rabouin; Speakers: Cristina Chimisso, Maarten Van Dyck, Katharina Kinzel, David Rabouin, Paolo Savoia

SYMPOSIUM: Thomas Kuhn's "The Structure of Scientific Revolutions": Interpretations and developments – Organizer: Friedrich Stadler; Speakers: Hans-Joachim Dahms, Michael Schorner, Christian Damböck, Christoph Limbeck-Lilienau

SYMPOSIUM: Carnap's linguistic pluralism and scientific methodology – Organizer: Richard Zach; Speakers: A. W. Carus, Richard Creath, Pierre Wagner, Richard Zach

C1. Methodological and Philosophical Issues of Logic, Mathematics and Computer Science

NICOLA ANGIUS – Corroborations of hypotheses and experimental computer science in software testing

JACOBO ASSE – Pluralism and mathematical objects

HERVÉ BARREAU – Aristotle's assertoric syllogistic

LIBOR BEHOUNEK – Fuzzy logics as the logics of linearly decomposable resources

FRANCESCA BOCCUNI – Plural logicism

JEAN-MARIE CHEVALIER – Are mathematics and logic sciences of observation? A semiotic approach to visual thinking

SORIN COSTREIE – Frege on "contentful mathematics"

LIESBETH DE MOL – The computer (as a medium) in mathematics. Mathematician-computer interactions, internalization, time and space squeezing

LUIS ESTRADA-GONZALEZ – On the meaning of connectives (A propos of a non-necessitarianist challenge)

SAMUEL C. FLETCHER, JASON HOELSCHER-OBERMAIER – Physical computability, efficiency, and the Church-Turing thesis

SIMON FRIEDERICH – Motivating Wittgenstein's perspective on mathematical sentences as norms

MICHELE FRIEND – Presenting pluralism in mathematics

PETER GABROVSKY – EXLOG: A non-standard logic programming language for experiment-based research

JOACHIM HERTEL – Frege on the iPad

JAAKKO HINTIKKA – Axiomatizing set theory

RALF KRÖMER – The duality of space and function, and category-theoretic dualities

ELAINE LANDRY – The genetic versus the axiomatic method: Resolving Feferman '77

AMIROUCHE MOKTEFI – Representing the "universe of discourse": Historical origin and philosophical relevance of a graphical convention in mathematics and logic

JARI PALOMAKI – Kant, Cantor, and the Burali-Forti's paradox

MARCO PANZA, ANDREA SERENI – On the indispensable premises of indispensability arguments

AHTI-VEIKKO PIETARINEN – A realist modal-structuralism

PAULA QUINON – Computational structuralism and Frege's constraint

PHILIPPE DE ROUILHAN – Remarks on recursive definitions of truth

ARNE SEEHAUS, MARTIN ZIEGLER – Raise and fall of scientific branches: On progress in mathematics

MARCOS SILVA – "Muss Logik für sich selbst sorgen?" On contrary propositions and material logical truth as problems to the neutrality of logic

MIN TANG – Ontology without abstract objects: A naturalistic defense of revolutionary fictionalism

GABRIEL TARZIU – What can science tell us about mathematical objects?

MARK VAN ATTEN – Kant and real numbers

SUSAN VINEBERG – Explanation and two kinds of investigation in the foundations of mathematics

PIOTR WILKIN – How formalized are informal proofs?

JAN WOLEŃSKI – Constructivism and metamathematics

SYMPOSIUM: Theories of continua: Logical and philosophical reflections – Organizer: Philip Ehrlich; Speakers: José Ferreirós, Maximo Dickmann, Philip Ehrlich, Paolo Giordano, Erik Palmgren, Geoffrey Hellman

SYMPOSIUM: Philosophy of mathematical practice – Organizers: Jose Ferreiros, Paolo Mancosu; Speakers: Tom Archibald, Kenneth Manders, Brendan Larvor, Valeria Giardino, Dirk Schlimm

SYMPOSIUM: Are aesthetic approaches in philosophy of mathematics topical? – Organizer: Caroline Jullien; Speakers: Roger Pouivet, Maria Giulia Dondero, Caroline Jullien

C2. Methodological and Philosophical Issues of Cognitive Science (including Linguistics and Psychology)

MICHELE ABRUSCI, CHRISTIAN RETORÉ – Quantification in ordinary language: From a critique of set-theoretic approaches to a proof-theoretic proposal

KENNETH AIZAWA – The autonomy of psychology in the age of neuroscience

NIMROD BAR-AM – Towards a rational theory of communication

KRYSTYNA BIELECKA – Are radical externalism and radical internalism the same?

DELPHINE BLITMAN – What linguistic nativism tells us about innateness

SHUSHAN CAI, HONGGUANG ZHANG – The role of abduction in learning and cognition

YANNICK CHIN-DRIAN – A naïve realist view of colour

IN-RAE CHO – Reassessing the rationality war

LIEVEN DECOCK – Reflectance physicalism and contrast colours

SIMONE DUCA – The suppositional Ramsey test in decision making

ERWIN ENGELER – Algebras of the mind and algebras of the brain

DINGZHOU FEI – Blackboard system as model of problem solving in Sudoku puzzles

YANXIA FENG – Emergent, mental causation and downward causation

SASCHA BENJAMIN FINK – What falsifies an NCC of specific content?

YAN GONG – Is rational-emotive behavior theory based on the methodology of critical rationalism?

LILIA GUROVA – The principle based explanations are not extinct in cognitive science: The case of the basic level effects

JOANA HOIS, OLIVER KUTZ – Towards linguistically-grounded spatial logics

OLIVER KUTZ, JOANA HOIS – Steering ontological blending

XIAOLI LIU – Representation and action: A theory of representation in the evolution-embodied cognition context

MANUEL LIZ, MARGARITA VAZQUEZ – Two approaches to the notion of Point of View

PAWEL LUPKOWSKI – Cooperative answering and inferential erotetic logic

HERNAN MIGUEL – The causal closure of the physical and the variable realization

MARCIN MIKOWSKI – Computational mechanisms and models of computation

GONZALO MUNEVAR – Damasio, self and consciousness

IGOR NEVVAZHAY – Dual nature of consciousness

Samuli Poyhonen – Should I split or should I lump? The epistemic-tool approach to scientific concept formation

Manuel Rebuschi – 'De dicto' versus 'de facto' attitudes

Valentine Reynaud – Can innateness assumptions avoid the tautology problem?

Dairon Rodriguez, Jorge Hermosillo, Bruno Lara – The Chinese room argument and the symbol grounding problem: A new perspective

Bazej Skrzypulec – The concept of "object" in the visual binding theories

Patrice Soom – On levels of mechanisms

Mariusz Urbanski, Joanna Urbanska – Abduction and rumormongering to the most coherent interpretation

Tomoyuki Yamada – Dynamic logics of speech acts as formal simulations of social interaction

Jerry Yang – In defense of a multiple content structure of self-representationalism

C3. Methodological and Philosophical Issues of Biology

Donato Bergandi – Ecology, evolution, ethics: In search of a meta-paradigm

Jonathan Birch – Is the concept of life response-dependent?

Jean-Sébastien Bolduc – Adaptationism: Behind criticisms and typologies, the tool

Hsien-I Chiu, Bo-Chi G. Lai – A moderate solution to the debate over the species concept

Ellen Clarke – Biological individuality in plants and beyond: A reconciliation for the genet-ramet dispute

Emmanuel D'Hombres – The Darwinian muddle on the 'Division of physiological labor': An attempt of clarification

Andreea Esanu – An argument against the evolutionary contingency thesis

Jean Gayon – Economic natural selection: What concept of selection?

Tarja Knuuttila, Andrea Loettgers – Modeling/experimenting? The combinatorial strategy in synthetic biology

Lukasz Lamza – Computational biophysics as a case against intertheoretic reduction

Paolo Lattanzio, Raffaele Mascella – On informational schemes in biology

Maël Lemoine – Function as a causal role in a biological model

Pablo Lorenzano – The status of the Hardy-Weinberg law

Jane Maienschein – Are embryos what we thought they were, and how do we know?

Hisashi Nakao, Edouard Machery – The evolution of punishment

Laura Nuño de la Rosa – Modelling development and evolution in three dimensions

Íñigo Ongay de Felipe – A materialist account of scientific reasoning in ethology

STEFAN PETKOV – The fitness landscape metaphor: Dead but not gone

MANUEL DE PINEDO GARCÍA – Individuation for holists: (physical) dispositions and (biological) affordances

THOMAS A. C. REYDON – Addressing a theory-practice gap: What can kind essentialism contribute to understanding classificatory practices in biology?

WALTER RIOFRIO – Cellular dynamics at the beginning of prebiotic world

CHRISTIAN SACHSE – Conservative reduction of biology

EDIT TALPSEPP – Essentialism, Darwinism and "theory theory"

JON UMEREZ – Epistemological reconstruction of the concept of level. Some preliminaries and a proposal

RONG-LIN WANG – On Rosenberg's Darwinian reductionism

CHARLES T. WOLFE –From substantival to functional vitalism and beyond in biomedical thought: Animas, organisms and attitudes

HSIAO-FAN YEH – Why the classical mendelian genetics are necessary? – A comparison of Lindley Darden's mechanism approach with C. Kenneth Waters' genetic approach

SYMPOSIUM: Evolution of biological complexity – Organizers: Matteo Mossio, Francesca Merlin; Speakers: Matteo Mossio, Francesca Merlin, Paul-Antoine Miquel, Pierre-Alain Braillard, Werner Callebaut, Arantza Etxeberria

C4. Methodological and Philosophical Issues of Chemistry

MARTÍN LABARCA, OLIMPIA LOMBARDI – The ontological autonomy of the chemical world: further arguments

JEAN-PIERRE LLORED Relational philosophy as a root for an epistemology of chemistry

ALEXANDER A. PECHENKIN – The paradigm changes in the study in the Belousov-Zhabotinsky reaction

REIN VIHALEMM – Philosophy of chemistry against standard scientific realism and anti-realism

C5. Methodological and Philosophical Issues of Physics

THOMAS D. ANGELIDIS – Special relativity prohibits spacelike causation and some implications

ARISTIDIS ARAGEORGIS – Spacetime as a causal set: Universe as a growing block?

JUAN SEBASTIÁN ARDENGHI, SEBASTIAN FORTIN, OLIMPIA LOMBARDI – The conceptual meaning of reduced states: Decoherence and interpretation

JULIEN BERNARD – From the hole argument (A. Einstein) to the ball of clay argument (H. Weyl)

TOMASZ BIGAJ – How to exchange quantum particles of the same type

ELENA CASTELLANI – Fundamentality, elementariness and scales

GRAZIANA CONTE – Information measures induced by partial Boolean algebras

MICHAEL CUFFARO – Many worlds, the cluster-state quantum computer, and the problem of the preferred basis

NAOUM DAHER – From independent models to a unified theory of dynamics

ROBERT DISALLE – Explanation, explication, and interpretation of space-time theories

JULIUSZ DOBOSZEWSKI – Specious present in branching space-times

STEFFEN DUCHEYNE – Testing universal gravitation in the laboratory, or the significance of research on the mean

MATT FARR – On the status of temporal unidirectionality in physics

MATHIAS FRISCH – Incantations of 'causation' and other philosophical sins, or: Rehabilitating Ritz

ALEXANDRE GUAY – Objectivity and physical symmetries

PANDORA HADZIDAKI – Bohr's model of the atom: Methodology, consistency and fruitfulness

CARL HOEFER – Mach's principle and the philosophy of space/time: What nature is trying to tell us

VASSILIOS KARAKOSTAS – Correspondence truth and quantum mechanics

MONTGOMERY LINK – Rotating universe

ARNAUD MAYRARGUE – An illustration of the importance of the epistemological point of view and of the context in Sciences: the astronomical refraction case during the 18th century

SANDRA MOLS – Going round the lack of time: Enforced entrusting and silent inter-expertise trading in time-short nanomagnetism knowledge making

F. A. MULLER – Circumveiloped by obscuritads

WAYNE C. MYRVOLD – The prospects for quantum state monism

PAUL NÄGER – Why quantum non-locality implies parameter dependence

GRAHAM NERLICH – Bell's Lorentzian pedagogy: A bad education

ARGYRIS NICOLAIDIS – Relational logic and modern science

EMBOUSSI NYANO – Einstein's philosophy and the origins of post-critical philosophy of sciences

KENT A. PEACOCK – Would superluminal influences violate the principle of relativity?

JOHANNES ROEHL – Forces—relations or dispositions?

IÑAKI SAN PEDRO – Causal relevance of measurement operations in the EPR paradox

RAQUEL ANNA SAPUNARU – Leibniz: Symmetry and harmony

ARIANNE SHAHVISI – The gravity of the past hypothesis: Lessons learnt from Earman and Wallace

LEBA SLEIMAN – Problems and promises of scientific method
ALBERT SOLÉ – Bohmian mechanics without wave function ontology
ADÁN SUS – Is inertia explained in general relativity?
MORGAN TAIT – The case for quantum state realism
MIKE TAMIR – Proving the principle: General relativity and geodesic universality
HAJIME TANAKA, KOJI NAKATOGAWA, HIROYASU NAGATA – A proposition called T0906 and Gödel's incompleteness theorems
MARKO URŠIC – Paradoxes of transfinite cosmology
PIERRE UZAN – Quantum theory beyond physics
LEV VAIDMAN – The past of a quantum particle
LOUIS VERVOORT – Probability is composed. The frequency interpretation of probability revisited
CHRISTIAN WÜTHRICH – How large is a structuralist universe?
SYMPOSIUM: Epistemological perspectives on the Large Hadron Collider – Organizer: Michael Stöltzner; Speakers: Michael Stöltzner, Arianna Borrelli, Koray Karaca

C6. Methodological and Philosophical Issues of Medicine

CHHANDA CHAKRABORTI – Vulnerability from infectious diseases and social determinants of health: In search for an ontology to guide health policy development
JAN DE WINTER – How to make the research agenda in the health sciences less distorted
SÉBASTIEN JANICKI – Between variability of the body and determinism of the care: a "mediated" relation
MARC KIRSCH – When society speaks to science: Politics, social representations and industrial interests in the medical definition of the concept of addiction, in the case of tobacco and nicotine
RENZONG QIU – Ontological and moral status of human-nonhuman animal mix organisms
NOEMÍ SANZ MERINO – How and why to epistemologically study applied bioethics to nanomedicine
STÉPHANIE VAN DROOGENBROECK – Experimental philosophy and evidence based medicine: Two criticized ways of doing science
XIAOMEI ZHAI – Philosophical and ethical issues in use or abuse of human body and its parts in biomedical technologies

C7. Methodological and Philosophical Issues of Environmental Sciences

JOEL KATZAV – Hybrid models, climate models and inference to the best explanation

RODOLFO HERNANDEZ PEREZ – Let the water flow to the city: A recent history of the water saving technologies for agriculture in China

MICHAEL POZNIC, RAFAELA HILLERBRAND – Climate science or climate fiction? The role of fictional elements in physics and in Earth sciences

CONSTANTIN STOENESCU – A new way of thinking in environmental sciences

C8. Methodological and Philosophical Issues of Economics and Social Sciences

MICHEL BOURDEAU – Two conflicting ideas upon the nature and the goals of man's action upon social phenomena

ROMULUS BRÂNCOVEANU – Weber's and Pareto's theories as methodological programs

ESA DIAZ-LEON – What is social construction?

DANIEL ECKERT – Guilbaud's reading of Arrow's theorem

SILVIA HARING, PAUL WEINGARTNER – On the conceptual clarification of "human environment", "action space" and "quality of life"

CHRYSOSTOMOS MANTZAVINOS – Which theory of explanation for the social sciences: Unificationist, mechanistic or manipulationist?

CARLO MARTINI – Modeling expertise in economics

ADRIAN MIROIU – Two approaches to representative voting

ANA MARÍA TALAK – Reconsidering values in assessing the progress of historiography of psychology

WEI WANG – Are there laws in the social sciences?

PAUL WEIRICH – Decisions without sharp probabilities

SYMPOSIUM: Duality within human sciences – Organizers & speakers: Antonella Corradini, Nicolò Gaj, Giuseppe Lo Dico

SYMPOSIUM: Business ethics and analytic philosophy – Organizer: Christoph Lütge; Speakers: Christian List, Lisa Herzog, Michael von Grundherr, César Canton, Christoph Lütge

SYMPOSIUM: Decision theory in economics: Between logic and psychology – Organizer: Samuel Ferey; Speakers: Jean-Sébastien Lenfant, Brian Hill, Michaël Cozic, Samuel Ferey, Nicola Gioccoli

D. Methodological and Philosophical Issues in Technology

Yamina Bettahar, Benoît Roussel – Production of intermediary objects in a collaborative network: Examples of impact on tools and methods in engineering innovation

Manjari Chakrabarty – An inquiry into the character of material artifacts

Luca Del Frate – Technical malfunction in terms of states and events

Nicolas Delhopital – Science, technology and society. An attempt to think their link through Maurice Blondel's (1861-1949) philosophy of action

Christopher Evans – Technoscience: Illuminating new blue skies

Marcello Frixione, Antonio Lieto – Formal ontologies and semantic technologies: A "dual process" proposal or concept representation

Pawel Garbacz – Artefacts and family resemblance

Sven Ove Hansson – What is so special with technological science?

Insok Ko – How to recycle Asimov's laws in roboethics: An intermediate suggestion

Baptiste Mélès – Nominalism of things and nominalism of events, from Turing Machines to functional programming

Susan G. Sterrett, Adrian Bejan – Geometric configuration in nature and in design: Is there a connection?

Margarita Vázquez, Manuel Liz – Models, commentaries, and theories

Martin A. Vezér – Comparing methodologies of classical, natural, field and computer experiments deployed in climate change studies

Symposium: Design as a challenge for the philosophy of science – Organizers: Maarten Franssen, Sjoerd D. Zwart; Speakers: Richard Buchanan, Maarten Franssen, Sjoerd D. Zwart, Riichiro Mizoguchi, Ibo van de Poel

Symposium: Artefact Functions – Organizer: Wybo Houkes; Speakers: Wybo Houkes, Françoise Longy, Pieter Vermaas

Appendix C: Affiliated Symposia

The Philosophy of Artificial Intelligence

    Conférences Pierre Duhem, Chairs: MAX KISTLER, CYRILLE IMBERT

Logic, Knowledge and Agency

    Beth Foundation, Chair: SONJA SMETS

The Philosophy of Mathematical Practice

    Association for the Philosophy of Mathematical Practice, Chairs: JESSICA CARTER, MARCO PANZA

Rebuilding Logic and Rethinking Language in Interactional Terms

    LOCI Project, Chair: ALAIN LECOMTE

Logical Modelling: The Interface between the Formal and the Empirical

    EuroCoRes programme LogICCC, Chairs: EVA HOOGLAND, BENEDIKT LÖWE

Climate Science and Climate Change: Epistemological and Methodological Issues

    Société de Philosophie des Sciences, Chairs: ANOUK BARBEROUSSE, CYRILLE IMBERT, STÉPHANIE RUPHY

Analysing Programs: Logic at Rescue

    LORIA, Chair: VÉRONIQUE CORTIER

Logical and Philosophical Foundations of Science and Technology – Historical Development, Contemporary Investigations, and Perspectives

    International Institute of Foundational, Interdisciplinary and Historical Problems of Science, Chair: BORIS CHENDOV

Methodological Problems of Technoscience

    Chairs: VITALY GOROKHOV, ARMIN GRUNWALD

The Logic of Opposition
  N.O.T., Chair: Fabien Schang

Science and Rationality
  International Academy for Philosophy of Science, Chair: Evandro Agazzi

Ontology between Philosophy and Computer Science
  Onto-Med, Chairs: Heinrich Herre, Roberto Poli

Intuitionistic Modal Logics and Application
  Chairs: Natasha Aleshina, Valeria de Paiva

Appendix D: Committees, Patronages, Supports and Partners

Executive Committee of the Division of Logic, Methodology and Philosophy of Science, 2007-2011

WILFRID HODGES (UK, President), SOSHICHI UCHII (Japan, 1st Vice-President), ANNE FAGOT-LARGEAULT (France, 2nd Vice-President), PETER CLARK (UK, Secretary-General), DAG WESTERSTÅHL (Sweden, Acting Secretary-General), RALF SCHINDLER (Germany, Treasurer), ADOLF GRÜNBAUM (USA, Past President)

General Program Committee

PETER SCHROEDER-HEISTER (Germany, Chair), PETER CLARK (UK, Representing Executive Committee), GERHARD HEINZMANN (France, Representing Local Organizing Committee), BERNADETTE BENSAUDE-VINCENT (France), ANDREAS BLASS (USA), THIERRY COQUAND (Sweden), DENNIS DIEKS (The Netherlands), JOHN DUPRÉ (UK), MARIA CARLA GALAVOTTI (Italy), ANTHONIE MEIJERS (The Netherlands)

Senior Advisors

MIC DETLEFSEN (USA/France), KOSTA DOŠEN (Serbia), ROBERT FRODEMAN (USA), FRANCESCO GUALA (Italy), MARCUS KRACHT (Germany), HANS RADDER (The Netherlands), FRIEDRICH STADLER (Austria), ALFRED TAUBER (USA)

Advisory Board

ALDO ANTONELLI (USA), SERGEI ARTEMOV (USA, Russia), JEREMY AVIGAD (USA), CRISTINA BICCHIERI (USA), HARVEY BROWN (UK), CHHANDA CHAKRABORTI (India), NICK CHATER (UK), CRISTINA CHIMISSO (UK), BRUNO COURCELLE (France), ANUJ DAWAR (UK), WILLIAM DEMOPOULOS (Canada), GILLES DOWEK (France), ROD DOWNEY (New Zealand), DONALD GILLIES (UK), FABRICE GZIL (France), WADE HANDS (USA), SVEN OVE HANSSON (Sweden), WYBO HOUKES (The Netherlands), COLIN HOWSON (Canada),

GÜROL IRZIK (Turkey), MATTHIAS KAISER (Norway), TARJA KNUUTTILA (Finland), PIETER KOK (UK), JIM LAMBEK (Canada), HANNES LEITGEB (UK), DANIEL LEIVANT (USA), OLIMPIA LOMBARDI (Argentina), HELEN LONGINO (USA), CHRYSOSTOMOS MANTZAVINOS (Germany), DAVID MARKER (USA), LUIS MARONE (Argentina), ITAY NEEMAN (USA), DAVID PEARCE (Spain), THOMAS PRADEU (France), DAG PRAWITZ (Sweden), JOACHIM SCHUMMER (Germany), JONATHAN SIMON (France), SUSAN STERRETT (USA), HELMUT SCHWICHTENBERG (Germany), PAUL B. THOMPSON (USA), MARCEL WEBER (Germany)

Local Organizing Committee

GERHARD HEINZMANN (Chair), CLAUDE DEBRU (Vice-Chair), PIERRE EDOUARD BOUR (Local Staff Chair), BERNADETTE CLASQUIN, LUCIE FLORENTIN, KATARZYNA GAN-KRZYWOSZYŃSKA, MARC HENRY, MARIE L'ÉTANG, CLÉMENTINE LE MONNIER, CINDY NEVÈS, MILICA PEJANOVIĆ

Local Scientific Committee

BERNARD ANCORI (France), DANIEL ANDLER (France), HOURYA BENIS-SINACEUR (France), PATRICK BLACKBURN (France), DENIS BONNAY (France), JACQUES BOUVERESSE (France), MICHEL BLAY (France), KARINE CHEMLA (France), GABRIELLA CROCCO (France), JACQUES DUBUCS (France), PASCAL ENGEL (Switzerland), ANNE FAGOT-LARGEAULT (France), JEAN GAYON (France), MARCEL GUILLAUME (France), MICHAEL HEIDELBERGER (Germany), CYRILLE IMBERT (France), PIERRE JACOB (France), JEAN-PIERRE KAHANE (France), PIERRE LIVET (France), THIERRY MARTIN (France), PHILIPPE MONGIN (France), PHILIPPE NABONNAND (France), MARCO PANZA (France), MICHEL PATY (France), ROGER POUIVET (France), JOËLLE PROUST (France), SHAHID RAHMAN (France), MANUEL REBUSCHI (France), LAURENT ROLLET (France), PHILIPPE DE ROUILHAN (France), LÉNA SOLER (France), JEAN-JACQUES SZCZECINIARZ (France), MARK VAN ATTEN (France), JOHANN VAN BENTHEM (The Netherlands), JOSEPH VIDAL-ROSSET (France)

Patronages

The 14th CLMPS was organized under the High Patronage of the President of the French Republic and under the patronage of the French Department for Research and Higher Education, UNESCO, the Academy of Technologies of France, and the French National Commission for UNESCO.

Supports and Partners

European Fund for Regional Development, Regional Delegation for Research and Technology (DRRT) Lorraine, Regional Council of Lorraine, Department of Meurthe-et-Moselle, Urban Community of Greater Nancy (Grand Nancy), Town of Nancy, Nancy-University Federation, University of Nancy 2, French Academy of Science, French National Centre for Scientific Research, French Society for

Philosophy of Science, French Institute for Research in Computer Science and Automation (INRIA), French Alternative Energies and Atomic Energy Commission (CEA), Lorraine Institute for Social Sciences and Humanities (MSH Lorraine), Henri-Poincaré Archives, Saint-Gobain PAM, CASDEN, Goethe-Institut Nancy.

14TH CLMPS
NANCY 2011

July, 19-26, 2011
Nancy, France

14th Congress of Logic, Methodology and Philosophy of Science

Logic and Science Facing the New Technologies

EXECUTIVE COMMITTEE OF DLMPS
Wilfrid HODGES (UK – Okehampton), Chair
Peter CLARK (UK – St Andrews)
Anne FAGOT-LARGEAULT (France – Paris)
Adolf GRÜNBAUM (USA – Pittsburgh)
Ralf SCHINDLER (Germany – Münster)
Soshichi UCHII (Japan – Kyoto)

ORGANIZING COMMITTEE
Gerhard HEINZMANN, Chair
Claude DEBRU, Vice-chair
Pierre-Edouard BOUR
Clémentine LE MONNIER

www.clmps2011.org

contact@clmps2011.org

GENERAL PROGRAM COMMITTEE
Peter SCHROEDER-HEISTER (Germany – Tübingen), Chair
Bernadette BENSAUDE-VINCENT (France – Paris)
Andreas BLASS (USA – Ann Arbor)
Peter CLARK (UK – St Andrews)
Thierry COQUAND (Sweden – Göteborg)
Dennis DIEKS (The Netherlands – Utrecht)
John DUPRÉ (UK – Exeter)
Maria Carla GALAVOTTI (Italy – Bologne)
Gerhard HEINZMANN (France – Nancy)
Anthonie MEIJERS (The Netherlands – Eindhoven)

INVITED SPEAKERS
Steve AWODEY (Carnegie Mellon)
Davis BAIRD (Clark University)
Yemima BEN-MENAHEM (Jerusalem)
Ulrich BERGER (Swansea)
Hans BRIEGEL (Innsbruck)
Jeffrey BUB (University of Maryland)
Craig CALLENDER (San Diego)
Anna CAMBON-THOMSEN (Toulouse)
Martin CARRIER (Bielefeld)
Cristiano CASTELFRANCHI (Roma)
Carlo CELLUCCI (Roma)
Roger COOKE (Washington)
Nachum DERSHOWITZ (Tel Aviv)
Heather E. DOUGLAS (Knoxville)
Kevin ELLIOTT (Columbia, South Carolina)
Michael FRIEDMAN (Stanford)
Roman FRIGG (London)
Etienne GRANDJEAN (Caen)
Martin GROHE (Humboldt, Berlin)
Yuri GUREVICH (Microsoft Research)
Ulrike HAHN (Cardiff)
Christopher HITCHCOCK (Caltech)
Paul HUMPHREYS (Charlottesville)
Julia KNIGHT (Notre Dame)
Saul KRIPKE (CUNY)
Peter KROES (Delft)
Hugh LACEY (Swarthmore)
William LAWVERE (Buffalo, New York)

SENIOR ADVISORS
Mic DETLEFSEN (USA/France)
Kosta DOSEN (Serbia)
Robert FRODEMAN (USA)
Francesco GUALA (Italy)
Marcus KRACHT (Germany)
Hans RADDER (Netherlands)
Friedrich STADLER (Austria)
Alfred TAUBER (USA)

Hannes LEITGEB (Bristol)
Tim LEWENS (Cambridge)
Christof LODING (Aachen)
Uskali MÄKI (Helsinki)
Jean-Pierre MARQUIS (Montréal)
Donald MCKENZIE (Edinburgh)
Dale MILLER (Ecole polytechnique, Palaiseau)
Joe MILLER (Wisconsin)
Justin MOORE (Cornell)
Michel MORANGE (Paris)
Yiannis MOSCHOVAKIS (UCLA / Athens)
Alfred NORDMANN (Darmstadt)
Paulo OLIVA (London)
Wendy PARKER (Ohio University)
Dag PRAWITZ (Stockholm)
Hans RADDER (Amsterdam)
Miklos REDEI (London)
Philippe SCHLENKER (Paris)
Philip SCOTT (Ottawa)
Brian SKYRMS (Irvine)
Miriam SOLOMON (Temple University)
Patrick SPEISSEGGER (McMaster)
Robert SPEKKENS (Ontario)
Wolfgang SPOHN (Konstanz)
Simon THOMAS (Rutgers)
Peter VANDERSCHRAAF (Merced)
Dag WESTERSTÅHL (Göteborg)
Marek ZUKOWSKI (Gdansk)

Under the High Patronage of the President of the French Republic

Under the High Patronage of   and the patronage of   and supported by

www.ingramcontent.com/pod-product-compliance
Lightning Source LLC
Chambersburg PA
CBHW051123230426
43670CB00007B/650